Advances in Industrial Control

For further volumes:
www.springer.com/series/1412

Other titles published in this series:

Digital Controller Implementation and Fragility
Robert S.H. Istepanian and James F. Whidborne (Eds.)

Optimisation of Industrial Processes at Supervisory Level
Doris Sáez, Aldo Cipriano and Andrzej W. Ordys

Robust Control of Diesel Ship Propulsion
Nikolaos Xiros

Hydraulic Servo-systems
Mohieddine Mali and Andreas Kroll

Model-based Fault Diagnosis in Dynamic Systems Using Identification Techniques
Silvio Simani, Cesare Fantuzzi and Ron J. Patton

Strategies for Feedback Linearisation
Freddy Garces, Victor M. Becerra, Chandrasekhar Kambhampati and Kevin Warwick

Robust Autonomous Guidance
Alberto Isidori, Lorenzo Marconi and Andrea Serrani

Dynamic Modelling of Gas Turbines
Gennady G. Kulikov and Haydn A. Thompson (Eds.)

Control of Fuel Cell Power Systems
Jay T. Pukrushpan, Anna G. Stefanopoulou and Huei Peng

Fuzzy Logic, Identification and Predictive Control
Jairo Espinosa, Joos Vandewalle and Vincent Wertz

Optimal Real-time Control of Sewer Networks
Magdalene Marinaki and Markos Papageorgiou

Process Modelling for Control
Benoît Codrons

Computational Intelligence in Time Series Forecasting
Ajoy K. Palit and Dobrivoje Popovic

Modelling and Control of Mini-Flying Machines
Pedro Castillo, Rogelio Lozano and Alejandro Dzul

Ship Motion Control
Tristan Perez

Hard Disk Drive Servo Systems (2nd Ed.)
Ben M. Chen, Tong H. Lee, Kemao Peng and Venkatakrishnan Venkataramanan

Measurement, Control, and Communication Using IEEE 1588
John C. Eidson

Piezoelectric Transducers for Vibration Control and Damping
S.O. Reza Moheimani and Andrew J. Fleming

Manufacturing Systems Control Design
Stjepan Bogdan, Frank L. Lewis, Zdenko Kovačić and José Mireles Jr.

Windup in Control
Peter Hippe

Nonlinear H_2/H_∞ Constrained Feedback Control
Murad Abu-Khalaf, Jie Huang and Frank L. Lewis

Practical Grey-box Process Identification
Torsten Bohlin

Control of Traffic Systems in Buildings
Sandor Markon, Hajime Kita, Hiroshi Kise and Thomas Bartz-Beielstein

Wind Turbine Control Systems
Fernando D. Bianchi, Hernán De Battista and Ricardo J. Mantz

Advanced Fuzzy Logic Technologies in Industrial Applications
Ying Bai, Hanqi Zhuang and Dali Wang (Eds.)

Practical PID Control
Antonio Visioli

(*continued after Index*)

Eduardo F. Camacho • Manuel Berenguel •
Francisco R. Rubio • Diego Martínez

Control of Solar Energy Systems

Springer

Prof. Eduardo F. Camacho
Departamento de Ingeniería de Sistemas y Automática
Escuela Superior de Ingenieros
Universidad de Sevilla
Seville, Spain

Prof. Francisco R. Rubio
Departamento de Ingeniería de Sistemas y Automática
Escuela Superior de Ingenieros
Universidad de Sevilla
Seville, Spain

Manuel Berenguel
Departamento de Lenguajes y Computación
Escuela Superior de Ingeniería
Universidad de Almería
Almería, Spain

Diego Martínez
Plataforma Solar de Almería
Centro Europeo de Ensayos de Energía Solar
CIEMAT
Tabernas, Spain

ISSN 1430-9491
e-ISSN 2193-1577
Advances in Industrial Control
ISBN 978-0-85729-915-4
e-ISBN 978-0-85729-916-1
DOI 10.1007/978-0-85729-916-1
Springer London Dordrecht Heidelberg New York

British Library Cataloguing in Publication Data
A catalogue record for this book is available from the British Library

Library of Congress Control Number: 2011944757

Printed on acid-free paper

Springer is part of Springer Science+Business Media (www.springer.com)

To Janet (E.F.C.)
To my daughters Laura and Ana (M.B.)
To my daughters María and Elena (F.R.R.)
To my family (D.M.)

Series Editors' Foreword

The series *Advances in Industrial Control* aims to report and encourage technology transfer in control engineering. The rapid development of control technology has an impact on all areas of the control discipline. New theory, new controllers, actuators, sensors, new industrial processes, computer methods, new applications, new philosophies, ..., new challenges. Much of this development work resides in industrial reports, feasibility study papers and the reports of advanced collaborative projects. The series offers an opportunity for researchers to present an extended exposition of such new work in all aspects of industrial control for wider and rapid dissemination.

Concerns about energy independence, CO_2 emissions and climate change and the engineering risks associated with nuclear power stations have all contributed to the political push to prioritise the exploitation of renewable energy resources in the developed nations. Key technologies in the renewable energy field for electrical power generation that can be considered commercially mature include hydro-power, tidal power, biomass digester technology and the use of wind turbines and the development of large-scale wind farms.

A technological field that is now finding wider terrestrial application is that of solar energy systems where an installation uses the Sun's radiant energy as the "fuel" of the system. Roof-top solar panels are a common sight in small-scale systems for domestic water and space heating, but what of *industrial*-scale solar energy systems? Progress on developing such systems and their control can be found in this very timely *Advances in Industrial Control* monograph, *Control of Solar Energy Systems* by Eduardo F. Camacho, Manuel Berenguel, Francisco R. Rubio and Diego Martínez. Given the very many specialised and small-scale applications of solar energy collectors, it cannot be stressed too strongly that the plants and installations described in this monograph are proposed for industrial-scale operation. Just as wind turbines have grown in size and have been integrated in large-scale wind farms to enable the generation of usefully significant amounts of electrical power, so too will the basic mechanisms of solar energy capture need to be developed to industrial dimensions to ensure a significant contribution.

In this monograph, we learn that there are really two candidate solar energy capture principles. The first of these uses photovoltaics to generate a voltage directly, and the second uses a solar thermal collector to capture the Sun's radiant energy using one or other of two main process architectures; either a heat transfer fluid (HTF) is taken to the vicinity of the locally focused radiant energy as in the parabolic trough, or the Sun's radiant energy is focused globally as in a central (tower) receiver system. This simple framework of concepts provides the structure of the monograph. Consequently, Chaps. 1 and 2 present the solar energy basics and some specific control fundamentals for solar energy systems. Chapter 3 looks at the control of photovoltaic plants, while Chaps. 4 and 5 focus on the control and the advanced control of parabolic trough systems. Chapter 6 examines central receiver systems with other possible systems and potential applications of solar energy (primarily as a furnace or for refrigeration) collected and described in Chap. 7. Finally, the critical topic of integrating these plants and installations into a larger-scale market-based network is considered in Chap. 8. This involves the control of plants with a diurnal operational characteristic or a highly variable generation profile and considers the issue of global plant control, so it is not surprising that upper-level supervisory process control makes an appearance in this chapter. The authors state quite early in the monograph "...the integration of solar energy plants in the electrical grid is a challenging problem"; in fact, it is a problem for many of the renewable energy systems, but one where the field of control systems theory and practice is capable of making significant contributions.

The monograph is a valuable compendium of process knowledge for the control of the various solar energy systems and can function as a reference book for those interested in learning more about these systems; it is especially valuable for the control research and engineering community working in the solar energy system field. Control researchers and students might note that although a wide range of advanced control solutions are studied in Chap. 5, there remains an obvious need for a common case study problem prescription to benchmark the different control schemes proposed. The monograph has the valuable attribute of describing many real plant installations and applications. These installations provided the test-beds for many of the control solutions described and results from control trials appear frequently in the monograph to amplify simulation results and to illustrate real-world implementations.

In 1997, the *Advances in Industrial Control* series published the first monograph, entitled *Advanced Control of Solar Plants* (978-3-540-76144-6) by the authors Eduardo F. Camacho, Manuel Berenguel, and Francisco R. Rubio. In the intervening years, the subject has matured and developed further. This new monograph shows how the authors and the control community are continuing to make a significant and substantial contribution to the control engineering involved in solar energy systems.

Industrial Control Centre
Glasgow, Scotland, UK

M.J. Grimble
M.A. Johnson

Preface

During the last 30 years, considerable research effort has been devoted to improve the efficiency of solar power plants from the control and optimization viewpoint. This book presents techniques to model and control solar energy systems. The book contains results obtained in several solar plants located at the Plataforma Solar de Almería (PSA), South-East Spain, which is the largest European center for research, development and testing of concentrating solar technologies.

The book is mainly aimed at practitioners, both from the solar energy community and the control engineering community, although it can be followed by a wide range of readers, as only basic knowledge of control theory and sampled data systems is required.

The book is organized as follows: Chap. 1 gives a brief introduction to solar energy fundamentals, including solar radiation related concepts and a classification of solar thermal technologies and energy storage systems. Chapter 2 presents control issues in solar systems, where the main Sun tracking mechanisms are studied, a brief overview of solar radiation estimation and forecast techniques is included, the control of fundamental variables is explained, as well as how the integrated control of solar energy systems should be addressed. Chapter 3 briefly introduces the photovoltaic plants, focused on automatic tracking strategies. Chapter 4 explains the basic modeling and control approaches related to thermal solar plants with parabolic trough distributed collectors. After reviewing different modeling approaches for these kinds of system, the basic control algorithms are explained, highlighting their main advantages and drawbacks: feedforward control, PID control and cascade control. In Chap. 5, parabolic troughs advanced control techniques are developed, covering a wide range of control schemes that have been tested at the PSA following a classification of these techniques. Some of these strategies are adaptive control, model-based predictive control, nonlinear control, fuzzy logic control and so on. Chapter 6 deals with the control of power towers with central receiver systems. After explaining the main control issues of these kinds of plant, including a general description of the control system, and types of receiver and model, both the heliostat field control and aiming strategies are explained, including basic control approaches. Chapter 7 briefly explains the main control issues related to other interesting solar energy applications, such as solar furnaces and solar refrigeration systems. Finally, Chap. 8 presents recent approaches to the integrated control of solar systems.

The text is composed of material collected from articles written by the authors, technical reports and lectures given to graduate students.

Seville, Almería
Spain

E.F. Camacho
M. Berenguel
F.R. Rubio
D. Martínez

Acknowledgements

The authors would like to thank a number of people and institutions who have made this book possible. Firstly, we thank Janet Buckley, who translated part of the book from our native language to English and corrected and polished the style of the rest. Our thanks go to Javier Aracil who introduced us to the exciting world of Automatic Control and to many other colleagues and friends from various universities, especially M.R. Arahal, C. Bordóns, F. Gordillo, M.G. Ortega, F.J. García-Martín, P. Lara (University of Seville, Spain), F. Rodríguez, J.L. Guzmán, J.C. Moreno, J.D. Álvarez, C.M. Cirre, A. Pawlowski, M. Pasamontes, D. Lacasa, J. González, M. Peralta, C. Rodríguez, M. Pérez (University of Almería, Spain), L. Valenzuela, E. Zarza, L.J. Yebra, M. Romero, L. Roca, J. Bonilla, A. Valverde, D. Alarcón, R. Monterreal (PSA—CIEMAT, Spain), S. Dormido (National Distance University, Spain), J.E. Normey-Rico (Federal University of Santa Catarina—Brazil) who developed with us many of the ideas appearing in the book and helped us to correct the manuscript.

Most of the material included in the book is the result of research work funded by the Spanish Ministry of Science and Innovation and EU-ERDF funds,[1] the Spanish Ministry of Education,[2] CIEMAT, the European Commission[3] and the Consejería de Innovación, Ciencia y Empresa de la Junta de Andalucía. We gratefully acknowledge these institutions for their support. This work has also been performed within the scope of the specific collaboration agreement between the PSA and the Automatic Control, Electronics and Robotics (TEP-197) research group of the Univer-

[1]DPI2001-2380-CO2-01/02, DPI2002-04375-C03-01/03, DPI2004-07444-C04-01/04, DPI2008-05818, DPI2004-06419, DPI2007-64697, DPI2010-19154, DPI2007-66718-C04-01/04, DPI2010-21589-C05-01/04.

[2]PHB2009-0008-PC.

[3]*Enhancement and Development of Industrial Applications of Solar Energy Technologies*, supported by EEC Program *Human Capital and Mobility—Large Installations Program*, EC-DGS XII Program *Training and Mobility of Researchers*, EC-DGS XII program *Improving Human Potential* and promoted by CIEMAT—PSA and the DISS project (contract No. JOR3-CT98-0277) within the framework of the E.U. JOULE Program.

sidad de Almería titled *Development of control systems and tools for thermosolar plants*.

The experiments described in the book could not have been carried out without the help of the Plataforma Solar de Almería (PSA) and its staff.

Finally, the authors thank their families for their support, patience and understanding of family time lost during the writing of the book.

Seville, Almería
Spain

E.F. Camacho
M. Berenguel
F.R. Rubio
D. Martínez

Contents

Abbreviations

Acronyms	
2DOF	*Two Degrees Of Freedom*
AC	*Alternating Current/Adaptive Control*
AI	*Artificial Intelligence*
AIC	*Akaike's Information Criterion*
ANFIS	*Adaptive Network-based Fuzzy Inference System*
ANN	*Artificial Neural Networks*
ARIMA	*Auto-Regressive Integrated Moving Average*
ARIMAX	*Auto-Regressive Integrated Moving Average with eXogenous inputs*
ARMA	*Auto-Regressive Moving Average*
ARMAX	*Auto-Regressive Moving Average with eXogenous inputs*
ARX	*Auto-Regressive model with eXogenous inputs*
BCS	*Beam Characterization System*
BPF	*Band Pass Filter*
B/W	*Black/White*
CARIMA	*Controlled Auto-Regressive Integrated Moving Average*
CARMA	*Controlled Auto-Regressive Moving Average*
CC	*Cascade Control*
CCD	*Charge Coupled Device*
CE	*Control Effort*
CESA-1	*Central Eléctrica Solar de Almería 1*
CIEMAT	*Centro de Investigaciones Energéticas, Medioambientales y Tecnológicas*
CF	*Comb Filter*
CPC	*Compound Parabolic Concentrator*
CRS	*Central Receiver System*
CSP	*Concentrating Solar Power*
CSR	*CircumSolar Radiation*
CST	*Concentrating Solar Thermal*
CV	*Control Volume*
DAE	*Differential Algebraic Equation*

DAPS	*Dynamic Aim Processing System*
DAS	*Data Acquisition System*
DC	*Direct Current*
DHA	*Discrete Hybrid Automata*
DIR	*Digital Image Radiometer*
DCS	*Distributed Collector System*
DES	*Double Exponential Smoothing*
DHA	*Discrete Hybrid Automata*
DISS	*Direct Solar Steam*
DLR	*Deutsches Zentrum für Luft- und Raumfahrt* *German Aerospace Center*
DSCF	*Distributed Solar Collector Field*
DSG	*Direct Steam Generation*
DTC	*Dead-Time Compensator*
ECMWF	*European Centre for Medium-range Weather Forecasts*
EHAC	*Extended Horizon Adaptive Control*
EPSAC	*Extended Prediction Self Adaptive Control*
ELC	*Extended Linear Complementary*
EWMA	*Exponential Weighting Moving Average*
FAM	*Fuzzy Associative Memory*
FF	*FeedForward*
FFFV	*Feedforward Function Feed Valve*
FFNN	*FeedForward Neural Network*
FIR	*Finite Impulse Response*
FIT	*Feed-In Tariff*
FL	*Feedback Linearization*
FLC	*Fuzzy Logic Control*
FM	*Flow Model*
FOPDT	*First Order Plus Dead Time*
FPPI	*Filtered Predictive-Proportional-Integral*
FRG	*Fuzzy Reference Governor*
FSM	*Finite State Machine*
FVM	*Finite Volume Method*
GA	*Genetic Algorithms*
GFS	*Global Forecast System*
GM	*Gain Margin*
GPC	*Generalized Predictive Control*
GPCIT	*Generalized Predictive Control Interactive Tool*
GPS	*Global Positioning System*
GS	*Gain Scheduling*
HAC	*Heliostat Array Controller*
HTF	*Heat Transfer Fluid*
IAE	*Integral of the Absolute Error*
IAS	*Image Analysis System*
IC	*Intelligent Control*

IFPIC	*Incremental Fuzzy PI Control*
IMC	*Internal Model Control*
IncCond	*Incremental Conductance method*
I/O	*Input–Output*
I-PD	*Integral-Proportional-Derivative*
IR	*Infra Red*
ISE	*Integral of the Square of the Error*
ITAE	*Integral of Time multiplied by the Absolute Error*
ITSE	*Integral of Time multiplied by the Squared Error*
JD	*Julian Day*
KBS	*Knowledge-Based System*
KF	*Kalman Filter*
KFDF	*Kalman Filter with Data Fusion*
LBL	*Lawrence Berkeley Laboratory*
LCP	*Linear Complementary Problem*
LQ	*Linear Quadratic*
LQG	*Linear Quadratic Gaussian*
LS	*Least Squares*
LTI	*Linear Time Invariant*
LTR	*Loop Transfer Recovery*
MBM	*Moving Boundary Model*
MED	*Multi-Effect Distillation*
MIMO	*Multiple-Inputs Multiple-Outputs*
MIP	*Mixed Integer Programming*
MISO	*Multiple-Inputs Single-Output*
MIN–MAX	*MINimum–MAXimum*
MLD	*Mixed Logical Dynamical*
MLP	*MultiLayer Perceptron*
MMPS	*Min–Max Plus Scaling*
MPC	*Model-based Predictive Control*
MPP	*Maximum Power Point*
MPPT	*Maximum Power Point Tracking*
MRAC	*Model-Reference Adaptive Control*
MSE	*Mean Squared Error*
MUSMAR	*MUltivariable Self-tuning Multipredictor Adaptive Regulator*
NARX	*Nonlinear Auto-Regressive model with eXogenous inputs*
NC	*Nonlinear Control*
NCEP	*National Centers for Environmental Prediction (USA)*
NDFD	*National Digital Forecast Database*
NEPSAC	*Nonlinear Extended Prediction Self-Adaptive Controller*
NI	*Numeric Inference*
NIP	*Normal Incidence Pyrheliometer*
NMP	*Non-Minimum Phase*
NN	*Neural Networks*
NNC	*Neural Networks Controllers*

NNLS	*Non-Negative Least Squares*
ODE	*Ordinary Differential Equation*
OPC	*Ole for Process Control*
OR	*Output Regulation*
ORG	*Optimizing Reference Governor*
PC	*Personal Computer*
PCM	*Phase Change Material*
PCS	*Power Conversion System*
PDE	*Partial Differential Equation*
PI	*Proportional-Integral*
PID	*Proportional-Integral-Derivative*
PM	*Phase Margin*
PPI	*Predictive-Proportional-Integral*
PRBS	*Pseudo Random Binary Sequence*
PSA	*Plataforma Solar de Almería* *Solar Platform of Almería*
PTC	*Parabolic Trough Collector*
PV	*PhotoVoltaics*
PWA	*PieceWise Affine*
P&O	*Perturbation and Observation method*
QFT	*Quantitative Feedback Theory*
QP	*Quadratic Programming*
RBFN	*Radial Basis Functions Network*
RC	*Robust Control*
RLS	*Recursive Least Squares*
RMSE	*Root Mean Squared Error*
RNN	*Recurrent Neural Network*
SAPS	*Static Aim Processing System*
SCADA	*Supervisory Control And Data Acquisition*
SEGS	*Solar Electricity Generating System*
SISO	*Single-Input Single-Output*
SP	*Smith Predictor*
SPGPC	*Smith Predictor Generalized Predictive Control*
SPSA	*Simultaneous Perturbation Stochastic Approximation*
SPT	*Solar Power Tower*
SSE	*Sum of Squared Errors*
SSPS	*Small Solar Power Systems*
STC	*Self-Tuning Control*
TDC	*Time Delay Compensation*
TDL	*Tapped Delay Lines*
TEP	*Thermosolar Electrical Plants*
TP	*Triangular-Partition*
TPE	*Triangular-Partition with Evenly spaced midpoints*
TS	*Training Set*
TSA	*Technology program Solar Air receiver*

UBB	*Unknown But Bounded*
UCM	*Unobserved Component Models*
UV	*Ultra Violet*
VS	*Validation Set*
VSCS	*Variable Structure Control System*
WRF	*Weather Research and Forecasting*
ZOH	*Zero Order Hold*
ZN	*Ziegler–Nichols*

Symbols

$\mathbb{N}$	Set of natural numbers
$\mathbb{R}$	Set of real numbers
$\arg(\cdot)$	Argument
$\det(\cdot)$	Matrix determinant
$(\cdot)^T$	Transpose of $(\cdot)$
$\mathbf{I}$	Identity matrix of appropriate dimensions
$\sin(\cdot)$	Sine function
$\cos(\cdot)$	Cosine function
$\exp(\cdot), e^{(\cdot)}$	Exponential function
$\log(\cdot)$	Natural logarithm function
$\min(\cdot)$	Minimum of a set
$\max(\cdot)$	Maximum of a set
s	Complex variable used in Laplace Transform
z^{-1}	Backward shift operator
z	Forward shift operator and complex variable used in Z-Transform
$\forall$	For all
$\Vert\cdot\Vert$	$C^{n\times m}$ norm
$\Vert\cdot\Vert_2$	$L_2^{n\times m}$ norm
$\Vert\cdot\Vert_\infty$	∞ norm
$\doteq$	Definition
$E[\cdot]$	Expectation operator
$\hat{\cdot}$	Expected value
$\bar{\cdot}$	Mean value
$\hat{x}(k+j \mid k)$	Expected value of $x(k+j)$ with available information at instant k
$\wedge$	Logical AND
$\vee$	Logical OR
$\neg$	Logical NOR
equalAB	A is the same as B
AgtB	A is greater than B

Variables and Parameters

Symbol	Description	Units	Chapter
a	discrete transfer function pole descriptive parameter		4, 5
	auxiliary transfer function (linear model of resonances)	–	4
A	shutter aperture	m	7
$\mathbf{A}$	state matrix		2
A_{ac}	accumulation tank area	m^2	7
a_f	discrete transfer function pole descriptive parameter		5
A_f	cross-sectional area for flow inside pipe	m^2	4, 5
a_{fv}	required DSG feed valve aperture	%	4
a_i	midpoint of fuzzy set A_i		5
A_i	fuzzy set		5
a_{iv}	required DSG injector valve aperture	%	4
A_m	cross-sectional area of pipe wall	m^2	4–6
	heliostat mirror area		6
a_n	heliostat-normal azimuth angle	rad, °	2
a_{pv}	DSG steam pressure valve aperture	%	4
a_s	azimuth	rad, °	1, 2
A_{st}	area if the storage tank	m^2	7
a_t	fixed tower azimuth angle	rad, °	2
b	auxiliary transfer function (linear model of resonances)	–	4
B	benefits	$	8
B_{spot}	benefits scheduled	$	8
$b_{..}$	discrete transfer function numerator parameter		5
b_D	estimated trend in the DES method		2
c	auxiliary parameter (linear model of resonances)	m	4–6

Symbol	Description	Units	Chapter
c_f	specific heat capacity of HTF	J/(kg K)	4–8
C_g	geometrical concentration factor	–	2
c_m	specific heat capacity of metal of pipe wall	J/(kg K)	4–6
C_{op}	operational cost	$	8
c_s	specific heat capacity of the sample	J/(kg K)	7
	thermal capacity per kg of the molten salt accumulator	J/K	8
d	dead-time in discrete time	samples	4, 5
D	Sun–Earth distance	km	1, 2
D_0	mean value of the Sun–Earth distance	km	1, 2
D_f	inside diameter of inner pipe	m	4, 5
D_m	outside diameter of pipe	m	2, 4, 5
D_{st}	diameter of the storage tank	m	8
dT_f	threshold that indicates if a situation is susceptible to generating an aimpoint adjustment	K, °C	6
dT_h	threshold that indicate if a situation is susceptible to generating heliostat adjustment	K, °C	6
e	generic tracking, identification or modeling error or discrete white noise with zero mean		5, 6 5
E	thermal energy	J	6
$\bar{e}$	average absolute error		2
$\bar{e}_{\%}$	average percent error	%	2
e_1	electricity price	$	8
e_b	highest electricity buying price	$	8
E_b	direct (beam) solar irradiance	W/m^2	2
E_c	solar constant	W/m^2	1
E_d	diffuse irradiance	W/m^2	2
E_{ext}	extraterrestrial solar irradiance outside the Earth's atmosphere	W/m^2	1
$E_{ext,h}$	extraterrestrial solar irradiance falling on a surface parallel to the ground	W/m^2	1
E_g	global irradiance	W/m^2	2
e_{m_in}	DSG inlet mass flow error	kg/s	4
e_s	lowest electricity selling price	$	8
e_{st}	price of energy stored	$	8
E_{st}	stored energy	kJ	8
e_T	DSG temperature error	K, °C	4
e_{xc}	eccentricity correction factor	–	2
f	forgetting factor of the identification algorithm	–	5
f_1	auxiliary parameter	1/s	6
F_1	filter applied to the plant input		7
F_2	filter applied to the plant output		7

Symbol	Description	Units	Chapter
f_{at}	attenuation factor due to turbidity of the atmosphere	–	6
f_d	focal distance of the parabolic mirrors	m	2
f_{di}	amount of diffuse irradiance intercepted	–	2
g	auxiliary parameter	1/s	6
G	collector aperture	m	4, 5
g_0	coefficient of discrete time PI transfer function		5
g_1	coefficient of discrete time PI transfer function		5
G_H	gas heater activation/deactivation signal	–	7
G_{Hoff}	gas heather off signal steps	–	7
G_{Hon}	gas heather on signal steps	–	7
$\mathbf{H}$	output matrix		2
H_{st}	storage tanks losses coefficient	W/(m^2 K)	7
h_{in}	specific enthalpy of HTF at collector row inlet	J/kg	2, 4
h_{in_c}	DSG specific enthalpy of water at collector inlet	J/kg	4
h_{inj}	DSG specific enthalpy of injection water at collector inlet	J/kg	4
H_l	convective heat transfer coefficient of pipe exterior	W/(m^2 K)	2, 4–6
	global coefficient of thermal losses		
$\widetilde{H}_l$	global thermal loss function	J/s	4
h_n	heliostat-normal altitude angle	rad, °	2
h_{out}	specific enthalpy of HTF at collector row outlet	J/kg	2, 4
h_{ref}	specific enthalpy reference of steam at collector outlet	J/kg	4
h_s	Sun altitude (solar elevation)	rad, °	1, 2
h_{st}	height of the storage tank	m	8
h_t	fixed tower altitude angle	rad, °	2
H_t	convective heat transfer coefficient of pipe interior	W/(m^2 K)	4–6
I	(corrected) direct solar irradiance	W/m^2	2, 4–7
	current of the PV cell	A	3
I_{CS}	intensity of the radiation of the solar halo	W/m^2	6
I_{MPP}	PV current to obtain MPP	A	3
I_{ref}	reference direct solar irradiance of FF controller	W/m^2	7
I_s	fraction of solar irradiance reflected by heliostats	W/m^2	7
I_{Sun}	intensity of the radiation of the solar disk	W/m^2	6
J	Julian day	–	1, 2, 6
	cost function		5
k	discrete time	sample	2–8
$K_{..}$	representative static gain of transfer function		4, 5, 7
k_f	thermal conductivity of the HTF	W/(m K)	4, 8
K_P	proportional gain of PID controller	–	4, 5, 7

Symbol	Description	Units	Chapter
l	length of tube segment	m	4
L	collector loop length or pipe length	m	2, 4–6
$\mathbf{L}$	gain of correction of the RLS algorithm		5
L_2	inlet–outlet loop two distance	m	4
L_{loc}	geographic longitude of the local meridian	rad, °	1
L_{PCS}	thermal losses inside the PCS	kW	8
L_{ref}	geographic longitude of the reference meridian	rad, °	1
L_{rs}	thermal power deliver to the refrigeration systems	kW	8
L_s	length of the shutter panel	m	7
ℓ	distance	m	4
m	positive integer—ahead forecast	samples	2
$\dot{m}$	HTF mass flow	kg/s	2, 5, 6
$\dot{m}_{12}$	difference between blowers G1 and G2 air mass flow	kg/s	6
$\dot{m}_1$	blower G1 air mass flow	kg/s	6
$\dot{m}_2$	blower G2 air mass flow	kg/s	6
$\dot{m}_{ff}$	DSG nominal mass flow	kg/s	4
$\dot{m}_{ff\_iv}$	injector valve DSG mass water flow	kg/s	4
$\dot{m}_{in}$	HTF mass water flow at collector row inlet	kg/s	4
$\dot{m}_{in\_c}$	DSG mass water/steam flow at collector inlet	kg/s	4
$\dot{m}_{inj\_dem}$	DSG injection water flow demanded at collector inlet	kg/s	4
$\dot{m}_{inj\_set}$	DSG injection water flow reference at collector inlet	kg/s	4
$\dot{m}_{inj}$	DSG injection water flow at collector inlet	kg/s	4
$\dot{m}_{in\_dem}$	DSG mass flow calculated by the master loop	kg/s	4
$\dot{m}_{out}$	HTF mass water flow at collector row outlet	kg/s	4
m_s	sample mass	kg	7
n	positive integer—number of PTC discretization elements		2, 4–6
$\mathbf{n}$	heliostat normal		2
N	prediction horizon or data window length	samples	2, 4–6
N_1	lower value of prediction horizon	samples	5
N_2	upper value of prediction horizon	samples	5
N_{CV}	number of CVs in the complete model of the DISS row	–	4
N_d	day number	–	1
N_D	parameter of the filter of the derivative action	–	4
n_{ope}	number of operative loops	–	4
n_P	number of patterns	–	4
N_r	reset velocity of repetitive control	–	5
N_u	control horizon	samples	5

Symbol	Description	Units	Chapter
p, $\mathbf{p}$	generic disturbance (vector) or noise	–	4, 5
P	output power of the PV array	W	3
	pressure	bar	4
	steam generator air input power	kW	6
$\mathbf{P}$	covariance matrix of the identification algorithm		5
	a posteriori estimate error covariance		2
$\mathbf{P}^-$	a priori estimate error covariance		2
P_{abs}	power absorbed by the receiver	kW	2
P_c	convection losses	W	7
$P_{contract}$	power scheduled	kW	8
P_{gross}	gross output power of the PCS	kW	8
P_h	total power reflected by an heliostat on a receiver	kW	6
P_{HTF}	power of the HTF at the PCS input	kW	8
P_i	input power	W	7
P_{load}	energy consumed by local electric loads	kW	5, 8
$P_{m\rightarrow a}$	power lost from the tube metal to the ambient	kW	2
$P_{m\rightarrow f}$	thermal power provided by a PTC	W/m	2
P_{MPP}	maximum power output	W	3
P_r	power delivered to the grid	kW	5, 8
	radiation losses	W	7
P_{ra}	Prandtl number	–	4
P_{rc}	convective thermal losses of pipe exterior per unit length	W/m	4
P_{ref}	power reference	kW	8
P_{s+}	energy stored at time interval $[k-1,k]$	kW	8
P_{s-}	energy extracted from the energy storage and delivered to the grid at time interval $[k-1,k]$	kW	8
P_{sol}	incident power on the receiver	W/m	2
P_{solar}	solar energy generated	kW	8
P_{start}	power needed to start up the plant	kW	8
P_{stored}	total energy stored	kW	8
P_{th}	thermal power provided by the solar field	kW	5
q	HTF pump volumetric flow rate	m^3/s, l/s	4, 5, 7, 8
$\mathbf{Q}$	process noise covariance matrix		2
q_f	filtered oil pump volumetric flow rate	m^3/s, l/s	5
q_{st}	HTF storage flow	m^3/s, l/s	8
q_T	HTF flow going from the storage system to the PCS	m^3/s, l/s	8
r	mirror specular reflectivity	–	2, 6
	generic reference		5–7
R	Earth radius	km	1
$\mathbf{R}$	measurement noise covariance matrix		2

Symbol	Description	Units	Chapter
R^2	coefficient of determination	–	2
r_c	reflectivity of the shutter secondary concentrator	–	7
r_h	heliostat's mirror reflectivity	–	7
s	complex variable used in Laplace transform	–	4, 5, 7
S	effective surface	m^2	2
S_{abs}	absorber pipe area of the collector loop	m^2	4
S_c	surface of the shutter secondary concentrator	m^2	7
$S_{f(90\%)}$	focus area where 90% of solar input energy concentrates	m^2	7
S_I	integral of the time varying discrete error signal		3
S_{ref}	reflecting surface of the collector field's mirrors	m^2	2
S_s	surface of the sample	m^2	7
S_{st}	storage tanks surface	m^2	7
t	time	s, min	4–8
T	temperature/sample temperature	K, °C	4–7
T_0	sample temperature at the equilibrium point	K, °C	7
T_2	temperature in an intermediate region in the sample	K, °C	7
T_3	temperature in the base of the sample	K, °C	7
T_a	ambient temperature	K, °C	2, 4–8
T_b	transmittance of the direct normal irradiance	–	2
t_d	input–output delay	s, min	4, 5
T_D	derivative time constant of PID controller	s, min	4, 5
T_f	temperature of HTF	K, °C	2, 4–6
$\bar{T}_f$	inlet–outlet average HTF temperature	K, °C	2, 4, 5
$T_{gh,in}$	input temperature of the gas heater	K, °C	7
$T_{gh,out}$	temperature at the gas heater output	K, °C	7
T_I	integral time constant of PID controller	s, min	3–5, 7
T_{in}	inlet HTF temperature	K, °C	4–8
T_{in1}	mean receiver air outlet temperature	K, °C	6
T_{in2}	thermal storage outlet temperature	K, °C	6
T_{in_c}	DSG water/steam temperature at collector inlet	K, °C	4
T_{inj}	DSG injection water temperature	K, °C	4
$T_{in_{Tur}}$	temperature of the inlet HTF of the power block	K, °C	8
T_m	temperature of metal of pipe wall	K, °C	2, 4, 6
$\bar{T}_m$	inlet–outlet average metal pipe temperature	K, °C	2
T_{max}	maximum temperature	K, °C	6
T_{min}	minimum temperature	K, °C	6
T_{out}	outlet HTF temperature	K, °C	4–8
$T_{out_{Tur}}$	temperature of the outlet HTF of the power block	K, °C	8
$t_p[\cdot]$	scaled temperature assigned to thermocouple i	K, °C	6
t_r	residence time	s, min	4, 5
T_r	tracking constant of anti-windup	s, min	5

Symbol	Description	Units	Chapter
T_{ref}	reference/set point HTF temperature	K, °C	4, 5, 7, 8
T_{rff}	reference temperature for the feedforward controller	K, °C	4, 5
t_{rt}	rise time	s, min	7
t_s	solar time	min, h	1, 2
T_s	sampling time	s, min	3–5, 7, 8
T_{sc}	inlet temperature to the absorption machine	K, °C	7
t_{st}	wall thickness of the storage tank	m	8
T_{st}	storage tank HTF mean temperature	K, °C	5, 7, 8
T_{stin}	storage tank inlet HTF temperature	K, °C	7, 8
T_{stout}	storage tank outlet HTF temperature	K, °C	7, 8
u	generic input		4–7
	perturbation of percentage of shutter rotated angle (linearized variable)	%	7
	position the tracking system assumes the Sun is in	rad, °	3
U	system input		5
	percentage of shutter rotated angle	%	7
$\hat{u}$	position obtained from the equations of Sun's movement	rad, °	3
$\tilde{u}$	correction of the position using estimated Sun position	rad, °	3
u'	virtual control signal		5
U_0	percentage of shutter rotated angle at the equilibrium point	%	7
u_{FF}	output of the feedforward controller		4
u_{max}	upper constraint on input amplitude		5, 6
u_{min}	lower constraint on input amplitude		5, 6
u_r	generic reference to input		6
U_∞	mean value of the input signal		5
v	velocity of HTF	m/s	4–7
	measurement noise		2
V	voltage of the PV cell	V	3
V_{st}	storage tank volume	m^3	7
$V1$	valve V1 state signal	–	7
$V3$	valve V3 state signal	–	7
V_f^i	volume of discrete element i of HTF	m^3	8
V_m^i	volume of discrete element i of metal	m^3	8
$VM1$	VM1 three-way valve state signal	–	7
V_{MPOP}	voltage of the maximum power operating point	V	3
V_{MPP}	voltage at which a PV array should operate to obtain MPP	V	3
V_{OOP}	voltage of optimal operating point	V	3

Symbol	Description	Units	Chapter
w	process noise		2
	DSG feed pump power	%	4
	weighting factor	–	5
W	aperture width of the parabolic mirrors	m	2
$\mathbf{W}$	matrix of weighting factors	–	5
x	generic state or time series variable		2, 4–7
	Cartesian coordinate	m	6
$\mathbf{x}$	input vector of an NN		4
x_{max}	upper constraint on state amplitude		6
x_{min}	lower constraint on state amplitude		6
x_r	generic reference to state		6
X_{target}	Cartesian coordinate of the target	m	6
y	position of the Sun	rad, °	3
	generic output variable		4–7
	Cartesian coordinate	m	6
Y	system output		5
$\mathbf{Y}$	vector of system outputs		5
y_m	generic model output		5
y_{max}	upper constraint on output amplitude		5, 6
y_{min}	lower constraint on output amplitude		5, 6
Y_{target}	Cartesian coordinate of the target	m	6
Y_{∞}	mean value of the output signal		5
z	generic measurement or auxiliary continuous variable		2, 6
	Cartesian coordinate	m	6

Variables and Parameters in Greek Symbols

Symbol	Description	Units	Chapter
α	shutter aperture angle	rad, °	7
	angle between the heliostats and the center of the target	rad, °	6
α_0	angle in which the shutter is completely closed	rad, °	7
$\alpha_{1,2,3}$	coefficients of discrete time bilinear model		5
α_a	absorption capacity of the sample	–	7
α_c	capacity of the sample to exchange heat with the air	–	7
α_d	discrete variable governing commutation between operation modes	–	6
α_e	emissivity of the sample	$W/(m^2\ K)$	7
α_{PID}	scale factor of the PID controller proportional gain	–	5
α_D	smoothing parameter for data of the DES method	–	2
$\alpha_{sol,A}$	solar radiation absorptance of absorber tube of PTC	–	2
α_W	H_∞ weighting function gain at high frequencies		7
α_{st}	proportion of the stored energy lost in each time period because of thermal losses	–	8
β	angle between the heliostats and the center of the target	rad, °	6
β_W	H_∞ weighting function gain at low frequencies		7
$\beta_{..}$	transfer function zero descriptive time constant	s, min	4, 5
β_s	smoothing parameter in data forecast	–	2
β_{st}	energy storage-conversion coefficient		8
γ	auxiliary parameter	$m^2\ °C/J$	4, 5
γ_A	PTC intercept factor	–	2
γ_{cl}	parameter modulating the closed-loop pole location		5
γ_D	smoothing parameter for trend in the DES method		2
γ_{te}	thermal to electrical power conversion factor	–	8
δ	weighting factor for future tracking errors	–	5
ΔE^a	energy change due to losses from wall to the ambient	J	8
ΔE^{cm}	energy change due to conduction among wall segments	J	8
ΔE^{co}	energy change due to conduction among HTF volumes	J	8
ΔE^t	energy change due to transport	J	8

Symbol	Description	Units	Chapter
ΔE^v	energy change due to convection	J	8
ΔE^w	energy change due to losses from HTF to wall	J	8
Δp_{fv}	DSG pressure drop across feed valve	bar	4
δ_s	declination	rad, °	1, 2
Δt	equation of time	min	1
ΔT	output–input temperature difference	K, °C	5
$\Delta \bar{T}$	difference in temperature between the average inlet and outlet temperature and the ambient temperature	K, °C	4
ε	characteristic time constant of a low pass filter	s, min	5
ζ	damping factor	–	5
η_a	thermal storage efficiency	–	5
η_b	thermal-to-electric power conversion	–	5
η_{col}	global collector efficiency	–	4, 5
η_{gross}	gross process efficiency	–	8
η_{net}	net process efficiency	–	8
η_{opt}	optical efficiency	–	2
$\bar{\eta}_{PCS}$	average efficiency of the PCS	%	8
$\bar{\eta}_{solar}$	average efficiency of the solar field	%	8
$\bar{\eta}_{st}$	average efficiency of the storage tanks	%	8
η_{th}	thermal efficiency	–	2
$\boldsymbol{\theta}$	vector of parameters for identification		5
θ_i	angle of incidence	rad, °	2, 4, 6
θ_f^i	estimation of HTF temperature in volume i	K, °C	8
θ_m^i	estimation of metal temperature in wall segment i	K, °C	8
θ_z	zenith angle	rad, °	1
κ	incidence angle modifier	–	2
κ_W	design parameter of the H_∞ method		7
λ	weighting factor for control effort	–	5, 6
	wavelength range	nm	2
μ_{A_i}	membership function of fuzzy set A_i		5, 7
μ_f	dynamic viscosity of HTF	Pa s	4
ξ	perturbation of sample temperature	K, °C	7
ρ_f	density of HTF	kg/m^3	4–8
ρ_m	density of metal of pipe wall	kg/m^3	4–6
ρ_t	tracking angle	rad, °	2
$\rho_{x,y}$	irradiance distribution centroid coordinates	m	6
ϱ	tracking specifications for QFT design		5
$\boldsymbol{\varrho}$	vector of parameters	–	4
σ	Stefan–Boltzmann constant	W/(m^2 K^4)	7
	standard deviation		2, 6
	discrete state		6

Symbol	Description	Units	Chapter
σ_{BQ}	optical or beam quality of an heliostat		6
σ_{total}	total beam dispersion error		6
ς	auxiliary binary variable		6
$\tau_{..}$	transfer function pole descriptive time constant	s, min	4, 5, 7
τ_1	T_f-related time constant against T_m variations	s, min	4–6
τ_{12}	T_m-related time constant against T_f variations	s, min	4–6
τ_2	T_m-related time constant against T_a variations	s, min	4–6
τ_f	time constant of a first order filter	s, min	5
$\tau_{sol,c}$	glass cover PTC transmissivity	–	2
ϕ	geographical latitude	rad, °	1, 2
$\boldsymbol{\Phi}$	matrix with regressors		5
$\boldsymbol{\varphi}$	vector with past inputs and outputs (regressor)		5
χ	unadjusted forecast of the DES method		2
ω	frequency	rad/s	4, 5, 7
ω_0	cut-off frequency in the design of PI controllers	rad/s	7
	crossover frequency of H_∞ weighting function	rad/s	
ω_n	natural frequency	rad/s	5
ω_s	hourly angle	rad, °	1, 2

Chapter 1
Solar Energy Fundamentals

1.1 Introduction

The use of renewable energy, such as solar energy, experienced a great impulse during the second half of the 1970s just after the first big oil crisis. At that time, economic issues were the most important factor and so interest in these types of process decreased when oil prices fell. There is renewed interest in the use of renewable energies nowadays, driven by the need of reducing the high environmental impact produced by the use of fossil energy systems.

As pointed out in [369], renewable spending has increased 30% in 2010 to a total of $243 billion. Nine-tenth of that is in the G-20 advanced industrial countries. The European region was the leading recipient of clean energy finance. Next comes Asia, while the Americas region is distant third in the race for clean energy investment. Globally, the solar sector grew fastest in 2010, attracting 53% more investment than the year before. Wind investment, in second place, grew 34%. Altogether, clean energy generating technology has doubled in the period 2008–2010 and now exceeds total global nuclear capacity. In terms of actual electricity produced, green energy still is only about a third or fourth of nuclear, but the progress in renewables is impressive indeed [369].

Global energy demand is about 16 TW approximately and is expected to double in the next twenty years. While increased energy production via traditional methods together with improved efficiency and conservation may alleviate some of the needs, such measures by themselves cannot meet the expected demand in the long run. The most abundant, sustainable source of energy is the Sun, which provides over 150000 TW of power to the Earth; about half of this energy reaches the Earth's surface while the other half gets reflected to outer space by the atmosphere (Fig. 1.1). Only a small fraction of the available solar energy reaching the Earth's surface would be enough to satisfy the global energy demand expected. Although most renewable energies derive their energy from the Sun, by solar energy we refer to the direct use of solar radiation. One of the greatest scientific and technological opportunities facing us today is to develop efficient ways to collect, convert, store and utilize solar energy at affordable costs. However, there are two main drawbacks

E.F. Camacho et al., *Control of Solar Energy Systems*, Advances in Industrial Control, DOI 10.1007/978-0-85729-916-1_1,

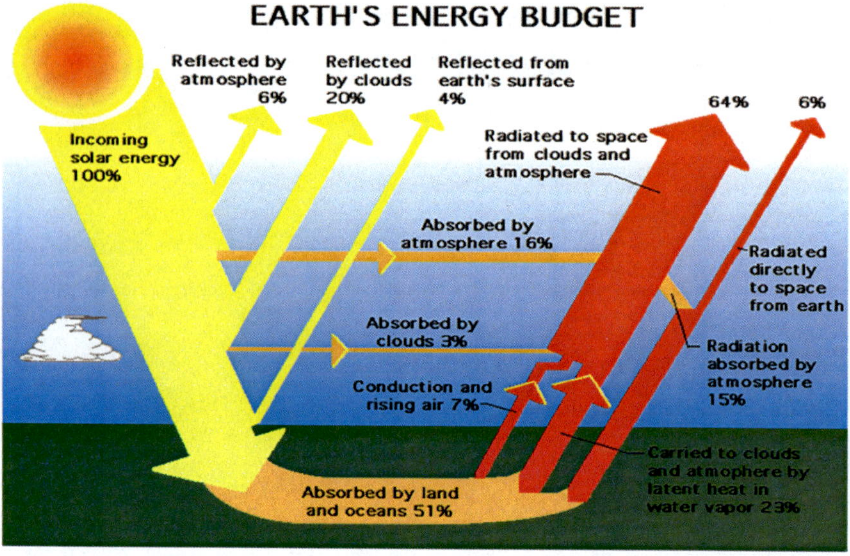

Fig. 1.1 Earth's energy budget (eosweb.larc.nasa.gov)

of solar energy systems: (i) the resulting energy costs are not yet competitive and (ii) solar energy is not always available when needed [89].

Considerable research efforts are being devoted to techniques which may help to overcome these drawbacks; control is one of these techniques. While in other power generating processes the main source of energy (the fuel) can be manipulated as the main control variable, in solar energy systems the main source of power, solar radiation, cannot be manipulated [85] and, furthermore, it changes in a seasonal and on a daily basis acting as a disturbance when considered from a control point of view. Solar plants have all the characteristics needed to use advanced control strategies able to cope with changing dynamics, non-linearities and uncertainties. The use of efficient control strategies resulting in better responses would increase the number of operational hours of solar plants and thus reduce the cost per kWh produced.

This chapter briefly describes solar radiation fundamentals, the main ways of collecting solar energy and related technologies, and a description of the solar plants of the Plataforma Solar de Almería (PSA, South-East Spain) that are used as test-bed plants throughout the text.

1.2 Solar Radiation

The Earth receives most of its energy from the Sun in the form of solar electromagnetic radiation. The Sun contains 99.9% of the total mass of the solar system. The average density of the Sun is surprisingly low (1.4 g/cm^3), the reason being that it is mostly composed of the lightest elements, hydrogen (70% by mass) and helium (27% by mass). The Sun's core is mostly composed of helium (65% by mass) while the hydrogen is reduced to 35% by mass because of being consumed in fusion reactions. Most of the other renewable sources of energy, such as wind energy, wave energy and bio-fuels depend on the Sun's energy. Furthermore, some of the non-renewable energy sources such as fossil fuels were originated by solar energy in the past. Solar energy originates from nuclear fusion reactions occurring at the core of the Sun where hydrogen atoms are fused into helium. The gravity at the core creates an intense pressure, which is high enough to force the fusion of hydrogen atoms. About 700 million tons per second of hydrogen is converted into helium. About half of the hydrogen found at the Sun's core has already been converted into helium. The remaining life expectation of the Sun is 5 billion years. The fusion reaction creates immense heat giving rise to temperatures at the core close to 15 million degrees Celsius (°C). At those temperatures, photons are emitted from the atoms and travel a very short distance before being absorbed by another atom causing the neighboring atom to heat and the subsequent emission of another photon. It takes a photon around 100000 years to reach the surface and about 8 minutes to travel the 149.5 million kilometers separating the Sun from the Earth.

The rate at which solar energy reaches a unit area at the Earth is called the solar irradiance or insolation and is measured in W/m^2. The integral of the solar irradiance during a period of time is called solar radiation or irradiation and is measured in J/m^2. Very frequently, solar irradiance is also denominated as solar radiation using the same units (W/m^2).

The total power emitted by the Sun's surface is about 63 MW/m^2. The photosphere has an average temperature of about 5800 K and delivers radiation with a spectrum similar to the spectrum of radiation emitted by a black body at 5800 K; most of the energy is emitted in the visible (450–700 nm) bands. A significant portion of solar energy is emitted in the infrared (IR) and ultraviolet (UV) and smaller quantities of energy are transmitted on radio, microwave, X-ray and gamma ray bands.

Solar radiation is absorbed by the atmosphere, soil and oceans. Wind and waves come from solar energy. A small portion of solar energy is transformed to kinetic energy of winds used by wind power systems. Wind also causes wave and currents which are used by waves or current generators.

Some of the solar energy is transformed by plants into biomass by photosynthesis, this is ultimately converted into heat energy by oxidation. The remaining part of the biomass produces organic sediments which will be transformed into fossil fuels in the future.

The Earth's disk intercepts solar radiation with an area πR^2; where R is the radius of the Earth. For a thermal equilibrium, the absorption of solar radiation must be equal to the energy emitted from the Earth to space. When the solar radiation hits the atmosphere, part of it is reflected, another part is scattered or absorbed by the air. The radiation that reaches the surface directly and is not absorbed, reflected or scattered, is called direct solar radiation. The reflected and scattered radiation reaching the ground is called diffuse radiation. The albedo factor is defined as the

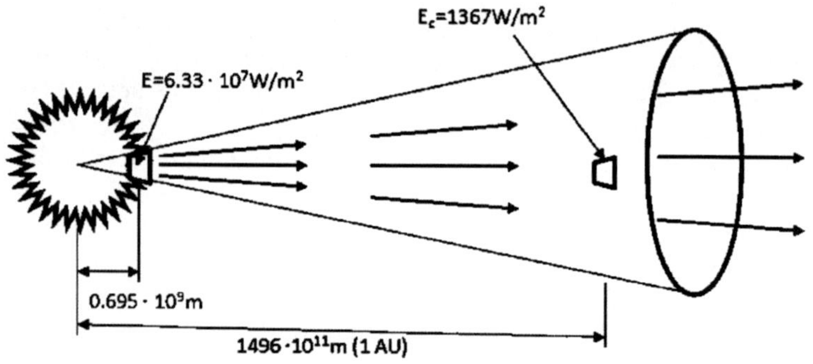

Fig. 1.2 The divergence of energy from the Sun to the Earth, [359]

fraction of radiation reaching the ground that is reflected back to the atmosphere. The power of direct solar irradiance on a clear day can be around 1 kW/m^2.

1.2.1 Solar Constant

The radiation intensity on the surface of the Sun is approximately $6.33 \cdot 10^7$ W/m^2. Since radiation spreads out as the distance squared, by the time it travels to the Earth ($1496 \cdot 10^{11}$ m or 1 AU is the average Earth–Sun distance, D_0), the radiant energy falling on 1 m^2 of surface area is reduced to 1367 W as depicted in Fig. 1.2 [359]. The intensity of the radiation leaving the Sun is relatively constant. Therefore, the intensity of solar radiation at a distance of 1 AU is called the solar constant E_c and has a currently accepted value of 1367 W/m^2 [359].

1.2.2 Extraterrestrial Solar Irradiance

As the Earth's orbit is elliptical, the intensity of solar radiation received outside the Earth's atmosphere varies as the square of the Earth–Sun distance D. The solar irradiance variation can be approximated by [359]

$$E_{ext} = E_c\left[1 + 0.034\cos\left(\frac{360N_d}{365.25}\right)\right] \quad [\mathrm{W/m^2}] \tag{1.1}$$

where E_{ext} is the extraterrestrial solar irradiance outside the Earth's atmosphere and N_d is the day number (starting at January 1st).

The extraterrestrial solar irradiance falling on a surface parallel to the ground (Fig. 1.3) is [359]

$$E_{ext,h} = E_{ext}\cos(\theta_z) \tag{1.2}$$

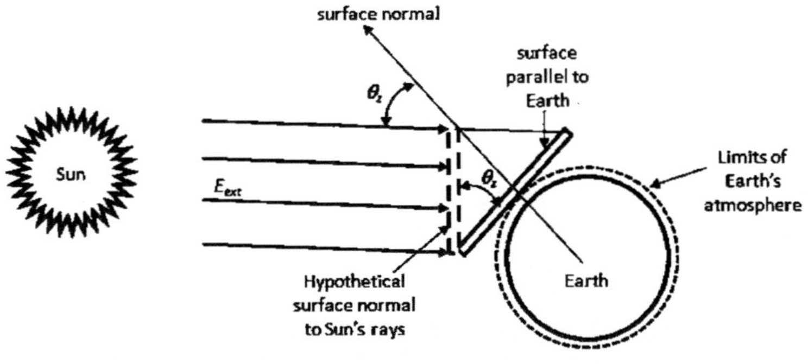

Fig. 1.3 The cosine effect as it relates to the concept of extraterrestrial horizontal irradiance, [359]

where θ_z the angle between the two surfaces, which is the solar zenith angle. Reduction of radiation by the cosine of the angle between the solar radiation and a surface normal is called the cosine effect.

Because of the cosine effect, the extraterrestrial solar irradiance on a horizontal plane varies cyclically as the Earth spins on its axis. The amount of solar radiation received on a horizontal surface outside the atmosphere forms an upper limit to the amount of radiation that will fall on a horizontal surface below the Earth's atmosphere. It also describes the cosine effect without the complication of air mass and cloud cover [359].

1.2.3 Measurement of Solar Irradiance

Pyranometers are the sensors used to measure global solar irradiance, that is, the Sun's energy coming from all directions in the hemisphere above the plane of the instrument. The measurement is of the sum of the direct and the diffuse solar irradiance.

To measure the direct normal component of the solar irradiance only, an instrument called a normal incidence pyrheliometer (NIP), or simply pyrheliometer is used.

The diffuse irradiance can be measured by modifying a pyranometer using a shadowing device large enough to block the direct irradiance onto this sensor.

The description of different pyranometers and pyrheliometers can be found in [359].

1.2.4 The Sun's Position

Along with the weather conditions, another factor that determines the amount of incident radiation on a solar collector is the apparent movement of the Sun across

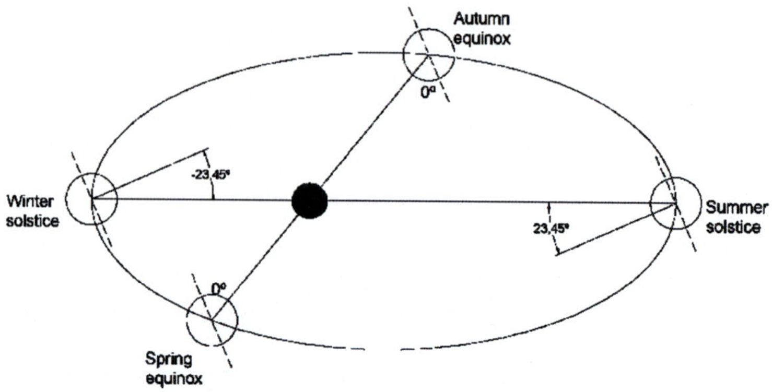

Fig. 1.4 Movement of the Earth's rotation around the Sun

the sky during the day. The Earth's orbit around the Sun has an elliptical trajectory, with 3% eccentricity. The imaginary line that represents the orbit described by the Earth is called the ecliptic. As it orbits, the axis of terrestrial rotation always forms the same 23.45° angle between the perpendicular and the ecliptic plane. The angle formed by the ecliptic plane and the equator varies during the year as shown in Fig. 1.4. This angle, known as the declination, varies from −23.45° on the winter solstice to 23.45° on the summer solstice. The Tropic of Cancer (23.45° north) and the Tropic of Capricorn (23.45° south) are the farthest latitudes at which the Sun is located on the perpendicular to the horizon one instant per year, at noon on the summer and winter solstices, respectively.

1.2.5 Geometry of the Sun's Movement

From the point of view of an observer on the Earth's surface, the Sun seems to describe an arc from sunrise to sunset. The local meridian plane by definition is located in the middle of this path at solar noon. The vertical direction of the observer's position on the Earth's surface intersects the sky vault at a point called the zenith. The Earth's axis forms an angle equal to the latitude of the location (ϕ) with the observer's horizontal plane.

The latitude angle ϕ is the angle between a line drawn from a point on the Earth's surface to the center of the Earth and the Earth's equatorial plane. The intersection of the equatorial plane with the surface of the Earth forms the equator and is designated as 0° latitude. The Earth's axis of rotation intersects the Earth's surface at 90° latitude (North Pole) and −90° latitude (South Pole). Any location on the surface of the Earth then can be defined by the intersection of a longitude angle and a latitude angle.

The Sun position may be referred to by two systems of coordinates centered on the observer, according to the system of reference selected: time coordinates

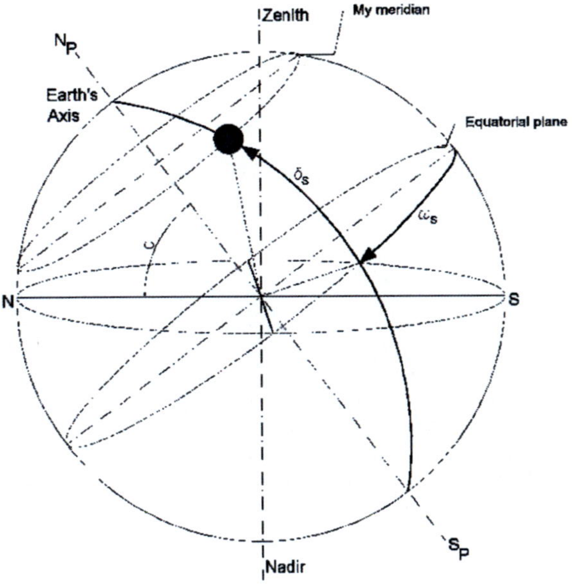

Fig. 1.5 Time celestial coordinates

(δ_s: declination, ω_s: solar hour) and horizontal coordinates (h_s: solar elevation, a_s: azimuth). These coordinates determine the solar vector, understood as a vector going from the observer to the Sun.

Figure 1.5 shows the time coordinates, where:

- δ_s is the solar declination [°]: Sun position angle with the terrestrial equatorial plane at solar noon.
- ω_s is the solar hour angle [°]: angular movement of the Sun on the plane of the solar trajectory. The origin of the hour angle is taken as solar noon and widens in the direction of movement of the Sun. Each hour is equal to 15°.

Figure 1.6 shows the horizontal coordinates, where:

- h_s is the solar elevation [°]: angle formed by the direct solar radiation and the horizontal plane. The complementary angle is the solar zenith angle θ_z.
- a_s is the solar azimuth [°]: angle formed by the direct solar radiation and the observer meridian. The origin of the azimuth is solar noon and is widens clockwise facing north from south of the location (in the northern hemisphere).

1.2.5.1 Calculating the Time Coordinates

The horizontal coordinates of the Sun vary depending on the time of day, day of year and latitude of the location. On the contrary, the time coordinates are easier to find, as the declination depends only on the day of the year and the hour angle at the time. The time coordinates are determined as a preliminary step for calculating the horizontal coordinates. Most solar calculations require solar time but our clocks

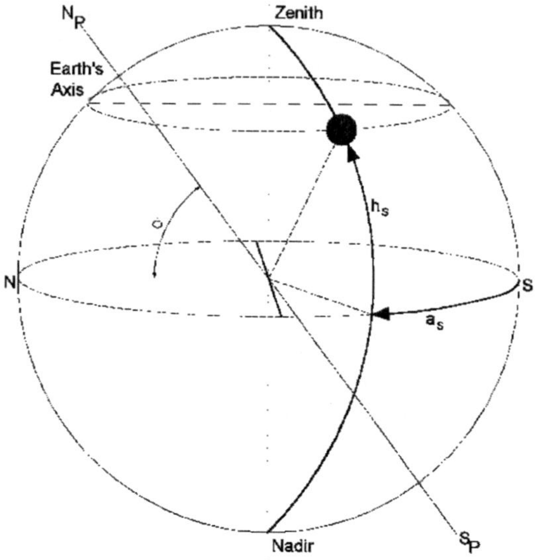

Fig. 1.6 Horizontal celestial coordinates

show the local time. There is a convention for using time, setting the civilian time in different cities with different solar times. The solar time (t_s) is calculated by

$$t_s = \text{local standard time} + 4(L_{ref} - L_{loc}) + \Delta t \quad [\text{min}] \tag{1.3}$$

where L_{ref} is the geographic longitude of the reference meridian [°] and L_{loc} the longitude of the local meridian [°]. A second correction is necessary because the Sun gets ahead of or behind the mean solar time. This equation is known as the equation of time:

$$\Delta t = 9.87\sin^2(B) - 7.52\cos(B) - 1.5\sin(B) \quad [\text{min}] \tag{1.4}$$

where $B = 360(J - 81)/364$ [°] and J the Julian day of the year. The equation of time is the difference between the local apparent solar time and the local mean solar time. There exist different approximations with different degrees of error [359].

When the solar time is known, the hour angle is calculated recalling that the origin of the coordinates is at the local meridian and that one solar hour is 15°.

$$\omega_s = 15(t_s - 12) \quad [°] \tag{1.5}$$

where t_s here is the solar time in hours.

The hour angle is negative in the morning and positive in the afternoon. The declination of the Sun depends on the day of the year as expressed by

$$\delta_s = 23.45\sin\left(360\frac{284 + J}{365}\right) \quad [°] \tag{1.6}$$

1.2.5.2 Calculating the Horizontal Coordinates

The horizontal coordinates (h_s, a_s) are calculated using spherical trigonometry based on the time coordinates, related by the equations

$$\begin{aligned} \sin(h_s) &= \sin(\phi)\sin(\delta_s) + \cos(\phi)\cos(\delta_s)\cos(\omega_s) \\ \sin(h_s)\cos(a_s) &= \sin(\phi)\cos(\delta_s) + \cos(\omega_s) - \cos(\phi)\sin(\delta_s) \\ \sin(h_s)\sin(a_s) &= \cos(\phi)\sin(\omega_s) \end{aligned} \quad (1.7)$$

From these equations the maximum solar elevation for a determined day and latitude can be found. At noon the hour angle and the solar azimuth angle are zero, $\omega_s = 0°$ and $a_s = 0°$, corresponding to the maximum elevation ($h_{s,max}$). Using these values in Eq. (1.7) gives

$$h_{s,max} = 90° - \phi + \delta_s \quad [°] \quad (1.8)$$

In [342], also meteorological, topographic and astronomic effects are studied.

1.3 Technology Classification

Solar energy has been used throughout time, mainly for heating and lighting but also for many other purposes such as refrigeration [413], detoxification [55], desalination [131, 415] and primarily for the generation of electricity. Solar powered electrical generation can be achieved either directly, by the use of photovoltaic (PV) cells, or indirectly, by collecting and concentrating solar power (CSP) to produce steam which is then used to drive a turbine to provide electrical power [139]. In [254] various advanced solar thermal electricity technologies are reviewed with an emphasis on new technology and new market approaches.

Figure 1.7 shows a diagram of basic solar energy conversion systems [359], able to convert the solar resource into a useful form of energy. The function of a solar collector is to intercept the incoming solar radiation and to change it into a usable form of energy. Storage or auxiliary systems are necessary if solar energy conversion systems are not connected to an electrical transmission grid [359].

In [205] various types of solar thermal collector and of application are presented, including flat-plate, compound parabolic, evacuated tube, parabolic-trough, Fresnel lens, parabolic dish and heliostat field collectors. This is followed by an optical, thermal and thermodynamic analysis of the collectors and a description of the methods used to evaluate their performance. Regarding the applicability of the collectors, some applications are related, like solar water heating (which comprise thermosyphon, integrated collector storage, direct and indirect systems and air systems), space heating and cooling (which comprise, space heating and service hot water, air and water systems and heat pumps, refrigeration), industrial process heat (which comprise air and water systems and steam generation systems), desalination, thermal power systems (which comprise the parabolic trough, power tower and dish systems), solar furnaces and chemistry applications. As can be seen, solar energy

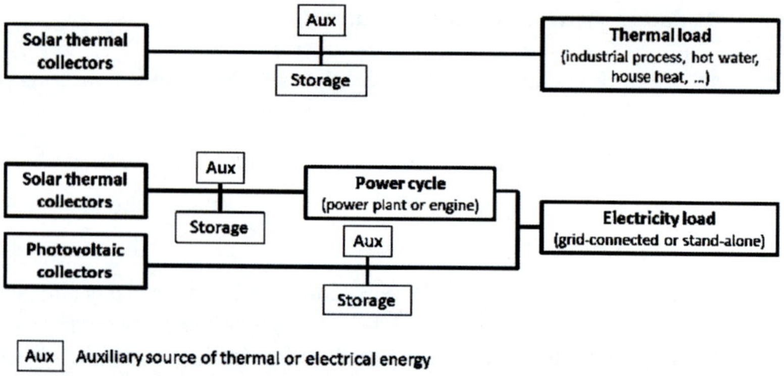

Fig. 1.7 Diagram of solar energy conversion systems, [359]

systems can be used for a wide range of applications and provide significant benefits.

In [342] a summary of solar energy sources and conversion methods is presented. Applications of solar energy in terms of low and high temperature collectors are given with future research directions. PV devices are discussed for future electric energy generations based on solar power site-exploitation and transmission by different means over long distances such as fiber-optic cables. Another use of solar energy is its combination with water and as a consequent electrolysis analysis generation of hydrogen gas, which is expected to be another form of clean energy sources. Combination of solar energy and water for hydrogen gas production is called solar-hydrogen energy. Possible future new methodologies are also mentioned in [342].

Most of the experiments described in this book have been performed at the Plataforma Solar de Almería (PSA, www.psa.es). This experimental facility belongs to the Centro de Investigaciones Energéticas Medioambientales y Tecnológicas (CIEMAT) and is the largest European center for research, development and testing of concentrating solar technologies. The PSA has two tower plants (CESA-1 and CRS), various linear focusing facilities and distributed solar collector fields (DISS and ACUREX), a solar furnace, a few solar dishes and a number of other experiments (Fig. 1.8).

1.3.1 *Electricity Generation*

1.3.1.1 Photovoltaic Plants

The direct generation of electricity from solar energy is based on the photovoltaic effect which refers to the fact that photons of light hitting certain materials will knock electrons into a higher state of energy producing an electrical current.

Although the first PV cells were used to generate electrical power for spacecrafts, there are many PV power generation systems for more normal applications such as

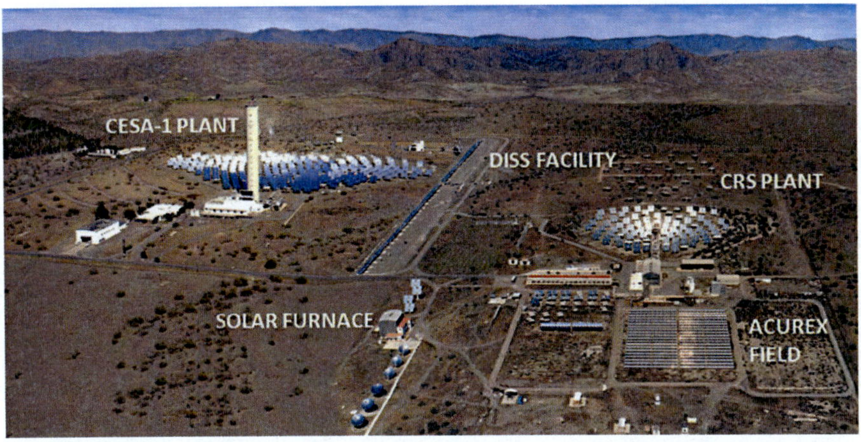

Fig. 1.8 Aerial view of the PSA facilities (courtesy of PSA, www.psa.es)

Fig. 1.9 Types of photovoltaic plant

houses isolated from the grid, pumps for water extraction, electric cars, roadside emergency telephones, remote sensing and cathodic protection of pipelines. When the power generated by the PV cells is transmitted to the grid, an inverter to convert the direct current (DC) to alternating current (AC) is required. PV cells are usually protected by a glass sheet and are connected together to form solar panels. Multiple solar panels or modules must be assembled as arrays when more power production is required. The prices of solar panels are still too high to compete with other non-renewable fuels. However, in most places, significant financial incentives such as the feed-in tariff (FIT) have triggered a huge increase in the demand for PV panels. Photovoltaic production is the fastest-growing energy technology. It has been increasing by almost 50% percent every year. At the end of 2010, the cumulative global amount of PV installations reached 15200 MW. Figure 1.9 shows the main types of photovoltaic plant according to the selected mechanism for tracking the Sun (in Chap. 2 the Sun tracking fundamentals are studied).

1.3.1.2 Thermosolar Plants/Concentrating Solar Thermal Systems

Concentrating solar thermal (CST) systems use optical devices (usually mirrors) and Sun tracking systems to concentrate a large area of sunlight onto a smaller receiving area. The concentrated solar energy is then used as a heat source for a conventional power plant. A wide range of concentrating technologies exists. The main concentrating concepts are: (a) parabolic troughs, (b) solar dishes, (c) linear Fresnels and (d) solar power towers. The main purpose of concentrating solar energy is to produce high temperatures and, therefore, high thermodynamic efficiencies.

Parabolic Troughs

Parabolic-trough systems are the most used CSP technology. A parabolic trough consists of a linear parabolic mirror that reflects and concentrates the received solar energy onto a tube (receiver) positioned along the focal line. The parabolic mirror follows the Sun by tracking along a single axis. The heat transfer fluid (HTF), typically synthetic oil or water, is pumped through the receiver tube and picks up the heat transferred through the receiver tube walls. Then, it is routed either to a heat exchanger when this fluid is oil, to produce steam that feeds an industrial process (for instance a turbine), to an ash tank when the fluid is pressurized water, to produce steam for an industrial process, or to a turbine when superheated and pressurized steam is produced directly in the solar field [416].

In [139] an overview of the parabolic-trough collectors that have been built and marketed during the past century, as well as the prototypes currently under development are studied. It also presents a survey of systems which could incorporate this type of concentrating solar system to supply thermal energy up to 400°C, especially steam power cycles for electricity generation, including examples of each application.

In order to provide viable power production, parabolic troughs have to achieve their task despite fluctuations in energy input, i.e. solar radiation. An effective control scheme is needed to provide operating requirements of a solar power plant. Most of the plants that are for the currently operational, such as the SEGS (Solar Electricity Generating System) plants in California [304], provide this energy in the form of oil heated by a field of parabolic-trough collectors. However, the processes usually connected to such fields for electricity generation [304, 398] or sea-water desalination [415] are most efficient when operated continuously. To do this, they must be provided with a constant supply of hot oil at some pre-specified temperature despite variations in the ambient temperature, the inlet temperature and the direct solar irradiance. This requirement prompted the use of a storage tank as a buffer between solar collection and the industrial process at early plants such as the SSPS system at the PSA, Spain [398] and SEGS I in California.

For this purpose, later plants (SEGS II–IX) operated a gas fired boiler running in parallel to the solar field in order to make up any shortfalls in the solar produced steam [304]. While these facilities enable the overall power output of the plant to be maintained during shortfalls, they do not remove the requirement for a fixed quality energy output from the field, in the form of tight outlet temperature control.

Maintaining a constant supply of solar produced thermal energy in the face of disturbances is not the purpose of this control because it is not a cost effective strategy; in theory, oversized collector fields could be used with parts only operating

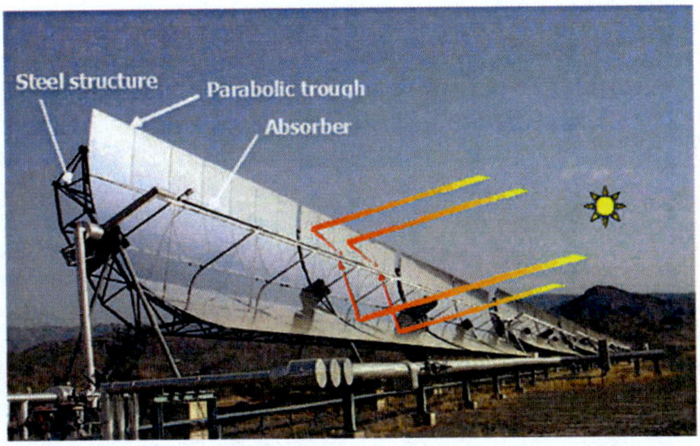

Fig. 1.10 Parabolic troughs at the PSA (courtesy of PSA)

during periods of low solar radiation. The aim of a control scheme should rather be to regulate the outlet temperature of the collector field by suitably adjusting the oil flow rate [398]. This is beneficial in a number of ways. Firstly, it supplies any available thermal energy in a usable form, i.e., at the desired operating temperature, which improves the overall system efficiency and reduces the demands placed on auxiliary equipment such as the storage tank. Secondly, the solar field is maintained in a state of readiness for the resumption of full scale operation when the intensity of sunlight rises once again; the alternative is unnecessary shutdowns and startup procedures which are both wasteful and time consuming. Finally, if the control is good, i.e., fast and well damped, the plant can be operated close to design limits, thereby improving productivity.

During the last thirty years considerable effort has been devoted by many researchers to improve the efficiency of solar thermal power plants with distributed collectors from the control and optimization viewpoints. Most of the work carried out and summarized in [85, 87, 88] (and extended in this book) has been devoted to improve the operation of the ACUREX field of the SSPS plant located in the PSA, Spain, which uses a parabolic-trough system using industrial oil as the HTF because commercial plants for electricity production [300] and the facilities available to tests automatic controllers are using this fluid. However, there are also some recent experiences of controlling parabolic-trough systems using water/steam as the HTF, like the DISS plant at the PSA shown in Fig. 1.10, [130, 226, 383, 384, 416, 419], or distributed collector solar fields using molten salt as the HTF [301]. The DISS plant consists of a parabolic-trough collector where the feed water is preheated, evaporated and converted into superheated steam as it circulates through the absorber tubes. The field consist of a single 665 m long row of parabolic-trough collectors with a total solar collecting surface of 3838 m^2. The modeling and control of these installations is dealt with in this book in Chaps. 4 and 5.

Fresnel Collectors

Linear Fresnel reflectors use various thin mirror strips to concentrate sunlight onto tubes where fluid is circulating (Fig. 1.11). Higher concentration can be obtained

Fig. 1.11 Fresnel linear reflector at the School of Engineers of Seville University

and these mirrors are cheaper than parabolic mirrors, but a more complex tracking mechanism is needed. Absorbers are located at the focal point of the mirrors and consist of an inverted air cavity with a glass cover enclosing insulated tubes. In some cases there are multiple absorbers to improve system efficiency.

Tower Plants with Central Receiver Systems

Solar power towers have an array of reflectors (heliostats) that are able to track the Sun's movement with two axis and concentrate solar radiation onto a central receiver at the top of a tower. The working fluid in the receiver is heated and then used for power generation. Power tower development is less advanced than trough systems but they offer higher efficiency and better energy storage capability.

A Solar Power Tower (SPT) plant consists of the heliostat field, receiver unit, heat transfer, exchange and storage unit, steam and electricity production units and the integrated control system. Usually, each of the units has its specific control device. The integrated control system communicates with the different subsystems to coordinate the different units in such a way that the plant operates in a safe and efficient way. Typically, a plant control system includes heliostat control and heliostat field dispatch optimization, water level control in receivers, main steam temperature control, steam supply pressure and temperature in heat storage system control under heat releasing condition and the main steam pressure control. At present, there are only two commercial tower power plants in operation, the 10 MW (PS10) and the 20 MW (PS20) plants, designed, built and operated by Abengoa Solar close to Seville in Southern Spain (Fig. 1.12).

The PSA has two exceptional facilities for testing and validating central receiver technology components and applications, the CRS and CESA-1. The SSPS-CRS plant was inaugurated in 1981 and it was originally an electricity production demonstration plant using a receiver cooled by liquid sodium which also served as a thermal storage fluid. The 43 m high metal tower has two test platforms. The CESA-I project, was built in 1983. Direct solar radiation is collected by the collector field

Fig. 1.12 PS20: 20 MW solar power tower (courtesy of Abengoa, www.abengoasolar.com)

of 300 heliostats (with 39.6 m^2 reflective surface each, Fig. 1.13). A maximum of 7 MW thermal power can be delivered by the field onto the receiver. The 80 m high concrete tower has three test levels, at 45 m, at 60 m and at 80 m. The modeling and control of these installations will be treated in Chap. 6.

Dish Collectors

A dish engine system consists of a parabolic collector dish that concentrates light onto a receiver positioned at the focal point of the reflector. The reflector is able to track the Sun along two axes. A Stirling engine positioned at the focal point of the parabola is used to generate power. Solar dish systems are modular and have

Fig. 1.13 CESA-1 tower plant at PSA (courtesy of PSA)

Fig. 1.14 Dish collector at PSA (courtesy of PSA)

comparatively high conversion efficiency and can be deployed either individually or grouped together.

Dish concentrators approximate a paraboloid of revolution. The mirrors require relatively high curvature because the focal length of the dish is only a few meters. Some concentrators use multiple mirrors to approximate the ideal concentrator shape. Other concentrators use a thin reflective membrane which is given the right shape by using vacuum. The concentration ratio of solar dish systems is usually around 2000 suns. The receiver, located at the focal point of the collector, absorbs energy reflected by the concentrator (Fig. 1.14).

The conversion of heat to mechanical energy in solar dish systems is done by compressing a working cold fluid, heating the compressed fluid and finally expanding it in a piston or turbine to produce work. An electric generator is used to convert the mechanical power to electrical power. The electrical power output in current Stirling dish systems is about 25 kWe.

1.3.2 Other Applications

1.3.2.1 Solar Furnaces

Solar energy has been used to produce high temperatures in solar furnaces. Solar radiation is collected by an array of mirrors which concentrate the solar radiation light onto a focal point. The largest solar furnace is Odeillo furnace built in 1968 in the Pyrenees, a mountain range between France and Spain. The furnace has a 16000 concentration factor and is capable of reaching temperatures of up to 3500°C. It uses an array of mirrors to collect and reflect solar radiation onto a larger curved mirror which focuses onto an area the size of a few square decimeters where samples are placed for material testing purposes (Fig. 1.15).

The PSA has a solar furnace facility which is able to reach concentrations of over 10000 suns. The main application field is materials testing. The solar furnace consists of a continuously solar-tracking heliostat field, a parabolic concentrator

Fig. 1.15 Solar furnace in Odeillo, France (www.sollab.eu/doctoralcolloquium6/)

mirror, an attenuator or shutter and the test zone located in the concentrator focus. In Chap. 7, Sect. 7.2, the modeling and control of the solar furnace of the PSA will be studied in detail.

1.3.2.2 Solar Cooling

Solar energy cannot only be used for heating but also for cooling purposes. Solar cooling also has the advantage of requiring little energy storage, because the needs of cold air match solar radiation very well.

There are various ways in which solar energy can be used for cooling [177, 209]. Absorption technology is the most popular and consists of using solar thermal energy to feed thermally driven chillers. Absorption chillers use a refrigerant which is absorbed by a liquid sorbent. The thermal solar energy is injected in the generator where the refrigerant is desorbed from the refrigerant–sorbent solution. The refrigerant vapor pressure generated is sufficient to condense the refrigerant in the condenser. The refrigerant vapor is absorbed back into the solution which is cooled in the absorber. The solution is pumped to the generator by a pump. Most absorption chillers use water as refrigerant and lithium bromide as solvent or ammonia as refrigerant and water as solvent. If water is used as the refrigerant (water/lithium bromide pair), the cold temperature is above 0°C. The absorption chillers using ammonia/water can generate evaporator temperatures as low as to −60°C and can be used for industrial cooling systems. Another type of chiller, adsorption chillers, uses water as a refrigerant which is adsorbed in a solid sorbent such as silica gel (among disposal of latent heat on the surface). The latent heat can decrease to zero when enough water molecules are added and then only evaporation heat is dissipated. The desorption of the stored water and the generation of enough pressure for the condensation can be done at temperatures below 70°C and the technology is appropriate for solar thermal technology with little concentration factors.

In a typical solar absorption cooler with a water/lithium bromide pair (Fig. 1.16), the solar energy collected is used at the generator to boil water from the lithium bromide and water solution. The water vapor is cooled down in the condenser and from there it goes to the evaporator where it is evaporated at low pressure providing the

Fig. 1.16 Solar cooling system at CIESOL R&D Center, University of Almería

required cooling. The heat of the strong lithium bromide solution is used to preheat the weak solution entering the generator. The water vapor leaving the evaporator is absorbed by the strong solution in the absorber. An auxiliary energy source, usually a gas boiler, is used to supply hot water to the chiller when the solar radiation is not sufficient.

Double or triple effect absorption cooling cycles have been used in different solar chillers to increase efficiency. Double effect absorption chillers require hot water temperatures of around 180°C. Since a flat-plate solar collector can only heat water up to 80°C, some concentration is needed. Fresnel and parabolic troughs are normally used. Solar cooling control systems will be treated in Chap. 7, Sect. 7.3.

1.3.2.3 Water Treatment and Solar Desalination Plants

One interesting field of solar chemistry applications is the solar photochemistry. Solar photochemical processes make use of the spectral characteristics of the incoming solar radiation to effect selective catalytic transformations which find application in the detoxification of air and water and in the processing of fine chemical commodities [205]. In solar detoxification photocatalytic treatment of non-biodegradable persistent chlorinated water contaminants typically found in chemical production processes is achieved. For this purpose parabolic-trough collectors (PTC) with glass absorbers are employed and the high intensity of solar radiation is used for the photocatalytic decomposition of organic contaminants. The process uses UV energy, available in sunlight, in conjunction with the photocatalyst, titanium dioxide, to decompose organic chemicals into non-toxic compounds. Another application concerns the development of a prototype employing lower concentration compound parabolic concentrators (CPC) [205].

An example are the photo-reactors placed in the PSA (Fig. 1.17(a)), where advanced oxidation processes allow to oxidize and mineralize almost any organic contaminant. Among them, the Photo-Fenton method is known for its high reaction rates and its applicability with natural sunlight, which is a cheap, sustainable photon source and an adequate solar technology developed to pilot-plant scale. The process cost may be considered the main reason that serious doubts are harbored as to its feasibility for commercial application. Research, therefore, focuses on ways to reduce these costs, through the application of sunlight instead of expensive UV lamps as a source of the required photons, study of the influence of process parameters on the reaction rates in order to increase reactor throughput, or the evolution

Fig. 1.17 Solar desalination plant at water detoxification loop at the PSA (courtesy of PSA)

of toxicity and biodegradability during treatment to determine the best point for discharging waste water or coupling with subsequent biological treatment. Experimental design approaches have been employed for pilot-plant experimentation to upscale the process and optimize pilot-plant experimentation strategies [7].

Water desalination can be achieved by using a number of techniques. These may be classified into the following categories: (i) phase-change or thermal processes and (ii) membrane or single-phase processes [205]. In the phase-change or thermal processes, the distillation of sea water is achieved by utilizing a thermal energy source. In the membrane processes, electricity is used either for driving high pressure pumps or for ionization of salts contained in the sea water. Desalination processes require significant quantities of energy to achieve separation. This is highly significant as it is a recurrent cost which few of the water-short areas of the world can afford.

Solar energy can be used for sea-water desalination either by producing the thermal energy required to drive the phase-change processes or by producing electricity required to drive the membrane processes. Solar desalination systems are thus classified into two categories, i.e. direct and indirect collection systems. As their name implies, direct collection systems use solar energy to produce distillate directly in the solar collector, whereas in indirect collection systems, two subsystems are employed (one for solar energy collection and one for desalination) [205].

In [318] a description of a technical development in desalination that combines a thermal desalination system and a solar field with a double effect absorption heat pump to reduce the cost of water production is presented. Figure 1.17(b,c) shows the AQUASOL plant at the PSA, where a solar distillation technology based on a multi-effect distillation (MED) plant with 3 m^3/h nominal distillate production. The MED plant (Fig. 1.17(b)) consists of 14 effectors in vertical arrangement at decreasing pressures from cell 1 to cell 14. Sea water is preheated and pumped to the first cell where it goes down to the following cells by gravity at the same time that part of the water is evaporated. For optimal operation, the feed-water inlet temperature in the first cell must be 66.5°C. It is possible to reach this temperature with heat from a solar field and also by steam generated by an auxiliary gas boiler coupled to a double effect absorption heat pump, which can work at variable loads of the steam (from 30% to 100%). Solar collector field consists of CPC collectors (Fig. 1.17(c)) coupled to two water storage tanks, 12 m^3 capacity each one, used as heat storage system. Energy supplied by the solar field and double effect absorption heat pump is transferred to a thermal storage tank using water as HTF. The AQUASOL plant can operate in three different modes; solar, fossil and hybrid (for more details see [318]).

1.4 Energy Storage

Intermittent energy sources are, in general, unpredictable as the generated power depends on the weather. In electrical grid energy sources, stored energy has to be used to balance the mismatch between the energy production from intermittent energy sources and the demand [185]. Energy is stored in fuels (coal, oil, gas) or other systems which store electrical, mechanical, thermal or chemical energy. Battery systems have been used in grid isolated small, usually domestic, systems. Batteries are expensive with high maintenance costs and limited lifespan. Hybrid and electric cars will also be able to store a substantial amount of energy if they are mass-produced. Electric utilities are planning to use plug-in vehicle batteries to store electricity. Mechanical energy can be stored as compressed air, flywheel or by pumped hydro storage. Water is pumped to a high storage reservoir using the excess capacity. Pumped storage is the most cost effective way of mass power storage, recovering about 75% of energy. Its main drawback is that two adjacent reservoirs with a significant difference in height are required. Energy from pumped water storage can be delivered very rapidly, normally in less than one minute and they are appropriate for grid regulation. Hydrogen can be produced by the electrolysis of water into hydrogen and oxygen or by reforming natural gas with steam and then converted back to electricity by fuel cells. Thermal energy storage is used by solar thermal plants which can use oil or molten salt as a heat store (Fig. 1.18) or directly store the steam produced. This stored energy can be used to generate electricity at night or when solar radiation is not sufficient to cover the production program. In thermal solar power plants, thermal storage and/or fossil backup act as the following [359].

- An output management tool to prolong operation after sunset, to shift energy sales from low revenue off-peak hours to high revenue peak demand hours and to contribute to guaranteed output.
- An internal plant buffer, smoothing out insolation charges for steadying cycle operation and for operational requirements such as blanketing steam production, component pre-heating and freeze protection.

(a) Oil storage (Acurex Plant at PSA)

(b) Molten salt storage (Andasol Plant in Spain)

Fig. 1.18 Examples of energy storage systems ((**a**) courtesy of PSA, (**b**) courtesy of ACS, http://www.protermosolar.com/boletines/boletin08.html)

In general, thermal storage is a physical or chemical process which takes place in the storage tank during system charging and discharging. The storage system consists of the storage tank, the storage medium and the specific charge and discharge devices. Thermal energy is transferred from the solar receiver to storage (charging) and from storage to the conversion system (discharging) by a HTF.

Thermal storage systems employed in solar thermal power plants, whether for power tower or parabolic-trough collector technologies, are based on:

- Sensible heat storage in saturated liquids.
- Sensible heat storage in solids.
- Latent heat storage.
- Thermochemical storage.

Both liquids and solids are used for storage of sensible heat. In addition to the thermophysical properties of the material employed, the cost of this material also becomes relevant to selection [185]. For temperatures up to 300°C, the most economical solution is mineral thermal oil at atmospheric pressure. For temperatures up to 410°C, synthetic or silicon oils are used, but they must work at low pressure and are more expensive. Molten salt and sodium are used between 300 and 550°C at atmospheric pressure, but they require antifreeze systems [402]. At higher temperatures, ceramic materials begin to be a competitive alternative.

Table 1.1 Overview of thermal storage systems that have been tested on large scale

Location	Storage	Primary coolant	Design temperature (°C)	Design
SEGS-1 Ca	Oil	Oil	240–307	2 tanks: hot/cold
SSPS Al	Oil	Oil	225–295	1 thermocline tank
SSPS Al	Oil/cast iron	Oil	225–295	1 hybrid tank
Solar-1 Ca	Oil/rock/sand	Steam	224–304	1 hybrid tank
CESA-1 Al	Molten salt	Steam	220–340	2 tanks: hot/cold
Solar-2 Ca	Molten salt	Molten salt	288–566	2 tanks: hot/cold
SSPS Al	Liquid sodium	Liquid sodium	270–530	2 tanks: hot/cold
TSA Al	Ceramic balls	Air	200–800	1 hybrid tank

Phase-change materials (PCM) offer important advantages, for example, high volumetric capacities and rather stable temperatures. Numerous latent heat storage systems using expendable metals or salt have been proposed and studied, but they have never been tested on large scale due to their difficult heat exchange [31, 124].

The use of reversible chemical reactions has great potential for offering a highly efficient, high-density thermal storage system. Chemical storage is especially appropriate for high temperatures and for long-term storage (several days or weeks). They generally require reactions that are completely reversible and for the equilibrium temperature to be the same as the charge/discharge temperatures in the specific system. The reactions may be homogeneous gas or heterogeneous solid–gas or liquid–gas reactions [222].

The basic operating strategy of a solar thermal plant is the one in which solar only operates the power block. If the solar field has exactly the right size to provide the energy the conversion unit needs at rated power, it is called a solar multiple equal 1. The solar multiple is closely related to the plant design point. If a solar multiple equal 1 is selected for June 21st, the plant will only operate at its rated power one day a year (northern hemisphere). If, on the contrary, it is selected for December 21st, the plant will operate all year at rated power but the extra energy will be wasted. Thermal storage systems provide the possibility of employing this excess energy at certain times of the year (summer) without increasing the solar field.

The specific solar plant technology, operating temperature and coolant chosen significantly affect thermal storage, which may either:

- use the same fluid as primary coolant and for thermal storage;
- use different fluids for coolant and storage.

In the first case only a storage tank or cold tank/hot tank design is used, while for the second there must be a hybrid storage system or regenerator. Table 1.1 gives an overview of thermal storage systems that have been tested on large scale.

In [185] the main characteristics of the different electricity storage techniques and their field of application (permanent or portable, long- or short-term storage, maximum power required, etc.) are analyzed. These characteristics serve to make comparisons in order to determine the most appropriate technique for each type of application.

1.5 Summary

In this chapter, solar energy fundamentals have been briefly introduced highlighting those aspects that are used in the following chapters. An overview of the main solar thermal technologies has also been performed, including general descriptions and examples, as well as an introduction to those facilities at the PSA that are used in this book to show illustrative examples of modeling and control techniques.

Chapter 2
Control Issues in Solar Systems

2.1 Introduction

This chapter is devoted to introduce the main issues involved in the control of solar energy systems. Four different levels can be distinguished: (i) the control of the solar collector units, (ii) solar radiation estimation and forecast, (iii) the control of the energy conversion systems and (iv) the overall control of the complete process.

The control of the solar collecting systems consist of controlling the solar collectors movements in such a way that the maximum solar energy is collected at any time. In the case of dish collectors with two degrees of freedom, the controls system keeps the collector surface perpendicular to the solar vector. In the case of solar trough or collectors with one degree of freedom, the mission of the controller is to maintain the solar collector surface normal as close as possible to the Sun vector. The controller has to compute the Sun vector, which depends of the geographical position of the collector and the date and time of day. The position of the collectors is then determined with a trigonometrical computation and this is sent to the servo controlling the collector axes. Fine tracking is obtained in some cases by using signals which depend on the angle formed by the collector surface normal and the solar vector.

In the case of solar power towers or Fresnel linear solar collectors, the reflecting mirrors have to be moved in such a way that the solar radiation is reflected on the central receiver or the receiver tube. The heliostats have to be controlled in such a way that the solar radiation is reflected on the central receiver. The heliostat surface normal should be in the bisectrix of the angle formed by the solar vector and the vector joining the center of the heliostat reflecting surface and the central receiver.

In order to control solar energy system, it is very important both to know the actual values of solar irradiance and even to be able to forecast this variable within different time windows to be used for control and operation planning purposes. Thus, adequate sensors to obtain values for solar irradiance are used in these kinds of plant (mainly pyranometers and pyrheliometers) and different algorithms to provide estimations of future values of solar irradiance where the solar plant is located.

The control of the variables associated to the solar conversion units depends very much on the type of system. In the case of photovoltaic (PV) systems, this involves

E.F. Camacho et al., *Control of Solar Energy Systems*, Advances in Industrial Control,
DOI 10.1007/978-0-85729-916-1_2, © Springer-Verlag London Limited 2012

the control of the voltage and intensity produced by the solar cells in order to operate at the maximum efficiency point and the controls of the associated DC/AC conversion power electronics. In the case of thermal solar plants, the solar energy heats up a fluid which is then used to produce steam necessary to drive the turbines. The variables to be controlled at this level are the temperatures and flows of the heat collecting fluid. The main difference with other power plants is that the main source of energy, the solar radiation, cannot be manipulated. An important part of this book is dedicated to this issue. The heated HTF is then used to produce steam and from this point, the controls at this level do not differ much from the control of other thermal plants.

The upper control level takes care of the operation of the complete solar system. The control decides what amount of energy is produced and delivered and what amount is stored and which are the set points of the main process variables at any given moment. Since the solar radiation varies along the day, the plant is rarely at a steady state condition and the determination of the optimal operating points should be done dynamically. Solar energy is intermittent and is not available when needed and where needed, and because of this the integration of solar energy plants in the electrical grid is a challenging problem. An efficient operation of solar plants with energy storage systems could allow a closer match between solar plants power generation and electrical power demand. Furthermore, proposals of new price mechanisms are appearing continuously with the liberalization of the energy prices and they are considered to be a fundamental part of the Smart Grid concept. The price fluctuations tend to correlate well with the demand and an efficient operation of a solar power plant installation requires all of these aspects to be considered.

2.2 Sun Tracking

2.2.1 The Need for Tracking the Sun

In order to collect solar energy on the Earth, it is important to know the angle between the Sun's rays and a collector surface (aperture). The collector aperture is defined as the surface of the plane normal to the Sun rays through which non-concentrated radiation enters the collector where it is reflected. The aperture angle in parabolic trough collectors (PTC) is the angle between the axis of the parabola and the line connecting the focus with one end of that parabola.

When a collector (its aperture normal) is not pointing directly at the Sun, some of the energy that could be collected is being lost [359].

In Sect. 1.2.5 the Sun's position angles relative to Earth-center coordinates (ϕ, δ_s, and ω_s) were defined. In the design of solar energy systems it is important to be able to predict the angle between the Sun's rays and a vector normal (perpendicular) to the aperture of the collector. This angle is called the angle of incidence θ_i. The tracking angle ρ_t is the amount of rotation that collectors require to place their aperture normal to the Sun's central ray.

A characteristic fundamental to the capture of solar energy is that the amount of energy incident on a collector is reduced by a fraction equal to the cosine of the angle between the collector surface and the Sun's rays. Knowing the position of the collector and the position of the Sun, equations may be used to predict the fraction of incoming solar energy that falls on the collector [359].

Concentrating solar technologies can make use of only direct solar irradiance, that is, the beam that come directly from the solar disk and not those that could be reflected from the surroundings. Therefore, solar collectors must be equipped with systems that enable the collecting surface to maintain the orientation necessary to reflect and concentrate the beam on the receiver. The Sun's position in the sky vault varies slowly during the day, in a trajectory above the horizon. Such a trajectory is different every day of the year.

2.2.1.1 The Solar Incidence Angle

Each of the CSP systems considered here tracks the Sun in a different way. As described in Chap. 1, the Sun's position with respect to the observer can be described by two angles, the azimuth angle and the elevation angle. A third angle is crucial to calculate the effective solar power usable by a solar collector; this is the incidence angle (θ_i). The effective collecting area is reduced by the cosine of the angle formed by the normal to the collecting surface and the Sun position vector at any instant. The incident solar power must by reduced by applying the factor $\cos(\theta_i)$ in the calculations. This is clearly a reducing factor as $-1 \leq \cos(\theta_i) \leq 1$. Most of the systems considered in this book are tracking the Sun, but most of them do not face it directly, so this *imperfect* tracking leads to the effect of the *cosine factor*. Then, the way to calculate the incident power on the receiver is

$$P_{sol} = E_b G \cos(\theta_i) \tag{2.1}$$

where E_b is the direct normal irradiance [W/m^2], G is the collector aperture area [m^2] and $\cos(\theta_i)$ is the cosine of the incidence angle.

2.2.1.2 The Cosine Factor Effect in the Different Solar-Tracking Systems

A PTC tracks the Sun on azimuth or elevation. The collector can be oriented East–West and track the Sun on elevation, or vice versa. The reflected beam is incident on a tube placed in the focal line of the collector, which moves along with it. Once the horizontal coordinates of the Sun are known, $\cos(\theta_i)$ can be calculated as follows [184]:

- For a collector with axis oriented East–West:

$$\cos(\theta_i) = \left(1 - \cos^2(\delta_s)\sin^2(\omega_s)\right)^{\frac{1}{2}} \tag{2.2}$$

- For a collector with axis oriented North–South:

$$\cos(\theta_i) = \left(\left(\sin(\phi)\sin(\delta_s) + \cos(\phi)\cos(\delta_s)\cos(\omega_s)\right)^2 + \cos^2(\delta_s)\sin^2(\omega_s)\right)^{\frac{1}{2}} \tag{2.3}$$

Linear Fresnel collectors are quite similar, except that the receiver, located well above a set of mobile flat mirrors, does not move, so tracking is on one-angle, or single-axis.

In a central receiver system, a heliostat tracks the Sun in such a way that the reflected rays are incident on a single receiver located on top of a tower. Although this tracking is two-axis, the cosine factor also appears, as the normal to the reflecting surface has an angle of deviation with respect to the Sun position vector.

The following formula can be used to calculate the cosine factor for a given heliostat [161]:

$$\cos(\theta_i) = \frac{\mathbf{n}^T \cdot \mathbf{H}}{\|\mathbf{n}\| \cdot \|\mathbf{H}\|} \tag{2.4}$$

$$\mathbf{n} = \left[\cos(\alpha_n)\sin(a_n), -\cos(\alpha_n)\cos(a_n), \sin(\alpha_n)\right]^T$$
$$\mathbf{H} = \left[\cos(h_t)\sin(a_t), -\cos(h_t)\cos(a_t), \sin(h_t)\right]^T$$
$$\mathbf{S} = \left[\cos(h_s)\sin(a_s), -\cos(h_s)\cos(a_s), \sin(h_s)\right]^T$$

where h_s and a_s are the solar altitude and azimuth angles, h_t and a_t are the fixed tower vector altitude and azimuth angles shown in Fig. 2.1 and h_n and a_n are the heliostat-normal altitude and azimuth angles, measured in the same way as h_s and a_s.

The most efficient tracking is done by the parabolic dish systems, in which tracking is two-axis and the parabola faces the Sun directly when in tracking mode, so there is no *cosine factor* effect. This is possible because the parabola has a single focal point, where the receiver may be placed.

2.2.2 Tracking Systems

A solar tracker is a device that points a solar collector mechanism toward the Sun or directs a reflector mechanism in such a way that it reflects the maximum energy onto a collector device. The relative position of the Sun in the sky changes both with the seasons and the time of day. As has been mentioned in the previous section, the solar power received by a solar collector is equal to the solar irradiance received at that location multiplied by the device surface and by the cosine of the angle formed by the Sun's rays and the normal surface. There are many types of solar-tracking mechanism with different accuracies. Solar tracking can be implemented by using one-axis and for higher accuracy, two-axis Sun-tracking systems.

Power tower heliostats need a good degree of accuracy to ensure that the power is reflected onto the receiver which can be situated hundreds of meters from the

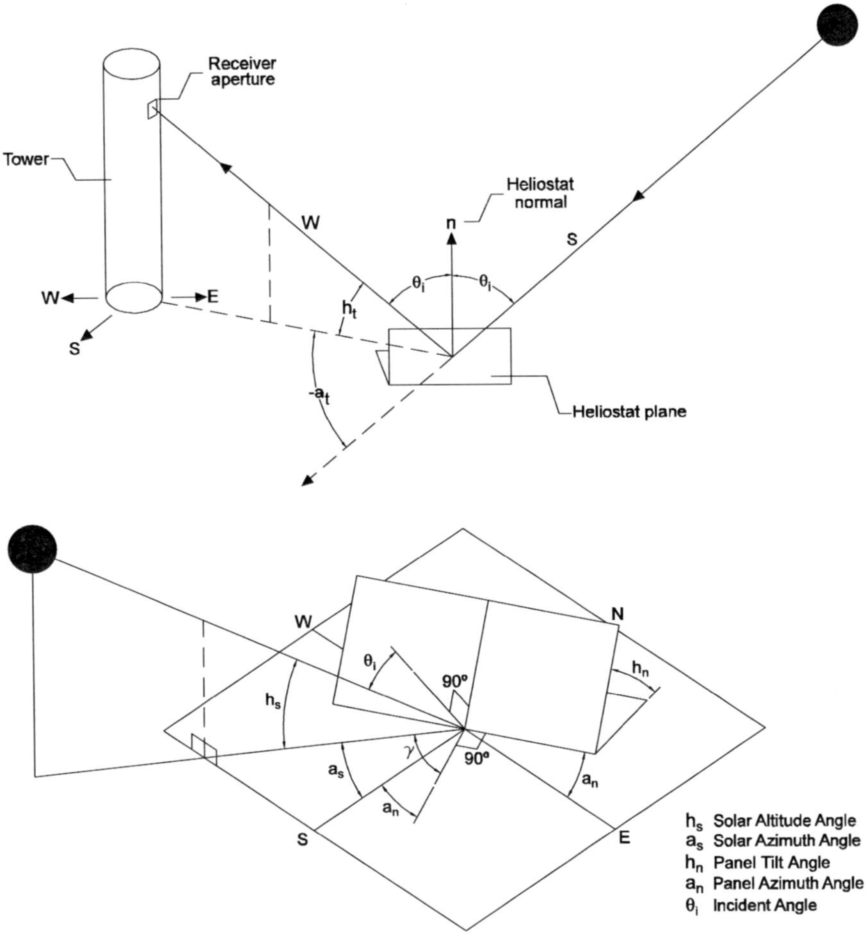

Fig. 2.1 Tracking and incidence angles for a heliostat

heliostat. Little accuracy is required for non-concentrating applications; in fact, most of these applications work without any solar tracking at all.

Tracking can significantly increase the amount of energy produced, especially in the early morning and late afternoon when the cosine of the angle of the direct solar irradiance with the surface normal is smaller.

In [263], a review of principle and Sun-tracking methods is developed focused on PV panels, but useful for other kinds of solar device. All tracking systems have all/some of the following characteristics:

- Single column structure or of parallel console type.
- One or two moving motors.
- Light sensing device.
- Autonomous or auxiliary energy supply.

- Light following or moving according to the calendar.
- Continuous or step-wise movement.
- Tracking all year or all year except winter.
- Orientation adjustment with/without the tilt angle adjustment.

Sun-tracking systems are usually classified into two categories [263]:

- Passive (mechanical) solar trackers are based on thermal expansion of a matter (usually Freon) or on shape memory alloys. Usually this kind of tracker is composed of couple of actuators working against each other which are, by equal illumination, balanced.
- Active (electrical) trackers can be categorized as microprocessor and electro-optical sensor based, computer-based controlled date and time, auxiliary bifacial solar cell based and a combination of these three systems. Electro-optical solar trackers are usually composed of at least one pair of anti-parallel connected photo-resistors or PV solar cells which are, by equal intensity of illumination of both elements, electrically balanced so that there is either no or negligible control signal on a driving motor. In auxiliary bifacial solar cell, the bifacial solar cell senses and drives the system to the desired position and in PC controlled date and time based, a computer calculates the Sun positions with respect to date and time with algorithms and create signals for the system control.

Many fast algorithms for calculating the Sun position used in engineering applications requiring little computation can be found in the literature. The well-known Spencer formula has a maximal error greater than 0.25°. There have been a number of algorithms proposed in literature [56, 102, 163] which increase precision without incurring high computational efforts. These algorithms work correctly for limited periods of time. There are also high-precision astronomical algorithms to compute the Sun's position with an error of less than 0.0003° over a very long period of time (2000 B.C.–6000 A.C.); they require a large amount of computation.

2.3 Solar Irradiance over a PTC

The global solar irradiance E_g over a horizontal surface that does not collect radiation due to reflection or diffusion is composed of the direct solar irradiance E_b (coming directly from the solar disc and measured over a plane normal to the direction of the solar irradiance) and the diffuse irradiance E_d (that subjected to absorption/diffusion processes in the atmosphere and measured over a horizontal plane) according to

$$E_g = E_b \cos(\theta_i) + f_{di} E_d \tag{2.5}$$

where f_{di} is the amount of diffuse irradiance intercepted by the surface. For concentrating collectors [305]:

$$f_{di} = \frac{1}{C_g} \quad \rightarrow \quad E_g = E_b \cos(\theta_i) + \frac{1}{C_g} E_d \tag{2.6}$$

where C_g is the geometrical concentration factor, that is, the ratio between the collector aperture surface and that of the absorbent tube (the part of the solar collector receiving the concentrated solar radiation and transforming it in thermal energy). For PTC with concentration factors over 14, the second term in the right hand of the equation can be neglected, so that $E_g \approx E_b \cos(\theta_i)$.

2.3.1 Optical and Geometrical Losses in a PTC

Optical efficiency η_{opt} is defined as the factor that takes into account energy losses between the reference solar power, P_{sol}, and the power absorbed by the receiver, P_{abs}. These energy losses are optical and geometric and are due to the following:

- Primary concentrator parabolic mirrors are not perfect reflectors, so their specular reflectivity, r, has to be taken into account.
- The glass cover is not perfectly transparent, but lets only part of the incident radiation through, which is determined by its transmissivity, $\tau_{sol,c}$.
- The absorber surface has a certain solar radiation absorptance $\alpha_{sol,A}$.
- There may be errors in the positioning of the receiver on the axis of the parabola which is formed by the mirrors, errors in the parabolic mirror shape itself, errors in solar tracking which cause reflected rays not to intercept the absorber, etc. All of these possible errors are included in the intercept factor, γ_A.

Power absorbed by the receiver, P_{abs}, may therefore be said to be given by

$$P_{abs} = GE_b \cos(\theta_i) r(\theta_i) \tau_{sol,c}(\theta_i, \lambda) \alpha_{sol,A}(\theta_i, \lambda) \gamma_A(\theta_i) \tag{2.7}$$

This equation emphasizes that the variables in it depend on the angle of incidence of the irradiance, θ_i, and for absorptance of the absorber and transmittance of the cover, on wavelength range, λ, which characterizes the incident radiation. Adding $\alpha_{sol,A}(\theta_i)$ and $\tau_{sol,c}(\theta_i)$ to the integrated absorptance of the absorber and transmittance of the cover, respectively, in relation to the solar radiation spectrum, the above expression may be simplified as

$$P_{abs} = GE_b \cos(\theta_i) r(\theta_i) \tau_{sol,c}(\theta_i) \alpha_{sol,A}(\theta_i) \gamma_A(\theta_i) \tag{2.8}$$

According to the definition of optical efficiency, η_{opt}, and according to the nomenclature used above, we have

$$\begin{aligned}
&\eta_{opt} = \frac{P_{abs}}{P_{sol}} \\
&\rightarrow \quad \text{if } P_{sol} = GE_b \cos(\theta_i) \\
&\qquad \rightarrow \quad \eta_{opt}|_{E_b \cos(\theta_i)} = r(\theta_i) \tau_{sol,c}(\theta_i) \alpha_{sol,A}(\theta_i) \gamma_A(\theta_i) \\
&\qquad \text{if } P_{sol} = GE_b \\
&\qquad \rightarrow \quad \eta_{opt}|_{E_b} = r(\theta_i) \tau_{sol,c}(\theta_i) \alpha_{sol,A}(\theta_i) \gamma_A(\pi) \cos(\theta_i) = \eta_{opt}
\end{aligned} \tag{2.9}$$

The definition of efficiency uses direct irradiance, E_b, as the reference, so the second equation in (2.9) is used. The incidence angle modifier, $\kappa(\theta_i)$, is the factor that takes all of the geometric and optical measurements into account, because the incident irradiance is at a certain angle θ_i, that is,

$$\kappa(\theta_i) = \frac{\eta_{opt}(\theta_i)}{\eta_{opt}(\theta_i = 0°)} = \frac{\eta_{opt}}{\eta_{opt,0°}} \tag{2.10}$$

Optical efficiency that considers normal incident irradiance ($\theta_i = 0°$) is called nominal optical performance, $\eta_{opt,0°}$. Optical performance can thus be expressed as

$$\eta_{opt} = \eta_{opt,0°}\kappa(\theta_i) \tag{2.11}$$

$$\eta_{opt,0°} = r(0°)\tau_{sol,c}(0°)\alpha_{sol,A}(0°)\gamma_A(0°) \tag{2.12}$$

and according to the second equation in (2.9),

$$\kappa(\theta_i) = \frac{r(\theta_i)\ \tau_{sol,c}(\theta_i)\ \alpha_{sol,A}(\theta_i)\ \gamma_A(\theta_i)}{r(0°)\ \tau_{sol,c}(0°)\ \alpha_{sol,A}(0°)\ \gamma_A(0°)}\cos(\theta_i) \tag{2.13}$$

The incidence angle modifier can also be provided as a polynomial expression on the incidence angle θ_i. Some authors, however, prefer to distinguish $\cos(\theta_i)$ in the modifier by the angle of incidence, $\kappa(\theta_i)$, writing Eq. (2.11) as

$$\begin{aligned} &\eta_{opt} = \eta_{opt,0°}\kappa(\theta_i)\cos(\theta_i) \\ &\text{considering}\quad \kappa(\theta_i) = \frac{r(\theta_i)\ \tau_{sol,c}(\theta_i)\ \alpha_{sol,A}(\theta_i)\ \gamma_A(\theta_i)}{r(0°)\ \tau_{sol,c}(0°)\ \alpha_{sol,A}(0°)\ \gamma_A(0°)} \end{aligned} \tag{2.14}$$

One of the geometric losses included by definition in the intercept factor, $\gamma_A(\theta_i)$, is the irradiance at the ends (because the collector has a finite length). Some authors [305] also prefer to distinguish this term in the modifier by the angle of incidence since it is merely a geometric reduction factor. They would therefore write Eq. (2.11) as

$$\eta_{opt} = \eta_{opt,0°}\kappa(\theta_i)\Gamma(\theta_i), \quad \text{where } \Gamma(\theta_i) = 1 + \frac{f_d}{L}\left(1 + \frac{G}{48 f_d^2}\right)\tan(\theta_i) \tag{2.15}$$

where f_d is the focal distance of the parabolic mirrors, G is the aperture width and L is the collector length. In any case, with a constant loss coefficient, the higher the optical efficiency, the higher collector performance is. Thus, the higher the variables that define it, the better the performance of the collectors will be. Mirror reflectivity and transmittance of the cover are largely determined by how clean the collector is. It is therefore important to have suitable collector maintenance and cleaning strategy in a PTC plant.

In this book, the measured direct solar irradiance and the corrected solar irradiance will be denoted I. The value of this variable is obtained directly from a pyrheliometer $I = E_b$ and it is going to be used for feedforward control purposes within a control algorithm $I = E_b \cos(\theta_i)S$, where $S = S_{ref}\eta_{opt}$, S_{ref} being the reflecting surface of the collector field's mirrors. A more detailed description of the model can be found in [38].

2.3.2 Thermal Losses in a PTC

Besides optical and geometric losses, thermal losses are important in PTC and they are occurred mainly in two places: in the absorber tube and in the HTF pipe.

Heat losses associated with the absorber tube are made of: heat loss by conduction through the pipe supports, losses by radiation, convection and conduction from metal tube into the glass cover and losses by convection and radiation from the tube crystal to the environment. In those absorber tubes in which vacuum exists between the metal tube and the glass, convection losses from the metal tube into the glass are eliminated and only losses by radiation and small ones by conduction between the glass-metal joint exist.

Although each of the heat losses could be calculated analytically, in practice a global heat losses coefficient from the absorber tube to the ambient is defined, in such a way that

$$P_{m\to a} = H_l \pi D_m L(\bar{T}_f - T_a) = H_l \pi D_m L \bar{\Delta T} \qquad (2.16)$$

where $\bar{T}_f = \frac{T_{out}+T_{in}}{2}$ is the mean temperature of the HTF (notice that some authors use instead the mean temperature of the metal absorber $\bar{T}_m$), T_a is the ambient temperature, T_{out} the outlet HTF temperature, T_{in} the inlet HTF temperature, D_m is the outside diameter of the absorber pipe and L is the length of the tube (length of the PTC). In (2.16) the global heat loss coefficient is given by surface unit of the absorber tube [W/(m^2 K)]. It can be defined by collector aperture surface unit $H_{l_{col}} = H_l/C_g$ [W/(m$^2_{col}$ K)]. In both cases, the coefficient can be obtained from experiments within the design temperature range. It can be expressed also in polynomial form as

$$H_l = a + b(\bar{T}_f - T_a) + c(\bar{T}_f - T_a)^2 \quad [\mathrm{W/(m^2K)}]$$

2.3.3 PTC Efficiency

The incident power on a PTC was given in Eq. (2.1). The thermal power given by a PTC can be computed in terms of enthalpy increment that the HTF experiments when flowing through the collector:

$$P_{m\to f} = \dot{m}(h_{out} - h_{in}) \quad [\mathrm{W}] \qquad (2.17)$$

where $\dot{m}$ is the mass flow rate [kg/s], h_{in} is the specific enthalpy of the HTF at the collector inlet [J/kg] and h_{out} is the specific enthalpy of the HTF at the collector outlet [J/kg].

Three different efficiencies can be defined in a PTC: global efficiency $\eta_{global} = \eta_{col}$, optical efficiency with an incidence angle of 0° ($\eta_{opt,0^\circ}$) and the thermal efficiency (η_{th}), plus the incidence angle modifier $\kappa(\theta_i)$. The global efficiency considers all the losses (optical, geometrical and thermal) and can be obtained as

$$\eta_{col} = \frac{P_{m\to f}}{P_{sol}} = \eta_{opt,0^\circ}\kappa(\theta_i)\eta_{th} \qquad (2.18)$$

The thermal efficiency η_{th} depends on the working temperature of the metallic absorber tube and can be given as a function of the ambient temperature and the absorber tube temperature for a determined value of the direct solar irradiance.

Considering the reference solar irradiance to be equal to the global irradiance, the efficiency of a PTC is given by

$$\eta|_{E_b \cos(\theta_i)} = \eta_{opt}|_{E_b \cos(\theta_i)} - \frac{H_l}{C_g}\frac{(\bar{T}_f - T_a)}{E_b \cos(\theta_i)} \tag{2.19}$$

Many authors [305] use for tracking PTC as reference solar irradiance the direct solar irradiance; the efficiency is given by

$$\eta|_{E_b} = \eta_{opt}|_{E_b} - \frac{H_l}{C_g}\frac{(\bar{T}_f - T_a)}{E_b} \tag{2.20}$$

2.4 Solar Irradiance Estimation and Forecast

Nowadays, there exist many systems where disturbance estimation would be needed to improve the overall performance of the control system. Some examples can be found in control systems for renewable power generation, especially for solar thermal and photovoltaic energy, where the solar radiation is used as the main energy source. Solar irradiance is a changing variable that can be perturbed by clouds, temporal dust concentration, vapor concentration, etc. Many of these changes have a temporal presence and can vary from one season to another. For this reason, some estimations about the future behavior of the solar radiation is required to optimize the process performance and minimize the use of auxiliary energy sources [288].

As a first approach, solar radiation estimation can be used for the following purposes:

1. Selection of the best location to build the solar plant (maximization of the solar resource).
2. Long-term operation planning, to estimate the power that can be supplied to the electrical network (daily and monthly prediction values [248]). At these higher decision-making levels, weather information is exploited indirectly by mapping it to economic variables such as user power demands [420].
3. Short-term operation planning, grid integration and operational control under FIT regulations (the regional or national electricity utilities have to buy renewable electricity at above-market rates set by the government). The prediction horizon in this case can be considered as short-term for prediction up to several hours. As pointed out in [311], short-term forecasts are needed for operational planning, switching sources, programming backup and short-term power purchases, as well as for planning for reserve usage and peak load matching. Also, actual measurements or estimations of the solar radiation are used to control the fundamental variables.

Table 2.1 Basic characteristics of solar radiation forecasting

Forecasting horizon:	Short-term	Long-term
Purpose (spatial nature):	Local	Global
Kinds of model:	Physics-based models	Statistical models

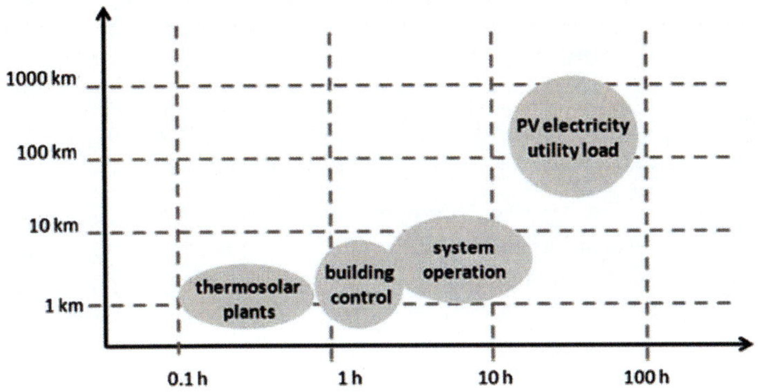

Fig. 2.2 Spatial and temporal scales of target applications, [175]

The previous objectives are also related to the existing methods to estimate solar radiation, which can be characterized by the prediction horizon length, their spatial character and the selected methodology (Table 2.1). Figure 2.2 shows typical target applications for solar radiation forecasting and their respective spatial and temporal scales [175]. Notice that, as pointed out in [420], major weather prediction centers, such as the European Centre for Medium-Range Weather Forecasts (ECMWF) and the US National Centers for Environmental Prediction (NCEP), are capable of producing high-precision weather forecasts several times a day, thanks to improved models of the atmosphere, greater availability of atmospheric data, increased computational power and the continued improvement of state estimation algorithms. Anyway, the models must be reconciled to the most recent observations. This state estimation problem is called in the weather forecast literature the data assimilation problem, where stochastic approaches as those treated in this section are used internally.

In the first case (location optimization), the most common methods to estimate the solar resource are based on the combination of local measurements (climate stations), satellite images and digital elevation models [62, 193, 374], where mainly physics-based models and artificial neural networks (ANN) are used for data fusion. In the second case (operation planning), the main approaches are based on the National Meteorological Institutes forecasts, historical datasets and satellite images to provide long-term solar radiation forecast. In [312] a comparison of different methods for two-days forecast is performed. The methods are based on the National digital forecast database (NDFD) of NOAA (http://www.weather.gov/ndfd/), the ECMWF (http://www.ecmwf.int) and the Weather research and forecasting (WRF) model (http://www.wrf-model.org) with first input of Global Forecast System (GFS) model. In the third approach (short-term operation planning and control), direct measurements from solar irradiance sensors (mainly pyrheliometers and pyranometers) are used both for feedforward control and short-term solar irradiance forecast, sometimes combined with fish-eye cameras (total sky imagers that take hemispherical sky photographs), ceilometers (recording cloud height) and satellite images [175, 214].

In this book the interest is focused on easy methods to perform a short-term solar irradiance forecast (daily solar radiation estimation). The estimation of solar irradiance on a horizontal surface can be mainly performed based on physical models or data-driven models. As pointed out by [378] and references therein, there exist mainly three methodologies for the estimation of horizontal global solar irradiance:

1. Estimation based on physical processes including complex radiative transfer models. The physical interactions between solar radiation and the terrestrial atmosphere (such as Rayleigh scattering, radiative absorption by ozone and water vapor and aerosol extinction) are considered. These models are also known as parametric models [399].
2. Approaches based on the traditional and long-utilized Ångström's linear approach and its modifications which is based on measurements of sunshine duration [14], providing empirical relationships between the ratio of the global solar radiation on a horizontal surface at the Earth's surface and at the top of the atmosphere and the percent possible sunshine (i.e. observed bright sunshine to maximum possible sunshine hours). These models are also known as decomposition models [399], as they usually use information only on global radiation to predict the beam and sky components.
3. Estimation based on data and statistical models represented by time-series or ANN, which can be based on sunshine duration measurements but also on other climatological parameters.

In [399] and cited references, solar radiation models for predicting the average daily and hourly global irradiance, beam irradiance and diffuse irradiance are reviewed. Examples of estimation based on the first two categories indicated previously are developed to predict the beam component or sky component based on other more readily measured quantities.

In what follows, selected examples of the mentioned models are developed based on the authors' experience.

2.4.1 Physics-Based Models

As pointed out by [399], parametric models require detailed information of atmospheric conditions. Meteorological parameters frequently used as predictors include the type, amount and distribution of clouds or other observations, such as the fractional sunshine, atmospheric turbidity and precipitable water content. Two well-known parametric models used for solar energy applications are:

- The ASHRAE algorithm [20], used by the engineering and architectural communities.
- The Iqbal model [188], which offers extra-accuracy over more conventional models.

Several approaches are now being developed based on NDFD created by different countries. In [291] a preliminary evaluation of a simple solar radiation forecast model using sky cover predictions from the NDFD of the USA as an input. The models developed by [165–168] provide predictions of clear-sky direct and diffuse broadband irradiance with great accuracy when detailed and accurate input data are available.

2.4.2 Decomposition Models

Development of correlation models that predict the beam or sky radiation using other solar radiation measurements is possible [399]. Decomposition models normally use information only on global irradiance to predict the beam and sky components. These relationships are usually expressed in terms of the irradiation which are the time integrals (usually over 1 h) of the radiant flux or irradiance. Examples of different decomposition models developed to estimate direct and diffuse irradiance from global irradiance can be found in [6, 399] and cited references.

2.4.2.1 An Example of a Model of the Solar Irradiance

Solar radiation undergoes changes due to its daily cycle and to passing clouds. In this subsection, the approximation given by [101] has been adopted to obtain a clear-day prediction of the solar irradiance.

The magnitude of the solar radiation at a determined point on the Earth's surface varies mainly with the geographical localization and the day of the year and also with the meteorological characteristics of the chosen instant. To calculate the global irradiance level on a surface, the extraterrestrial irradiance level E_c (solar constant[1]) must be known. This value reaches the atmosphere of the Earth with a small modification E_{ext}, due to the elliptical trajectory of the Earth around the Sun.

As the radiation heats the atmosphere, fragmentation is produced, one part is reflected outside and another part is absorbed by the atmosphere; the rest penetrates the atmosphere. One part of this is dispersed, the other is reflected and the rest reaches the Earth in the form of direct irradiance E_b.

---

[1]This is the amount of total energy that contains the extraterrestrial solar irradiance, integrated in all the spectrum of wave lengths. The value used is $E_c = 1367\ \mathrm{W/m^2}$.

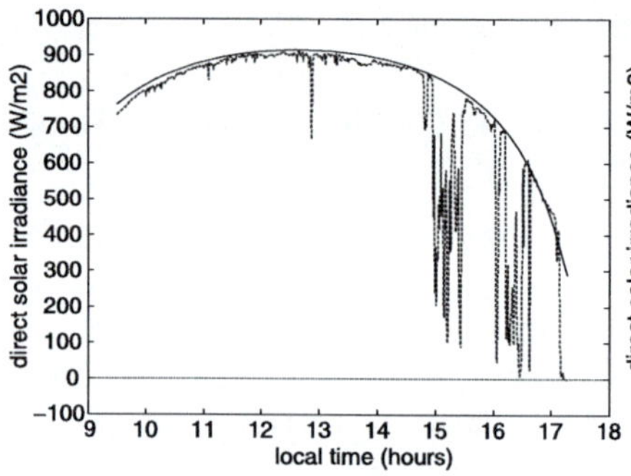

(a) November 26th, 1990

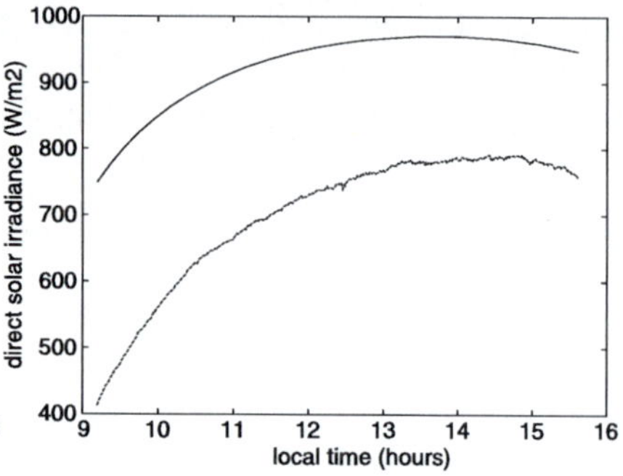

(b) September 19th, 1991

Fig. 2.3 Direct solar irradiance and clear-day prediction

Part of the amount of irradiance reflected and dispersed reaches the Earth in the form of diffuse irradiance. The irradiance that reaches the atmosphere is the solar constant multiplied by the eccentricity correction factor e_{xc}:

$$E_{ext} = E_c \left(\frac{D_0}{D} \right)^2 = E_c e_{xc}; \quad e_{xc} = 1 + 0.033 \cos \left(\frac{2\pi J}{365} \right) \tag{2.21}$$

D being the Sun–Earth distance at a determined instance and D_0 its mean value and J is the Julian Day.

An approximation to the instantaneous value of direct solar irradiance is the one obtained by [101]:

$$E_b = 0.9751 E_{ext} T_b \tag{2.22}$$

where T_b is the transmittance of the direct normal irradiance, which depends on factors such as the ozone cover thickness, the thickness of the water steam cover, two coefficients of atmospheric cloudiness and the air mass at standard pressure.

Expression (2.22) has been used to predict the evolution of the solar irradiance and then compared to the actual direct solar irradiance measured at the PSA. Figure 2.3(a) corresponds to a day with high solar radiation levels in which, at the end of the daily operation, large clouds appeared, while Fig. 2.3(b) corresponds to a day with low solar radiation levels, due to the presence of contamination in the air, but without the presence of passing clouds. As can be seen, lines have been drawn depending on the local hour. By using the prediction model, a filtered value of the solar radiation evolution is directly obtained.

2.4.3 Statistical Models

The perturbation variables are usually represented as time-series structures due mainly to their stochastic behavior. Using time-series models is one of the ways to estimate future values of disturbances. These models are obtained using past data and are used to estimate the future behavior along a prediction horizon [311]. Time-series models are based on the assumption that the modeled data are autocorrelated and characterized by trends and seasonal variations. Thus, well-known autocorrelated models (ARMA, ARIMA, ARMAX, ARIMAX) could be also used for disturbance estimation [261, 311, 410]. On the other hand, ANN provides also a good solution to perform estimations because its design is based on training and no statistical assumptions are needed for the source data. Neural networks are widely accepted as a technology to predict time-series offering an alternative way to solve complex problems [283, 409]. In [311] a comparison between several types of time-series model was performed: regression models, unobserved component models (UCM), ARIMA models, transfer function models, neural networks and hybrid models (combining regressions and neural nets). In [132] a total of 18 empirical models in linear, quadratic, cubic, logarithmic, exponential and hybrid forms using only sunshine hours, latitude and altitude were compared to estimate monthly average daily global solar radiation on a horizontal surface for 159 weather stations in Turkey.

In this section, a summary of the results in [288] is included, where four different well-known time-series methods are analyzed and compared, namely, Discrete Kalman Filter [337], Discrete Kalman Filter with Data Fusion [279], Exponentially Weighted Moving Average [381] and Double Exponential Smoothing [268]. These methods are used to obtain forecasts of the direct solar irradiance using real data. The nomenclature that will be used in next sections is the following: $z(k)$ is the measurement in the discrete-time instant k and $\hat{x}(k+1)$ is the predicted value one sample ahead.

2.4.3.1 Discrete Kalman Filter

The Kalman Filter (KF) addresses the general problem of trying to estimate the state $x \in \mathbb{R}^n$ of a discrete-time process that is governed by a linear stochastic difference equation. Matrix $\mathbf{A}$ in the difference equation, Eq. (2.23), relates the state at the previous time step $k-1$ to the state at the current step k and matrix $\mathbf{H}$ in (2.24) relates the state to the measurement value $z(k) \in \mathbb{R}^n$ [337].

$$x(k) = \mathbf{A}x(k-1) + w(k-1) \tag{2.23}$$

$$z(k) = \mathbf{H}x(k) + v(k) \tag{2.24}$$

The random variables $w(k)$ and $v(k)$ represent the process and measurement noise, respectively. They are assumed to be independent (of each other), white and with normal probability distributions. The KF estimates a process by using a form of feedback control: the filter estimates the process state at some time and then obtains feedback in the form of noisy measurements. As such, the equations for the KF fall into two groups: time update equations and measurement update equations. The discrete KF time update equations are [337]

$$\hat{x}(k)^- = \mathbf{A}\hat{x}(k-1) \tag{2.25}$$

$$\mathbf{P}(k)^- = \mathbf{A}\mathbf{P}(k-1)\mathbf{A}^T + \mathbf{Q} \tag{2.26}$$

and the discrete KF measurement update equations:

$$\mathbf{K}(k) = \mathbf{P}(k)^{-}\mathbf{H}^{T}\left(\mathbf{H}\mathbf{P}(k)^{-}\mathbf{H}^{T} + \mathbf{R}\right)^{-1} \tag{2.27}$$

$$\hat{x}(k+1) = \hat{x}(k)^{-} + \mathbf{K}(k)\left(z(k) - \mathbf{H}\hat{x}(k)^{-}\right) \tag{2.28}$$

$$\mathbf{P}(k) = \left(\mathbf{I} - \mathbf{K}(k)\mathbf{H}\right)\mathbf{P}(k)^{-} \tag{2.29}$$

where $\mathbf{P}(k)^{-}$ is the a priori estimate error covariance, $\mathbf{P}(k)$ is the a posteriori estimate error covariance, $\hat{x}(k)^{-}$ is the a priori estimate, and $\mathbf{I}$ is the identity matrix of the appropriate size. In practice, the process noise covariance $\mathbf{Q}$ and measurement noise covariance $\mathbf{R}$ matrices might change with each time step or measurement, however here we assume they are constant. Detailed information about KF implementation can be found in [396].

2.4.3.2 Discrete Kalman Filter with Data Fusion

The main structure of the discrete Kalman Filter with Data Fusion (KFDF) has the same formulation as in the previous section, with the exception that data fusion is performed. Assume now that we are dealing with two different measurements that provide a reading for some variable of interest z. We call z_1 the measure from the first source and z_2 the measure from the second source. If both measurements are equally good (the quadratic standard deviation $\sigma_1^2 = \sigma_2^2$), we just take the average of both numbers. In this case, a weighted average of both measurements can be obtained to generate an estimation of z, $\hat{x}(k+1)$. One possibility is weighting each measurement inversely proportional to its precision, that is [279],

$$\hat{x}(k+1) = \frac{z_1(k)\sigma_2^2 + z_2(k)\sigma_1^2}{\sigma_1^2 + \sigma_2^2} \tag{2.30}$$

Note that the above estimation can be also rewritten as

$$\hat{x}(k+1) = z_1(k) + K\left(z_2(k) - z_1(k)\right) \tag{2.31}$$

where now $K = \sigma_1^2/(\sigma_1^2 + \sigma_2^2)$. The update equation has the same general form as in the previous paragraph (see [279] for further information). In what follows, the measurement z_1 is a real value from a sensor and z_2 is an estimated value calculated using a theoretical model [288].

2.4.3.3 Exponentially Weighted Moving Average

Exponentially Weighted Moving Average (EWMA) is commonly used with time-series data to smooth short-term fluctuations and highlight longer-term trends or cycles. The threshold between short and long terms depends on the application and the parameters of the moving average will be accordingly set. Mathematically, a moving average is a type of convolution and so it is also similar to a low-pass filter used in signal processing [68]. The arithmetic assumption of moving average assumes that only the latest values have an effect on the current value. However, the moving average assigns equal weight to each value as compared to the trend line that gives weight to only the first and the last values for determination. The long-term moving averages can be useful for measuring the trend whereas the short-term averages can be utilized for measuring the changes in the trend [63, 381]:

$$\hat{x}(k+1) = \frac{z(k) + (1-\beta_s)z(k-1) + \cdots + (1-\beta_s)^N z(k-N)}{1 + (1-\beta_s) + \cdots + (1-\beta_s)^N} \tag{2.32}$$

where N is the sample window size and $\beta_s \in (0, 1)$ is the smoothing parameter. The sample window is composed by the N most actual samples of analyzed variable from $z(k)$ to $z(k-N)$. When the prediction for time instant $k+1$ is done, its value is located at first place in a sample window which is organized as $[\hat{x}(k+1), z(k), z(k-1), \ldots, z(k-N+1)]$. This sample window is used to make the prediction of value in time instant $k+2$ and so on. When the process reaches the prediction horizon, a new sample window is formed from the last N measurements.

2.4.3.4 Double Exponential Smoothing

The Double Exponential Smoothing (DES) is described by the following equations [268]:

$$\chi(k) = \alpha_D z(k) + (1-\alpha_D)\big(\chi(k-1) + b_D(k-1)\big) \tag{2.33}$$

$$b_D(k) = \gamma_D\big(\chi(k) - \chi(k-1)\big) + (1-\gamma_D)b_D(k-1) \tag{2.34}$$

where $\chi(k)$ is the unadjusted forecast, $b_D(k)$ is the estimated trend, α_D is the smoothing parameter for data and γ_D is the smoothing parameter for trend. Note that the current value of the series is used to calculate its smoothed value replacement in double exponential smoothing. The one-period and m-periods ahead forecasts are given, respectively, by

$$\hat{x}(k+1) = \chi(k) + b_D(k) \tag{2.35}$$

$$\hat{x}(k+m) = \chi(k) + m b_D(k) \tag{2.36}$$

There are a variety of schemes to set initial values for χ and b_D in double smoothing, but in [288] $\chi(1) = z(1)$ and $b_D(1) = (z(1) + z(2) + z(3))/3$ as suggested in [268]. The first smoothing equation adjusts χ directly for the trend of the previous period, $b_D(k-1)$, by adding it to the last smoothed value, $\chi(k-1)$. This helps to eliminate the lag and brings χ to the appropriated base of the current value. Then, the second smoothing equation updates the trend, which is expressed as the difference between the last two values. The equation is similar to the basic form of single smoothing, but here it is applied to the updating of the trend. The values for α_D and $\gamma_D \in (0, 1)$ can be obtained via optimization techniques as described in [268].

Table 2.2 Performance indices (courtesy of A. Pawlowski et al., [288])

MSE	RMSE	nRMSE	R^2
$\frac{1}{N}\sum_{k=1}^{N}(\hat{x}(k)-z(k))^2$	$\sqrt{\frac{1}{N}\sum_{k=1}^{N}(\hat{x}(k)-z(k))^2}$	$\frac{\sqrt{\frac{1}{N}\sum_{k=1}^{N}(\hat{x}(k)-z(k))^2}}{\sqrt{\frac{1}{N}\sum_{k=1}^{N}(z(k))^2}}$	$\frac{\sum_{k=1}^{N}(\hat{x}(k)-\bar{z})^2}{\sum_{k=1}^{N}(z(k)-\bar{z})^2}$

2.4.3.5 Illustrative Results

The predicted results for each combination were compared statistically using four different parameters [283]: the Mean Square Error (MSE), the Root Mean Square Error (RMSE), the normalized RMSE (nRMSE), and the Coefficient of Determination R^2 (Table 2.2). When a zero value is obtained for MSE, it means that the estimator $\hat{x}(k)$ predicts observations of the parameter $z(k)$ with perfect accuracy. The unbiased model with the smallest MSE is generally interpreted as the one best explaining the variability in the observations. RMSE is a frequently used measurement of the differences between root mean square error values predicted by a model (or an estimator) and the values actually observed from the variable being modeled or estimated. The nRMSE measurement is useful for comparison while R^2 is used in the context of statistical models, where its main purpose is the prediction of future outcomes on the basis of other related information (it is the proportion of variability in a data set that is accounted for by the statistical model). It provides a measurement of how well future outcomes are likely to be predicted by the forecast method.

Additionally, two measurements of the error are used to show the quantitative variation of the forecasts. These measurements are the average absolute error $\bar{e}$ and the average percent error $\bar{e}_{\%}$:

$$\bar{e} = \frac{1}{N}\sum_{k=1}^{N}\left|z(k)-\hat{x}(k)\right|, \qquad \bar{e}_{\%} = \frac{1}{N}\sum_{k=1}^{N}\frac{|z(k)-\hat{x}(k)|}{|z(k)|}100\% \tag{2.37}$$

The meteorological data used in this section have been recollected during one year by a meteorological station located at the Almería, Spain, 36°46′N2°48′W. For this database, solar irradiance samples are acquired every minute. As commented above, the KFDF technique requires a theoretical model to obtain data for fusion purposes. In this case, the model for direct irradiance explained in Sect. 2.4.2.1 has been used [42, 85].

All the presented methods have some parameters to be set and which have a great influence on the overall performance and accuracy of the forecast. Therefore, a calibration stage with 10 days was performed in order to set up the appropriated values for these parameters, where the minimization of the RMSE error was selected as a reference measurement. The selected days for this calibration process include diverse situations of the solar radiation from different year seasons.

Figures 2.4 and 2.5 present the results for a day with clouds alteration for two prediction horizons of 5 and 15 samples, respectively. The original solar irradiance time-series is characterized by many local changes of trend caused by passing clouds. For the 5-sample horizon, the DES, KF and EWMA methods produce a

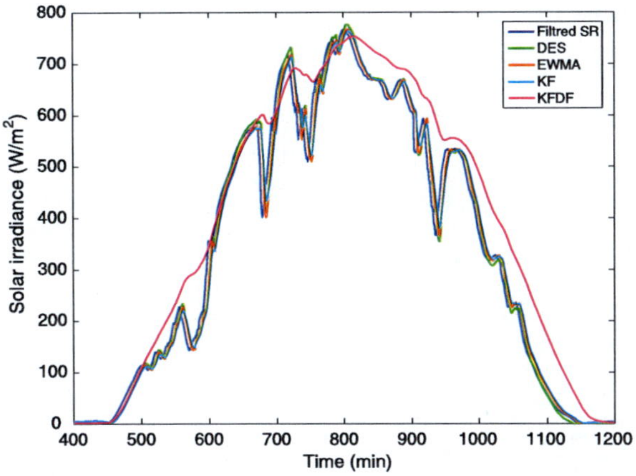

Fig. 2.4 Day with passing clouds and a 5-sample horizon (courtesy of A. Pawlowski et al., [288])

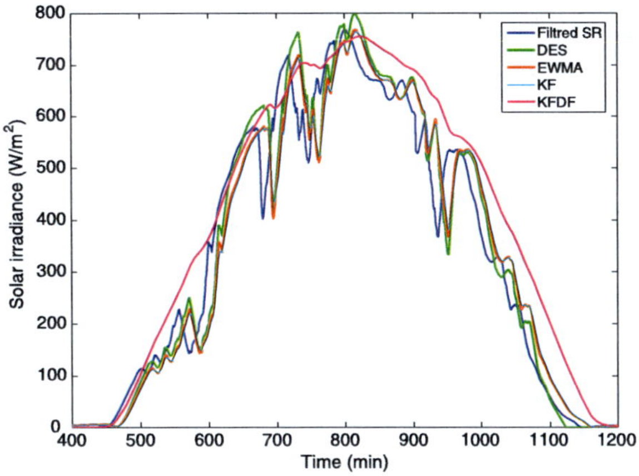

Fig. 2.5 Day with passing clouds and a 15-sample horizon (courtesy of A. Pawlowski et al., [288])

small predicted signal delay. The KFDF technique keeps the theoretical data trend despite of changes in real signal and produces only a slow decrease of the forecasted value. Comparing Figs. 2.4 and 2.5, it can be observed that the signals are characterized by a delay, which is more visible for the last case. In comparison with the rest of the methods, the DES technique produces forecasts that slightly surpass the magnitude of real solar irradiance signal.

Table 2.3 compares the performance indices for the day with passing clouds. The RMSE and nRMSE measurements indicate that the best results for all techniques are generated by the DES method. The second one according to this measurement is the EWMA method. The R^2 for the maximal horizon reaches the highest value 0.9848 for the DES method. However, the worst value is obtained by the KFDF. Furthermore, for this analysis, the lowest value of the average absolute error $\bar{e}$ for 15-sample horizon is obtained for the DES technique with a value of 26.22 W/m^2. The outcome of average percentage error $\bar{e}_{\%}$ is 32% for the DES method and for the

Table 2.3 Performance comparison for a day with passing clouds (courtesy of A. Pawlowski et al., [288])

Index	Horizon	KF	KFDF	EWMA	DES
MSE	5	676.7	2966	470.5	461.8
	10	1563	3700	1323	1280
	15	2508	4296	2260	2167
RMSE	5	26.01	54.46	21.69	21.49
	10	39.54	60.83	36.38	35.78
	15	50.08	65.54	47.54	46.55
nRMSE	5	0.080	0.168	0.067	0.066
	10	0.122	0.188	0.112	0.110
	15	0.155	0.202	0.147	0.144
R^2	5	0.9949	0.9858	0.9965	0.9966
	10	0.9883	0.9827	0.9901	0.9909
	15	0.9812	0.9805	0.9831	0.9848
$\bar{e}$ [W/m^2]	5	13.86	31.43	11.09	10.70
	10	21.90	35.31	19.59	18.59
	15	28.56	38.37	26.45	24.69
$\bar{e}_{\%}$ [%]	5	14.89	77.37	10.89	17.21
	10	26.07	89.55	22.12	26.22
	15	37.18	100.76	33.42	32.37

EWMA and KF it is 33% and 37%, respectively. The KFDF technique produces an unacceptable percentage error for the 15-sample horizon.

2.4.3.6 Computational Intelligence-Based Models

Several methods for forecasting solar irradiance in different time scales have appeared recently [250] based on ANN, fuzzy logic and hybrid system such as ANFIS, ANN-wavelet and ANN-genetic algorithms (GA). These approaches can be classified into three different types [249, 250].

- The first one can estimate the solar irradiance (in different scale times) based on some meteorological parameters such as air temperature, relative humidity, wind speed, wind direction, cloud, sunshine duration, clearness index, pressure and geographical coordinates as latitude and longitude. Multilayer Perceptron (MLP) network, Radial Basis Function (RBF) network and fuzzy logic can resolve this problem [249, 250].
- A second approach allows predicting the future solar irradiance (in different scale times) based on the past observed data [250]. Therefore, in this case recurrent

neural networks (RNN), wavelet-networks and wavelet-networks-fuzzy are very suitable.
- The last one combines the two previous approaches, so that different ANN-architectures and ANFIS are adequate.

Other examples of these approaches can be found in [250] and the cited references. Representative cases can be found in [409], where the thermal energy collection of solar heat energy utilization system based on solar radiation forecasting at one-day ahead 24-hours thermal energy collection is performed using three different ANN models trained by weather data based on tree-based model and tested according to forecast day. Selected models are feedforward neural network (FFNN), radial basis function neural network (RBFNN) and RNN. In [250] a practical method for solar irradiance forecast using ANN is presented. The proposed MLP model makes it possible to forecast the solar irradiance on a base of 24 h using the present values of the mean daily solar irradiance and air temperature. In [123] ANN are applied to multi-step long-term solar radiation prediction. The networks are trained as one-step-ahead predictors and iterated over time to obtain multi-step longer-term predictions. Auto-regressive solar radiation models and solar radiation models auto-regressive with exogenous inputs are compared, considering cloudiness indices as inputs in the latter case. These indices are obtained through pixel classification of ground-to-sky images. The input–output structure of the ANN models is selected using evolutionary computation methods.

As pointed out by [311], several problems with ANN have been noted in the literature, related with the existence of local minima, noise over-fitting and others. The problems with neural nets have been one motivating factor in the development of new classes of model combining nets with other techniques. One idea is to combine ANN with wavelets [91, 92], and regressions and ANN [180, 421]. Typically, in these hybrid models, an initial regression or ARIMA is estimated and the residuals are then processed using a neural net. The forecasts from the two separate stages are then combined.

2.5 Control of the Energy Conversion Units

As pointed out in the introduction of the chapter, the control of the variables associated to the solar conversion units depends very much on the type of system. In the case of photovoltaic (PV) systems, this involves the control of the voltage and intensity produced by the solar cells in order to operate at the maximum efficiency point and the control of the associated DC/AC conversion power electronics.

In the case of thermal solar plants, a HTF is heated and thus associated temperatures and flows have to be controlled, as well as temperature, flow, and pressure of the produced steam. Values of solar irradiance are used to compensate for its variations. Once the steam is produced, the controls at this level do not differ much from the control of other thermal plants. In most cases oil is used as energy conversion fluid and because oils decompose and generate dangerous inflammable fumes at relative low temperatures, the maximum operating temperatures have to be kept below these temperatures, which are rather low. In order to increase the performance of the steam generating units, the solar field has to be operated close to the maximum operating temperatures.

The control of energy conversion units will be treated in detail in the following chapters.

2.6 Integrated Control

Solar energy is intermittent and is not always available when needed and where needed. Because of that, the integration of solar energy plants in the electrical grid is a challenging problem. Energy storage is a potential solution to the integration problems. An efficient operation of solar plants with energy storage systems could allow a closer match between solar plants power generation and electrical power demand. In thermal solar power plants, energy is usually stored as heat. Heat storage systems allow solar thermal plants to produce electricity at night and on cloudy days. Additionally, the equipment (steam generators turbines and electrical generator) does not need to be designed for the peak power but for the average power and thus installation costs are reduced. Furthermore, equipment can operate continuously with smaller payback periods. When collected solar power is higher than the demanded energy, excess heat is transferred to a thermal storage medium and stored in an insulated reservoir. The stored heat can be withdrawn for power generation when needed. Different systems have been used in solar power plants such as molten salts, pressurized steam, change materials and concrete. Energy storage tanks are usually well insulated and can store energy for a few days and provide enough thermal storage for a few plant operational hours.

The price of electricity is determined by a variety of methods which change not only from place to place, but also with the commissioning date of the plants. Furthermore, proposals of new price mechanisms are appearing continuously with the liberalization of the energy prices and they are considered to be a fundamental part of the Smart Grid concept. In the electrical energy market, the price fluctuations will tend, usually, to correlate well with the hourly demand because the prices will try to cover the marginal cost of generation which will normally increase with demand. Risk factors are also considered for establishing prices, for example, the available generators to cope with unexpected demand peaks will affect the electricity price.

The use of renewable energies is encouraged in many countries by different policies such as FIT which guarantees grid access and long-term contracts for the electricity produced by renewable sources at the cost of renewable energy generation and which tends toward market prices with time. The electrical utilities or regional or national authorities are obliged to buy the renewable electricity from the renewable energy producer. The FIT are determined to make it possible for an installation to be operated cost-effectively if the installation is managed efficiently and state of the art techniques are used. FIT may therefore differ depending on the sources, location, project size and other important factors.

Solar plant operators can benefit from energy storage systems by shifting generation from valley periods to peak periods taking advantage of the corresponding price variations. Two different scenarios will be analyzed in Chap. 8: (a) non-committed production, where the electrical utilities buy all the production of the solar power plant at prices at fixed prices and, (b) committed production, where the solar plant and the utility agree to a production and price in advance. The electrical utility penalizes the deviation with the agree contract.

If the prices are constant, the best option would be to produce as much electrical energy as possible. Storage can be used in those situations where more solar energy can be collected than energy converted and delivered. This may be due to maintenance operation or turbines or generator having been designed for nominal conditions below the maximum solar conditions.

If prices vary with the time of day or if the production is committed with penalization errors, the energy storage can be used to maximize profits. The electrical energy produced depends basically on the solar radiation during that sampling period and on the operating conditions of the plant. For example, when the plant is starting up, some of the solar energy will need to be used to warm up the plant systems to the operative stage while if the plant was already generating electricity in the previous period, no energy will be needed for warming up. Furthermore, the warming up of the plant will require some time.

The electricity market works with different models. In many of them, the hourly daily price is fixed by a negotiation procedure. First, usually in the morning, a period opens for filing buying and selling offers for the energy for every hour of the day. There are two types of decision that have to be taken. First, at the beginning of the day, to determine the offer for the next 24 hours and once this have been agreed, the next problem is to determine, for each time period, the power to be delivered and the power to be stored in order to maximize profits and fulfill all operational objectives. Since solar energy cannot be exactly predicted in advance, the optimization problems involved have to be formulated in an uncertain framework. The performance of solar plants depends on the environmental and plant operating conditions. Plant operation conditions (modes and set points of fundamental variables) have to be determined by the control system in order optimize plant performance.

All these issues will be discussed in Chap. 8.

2.7 Summary

This chapter has been devoted to describing the control problems found in solar energy systems. Four different levels can be distinguished: (i) the control of the solar collector units, (ii) solar radiation estimation and forecast, (iii) the control of the energy conversion systems and (iv) the overall control of the complete process. In the case of the control of the solar collector units and solar radiation estimation and forecast, an overview of the fundamentals and techniques has been carried out, including some illustrative results. The next chapter will deal with the modeling and control of the energy conversion systems and the overall control of the complete process.

Chapter 3
Photovoltaics

3.1 Introduction

Electrical energy can be obtained directly from sunlight using photovoltaic (PV) devices (cells). The cells convert solar energy into electrical energy via the photovoltaic effect. PV cells are large area p–n diodes which are assembled in modules (panels). PV installations have no moving parts and do not vibrate, produce noise, or require cooling or tall towers. Since the production of the actual PV cells requires crystalline silicon, their cost is high. However, the cost of PV cells has been decreasing steadily from US $20 per W in 1990 to less than US $5 per W nowadays (2011). The world wide PV generating capacity has been increasing steadily in the last 20 years. The process of manufacturing PV cells is energy intensive. Every cell consumes a few kWh before it is placed facing the Sun and produces a kWh of energy. The first application of photovoltaic cells was to provide operating power to satellites and other spacecrafts; after this, many applications of a PV system able to provide power to small electrical appliances were developed.

The power that one cell, or module, can produce is not usually enough to cover the needs of a home or a business. Modules are normally connected together to form an array. Most PV arrays use an inverter (see Fig. 3.1) to convert the direct current (DC) produced by the cells into alternating current (AC) that can be injected into the existing lines to power domestic or business loads such as lights and motors. The modules in a PV array are usually connected in series to obtain the appropriate voltage and then in parallel to produce the desired current.

A grid-connected PV system is shown in Fig. 3.2 [231]. The load can be fed by the electrical utility and by a PV system. The power not consumed by the load can be fed to the electrical utility.

3.2 Power Point Tracking

In PV power systems, maximum power point tracking (MPPT) is essential because it takes full advantage of the available solar energy. A MPPT is a high efficiency DC

E.F. Camacho et al., *Control of Solar Energy Systems*, Advances in Industrial Control,
DOI 10.1007/978-0-85729-916-1_3, © Springer-Verlag London Limited 2012

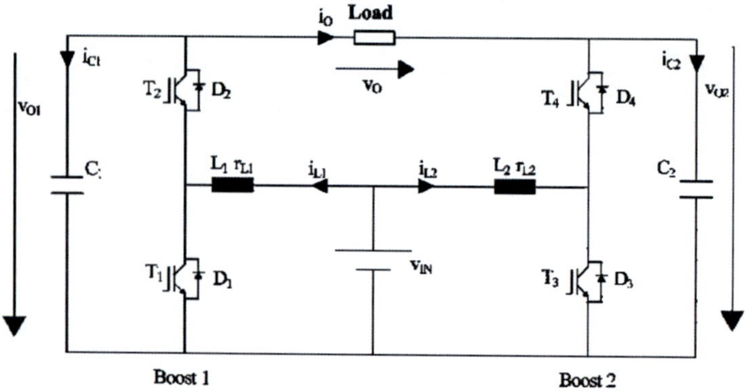

Fig. 3.1 Boost DC/AC inverter

Fig. 3.2 Topologies of a grid-connected PV system

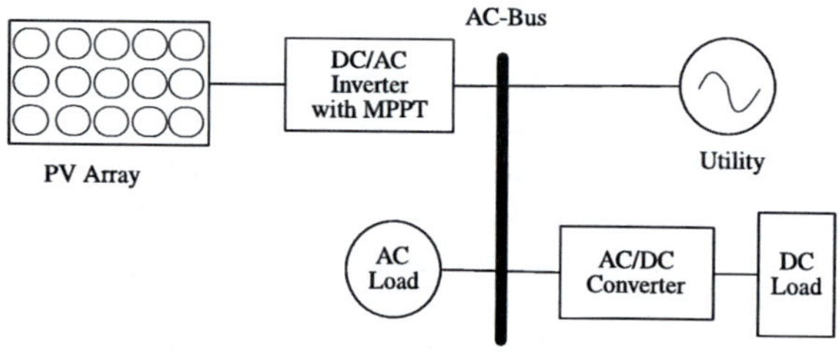

to DC converter that presents an optimal electrical load to a solar panel or array and produces a voltage suitable for the load.

PV cells have a single operating point where the values of the current (I) and voltage (V) of the cell result in a maximum power output. These values correspond to a particular resistance, which is equal to V/I. A PV cell has an exponential relationship between current and voltage and the maximum power point (MPP) occurs at the knee of the curve, where the resistance is equal to the negative of the differential resistance ($V/I = -dV/dI$), see Fig. 3.3. Maximum power point trackers use some type of control system to search for this point and, thus, allow the converter to extract the maximum power available from a cell.

Fig. 3.3 Characteristic PV array power curve

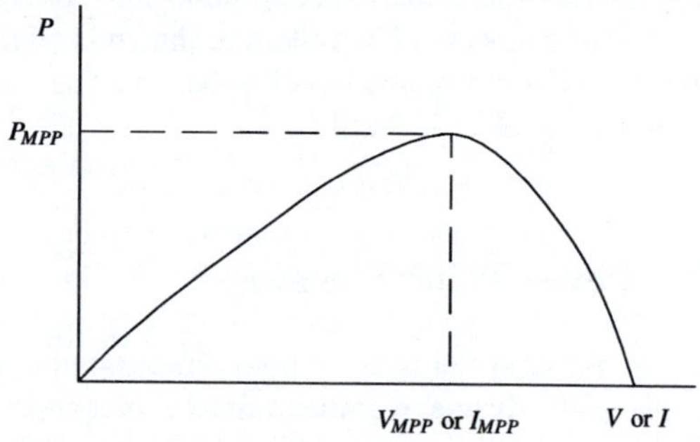

Traditional solar inverters perform MPPT for an entire array as a whole. In such systems, the same current, dictated by the inverter, flows though all the panels in the string. However, because different panels have different I–V curves, i.e. different MPP (due to manufacturing tolerance, partial shading, etc.), this architecture means some panels will be performing below their MPP, resulting in a loss of energy. Some companies are now placing peak power point converters into individual panels, allowing each to operate at peak efficiency despite uneven shading, soiling or electrical mismatch.

At night, an off-grid PV power system uses batteries to supply its loads. Although the battery pack voltage when fully charged may be close to the peak power point of the PV array, this is unlikely to be true at sunrise when the battery is partially discharged. Charging may begin at a voltage considerably below the array peak power point and a MPPT can resolve this mismatch.

When the batteries in an off-grid system are full and PV production exceeds local loads, MPPT can no longer operate the array at its peak power point as the excess power has nowhere to go. The MPPT must then shift the array operating point away from the peak power point until production exactly matches demand. An alternative approach commonly used in spacecraft is to divert surplus PV power into a resistive load, allowing the array to operate continuously at its peak power point.

In a grid-tied photovoltaic system, the grid is essentially a battery with infinite capacity. The grid can always absorb surplus PV power and it can cover shortfalls in PV production (e.g. at night). Batteries are thus only needed for protection from grid outages. The MPPT in a grid-tied PV system will always operate the array at its peak power point, unless the grid fails when the batteries are full and there are insufficient local loads. It would then have to move the array away from its peak power point, as in the off-grid case (which it has temporarily become).

MPPTs can be designed to drive an electric motor without a storage battery. They provide significant advantages, especially when starting a motor under load. This can require a starting current that is well above the short-circuit rating of the PV panel. A MPPT can step the relatively high voltage and low current of the panel down to the low voltage and high current needed to start the motor. Once the motor is running and its current requirements have dropped, the MPPT will automatically increase the voltage to normal. In this application, the MPPT can be seen as an electrical analogue to the transmission in a car; the low gears provide extra torque to the wheels until the car is up to speed.

Figure 3.3 shows the characteristic power curve for a PV array. The problem considered by MPPT techniques is to automatically find the voltage V_{MPP} or current I_{MPP} at which a PV array should operate to obtain the maximum power output P_{MPP} under a given temperature and irradiance. It is noted that under partial shading conditions, it is possible in some cases to have multiple local maxima, but overall there is still only one true MPP. Most techniques respond to changes in both irradiance and temperature, but some are specifically more useful if temperature is approximately constant. Most techniques would automatically respond to changes in the array due to aging, though some are open-loop and would require periodic fine-tuning. In our context, the array will typically be connected to a power converter that can vary the current coming from the PV array.

Fig. 3.4 Block diagram

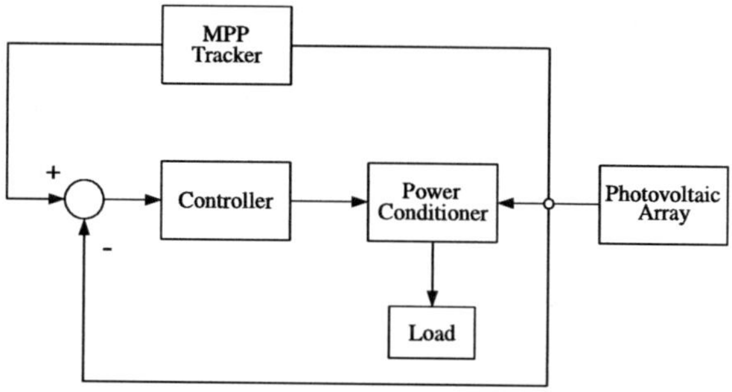

There are several MPPT techniques. Two of them are popular tracking methods based on power measurement which are widely adopted in PV power systems: the perturbation and observation method (P&O) [401] and the incremental conductance method (IncCond) [134]. In fact, P&O and IncCond are based on the same technology, regulating the voltage of the PV array to follow an optimal set point, which represents the voltage of optimal operating point V_{OOP}, as shown in Fig. 3.4. In some of the literature, V_{OOP} is also symbolized by V_{MPOP} which stands for the voltage of the maximum power operating point.

The local maximum power operating point is continuously tracked and updated by the MPP tracker to satisfy $dP/dV = 0$. By investigating the power–voltage relationship, which is the P–V curve of a typical PV module, the MPP can always be tracked if we keep dP/dV equal to zero for any solar insolation or temperature since all local MPP have the same mathematical attribute.

3.3 Solar Tracking

As has been analyzed in Chap. 2, a solar tracker is a device for orienting a PV panel toward the Sun, thus increasing the effectiveness of such equipment over any fixed position, at the cost of additional system complexity. There are many types of solar tracker of varying cost, sophistication and performance.

The application of high-concentration solar cell technology allows a significant increase in the amount of energy collected by solar arrays per area unit. However, to make this possible, stricter specifications on the Sun's pointing error are required. In fact, the performance of solar cells with concentrators decreases drastically if this error is greater than a small value. These specifications are not fulfilled by simple tracking systems, due to different sources of errors (e.g. small misalignments of the structure with respect to geographical North) that appear in practice in low-cost, domestic applications.

Thanks to technical advances, reasonably priced high-concentration solar PV arrays should soon be available. However, the future use of this kind of solar PV array in low-cost installations will bring with it a new type of problem: the necessity for high-accuracy solar pointing. High-concentration solar PV arrays require greater solar tracking precision than conventional photovoltaic arrays and therefore, a relatively low pointing error must be achieved for this class of installation. Since design and installation are optimized in large plants, this error requirement is usually achieved. Nevertheless, the cost of such optimization is prohibitive for low-cost installations.

Several classes of structure can be distinguished, depending on the classification criteria:

- Regarding movement capability, three main types of Sun tracker exist [176]: fixed surface, one axis trackers [303] and two axis trackers [1]. The main difference among them is their ability to reduce the pointing error, increasing the daily radiation that the solar cells receive and, thus, the electric energy that they produce. A theoretical comparative study between the energy available to a two axis tracker, an East–West tracker and a fixed surface was presented in [267]. As main result, it concluded that the annual energy available to the ideal tracker is 5–10% higher than the East–West tracker and 50% higher than the fixed surface.
- With regards to the control units, the main types of solar tracker are [328]: passive, microprocessor and electro-optical controlled units. The first ones do not use any electronic control or motor [114]. The second ones use a mathematical formula to predict the Sun's movement and do not need to sense the sunlight. An example of this kind of unit can be found in [1]. Finally, the electro-optical controlled units use information from some kind of sensor (e.g., auxiliary bifacial solar cell panel, pyrheliometer) to estimate the Sun's real position which is used in the control algorithm [303, 327].

3.4 Automatic Tracking Strategy

A control strategy for two-axis trackers is presented in this section, which is executed in a microprocessor. The correct pointing is inferred from the electrical power generated, which must be sensed on-line. The control strategy consists of a combination between:

1. An open-loop tracking strategy which corresponds to a *microprocessor controller* in the classification [328]. This controller is based on the solar movement model.
2. A closed-loop strategy, which corresponds to *electro-optical controller* in the previous classification. This strategy consists of a dynamic controller that feeds back generated power measurements. Furthermore, in order to make the system autonomous, a search mode is included that operates when the tracking error is too high. To prevent the system going into search mode too often, a reduced table of errors is also stored that is updated every half an hour if there is enough radiation.

The control algorithm takes into account the different types of error that can appear in practice in low-cost, domestic systems, e.g. the placement of and problems

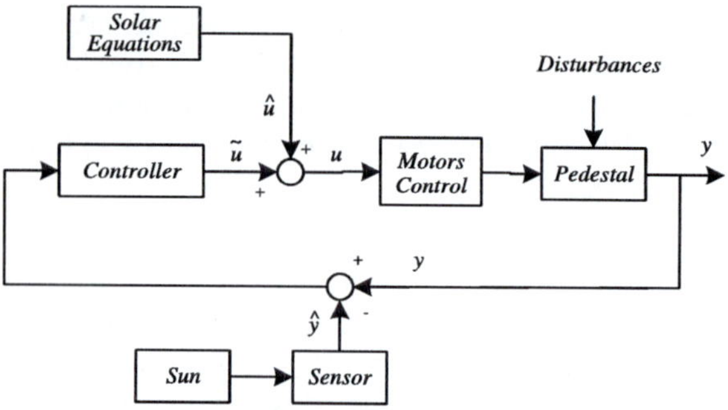

Fig. 3.5 Operation in tracking mode [335]

with the mechanical structure and errors of time and location. As a result, whatever the type of error, the controller can make the tracker follow the Sun. In fact, the proposed algorithm is also valid for large high-precision trackers since it contributes to reducing these errors.

A hybrid tracking strategy [335] that basically consists of two modes is used: in one mode, normal Sun tracking is carried out, maintaining a tracking error of less than a pre-specified value. In the other, a Sun search is undertaken by means of an ever-widening rectangular spiral; this is necessary when the Sun needs to be located because of some external disturbance (for example, a period of prolonged cloud cover). Each of these modes is described in greater detail below.

3.4.1 Normal Tracking Mode

This mode is effective whenever the Sun tracking error is smaller than a specified bound and the solar radiation large enough for the system to produce electric energy. It is a hybrid tracking system that consists of a combination of open-loop tracking strategies based on solar movement models (feedforward control) and closed-loop control strategies using a feedback controller. The feedback controller is designed to correct the tracking errors made by the feedforward controller in the open-loop mode. The operation in this mode is shown schematically in Fig. 3.5.

In this figure, u represents the position (azimuth and elevation) the tracking system assumes the Sun is in. It can be seen that this estimated position of the Sun is obtained by adding two values: $\hat{u}$, which is the position obtained from the equations that model the Sun's movement and $\tilde{u}$, which is a correction of that position based on the estimated position of the Sun, $\hat{y}$.

As has been explained in Chap. 2, there are several algorithms for calculating the position of the Sun ($\hat{u}$) based on the date and time provided by an auxiliary clock and geographical data (longitude and latitude of the point used to estimate the position of the Sun). In this section the PSA algorithm, developed by the Plataforma Solar de Almería [56], has been used. This algorithm has improved the calculation of universal time as well as the treatment of leap years and it also makes the calcula-

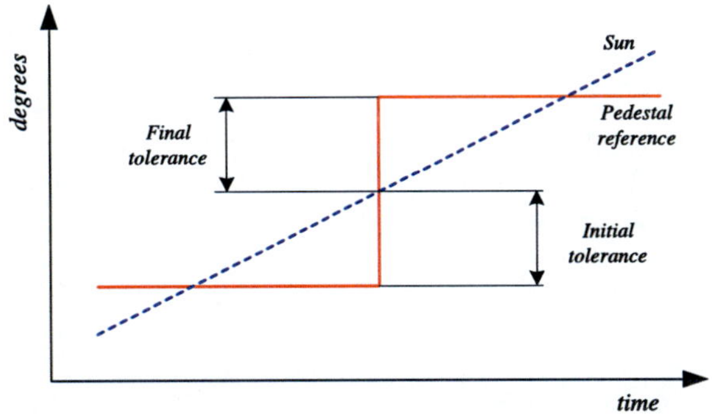

Fig. 3.6 Basic scheme for the movement of the mechanical structure [335]

tion more quickly and robustly, eliminating unnecessary operations by using simple, valid equations in both hemispheres.

However, despite the precision of this algorithm, errors in estimating the Sun's position are possible for several reasons, such as variations in the time given by the auxiliary clock with regard to actual solar time, lack of precision in the geographical location of the driver (errors in the estimation of latitude and longitude, although usually small if global positioning systems–GPS—technology is used) and errors in the alignment of the mechanical structure with respect to geographical North (different from magnetic North). In fact, this last kind of error is very frequent in low-priced installations if no specialized staff is employed for set-up adjustment, or if the wind causes any misalignment.

This fact justifies the necessity of including a correction ($\tilde{u}$) for the Sun's feed-forward position ($\hat{u}$) in order to obtain a better estimate of its real position (u), especially when it is important for the tracking error to be very low, as is in this case. This correction is provided by the *Controller* block, which will be analyzed later.

Once a realistic estimate of the Sun's position is obtained (u), the *Motor Control* block gives the necessary commands to the motor driver in order to move the platform according to the solar trajectory. For energy reasons, as the main objective of the strategy is the generation of energy using the Sun as a source, the tracker is not commanded to follow the Sun at all times, because this would cause continuous movement of the driver motors, which would in turn result in excessive energy consumption. Instead, to prevent unnecessary movement of the mechanical structure (see [27, 336]), the strategy implemented in the controller is the following (Fig. 3.6): the structure does not move so long as the tracking error is less than a certain tolerance assuming that the Sun is where the u signal says it is. When this error is greater than this tolerance, the controller orders the driver to move to a point which the Sun will arrive at in a certain amount of time. Thus, the tracker *waits* for the Sun. This process is identical and independent for each axis. However, the two axes never move at the same time: before ordering the movement of one of the axes, a check is made to ensure that the other axis is not moving. The two axes are not allowed to move simultaneously because of the type of sensor used to indicate whether the tracking of the Sun is correct, as will be seen later.

As the Sun moves along its trajectory throughout the day, signals are sent so that the driver moves appropriately, thus generating the electrical energy that the system was designed for. The power is used as a *Sensor* to confirm that the driver is tracking the Sun correctly, so a decrease in the power generated (under normal external conditions, i.e. without taking into account extended cloudy periods, for example) indicates tracking problems. It is known that the greater the error in either of the two coordinates (azimuth and elevation) is, the less the power generated. As a result, if the driver moves on either of the coordinates (while the other remains fixed) it can be assumed that the real position of the Sun for that coordinate corresponds to the point where the maximum power was produced during that movement. This is why both motors cannot move simultaneously. In this way, the power generated is used as a sensor to determine the Sun's position.

Finally, the *Controller* block was designed in order to close the feedback control loop; this implements a proportional and integral (PI) control strategy for each coordinate independently and its purpose is to bring about a difference of zero between the u signal and the real position of the Sun. This controller uses an estimate of this difference, calculated as follows: as the system moves from one position to the next (keep in mind that the system moves ahead to wait for the Sun), the control system samples the power generated by the power sensor. It is assumed that the point at which the maximum amount of power is produced is equal to the position of the Sun at the coordinate where the mechanical structure is being oriented (as explained in the preceding paragraph). Thus, by comparing the Sun's position according to this system with the value given by the corrected solar equations, the tracking error for each axis is obtained.

The error estimate is computed taking past and present error measurements into account, it is defined by the PI controller, applied in a discrete way:

$$\tilde{u}(k+1) = K_P\left(\tilde{y}(k) + \frac{1}{T_I}S_I(k)\right) \tag{3.1}$$

$$S_I(k) = S_I(k-1) + T_s\tilde{y}(k) \tag{3.2}$$

$\tilde{u}(k+1)$ being the present error estimate, $\tilde{y}(k)$ the last tracking error measurement, $S_I(k)$ the integral of the time varying discrete error signal, T_s the present sampling period and K_P and T_I constants tuned to give an adequate relative weight to the proportional and integral parts in the computation of the error estimate.

Regarding the sampling process for the power signal generated, it should be noted that it needs to be quick enough to have enough points to estimate the real position of the Sun for each coordinate with a certain amount of precision.

With regard to the PI control strategy employed in the *Controller* block, it is worth noting that the control laws (one for each coordinate) will not be carried out with a constant sampling time, as is usually the case with conventional discrete-time control systems, even though changes in the sampling time are small. In this case, the PI controllers will only operate when each incremental movement of the structure has finished and the structure has reached its final reference point. Given

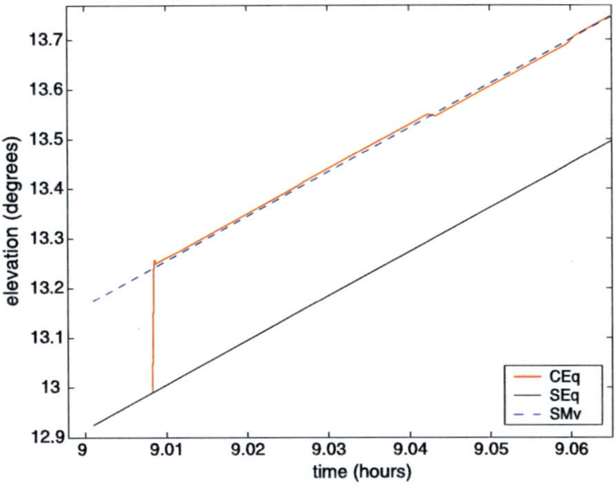

Fig. 3.7 Evolution of the coordinate elevation [335]

that these movements are determined by a certain tolerance in the orientation error (taking into account the strategy followed by the *Controller*) and that the velocity of the Sun (for both coordinates) is not constant throughout the day, the time that the structure must wait for the Sun and, consequently, the corresponding PI sampling time will vary depending on the position of the Sun. Furthermore, since both motors cannot move simultaneously, there is yet another delay in the movement of one of the motors if the other one is already moving.

Given the proposed control law, that the controller has an integral effect regardless of the sampling time used and that the variations of the sampling time are small, this controller will incorporate the error measured between the Sun's real position and the estimated position, u, providing a correction, $\tilde{u}$, that will cause the estimated position of the Sun to move toward the real position. Also note that this controller provides a continuous error of almost zero when taking into account that the usual time of the correction carried out by the controller is well below the characteristic time of the variations in the Sun's position throughout the day.

This tracking strategy obtains a close approximation to the evolution of the Sun's elevation and azimuth, even if the solar equations yield quite large errors. Figures 3.7 and 3.8 show a simulated example of the evolution of the three variables (the Sun's real movement (SMv), the progression of the values yielded by the solar equations (SEq) and the evolution obtained after the corrections (CEq)).

From the result of these simulations, it can be seen that the correction provided by the PI causes the corrected trajectory, which was initially the same as the trajectory calculated based on the solar movement equations (feedforward control), to tend toward the Sun's real trajectory, both with regard to the azimuth and the elevation coordinates. It can also be seen that the update time of the corrected trajectory (the PI sampling time) varies because of the type of strategy employed in the *Controller* that moves the mechanical structure.

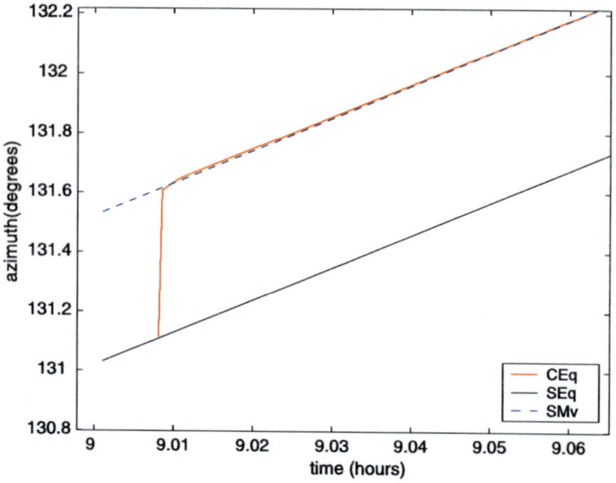

Fig. 3.8 Evolution of the coordinate azimuth [335]

3.4.2 Search Mode

As was mentioned above, the normal tracking mode operates so long as the Sun tracking error is small enough and the solar radiation great enough for the system to produce electric energy, and then the previous strategy is possible. If the tracking error is larger, no electric energy is produced and the sensor strategy will not work. Thus, for the system to function autonomously, how the system will react when one of these premises ceases to be true has also to be taken into account.

This section will describe how the search mode was designed. This mode will only operate when the tracking error is not small enough but solar radiation is great enough for the system to produce electric energy. Note that for this to occur, there must be an additional solar irradiance sensor (e.g., a pyrheliometer) that indicates whether irradiance exceeds the minimum threshold required.

Thus, if the Sun tracking error is too great (greater than a given upper bound) due to a combination of time errors, errors in the alignment of the mechanical structure and external disturbances, the solar arrays will not produce electric power and it will not be possible to feedback any information about the tracking errors. A clear example of this problem would be the presence of clouds for a prolonged period of time. During this time interval, no corrections are produced as a result of feedback from the system and the reference will be that provided by the equations (in open loop) that are available when the clouds disappear. If it remains cloudy for a long period of time and the errors associated with the equations are great, the misalignment between the Sun and the position sought by the positioner when the structure begins to move again may be rather great; consequently, the power sensor will not provide adequate information with which to correct this problem.

It is thus necessary to create a procedure which allows the Sun to be found when, for whatever reason, feedback does not occur. This is the search mode algorithm. An exception is the case in which the lack of energy produced by the inverter is a result of low solar irradiance (caused, for instance, by the presence of clouds); this case

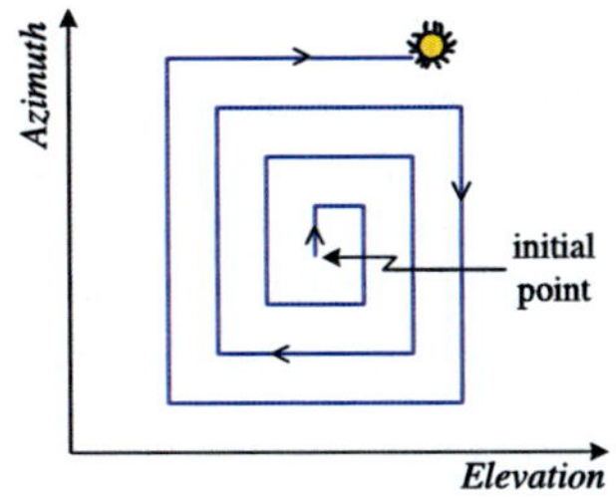

Fig. 3.9 Search mode [335]

will be analyzed in the next section. In this case, no matter how great the tracking error, the search mode should not be used, because the low solar irradiance keeps us from detecting when the mechanical structure is at the correct tracking point.

In search mode, the movement of the structure follows a square spiral in the azimuth–elevation plane in order to try to detect the position of the Sun (see Fig. 3.9). As the movement takes place, a check is made as to whether the system is generating electric power. As soon as electric power is produced, this mode is abandoned and the controller enters into the normal tracking mode.

As Fig. 3.9 shows, the structure movement in the azimuth–elevation plane is completely rectangular, because of the alternation in the movement of both mobile axes.

The amount by which the range of the movement is increased with each step is important. Special care must be taken in order not to increase the distance so much as to not detect the Sun between one movement and the next along the same side of the spiral. If the specifications require the Sun to be found within, for example, one degree, an increment in the spiral cannot be allowed too close to this value, or the risk of not detecting the Sun in the first step and going too far with regard to the Sun's position in the second step is taken. An example of this is shown in Fig. 3.10.

Thus, special care has to be taken when choosing how much to increment the step for the spiral. The most delicate point is at the end or the beginning of each step

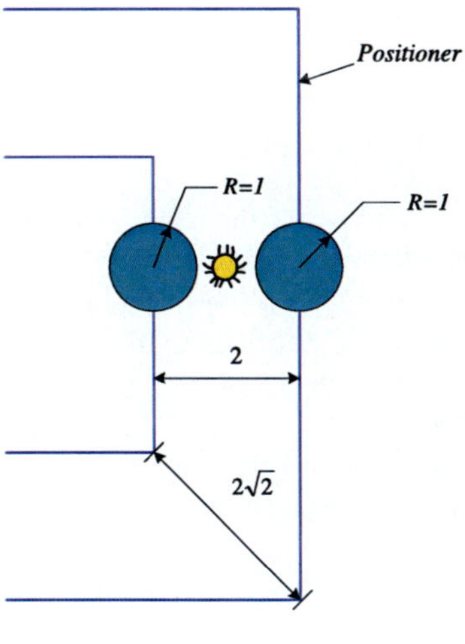

Fig. 3.10 Incorrect search step [335]

because at these points, the distance between one trajectory and the next is the value of the step multiplied by the square root of two, as is shown in the figure above. Furthermore, it should be taken into account that during this search the Sun does not remain fixed, but rather continues along its trajectory. These two circles should overlap enough to ensure that the Sun is not by-passed without realizing it.

3.4.3 *Other Situations*

This subsection briefly presents the actions taken when the solar irradiance level is lower than the minimum level established for generating electrical energy by means of the system.

There are different possible actions in this case, each one with advantages and disadvantages. However, the option of not moving the mechanical structure to follow the movement of the Sun so long as the irradiance threshold is not met has been chosen. The main advantage of this option is that no energy is consumed by moving the structure; the main disadvantage is that no energy is produced until there is sufficient irradiance again and the structure has tracked the Sun.

With this strategy, the main problem comes when it is time to track the mechanical structure, because during the cloudy period there has been no feedback on the actual solar position, i.e. the *Controller* block has not been operating.

This means that, at first, the estimated position of the Sun would be the position generated using the solar equation along with the last correction generated by the PI. Theoretically, there should not be any problems with the automatic operation of the system; if the position given by the solar equations is not precise enough, the system will go into search mode until proper tracking is achieved and then the system will return to normal tracking mode.

However, in practical implementation, the system might go into search mode too often and thus waste energy. To prevent this, a small table was incorporated, where every so often (in this case, every half hour), the calculated errors from the periods when there are no clouds are stored. Simulations have shown that using this table prevents the excessive use of the search mode after prolonged cloudy periods.

The main difference between this tracking system and other systems (e.g., [327, 328, 336]) is that it is not necessary to store the errors with regard to each of the structure positions.

To illustrate the behavior of this strategy, the result of several simulations which were carried out is shown in the following. In the simulations, there is a cloudy period between 9:00 and 11:00 (solar time) approximately. Furthermore, different sources of error were introduced: time error, tracking errors, etc. The results of two simulations are shown: one without and the other with the error table mentioned above. It can be noticed that without the table, the behavior is worse than when it is used. This is because without the table, the corrections made in the equations are not good enough to prevent a considerable misalignment.

Figures 3.11 and 3.12 show simulations of the evolution of the Sun (SMv) and the corrected equation (CEq), which corresponds to signal u in Fig. 3.5 as well as the

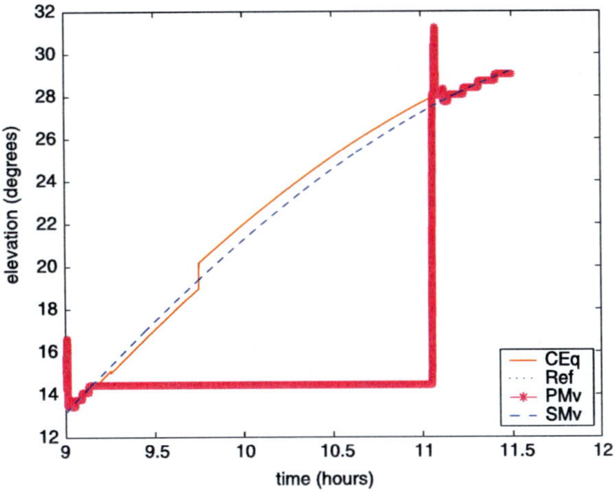

Fig. 3.11 Mechanical structure movement without using error table (elevation) [335]

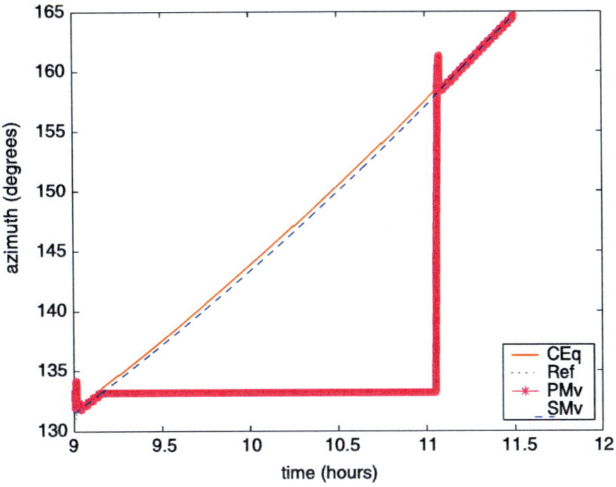

Fig. 3.12 Mechanical structure movement without using error table (azimuth) [335]

movement of the mechanical structure (PMv) when the error table is not used. Since, for the sake of clarity, the structure was kept still during the cloudy period (from about 9:00 to 11:00), there is a great discrepancy among the curve for mechanical structure movement and the other curves. This discrepancy will not have much of an effect because the system cannot generate energy when the Sun is hidden by the clouds. At the end of the cloudy period, the structure moves toward the Sun and, except for an overshoot evident in the graphs, the curves appear to overlap the rest of the time. The overshoot is a result of the search mode, as will be seen in Figs. 3.15 and 3.16.

When the error tables are used, the corrections as a result of feedback are much better and a search is unnecessary. Thus, a series of movements is avoided that would consume part of the energy generated. This can be observed in Figs. 3.13 and 3.14.

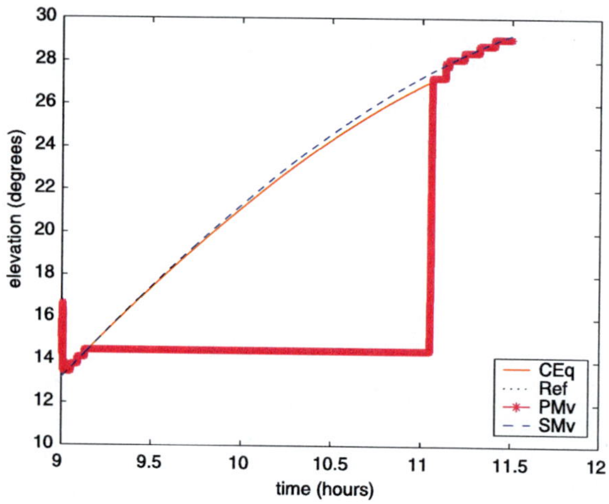

Fig. 3.13 Mechanical structure movement using error table (elevation) [335]

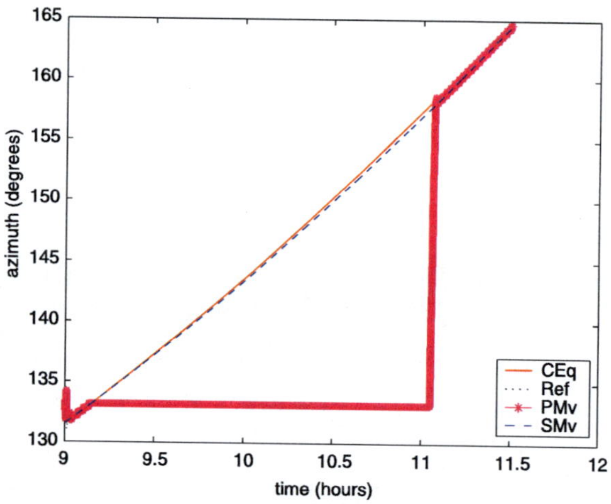

Fig. 3.14 Mechanical structure movement using error table (azimuth) [335]

Figures 3.15 and 3.16 show the same movements illustrated in Figs. 3.11–3.12 and 3.13–3.14 in the azimuth–elevation plane. The system starts from similar initial conditions (i.e. from the search mode) in both cases. It can be seen that when the error table is not included, there is a great discrepancy between the value of the corrected equations and the position of the Sun at the moment when the clouds disappear. This fact makes the search mode necessary again. However, when the error table is used, these discrepancies are not significant, which implies that a second call to the search mode is no longer necessary.

Obviously, the tables of error have limited validity in time. Apart from possible changes in origin or values of error, it has to be taken into account that the trajectory that the Sun follows varies with time and this variation is different, depending on the time of year.

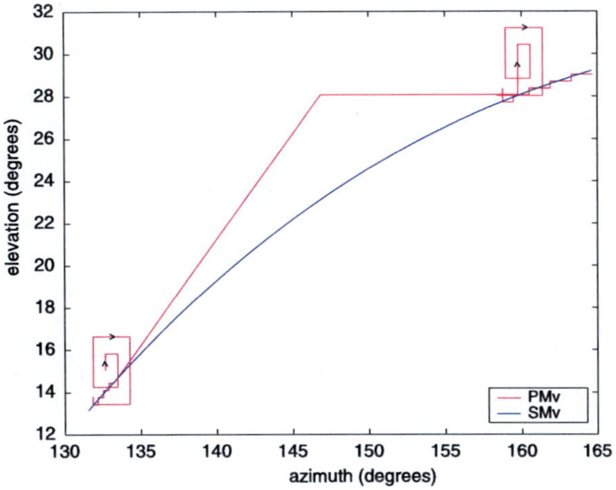

Fig. 3.15 Mechanical structure movements in elevation and azimuth coordinates (without using the error table) [335]

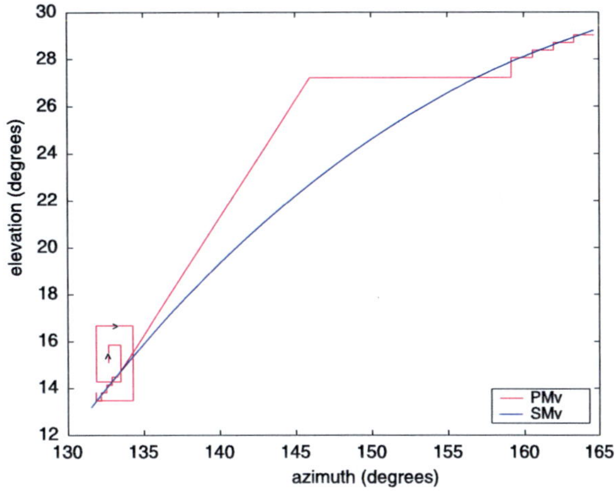

Fig. 3.16 Mechanical structure movements in elevation and azimuth coordinates (using the error table) [335]

Having carried out different simulations, if the table for when there is no need for a search to resume tracking after an extended cloudy period is considered valid, it is possible to consider about a margin of 25 to 30 days valid when the date is near the winter or summer solstices and 15 to 20 days when the date is near the spring or autumn equinoxes.

This means that after 15 to 20 days of continuous, total cloud cover, when the Sun comes back out the operation will be normal, with no need for a search. For longer periods of cloudiness, a search would be necessary before resuming normal tracking but, in general, the system will be adequate.

Finally, it should be noted that additional questions, such as safety routines to preserve the mechanical structure, e.g. the inclusion of a predetermined defense position in which the structure would offer minimal aerodynamic resistance, were not considered. The mechanical structure could adopt this position for instance, if

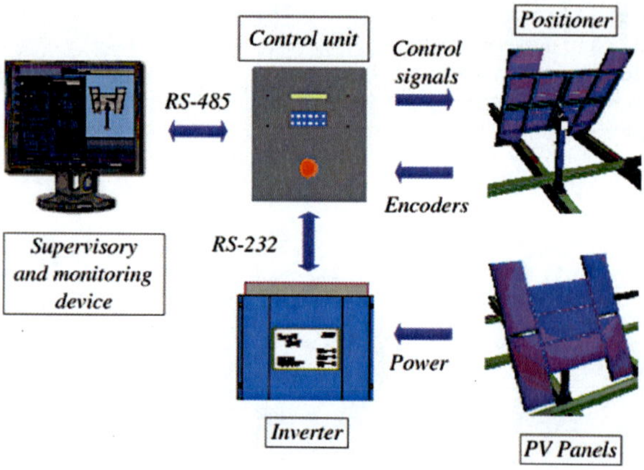

Fig. 3.17 Control loop scheme [335]

wind speeds endangered the structure itself (which would also imply the inclusion of an anemometer in the sensor system) or at night, when the Sun is not out.

3.4.4 Experimental Results

Operation of the prototype of the control unit (Fig. 3.17) developed was tested using the mechanical structure shown in Fig. 3.18. This low-cost positioning system is located on the roof of the *Department of Systems and Automatic Control Engineering Laboratories* at the *University of Seville*, in *Spain*. It was checked to make sure that it worked correctly with regard to both hardware (movement of both axes, decoding, etc.) and software (execution of basic programs, monitoring, etc.).

As can be seen in Fig. 3.18, the positioning system supports flat plate PV arrays instead of a concentrating PV system. Since high-concentration solar arrays were not available, several cells of slender-built tubes were mounted on the arrays. These cells guarantee that no solar irradiance gets to the arrays when the tracking error is greater than some degrees.

The different strategies tested in simulation were carried out in the control unit in order to fine-tune the controllers. To check the robustness of the algorithms, several error sources were included in the experiments, such as an offset in the time given by the auxiliary clock and a misalignment on the mechanical structure orientation with respect to the geographical North.

Figures 3.19 and 3.20 show the experimental power attained using the open-loop tracking strategy, as well as that obtained using the proposed hybrid strategy. Solar irradiance graphs are also included, showing that the experiments were performed under similar solar irradiance conditions during the first four hours. In this period of time the electric power generated using the hybrid strategy is, in mean values, 55% higher than the open-loop one.

It can be observed that the power generated by the solar arrays using the open-loop strategy has a maximum level of about 65 W for almost two hours at noon.

Fig. 3.18 Mechanical structure of the solar tracker (front face) [335]

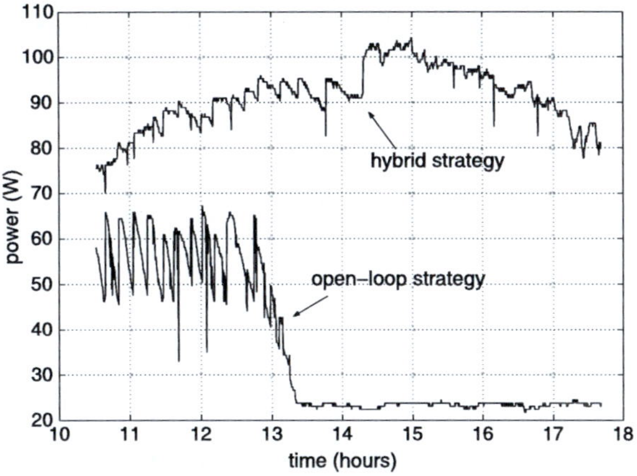

Fig. 3.19 Experimental results using an open-loop and the proposed hybrid strategy (electric power generated) [335]

However, the positioner loses the Sun from the afternoon onwards, which means that the cells of slender-built tubes throw shadow upon some of the solar cells with consequent decrease in the level of electric power generated.

The above mentioned fact does not occur when the proposed hybrid tracking strategy is used. Besides the fact that the level of power generated is about 90 W at noon (a benefit of about 40% with respect to the open-loop strategy, despite the low quality of the mechanical structure), the arrays not only do not lose the Sun but also their alignment is corrected, with the consequent increment of electric power generated (with a maximum greater than 100 W).

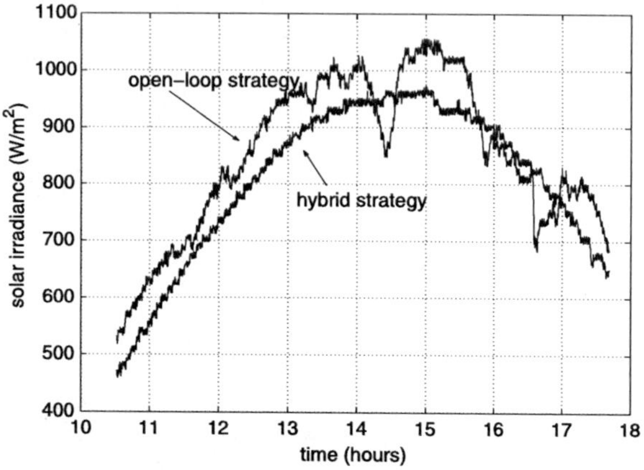

Fig. 3.20 Experimental results using an open-loop and the proposed hybrid strategy (solar irradiance) [335]

The reader may be surprised that the amount of generated power is so low. This is due to the fact that the experiments were carried out using a constant non-optimal electric load. Of course, the level of power could have been increased if an MPPT device had been used (see, for example, [183] and references therein). However, since the goal of these experiments is to evaluate the Sun's pointing error using different tracking strategies, the comparison between the level of generated power was taken into consideration, but not the level of power per se.

3.5 Summary

This chapter has dealt with the fundamentals of PV systems. After a general overview of the related concepts, a new Sun tracking strategy which provides small Sun tracking errors (needed by high-concentration solar arrays) has been introduced. The algorithm consists of two tracking modes: a *normal tracking mode*, used whenever the Sun tracking error is small enough and the solar irradiance is great enough; and a *search mode*, which operates when the first of the above conditions is not fulfilled but there is sufficient solar radiation to produce a minimum amount of electric power. Energy saving factors have been taken into account in the tracking strategy design. Simulated and experimental results have been shown, which demonstrate the benefits of the strategy with respect to a classical open-loop one, when errors in the estimation of the Sun's position (such as variations in the time given by the auxiliary clock or lack of precision in the alignment of the mechanical structure with respect to geographical North) are included.

Chapter 4
Basic Control of Parabolic Troughs

4.1 Introduction

Parabolic trough collectors (PTC) concentrate sunlight onto a receiver pipe located along their focal line. A heat transfer fluid (HTF), typically synthetic oil, water or molten salt, is heated as it flows along the receiver pipe and is routed either to a heat exchanger when this fluid is oil or molten salt to produce steam that feeds an industrial process (for instance a turbine), or to a flash tank when the fluid is pressurized water, to produce steam of up to 200°C for an industrial process, or get again to a turbine when superheated and pressurized steam is produced directly in the solar field [416–418]. In order to provide viable power production PTC have to achieve their task despite fluctuations in energy input, i.e. the sunlight. An effective control scheme is needed to provide the operating requirements of a solar power plant. A prototype that has been used as a test-bed plant for the development of the new generation of solar plants with PTC is the ACUREX field of the PSA that will be used to briefly describe the technology and associated subsystems. Information about modeling and control of the new generation of collectors using direct steam generation (DSG), represented by the DISS plant of the PSA, is included at the end of the chapter.

As was commented in Chap. 1, the main difference between a conventional power plant and a solar plant is that the primary energy source, although variable, cannot be manipulated. The intensity of the solar radiation, in addition to its seasonal and daily cyclical variations, also depends on atmospheric conditions such as cloud cover, humidity and air transparency. Due to this fact, a solar plant is required to cope with certain problems that are not encountered in other thermal power plants. The objective of the control system in a distributed solar collector field (DSCF) is to maintain the outlet HTF temperature of the field (or the highest outlet HTF temperature reached by one of the collectors at each sampling time) at a desired level in spite of disturbances such as changes in the solar irradiance level (caused by clouds), mirror reflectivity or inlet HTF temperature. This is achieved by adjusting the fluid flow and the daily solar power cycle characteristics are such that the HTF flow has to change substantially during operation. This leads to significant variations in the

E.F. Camacho et al., *Control of Solar Energy Systems*, Advances in Industrial Control, DOI 10.1007/978-0-85729-916-1_4,

dynamic characteristics of the field such as the response rate and the dead time, which cause difficulties in obtaining adequate performance over the operation range with a fixed parameter controller. Thus, from the control point of view, the main characteristics of this kind of plant are:

- Non-linearities, complexities, requiring modeling simplifications, changing dynamics and changing environmental conditions: (i) the solar radiation acts as a fast disturbance in respect to the dominant time constant of the process; (ii) the existence of time-varying input/output transport delay, since the delay in action depends on the manipulated variable (HTF flow rate); this type of delay appears both in the field and in the pipe connecting the loops to the storage tank; (iii) when modeling simplifications are made, there are strong unmodeled dynamics and the linearized dynamics vary with the operating point; indeed, the plant is best modeled as a distributed parameter system and, furthermore, there are resonance modes (frequencies at which the magnitude of the frequency response has a maximum or minimum value) in the frequency response of the collector field within the control bandwidth, such that when the system is excited by a signal (HTF flow or solar irradiance) with principal frequency components corresponding to those of the resonance modes, oscillations may appear at the system output.
- A PTC is essentially a very large heat exchanger and these types of system are quite common in the process industry; thus, most of the experience gained by the control of solar collector fields can be used for other, more common, industrial processes.

These aspects render the control problem in question a difficult one and call for the use of carefully designed control algorithms, robust enough to cope with the high levels of uncertainty present in the plant. The activities performed by the control groups related to this field [87, 88] cover modeling, identification and simulation, classical proportional-integral-derivative control (PID), feedforward control (FF), cascade control (CC), adaptive control (AC), gain-scheduled control (GS), internal model control (IMC), time delay compensation (TDC), optimal control (LQG), non-linear control (NC), robust control (RC), model-based predictive control (MPC), fuzzy logic control (FLC), neural network controllers (NNC) and hierarchical control (HC). The basic control approaches (PID, CC and FF) are briefly commented on this chapter, while the rest are described in the following one.

4.2 Description of the Technology and Subsystems

Most of the DSCF plants are composed of a distributed collector system (DCS), a thermal storage system and the power block (Fig. 4.1) governed by a control system.

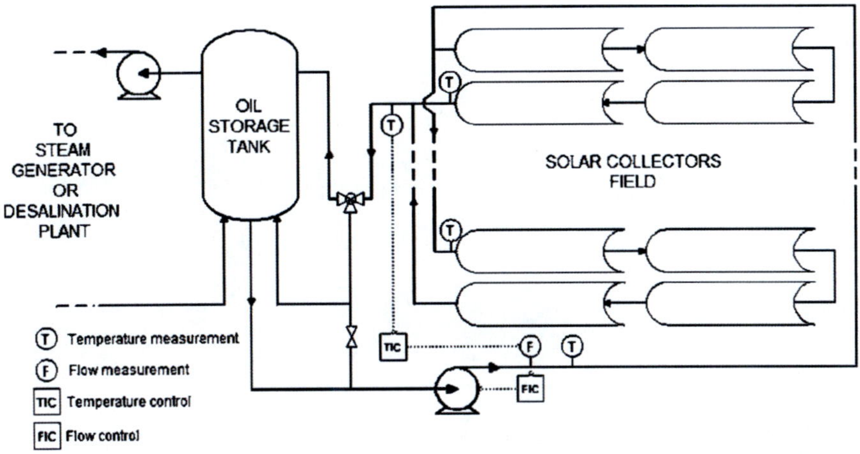

Fig. 4.1 Schematic diagram of the ACUREX solar collector field

4.2.1 Distributed Collector Field

The ACUREX DSCF consists of 480 East–West aligned single axis tracking collectors forming 10 parallel loops with a total mirror aperture area of 2672 m^2. Each of the loops is formed by four 12-module collectors suitably connected in series. The loop is 172 m long, the active part of the loop (those parts receiving beam irradiance) measuring 142 m and the passive part (those not receiving beam irradiance) 30 m. The HTF used is Therminol® 55 thermal oil, capable of supporting temperatures of up to 300°C; this is pumped from the bottom of a storage tank through the solar field where it picks up the heat transferred through the receiver tube walls to the top of the tank. The heated HTF stored in the tank is used to boil water, which is utilized in a steam turbine to drive an electricity generator or to feed the heat exchanger of a desalination plant (Figs. 4.1 and 4.2). The operation limits for the HTF pump are between 2.0 and 12.0 l/s. The minimum value is there for safety, mainly to reduce the risk of the HTF decomposing, which happens when the HTF temperature exceeds 305°C. Another important restricting element in this system is the difference between the inlet and outlet HTF temperatures of the field that must be less than 80°C. If the difference is higher than 100°C there is a significant risk of HTF leakage due to high HTF pressure in the pipe system. The field is also provided with a Sun tracking system (as those commented on in Chap. 1) which causes the mirrors to revolve around an axis parallel to that of the pipe. The seeking mechanism can reach three possible states: (i) Track: The mechanism seeks the Sun and the collectors focus on the pipe. (ii) Desteer: The mechanism steers the collector several degrees away from the Sun and continues tracking with the receiver out of focus. This protects the field from over-heating in case of a pump failure. (iii) Stow: The mechanism moves the collector to an inverted position at the end of each day or in the event of a serious alarm. The system takes about 5 min to take

Fig. 4.2 ACUREX distributed solar collector field (courtesy of PSA)

the field from stow to track. A more detailed description of the field can be found in [206].

4.2.2 Storage

The storage tank was included in order to allow flexible electricity production and to provide a buffer between electricity generation and fluctuating solar input [206]. For the initial start-up of the plant, the system is provided with a three-way valve, which allows the HTF to circulate in the field until the outlet temperature is adequate for entering the storage tank.

The tank used in this collector field has a capacity of 140 m^3 which allows for storing of 2.3 thermal MWh for an inlet field temperature of 210°C and an outlet field one of 290°C. The good thermal stratification of this HTF allows it to be stored for various days.

4.2.3 Control System

In the case of the ACUREX field, a supervisory control and data acquisition (SCADA) system runs in a computer, working automatically except in certain operations at the start and at the end of the operation. Every 3 s more than 150 data are registered and processed. The SCADA provides a supervisory module which controls the various alarms which may be produced during daily operation.

The direct control system of the pump which impels the fluid is an analog PI adjusted to the design point corresponding to 12 noon (solar time), March 21st with direct beam irradiance of 920 W/m^2. Evidently, the environmental conditions under which a solar plant has to operate are very different from the design conditions.

Typical sample times used in this installation are either 15 or 39 s, being a good trade-off both for set point tracking and disturbance rejection.

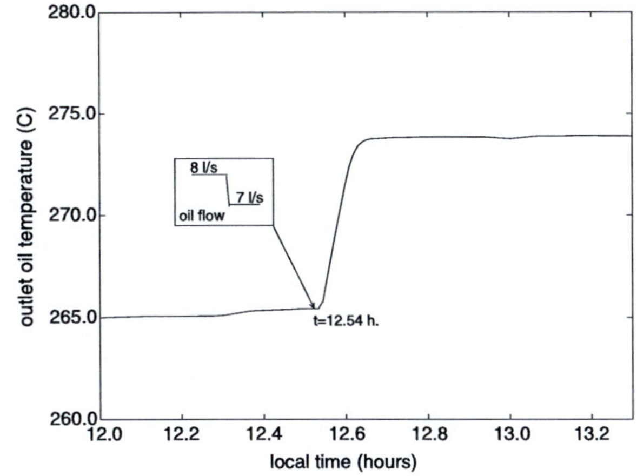

Fig. 4.3 Response to an open loop oil flow step [85]

4.3 Modeling and Simulation Approaches

Several classifications of modeling approaches can be found in the literature, that presented in [69] being widely accepted. The hierarchy of process models has been used for different purposes in this type of solar plant: control models, simulation models, set point optimization models, fault tolerance, etc. Models for control purposes range from the simplest ones based on steady-state relationships or on linear low-order approaches to non-linear empirical or first principles-based ones. In practice, the DCS has been modeled both by using first principles or empirically by conducting practical tests. In this second case, when introducing a step input signal in the HTF flow in an open loop configuration (reaction curve method) while the rest of disturbances are in quasi-steady state, the response can be approximated by that of a first-order system or overdamped second-order system with a delay depending on the fluid velocity, as will be seen in Fig. 4.3. This kind of step response suggests the use of low-order linear descriptions of the plant (as is the norm in the process industry) to model the system and design diverse control strategies. When using persistent excitation signals (e.g. random binary sequences) or analytically examining the dynamics of the system [241, 242] it can be seen that the plant exhibits a number of resonance modes within the control bandwidth. Thus, non-linear models (both mechanistic and empirical ones) or high-order linear models around different operating points have to be used [85, 87]. All these modeling approaches are discussed in this chapter.

4.3.1 Fundamental Models

A DSCF, under general assumptions and hypotheses, may be described by a distributed parameter model of the temperature [38, 81, 85, 97, 99, 100, 213, 280, 325], using the following system of partial differential equations (PDE) describing the energy balance:

$$\rho_m c_m A_m \frac{\partial T_m}{\partial t}(t,\ell) = \eta_{col} G I(t) - P_{rc} - D_f \pi H_t \big(T_m(t,\ell) - T_f(t,\ell)\big) \quad (4.1)$$

$$\rho_f c_f A_f \frac{\partial T_f}{\partial t}(t,\ell) + \rho_f c_f q(t) \frac{\partial T_f}{\partial \ell}(t,\ell) = D_f \pi H_t \big(T_m(t,\ell) - T_f(t,\ell)\big) \quad (4.2)$$

where the subindex m refers to the metal and that of f to the fluid; $\rho_{m,f}$ is the density, $c_{m,f}$ is the specific heat capacity, $A_{m,f}$ is the cross-sectional area, $T_{m,f}$ the temperature, t the time, ℓ the space, η_{col} is the global collectors' efficiency, G is the collectors' aperture, I is the corrected direct solar irradiance, P_{rc} are convective thermal losses of pipe exterior per unit length, $D_{m,f}$ is the pipe diameter, H_t is the convective heat transfer coefficient of pipe interior and q is the HTF pump volumetric flow rate (in the table of variables and parameters defined at the beginning of the book, the corresponding units are included). A simplified energy balance neglecting heat losses and assuming incompressibility of the fluid and no diffusion has also been used by several authors [135–138, 195, 196, 351, 352, 388, 389], described by

$$A_f \frac{\partial T_f}{\partial t}(t,\ell) + q(t) \frac{\partial T_f}{\partial \ell}(t,\ell) = \frac{\eta_{col} G}{\rho_f c_f} I(t) \quad (4.3)$$

where $T_f(t,\ell)$ is the HTF temperature at position ℓ along the tube, with boundary condition $T_f(t,0) = T_{in}(t)$ (inlet HTF temperature to the DSCF).

The objective is to control the outlet HTF temperature variable $T_{out}(t) = T_f(t,L)$ to its specified set point. The incoming energy depends on the global collectors' efficiency and thus, on the peak optical efficiency of the collectors, on the mirror reflectivity, on the effective reflecting surface and on the effective irradiance onto the collector $I(t)$. These last two variables depend on the incidence angle between solar rays and the vector normal to the collector surface (θ_i), this angle being a function of the solar hour and date (see Chap. 2).

Both lumped and distributed parameter versions of the models obtained from Eqs. (4.1), (4.2) and (4.3) have been used both for control and simulation purposes. Depending on the applications, the properties of the HTF are considered constant or functions of the temperature. The development of numerical simulation models of the plant has played an important role in the design of different control strategies avoiding a number of expensive and time consuming controller tuning tests at the solar power plant. Based on Eqs. (4.1) and (4.2), a distributed parameter model of the ACUREX field was developed [81, 97] and implemented [38, 85] and has been used for simulation purposes by many researchers. This model is summarized in the following section.

4.3.2 Distributed Parameter Model

The dynamic behavior of a collector of the DSCF is simulated by 100 lumped-parameter submodels. Temperatures of the HTF and the tube walls are modeled separately. The model also takes Sun position, field geometry, mirror reflectivity, solar irradiance and the inlet field HTF temperature into account [38, 81]. The collector loop is the basic subsystem which determines the behavior of a collector field. If it is possible to model a loop, the behavior of the whole field can be determined by simply adding the parallel loops and allowing for transport delays in the interconnecting tubes. The present model has been developed to fulfill the following objectives:

- Simulation of the field behavior in order to optimize the temperature regulation system.
- A study of the behavior of the system under specific operational conditions such as passing clouds.
- A study of extreme situations by simulation of failures, desteer mechanisms, etc.
- Application to other collector fields by modifying the corresponding parameters.

The model simulates the temperature distribution in the absorption tube and in the HTF along the collector loop at a point in time, as well as the temporary variation of the temperatures at determined points of the collector. The following hypotheses have been made:

- The properties of the HTF are considered as functions of the temperature.
- The flow in each section is presumed to be circumferentially uniform and equal to the average value.
- Variations in the radial temperature of the tube wall are not taken into account. This assumption is reasonable in the case of a thin wall with good thermal conductivity.
- The HTF flow and the irradiance are considered as time functions and are always the same for each element (an incompressible fluid is presumed).
- Losses caused by the conduction of axial heat on both sides of the wall and from the fluid are negligible. Axial conduction in the tube should be slight given that the wall is thin, having high heat resistance. Axial conduction in the fluid is relatively slight as the HTF conductivity is poor.

Using the above hypotheses and applying the conservation of energy in the metal tube of a length control volume (CV) $\Delta\ell$ over a time interval Δt described by Eqs. (4.1) and (4.2) and supposing $P_{rc} = D_m \pi H_l (T_m - T_a)$, where H_l is the convective heat transfer coefficient of pipe exterior, D_m the outside diameter of pipe, T_m the temperature of metal of pipe wall and T_a the ambient temperature (see the table of variables and parameters defined at the beginning of the book), the simulation model for the active zones (those receiving concentrated solar irradiance) can be obtained. The equations which describe performance in a passive element (non-irradiated) are similar except that solar energy entrance is nil and the heat loss coefficient is much less because of thermal insulation. Thus, the model for the complete field is built of a series of active and passive elements. Equations generated can be solved using an iterative process with finite differences. The temperatures of the fluid and of the absorbency tube are calculated for each time interval and for each element. Each segment is 1 meter long and the integration interval is 0.5 s.

A two stage algorithm has been chosen to solve the temperature equations. At the first stage, the temperatures of the fluid and of the metal are calculated presuming the fluid to be in a steady state. At the second stage the fluid temperature is corrected in function of the net energy transported by the fluid.

- First Stage

$$\begin{aligned} T_m(k,n) = T_m(k-1,n) + \frac{\Delta t}{\rho_m c_m A_m}\big(\eta_{col} G I(k) \\ - D_m \pi H_l\big(T_m(k-1,n) - T_a(k)\big) \\ - \pi D_f H_t\big(T_m(k-1,n) - T_{1f}(k-1,n)\big)\big) \end{aligned}$$

$$T_f(k,n) = T_{1f}(k-1,n) + \frac{\pi D_f H_t \Delta t}{\rho_f c_f A_f}\big(T_m(k-1,n) - T_{1f}(k-1,n)\big)$$

- Second Stage

$$T_{1f}(k,n) = T_f(k,n) - \frac{q(k)\Delta t}{A_f \Delta \ell}\big(T_f(k,n) - T_f(k,n-1)\big)$$

In these difference equations, $T_f(k,n)$ and $T_m(k,n)$ are the temperatures of fluid and metal in segment n during the k time interval. The different constants and coefficients used in the previous equations have been determined using real data from the plant. Many of them are adjusted to polynomial functions of the temperature by a least squares (LS) method [97].

Properties of Thermal Fluid One of the main characteristics of Therminol® 55 oil is its low thermal conductivity. Furthermore, its density is highly dependent on its temperature, which permits the use of just one storage tank to contain both hot and cold oil in thermal stratification (the thermocline effect). From data supplied by the oil producer [206], its physical properties have been obtained as the following polynomial function of the temperature [°C]: density: $\rho_f = 903 - 0.672 T_f$ [kg/m^3], specific thermal capacity: $c_f = 1820 + 3.478 T_f$ [J/(kg °C)], thermal conductivity: $k_f = 0.1923 - 1.3 \cdot 10^{-4} T_f$ [W/(m °C)], dynamic viscosity: $\mu_f = 1.41 \cdot 10^{-2} - 1.6 \cdot 10^{-4} T_f + 6.41 \cdot 10^{-7} T_f^2 - 8.66 \cdot 10^{-10} T_f^3$ [Pa s], Prandtl number: $P_{ra} = 212 - 2.2786 T_f + 8.97 \cdot 10^{-3} T_f^2 - 1.2 \cdot 10^{-5} T_f^3$.

Convective Heat Transfer Coefficient of Inside Pipe This coefficient can be expressed as a function $H_t = H_v q^{0.8}$, where

$$H_v(T_f) = 2.17 \cdot 10^6 - 5.01 \cdot 10^4 T_f + 4.53 \cdot 10^2 T_f^2 - 1.64 T_f^3 + 2.1 \cdot 10^{-3} T_f^4$$

Global Coefficient of Thermal Losses The losses have been evaluated by various tests carried out on the ACUREX field with the oil circulating. In the steady state losses are calculated by multiplying the enthalpy lost in the oil by the mass flow. Using the data, the loss coefficient is given by

$$H_l = 0.00249 \bar{\Delta T} - 0.06133 \quad \big[\mathrm{W/(m^2\,°C)}\big]$$

where $\bar{\Delta T}$ is the difference in temperature between the average inlet and outlet temperature and the ambient temperature ($\bar{\Delta T} = (\bar{T}_f - T_a) = (\frac{T_{out}+T_{in}}{2} - T_a)$). In

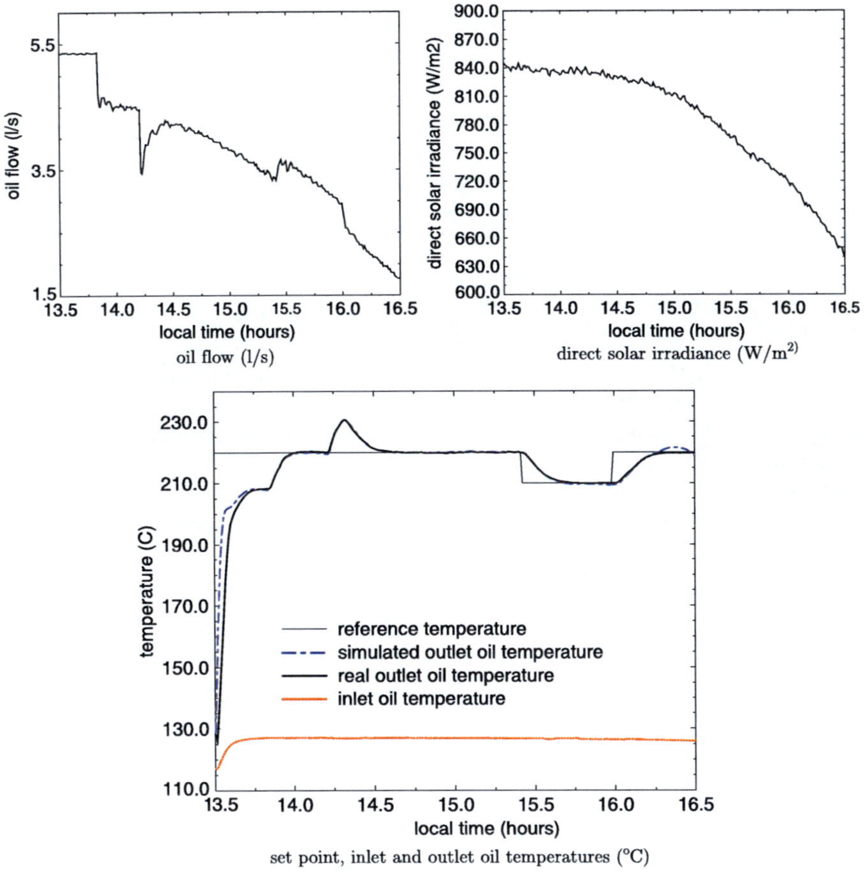

Fig. 4.4 Comparison simulator-real plant [85]

the model the thermal loss coefficient is calculated by applying the above equation to each element of length and having $\bar{\Delta T}$ equal to the temperature of this element minus the ambient temperature.

Model Validation A complete study in which results obtained by the simulation model are compared to real experimental data can be found in [81]. In these experiments, the response of the simulator is compared to that of the real plant under various different conditions (changes from track to desteer position, from desteer to track position, response to HTF flow steps, etc.). As an illustrative example, Fig. 4.4 shows the output of the model in an open loop configuration and that obtained by the real system under the same conditions of HTF flow, solar irradiance and inlet HTF temperature (without any intermediate correction using the real plant output). As can be seen, after an initial transient in which the temperatures of the model and the real system differ substantially (mainly due to the fact that it is impossible to know the initial temperature profile both in the tubes and at the lower thermocline of the storage tank and to the degradation of the thermal properties of the HTF at low temperatures), the model tends to follow the system behavior with a logical difference which tends to zero as the system approaches the steady state. At the end of the operation (which normally finishes at four pm local time), a discrepancy appears between the outlet of the model and that of the plant due to the progressive decrease in solar irradiance, a case in which the coefficients of the model are not as exact as could be desired.

The temperature increase experimented with by both the model and the field at about 14.25 was due to the inclusion of a loop that was closed at the beginning of the operation. Due to this fact, the HTF flow circulating through each loop experimented a sudden decrease. The second significant change in the HTF flow signal and, indeed, in the output of the system was produced at about 15.4, due to a set point change. As can be seen, the behavior of the non-linear distributed parameter model in both situations is very similar to that of the real plant. From the control viewpoint, in order to ensure a secure operation, the controlled variable is the outlet temperature of the loop with the highest value at each sampling time and not the global outlet HTF temperature. This fact can introduce additional dynamics if the highest temperature loop changes occur during operation, but avoids any of the loops exceeding the maximum allowable temperature, in which case the mirrors are sent to stow position by the supervisory system, causing heavy losses in daily operation.

Some authors have modified this original simulation model or performed different implementations using other numerical methods or taking into account the dynamics of the tubes connecting the outlet of the DSCF with the storage tank. As shown in [309], the dynamic characteristics of the tube joining the output of the loops to the top of the storage tank are given by a gain of less than one, a time constant and a variable delay. This approximation has been adopted in order to modify the basic formulation of the non-linear model to account for dynamic characteristics introduced by the tube. The modified model has been validated with data obtained at the plant in closed-loop operation [309]. In [274] a modification was performed on this non-linear model of parabolic trough collectors in [38, 85] to include varying transport delay. In [241], a modification of the original model was also developed, as it was limited by being unable to adequately represent transport delay effects and the inconvenience of not having a steady-state finder. When using the model for transient studies, the initial conditions are found simply by running the model over a period to permit initial transients to decay. To overcome this, the discrete model equations were reformulated to provide the capability of direct calculation of steady-state conditions using an implicit trapezoidal approximation instead of a two-step Euler approximation such as that used by [38, 85]. All the models mentioned are based on standard fluid flow and thermodynamic considerations, but considering incompressible fluid. Nowadays, emphasis is placed on modeling PTC with direct steam generation, as will be seen in Sect. 4.5.2.

Dynamic validation of the models has been carried out in various ways. Most of the authors have used typical step-response test performed at the plant. In [247] dynamic validation was achieved by making a comparison between the plant and a model in the frequency domain. The frequency response of the plant was obtained by a Fourier analysis of measured input and output data during transients. The method of excitation used was the simple pulse test. This was chosen in preference to periodic signals such as the common pseudo random binary sequence (PRBS) simply because it extracts dynamic information very quickly. In comparison, a PRBS signal takes well over an hour to extract the relevant data with sufficient accuracy, suffering from the influence of solar irradiance drifts. In order to use PRBS type signals, computer models have to be used, as in [84]. Equations (4.1) and (4.2) have also been used for control purposes [85, 87] in the development of feedforward controllers [38, 82, 85, 100, 195, 196, 247, 325, 329, 331, 334, 348, 382], non-linear PID controllers including real-time numerical integration of the distributed plant model [195, 196], non-linear model-based predictive controllers [16–18, 34, 35, 40, 42, 84, 295, 297, 298], internal model control [135–138], time delay compensation [274], feedback linearizing controllers [29, 99, 107, 109, 186, 349], multirate controllers [350, 351] (all these strategies are treated in other sections) and for set point optimization purposes.

4.3.3 Analysis of the Dynamic Response of the Plant

An analysis of both the time and frequency responses was carried out to characterize the system. A summary is included in the following subsections.

4.3.3.1 Analysis of the Time Response

One of the tests carried out to characterize the field dynamically consists of introducing a step input signal in the HTF flow in an open loop configuration (without flow recirculation). The response to a change in HTF flow from 8 to 7 l/s can be seen in Fig. 4.3 (at solar midday). As can be appreciated, the response can be approximated by that of a first-order system or an overdamped second-order system with a delay. This kind of step response suggests the use of low-order linear descriptions of the plant (as is usual in the process industry) to model the system and to design diverse control strategies.

4.3.3.2 Analysis of the Frequency Response

To achieve a greater knowledge of the plant dynamics, PRBS tests were carried out in order to obtain input–output data to calculate the frequency response of the plant corresponding to different operating conditions. Figure 4.5 shows the frequency response[1] obtained by a spectral analysis of the input–output data for a PRBS test on

[1]The frequency response corresponds to the system constituted by the feedforward studied in Sect. 4.4.1 in series with the plant.

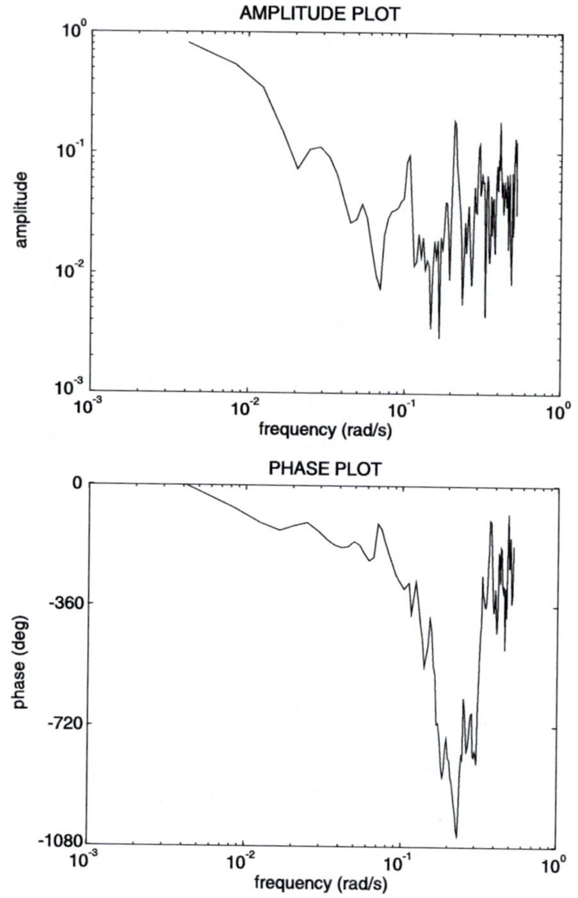

Fig. 4.5 Frequency response of the plant under a fixed operating condition

the plant corresponding to a mean flow condition of 6 l/s. As can be seen, the plant exhibits a number of resonance modes within the control bandwidth.

Figure 4.6 shows the theoretical (without taking into account thermal losses) and real frequency response at one of these operating points. As can be seen, the amplitude decreases at certain frequencies but does not reach zero because the field thermal losses dampen this decrease.

The resonance behavior can be analyzed by considering an approximate simple system which consists of a tube through which fluid is pumped. If this tube is divided into n elements measuring $\Delta\ell$ meters long, considering fluid properties to be constant (in reality they are functions of the temperature) and the thickness of the tube walls equal to zero (or infinite thermal conductivity), by making an energy balance in each of these elements:

$$\frac{\partial(\rho_f A_f \Delta\ell c_f T_f(t,\ell))}{\partial t} = I(t) - \rho_f c_f q(t)\big(T_f(t,\ell+\Delta\ell) - T_f(t,\ell)\big) \qquad (4.4)$$

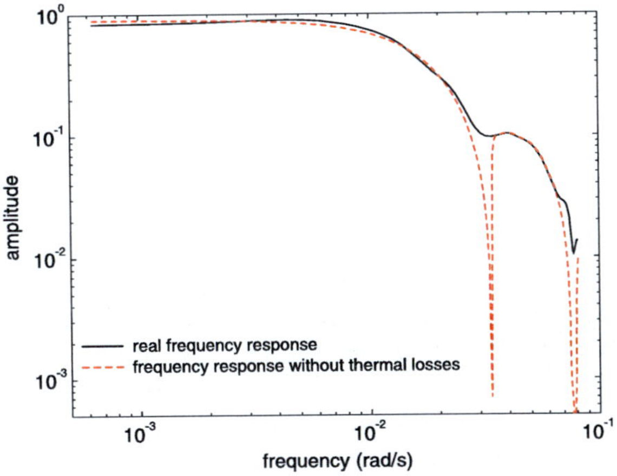

Fig. 4.6 Real and theoretical frequency responses [85]

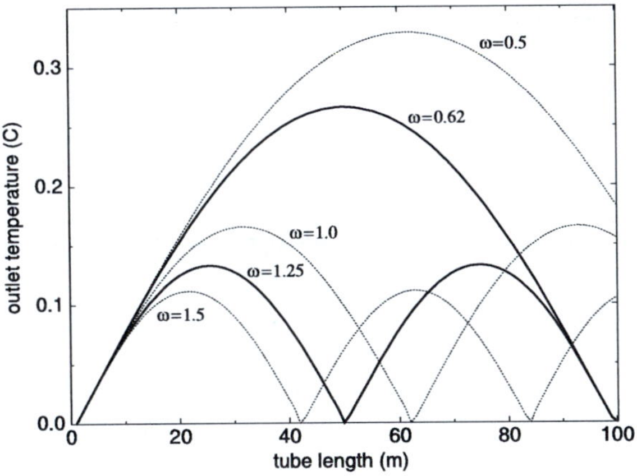

Fig. 4.7 Amplitude of temperature wave in each one of the tube elements (ω rad/s) [85]

Considering a small disturbance model around an equilibrium point $(q_0, T_{f0}(\ell))$ and neglecting second-order terms $(T_f(t,\ell) \approx T_{f0}(\ell) + \tilde{T}(t,\ell); q(t) \approx q_0 + \tilde{q}(t))$, if the flow perturbation is made equal to a sine wave $\tilde{q}(t) = Q\sin(\omega t)$ and Eq. (4.4) is solved numerically for different values of the excitation frequency, the results can be seen in Fig. 4.7, which shows the amplitude of the sine wave obtained for the temperature of the HTF $\tilde{T}(t,\ell)$ along the pipe. As can be seen, the outlet HTF temperature has a zero response in a pipe of length $L = 100$ m for a frequency of 0.62 and 1.25 rad/s. In practice, these frequencies are lower and the drops are attenuated by the field losses.

Another way of analyzing the resonance modes is by considering a sinusoidal solar irradiance profile and studying variations produced in the outlet temperature, as in [242]. In order to summarize the physical mechanism which quantitatively accounts for resonance modes, solar irradiance is presumed to be a sinusoid signal to study its effect on the outlet temperature when steady-state conditions in the HTF flow are considered. The variation in outlet HTF temperature $\tilde{T}_{out}$ of one of the loops is a function of the energy accumulated by the fluid when pumped through the tube. If air thermal losses and the thermal capacity of the tube are neglected, the variation in the outlet temperature due to a change in solar irradiance can be expressed as the integral of the energy absorbed by the fluid ($\tilde{h}$) when circulating through the tube:

$$\tilde{T}_{out} \approx K_h \int_0^{t_r} \tilde{h}\, dt$$

where t_r is the residence time of the fluid in the field ($t_r \approx L/v$), with L being the tube length and v the fluid velocity. If solar irradiance is a sinusoid of frequency ω, when $\omega = 2\pi/t_r$ (period equal to the residence time), the value of $\tilde{T}_{out}$ is zero and so, no changes are observed in the outlet temperature. Extending this result, if the period is considered to be 1.5 times the residence time ($\omega = 3\pi/t_r$), the integral and thus the variations, in the output amplitude, will be maximum for a determined magnitude of the input signal. These situations are valid for high frequencies. In general, if thermal losses are neglected, a minimum will be obtained for $\omega = 2n\pi/t_r$ and a maximum for $\omega = 2\pi(n + 1/2)/t_r$, with n being a positive integer. However, thermal losses cause these maximum and minimum to be not exactly in these points. In practice, air thermal losses and the thermal capacity of the tube cannot be neglected and attenuate the variations in the outlet temperature.

From the control viewpoint, the plant can be approximated by simplified linear models when considering the operation around a set point (small signal model). As has been seen in the analysis of the step response, the plant can be approximated by a simplified first-order model and a dead time, as in [82, 84]. This is a good approximation if only low frequencies are excited. If this is not the case, as when more demands are made on the plant response time, the resonance modes (unmodeled dynamics) may give rise to unacceptable oscillatory behavior.

4.3.4 *Simplified Fundamental Models*

4.3.4.1 Lumped-Parameter Model

Supposing a lumped description of the plant, the variation in the internal energy of the field can be described by

$$c_1 \frac{dT_{out}(t)}{dt} = c_2 I(t) - c_3 q(t)\big(T_{out}(t) - T_{in}(t)\big) - c_4 \bar{\Delta T}(t) \qquad (4.5)$$

where all the variables are known and the coefficients c_i, $\forall i \in \{1, \ldots, 4\}$ can be determined by experimental tests (for the ACUREX field they were determined in [329]).

4.3.4.2 Bilinear Models

As has been pointed out in Sect. 4.3.1, many authors use a simplified PDE model of the PTC such as that shown in Eq. (4.3). Rearranging terms [29], this can be expressed as

$$\frac{\partial T_f}{\partial t}(t,\ell) + v(t)\frac{\partial T_f}{\partial \ell}(t,\ell) = \gamma I(t) \tag{4.6}$$

where $\gamma = \frac{\eta_{col} G}{\rho_f c_f A_f}$. As in [29], assuming a smooth variation of HTF temperature along the pipe, the temperature distribution can be approximated by

$$\left.\frac{\partial T_f}{\partial \ell}\right|_{\ell\in(\ell_{i-1},\ell_i]} \approx \frac{T_{f_i} - T_{f_{i-1}}}{l}, \quad i = 1,\dots,n \tag{4.7}$$

where l is the length of each segment, n is the number of segments, $\ell_i = il$, $L = nl$ is the pipe length and $T_{f_i} = T_f(t, il)$. Defining the state variables $x_i = T_f(t, il)$, $i = 1, \dots, n$, the process dynamics can be described by a system of bilinear ordinary differential equations:

$$\frac{dx_i}{dt} = -v\frac{(x_i - x_{i-1})}{l} + \gamma I, \quad i = 1,\dots,n,\ x_0 = T_{in} \tag{4.8}$$

This model can be written as

$$\frac{dx}{dt} = f(x) + g(x)u \tag{4.9}$$

where $x = [x_1, \dots, x_n]^T$ and

$$f(x) = \gamma I \begin{bmatrix} 1 \\ 1 \\ \vdots \\ 1 \end{bmatrix}, \qquad g(x) = -\frac{1}{l}\begin{bmatrix} x_1 \\ x_2 - x_1 \\ \vdots \\ x_n - x_{n-1} \end{bmatrix}$$

$f(x)$ is independent of x, $f(0) \neq 0$ and $g(x) = Bx$ where

$$B = -\frac{1}{l}\begin{bmatrix} 1 & 0 & \cdots & 0 \\ -1 & \ddots & & \vdots \\ \vdots & \ddots & \ddots & 0 \\ 0 & \cdots & -1 & 1 \end{bmatrix}$$

For the number of states n high enough, this model reasonably describes the transport and heating phenomena inside the pipe and can be used for feedback linearization control purposes [29].

It is also possible to use a lumped-parameter physical model of the plant for control purposes [109, 318, 319], obtained from the distributed parameter non-linear model based on partial differential equations (PDE) to model transport phenomena. Equation (4.10) describes a simplified balance of the variation of the internal energy of the plant. This model has been obtained by removing the dependence on space in the original PDE model. Moreover, some experiments to obtain parameters such as the thermal loss coefficient given by Eq. (4.11) and the time delay of

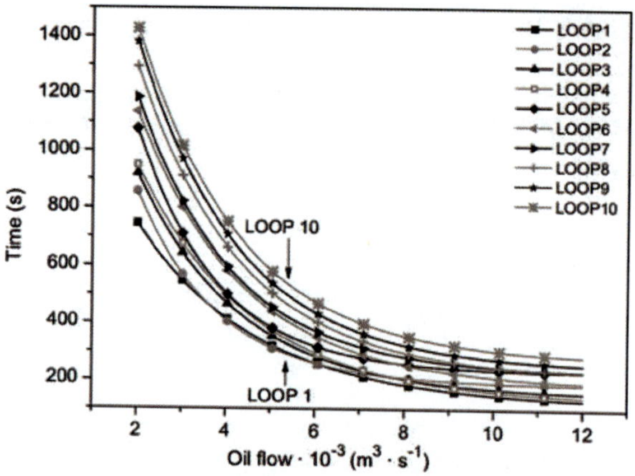

Fig. 4.8 Time delay between the inlet temperature sensor measurement and the ten outlet temperatures of the field as function of the flow (courtesy of C.M. Cirre *et al.*, [109])

the inlet temperature were performed at the actual plant. Equation (4.11) shows an empirical function that provides the global thermal losses [W], with a coefficient of determination of 0.98. Depending on the loop, the coefficients will vary. In this illustrative case, the coefficients related to loop number 2 of the ACUREX field are shown.

$$\rho_f(\bar{T}_f)c_f(\bar{T}_f)A_f\frac{dT_{out}}{dt} = \eta_{col}GI - \rho_f(\bar{T}_f)c_f(\bar{T}_f)A_f v(t-t_d)\frac{T_{out}-T_{in}(t-t_r)}{L} - \frac{\widetilde{H}_l(\bar{T}_f, T_a)}{L_2} \tag{4.10}$$

$$\widetilde{H}_l = c_1 \Delta T - c_2 \tag{4.11}$$

where v is the HTF velocity, t_r is the inlet–outlet temperature transport delay or residence time, t_d the flow–outlet temperature HTF related dead time, L is the collector length, L_2 the inlet–outlet loop number 2 distance, $\widetilde{H}_l$ a global thermal loss function and c_1 and c_2 coefficients obtained from experimental tests. Previous lumped-parameter models [85, 97] designed for control did not take into account the variable delay between the measured inlet and outlet temperatures due to HTF transport from the inlet (near the storage tank) to the solar field outlet. Experiments at the real plant in steady state show that a relationship between flow rate, loop number and delay can be approximated as an exponential expression (see Fig. 4.8). Equation (4.12) gives an expression obtained experimentally from the inlet–outlet transport delay t_r [s] which is dependent on the HTF flow rate. This equation includes residence time calculations based on different sections of pipe and approximates a per-piece expression. Coefficients A_1, q_1 and y_0 vary, depending on the collector loop in which the temperature is being controlled. For loop 2, these coefficients are $A_1 = 2008$ [s], $q_1 = 0.00182$ [m^3/s] and $y_0 = 184$ [s]. The maximum delay occurs in loop 10 with minimum flow and the minimum delay (maximum flow) in loop 1. Notice that the delay (t_d) between a change in HTF flow rate and the outlet temperature response is also taken into account. This delay is due to several factors: delays in pump response, delays associated to sensors, and the residence time between the loop outlet and outlet temperature sensor measurement. At present, technology is insufficient to evaluate actuator uncertainty, so based on experimental measurements, this delay is around 30 s.

$$t_r = A_1 e^{-(q(t-t_d)/q_1)} + y_0 \tag{4.12}$$

This stationary approximation of the input–output temperature delay can be used for control if the HTF flow rate does not change much on the corresponding time scale. If a wide range of operating conditions has to be covered, the variable time delay can be estimated by numerical integration using the method proposed by [274] for this type of plant. As the flow rates change at each sampling time, the system transport delay can be estimated as an integer multiple n of the sampling time. At each sampling time k, the new required flow $q(k)$ is calculated by the controller. The distance $\Delta\ell_k$ that can be covered by the fluid during one sample T_s with a corresponding flow rate $q(k)$ is given in Eq. (4.14). Using a discrete-time approximation of the equation which accounts for the flow rates during different sampling intervals, the value of n can be computed for each sample as follows:

$$\Delta\ell_k = T_s q(k)/A_f \tag{4.13}$$

$$L = \int_0^{t_r} v(t)\,dt \quad \rightarrow \quad \frac{T_s}{A_f}\sum_{i=0}^{i=n-1} q(k-i) = L \tag{4.14}$$

where the values of $q(i)$ are found from those measured previously ($q(k)$ is the current flow rate, $q(k-1)$ was the flow demanded at the previous sampling time and consequently, $q(k-n+1)$ was the flow demanded at n previous sampling times, so that $nT_s \approx t_r$). Thus, the transport delay can be estimated because the digital implementation of Eq. (4.14) is quite simple. This approximation is very useful for finding an approximation of the input–output temperature transport delay t_r and for modeling one of the causes producing input flow–output temperature delay, as will be commented on in the following paragraphs. This method was thus implemented and improved to take the number of loops working during the test, the different cross-sections of the pipes, and the loop being controlled into account. Figure 4.9 shows the comparison of the results obtained both in experiments (4.12) and by numerical integration (4.14) under steady-state conditions. The sample time was 39 s. In this figure it can be seen that for HTF flow levels of $8 \cdot 10^{-3}$ m^3/s and $9 \cdot 10^{-3}$ m^3/s the transport delay is almost the same, with greater differences appearing at low flow rates. It must be kept in mind that the plant is composed of different parts and pipes with different cross-sections and that to implement the method proposed in [274] only the most important parts have been used. Notice that the method accuracy depends on the sample time. As the sample time increases, the transport delay calculation becomes less accurate. However, for typical sample times used in the facility (between 15 and 39 s), this approach was shown to work well. The inclusion of the variable input–output temperature transport delay in the model plays an important role, mainly during operation start-up. The delay (t_d) is added to the residence time (t_r), which varies with the HTF flow and can be estimated using (4.14).

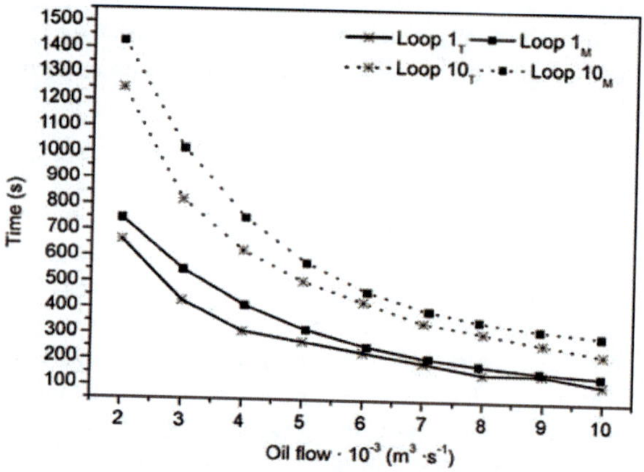

Fig. 4.9 Comparison between the transport delay obtained by experimental methods (subscript '*M*') with the time delay obtained by theoretical calculations (subscript '*T*') (courtesy of C.M. Cirre et al., [109])

For a single collector loop, the flow through it is given by $q(t-t_d)/n_{ope}$, where n_{ope} is the number of operative loops. The system controllability canonical form [354] is shown in Eq. (4.15) and can be used for feedback linearization control purposes, as will be shown in Chap. 5:

$$\frac{dT_{out}(t)}{dt} = \frac{\eta_{col} G I(t)}{\rho_f(\bar{T}_f)c_f(\bar{T}_f)A_f} - \frac{q(t-t_d)(T_{out}(t)-T_{in}(t-t_r))}{A_f n_{ope} L} - \frac{\widetilde{H}_l(\bar{T}_f, T_a)}{\rho_f(\bar{T}_f)c_f(\bar{T}_f)A_f L_2} \tag{4.15}$$

4.3.4.3 Models of Resonances

This section outlines the gray-box models developed in [8, 11, 12]. A gray-box model is a hybrid physical and empirical modeling approach, which can set adjustable parameters that can be physically interpreted [230]. The gray-box model for a DSCF is equivalent to that of a tubular heat exchanger in which the outlet temperature does not vary with space but with time, i.e., condensers. It is important to highlight that, in contrast to heat exchangers where the fluid outlet temperature can be controlled either by the fluid velocity or the output power supply, in a DSCF, the fluid outlet temperature can only be controlled using the fluid velocity, because the main energy source, solar radiation, cannot be manipulated and is, therefore, treated as a system disturbance.

Figure 4.10 shows a cross-section of a PTC in which warm fluid flows at a velocity $v(t)$ which could vary over time. At the same time, the fluid is heated by solar radiation incident on the outside of the tube. The hot fluid leaves the solar collector at the end of tube $\ell = L$ and the temperature of the fluid varies with time and space $T_f(t,\ell)$, while direct solar irradiance $I(t)$, the ambient temperature $T_a(t)$ and the tube temperature $T_m(t)$ are taken to be only time dependent. Some more simplifying assumptions are that the fluid is incompressible, its specific heat capacity and

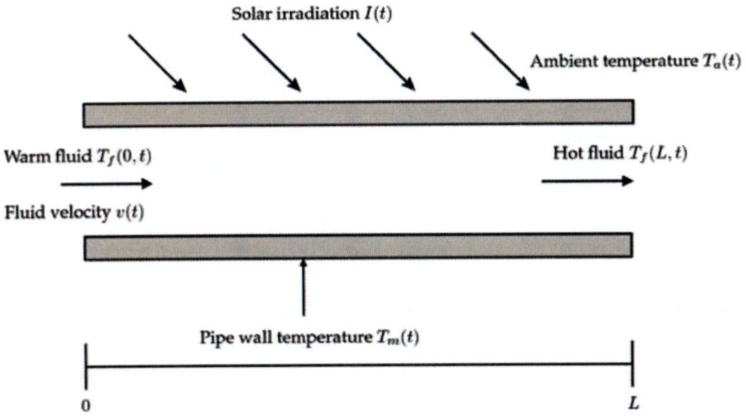

Fig. 4.10 PTC cross-section

density are considered constant and the thermal resistance of the tube wall is ignored. Only fluid velocity, solar irradiance, ambient temperature and incoming fluid temperature are considered to dynamically influence the outgoing fluid temperature. The aim of the mathematical analysis is to find transfer functions which relate the controlled variable $T_f(t, L) = T_{out}(t)$ to the manipulated variable $v(t)$ and disturbance variables $I(t)$, $T_a(t)$, $T_f(t, 0) = T_{in}(t)$. The table of variables and parameters defined at the beginning of the book summarizes model variables and parameters described in Fig. 4.10, while the mathematical analysis for the model is given in the following.

Based on Eqs. (4.1) and (4.2), after simplification and Taylor series expansions to arrive at linear approximations of the non-linear terms, Eqs. (4.16) and (4.17) are found [8, 11]:

$$\frac{\partial T_m}{\partial t} = I\gamma - \frac{1}{\tau_2}(T_m - T_a) - \frac{1}{\tau_{12}}(T_m - T_f) \tag{4.16}$$

$$\frac{\partial T_f}{\partial t} = -(v - v_s)\frac{dT_{f_s}}{d\ell} - v_s\frac{\partial T_f}{\partial \ell} + \frac{1}{\tau_1}(T_m - T_f) \tag{4.17}$$

The subscript s denotes the steady-state value and $\tau_1 = \frac{A_f \rho_f c_f}{\pi D_f H_t}$, $\tau_{12} = \frac{A_m \rho_m c_m}{\pi D_f H_t}$, $\tau_2 = \frac{A_m \rho_m c_m}{\pi D_m H_l}$ and $\gamma = \frac{\eta_{col} G}{A_m c_m \rho_m}$. Equations (4.16) and (4.17) can be expressed in terms of deviation variables around steady-state conditions:

$$0 = I_s\gamma - \frac{1}{\tau_2}(T_{m_s} - T_{a_s}) - \frac{1}{\tau_{12}}(T_{m_s} - T_{f_s}) \tag{4.18}$$

$$0 = -v_s\frac{dT_{f_s}}{d\ell} + \frac{1}{\tau_1}(T_{m_s} - T_{f_s}) \tag{4.19}$$

Thus, the steady-state temperature T_{f_s} is only a function of ℓ:

$$T_{f_s} = I_s\gamma\tau_2 + T_{a_s} + (T_{f_{so}} - I_s\gamma\tau_2 - T_{a_s})e^{-\frac{\ell}{c}} \tag{4.20}$$

where $T_{f_{so}} = T_{f_s}(\ell = 0) = T_f(t = 0, \ell = 0)$ and $c = v_s \tau_1 (1 + \frac{\tau_2}{\tau_{12}})$. By subtracting Eq. (4.18) from Eq. (4.16) and Eq. (4.19) from Eq. (4.17), and introducing deviation variables:

$$\frac{\partial \tilde{T}_m}{\partial t} = \tilde{I}\gamma - \frac{1}{\tau_2}(\tilde{T}_m - \tilde{T}_a) - \frac{1}{\tau_{12}}(\tilde{T}_m - \tilde{T}_f) \tag{4.21}$$

$$\frac{\partial \tilde{T}_f}{\partial t} = -\tilde{v}\frac{dT_{f_s}}{d\ell} - v_s \frac{\partial \tilde{T}_f}{\partial \ell} + \frac{1}{\tau_1}(\tilde{T}_m - \tilde{T}_f) \tag{4.22}$$

where $\tilde{T}_f = T_f - T_{f_s}$, $\tilde{v} = v - v_s$, $\tilde{I} = I - I_s$, $\tilde{T}_m = T_m - T_{m_s}$ and $\tilde{T}_a = T_a - T_{a_s}$. Through a Laplace transform, previous PDE are converted into ordinary differential equations (ODE):

$$s\tilde{T}_m(s) = \tilde{I}(s)\gamma - \frac{1}{\tau_2}\big(\tilde{T}_m(s) - \tilde{T}_a(s)\big) - \frac{1}{\tau_{12}}\big(\tilde{T}_m(s) - \tilde{T}_f(s)\big) \tag{4.23}$$

$$s\tilde{T}_f(s) = -\tilde{v}(s)\frac{dT_{f_s}}{d\ell} - v_s \frac{d\tilde{T}_f(s)}{d\ell} + \frac{1}{\tau_1}\big(\tilde{T}_m(s) - \tilde{T}_f(s)\big) \tag{4.24}$$

Rewriting Eqs. (4.23) and (4.24) to remove $\tilde{T}_m(s)$:

$$\frac{d\tilde{T}_f(s)}{d\ell} + \frac{a}{v_s}\tilde{T}_f(s) = -\frac{\tilde{v}(s)}{v_s}\frac{dT_{f_s}}{d\ell} + \frac{\tau_2 b\gamma}{v_s}\tilde{I}(s) + \frac{b}{v_s}\tilde{T}_a(s) \tag{4.25}$$

where $a(s) = s + \frac{1}{\tau_1} - \frac{\tau_2}{\tau_1(\tau_{12}\tau_2 s + \tau_{12} + \tau_2)}$ and $b(s) = \frac{\tau_{12}}{\tau_1(\tau_{12}\tau_2 s + \tau_{12} + \tau_2)}$. Equation (4.25) is an ODE, with boundary condition $\tilde{T}_f(s, \ell) = \tilde{T}_f(s, 0)$ at $\ell = 0$, which is solved giving

$$\begin{aligned} \tilde{T}_f(s)e^{\frac{a}{v_s}\ell} &= -\frac{\tilde{v}(s)}{v_s}\int \frac{dT_{f_s}}{d\ell}e^{\frac{a}{v_s}\ell}\,d\ell + \frac{\tau_2 b\gamma}{v_s}\tilde{I}(s)\int e^{\frac{a}{v_s}\ell}\,d\ell \\ &\quad + \frac{b}{v_s}\tilde{T}_a(s)\int e^{\frac{a}{v_s}\ell}\,d\ell + \tilde{T}_f(s,0) \end{aligned} \tag{4.26}$$

It is easy to find $\frac{dT_{f_s}}{d\ell}$ from Eq. (4.20), in such a way that

$$-\frac{\tilde{v}(s)}{v_s}\int \frac{dT_{f_s}}{d\ell}e^{\frac{a}{v_s}\ell}\,d\ell = -\tilde{v}(s)\frac{T_{a_s} + I_s\gamma\tau_2 - T_{f_{so}}}{ac - v_s}\big[e^{\ell(\frac{a}{v_s} - \frac{1}{c})} - 1\big] \tag{4.27}$$

Equation (4.26) can be solved providing a general equation:

$$\begin{aligned} \tilde{T}_f(s) &= -\tilde{v}(s)\frac{T_{a_s} + I_s\gamma\tau_2 - T_{f_{so}}}{ac - v_s}\big[e^{-\frac{\ell}{c}} - e^{-\frac{a}{v_s}\ell}\big] + \tilde{I}(s)\tau_2\gamma\frac{b}{a}\big[1 - e^{-\frac{a}{v_s}\ell}\big] \\ &\quad + \tilde{T}_a(s)\frac{b}{a}\big[1 - e^{-\frac{a}{v_s}\ell}\big] + \tilde{T}_f(s,0)e^{-\frac{a}{v_s}\ell} \end{aligned} \tag{4.28}$$

It is possible to extract a single-input single-output (SISO) transfer function which relates the fluid outlet temperature to the only manipulated variable, its velocity, when solar irradiance, ambient temperature, and incoming fluid temperature do not vary:

$$\frac{\tilde{T}_f(s)}{\tilde{v}(s)} = \underbrace{-\frac{T_{a_s} + I_s\gamma\tau_2 - T_{f_{so}}}{ac - v_s}}_{L_e(s)} \underbrace{\left[e^{-\frac{\ell}{c}} - e^{-\frac{a}{v_s}\ell}\right]}_{R_i(s)} \tag{4.29}$$

For a better understanding, Eq. (4.29) can be split into two transfer functions which are labeled as $L_e(s)$ (for the left side) and $R_i(s)$ (for the right side).

Focusing attention on the left side, $L_e(s)$:

$$\begin{aligned} L_e(s) &= -\frac{T_{a_s} + I_s\gamma\tau_2 - T_{f_{so}}}{ac - v_s} \\ &= -\frac{T_{a_s} + I_s\gamma\tau_2 - T_{f_{so}}}{\frac{(\tau_1\tau_{12}\tau_2 s^2 + (\tau_1(\tau_{12}+\tau_2)+\tau_{12}\tau_2)s + \tau_{12})c - v_s(\tau_1\tau_{12}\tau_2 s + \tau_1(\tau_{12}+\tau_2))}{\tau_1\tau_{12}\tau_2 s + \tau_1(\tau_{12}+\tau_2)}} \end{aligned} \tag{4.30}$$

Expanding terms and after a few simplifications [11], it is possible to obtain

$$\begin{aligned} L_e(s) &= -\frac{(T_{a_s} + I_s\gamma\tau_2 - T_{f_{so}})}{c} \frac{(\tau_1\tau_2 s + \tau_{12} + \tau_2)}{\tau_{12}\tau_2 s^2 + \left(\tau_{12} + \tau_2 + \frac{\tau_{12}\tau_2^2}{\tau_1(\tau_{12}+\tau_2)}\right)s} \\ &= -\frac{(T_{a_s} + I_s\gamma\tau_2 - T_{f_{so}})}{c} \frac{\left(s + \frac{\tau_{12}+\tau_2}{\tau_1\tau_2}\right)}{s^2 + \left(\frac{\tau_{12}+\tau_2}{\tau_{12}\tau_2} + \frac{\tau_2}{\tau_1(\tau_{12}+\tau_2)}\right)s} \end{aligned} \tag{4.31}$$

On the other side, the term $e^{-\frac{a}{v_s}\ell}$ in $R_i(s)$ can be expanded in [11]

$$e^{-\frac{a}{v_s}\ell} = e^{-\frac{\ell}{c}} e^{-\frac{\ell}{v_s}\left(s + \frac{\tau_2}{\tau_1(\tau_{12}+\tau_2)}\left(\frac{s}{s+\frac{\tau_{12}+\tau_2}{\tau_{12}\tau_2}}\right)\right)} \tag{4.32}$$

Obtaining the common factor $e^{-\frac{\ell}{c}}$ in $R_i(s)$:

$$R_i(s) = \left[e^{-\frac{\ell}{c}} - e^{-\frac{a}{v_s}\ell}\right] = e^{-\frac{\ell}{c}}\left[1 - e^{-\frac{\ell}{v_s}\left(s + \frac{\tau_2}{\tau_1(\tau_{12}+\tau_2)}\left(\frac{s}{s+\frac{\tau_{12}+\tau_2}{\tau_{12}\tau_2}}\right)\right)}\right] \tag{4.33}$$

Finally, if Eqs. (4.31) and (4.33) are put together and the substitution $\ell = L$ is made, the transfer function relating fluid outlet temperature to its velocity (assuming that the rest of the disturbance variables do not vary) is obtained:

$$\begin{aligned} G(s) = \frac{\tilde{T}_f(s, L)}{\tilde{v}(s)} &= -\frac{T_{a_s} - T_{f_{so}} + I_s\gamma\tau_2}{c} e^{-\frac{L}{c}} \frac{\left(s + \frac{\tau_{12}+\tau_2}{\tau_{12}\tau_2}\right)}{s^2 + \left(\frac{\tau_{12}+\tau_2}{\tau_{12}\tau_2} + \frac{\tau_2}{\tau_1(\tau_{12}+\tau_2)}\right)s} \\ &\quad \times \left[1 - e^{-\frac{L}{v_s}\left(s + \frac{\tau_2}{\tau_1(\tau_{12}+\tau_2)}\left(\frac{s}{s+\frac{\tau_{12}+\tau_2}{\tau_{12}\tau_2}}\right)\right)}\right] \end{aligned} \tag{4.34}$$

Equation (4.34) is the transfer function relating the controlled variable to the only manipulated variable and it can be used to design a feedback controller for the plant, as will be seen in Chap. 5. In [11], similar transfer functions relating the fluid outlet temperature to the disturbance variables are obtained, assuming the remaining variables are in steady state:

$$\frac{\tilde{T}_f(s,L)}{\tilde{I}(s)} = \tau_2 \gamma \frac{b}{a}\left[1 - e^{-\frac{a}{v_s}L}\right] \tag{4.35}$$

$$\frac{\tilde{T}_f(s,L)}{\tilde{T}_a(s)} = \frac{b}{a}\left[1 - e^{-\frac{a}{v_s}L}\right] \tag{4.36}$$

$$\frac{\tilde{T}_f(s,L)}{\tilde{T}_f(0,s)} = e^{-\frac{a}{v_s}L} \tag{4.37}$$

where $a(s)$ and $b(s)$ are transfer functions, as explained previously. Equations (4.35), (4.36), and (4.37) can be used to find simplified feedforward controllers to compensate for measurable disturbances. Notice that Eqs. (4.34) to (4.37) are simplifications, as each was found assuming that the rest of the input/disturbance variables are in steady state which is never true in a solar plant. Nevertheless, this kind of simplification allows fundamental transfer functions capturing the main system dynamics and the influence of disturbances to be found that can be used for implementing high-performance control schemes [8, 11, 12]. Equations (4.36) and (4.37) yield the same results as those in [115], whereas Eq. (4.34) is similar to that in [8], Eq. (4.37) shows almost a total delay between the fluid outlet temperature and its input temperature. Equations (4.34), (4.35) and (4.36) predict a resonance effect when the fluid outlet temperature is related to its velocity, solar irradiance or ambient temperature. In these three equations, the function in brackets represents system resonance dynamics. Equations (4.34) to (4.37) have a complex expression in Laplace transform s of:

$$e^{\frac{-L}{v_s}Q(s)} = \begin{cases} e^{\frac{-L}{v_s}(x_b(\frac{x_a s}{x_b s+1}))} & \text{if Eq. (4.34)} \\ e^{\frac{L}{v_s}(\frac{x_a}{x_b s+1})} & \text{if Eqs. (4.35), (4.36) or (4.37)} \end{cases} \tag{4.38}$$

with variables $x_a = \frac{\tau_2}{\tau_1(\tau_{12}+\tau_2)}$ [1/s] and $x_b = \frac{\tau_{12}\tau_2}{\tau_{12}+\tau_2}$ [s].

The complex expression $e^{-\frac{L}{v_s}Q(s)}$ is not easy to invert but, with the help of its frequency response and taking into account that the transfer function can be closely approximated by another transfer function with one pole and one zero [8], the expression can be approximated by

$$e^{\frac{-L}{v_s}Q(s)} \approx K\frac{-\beta s+1}{\tau s+1} \tag{4.39}$$

which can be considered a first-order Padé approximation of Eq. (4.38), where parameters β and τ are the same for Eqs (4.34) to (4.37), whereas static gain K of the transfer function has different values depending on the equation

$$K = \begin{cases} 1 & \text{if Eq. (4.34)} \\ e^{\frac{L}{v_s}(\frac{\tau_2}{\tau_1(\tau_{12}+\tau_2)})} & \text{if Eqs. (4.35), (4.36) or (4.37)} \end{cases} \tag{4.40}$$

Therefore, the transfer functions in Eqs. (4.34) to (4.37) could be represented as Eqs. (4.41) to (4.44):

$$G(s) = \frac{\tilde{T}_f(s,L)}{\tilde{v}(s)} = -\frac{T_{a_s} - T_{f_{so}} + I_s\gamma\tau_2}{c}e^{-\frac{L}{c}}\frac{(s+\frac{\tau_{12}+\tau_2}{\tau_{12}\tau_2})}{s^2+(\frac{\tau_{12}+\tau_2}{\tau_{12}\tau_2}+\frac{\tau_2}{\tau_1(\tau_{12}+\tau_2)})s}$$

$$\times \left[1 - e^{-\frac{L}{v_s}s}\left(\frac{-\beta s + 1}{\tau s + 1}\right)\right] \tag{4.41}$$

$$I(s) = \frac{\tilde{T}_f(s, L)}{\tilde{I}(s)} = \tau_2 \gamma \frac{b}{a}\left[1 - e^{-\frac{L}{v_s}s}\left(e^{\frac{-L}{v_s}\frac{1}{\tau_1}} e^{\frac{L}{v_s}(\frac{\tau_2}{\tau_1(\tau_{12}+\tau_2)})}\left(\frac{-\beta s + 1}{\tau s + 1}\right)\right)\right] \tag{4.42}$$

$$H(s) = \frac{\tilde{T}_f(s, L)}{\tilde{T}_a(s)} = \frac{b}{a}\left[1 - e^{-\frac{L}{v_s}s}\left(e^{\frac{-L}{v_s}\frac{1}{\tau_1}} e^{\frac{L}{v_s}(\frac{\tau_2}{\tau_1(\tau_{12}+\tau_2)})}\left(\frac{-\beta s + 1}{\tau s + 1}\right)\right)\right] \tag{4.43}$$

$$T(s) = \frac{\tilde{T}_f(s, L)}{\tilde{T}_f(0, s)} = e^{-\frac{L}{v_s}s}\left(e^{\frac{-L}{v_s}\frac{1}{\tau_1}} e^{\frac{L}{v_s}(\frac{\tau_2}{\tau_1(\tau_{12}+\tau_2)})}\left(\frac{-\beta s + 1}{\tau s + 1}\right)\right) \tag{4.44}$$

4.3.5 Data-Driven Models

Linear black-box models have been obtained from parameter identification by many authors for control purposes [87]. Low-order linear models have commonly been used for adaptive control [77, 82, 83, 274, 292, 334], while high-order linear models are used for gain-scheduled controllers [73, 75, 76, 84, 85, 197, 264, 295, 296, 308], and all these are dealt with in the next chapter.

Regarding non-linear models, several methodologies, including numerous types of artificial neural network (ANN), have been proposed for building a non-linear model of the solar power plant, which was later used for simulation purposes or as a core element in various model-based prediction schemes.

In [70] a black-box identification of the solar collector field of an air conditioning plant is carried out. As the collectors have non-linear dynamics and flow variant time delay, a method for compensating the flow variant time delay similar to that of [274] is proposed and thereafter, a black-box method for non-linear systems without time delays can be applied. The obtained non-linear model describes the collector dynamics well.

In [203, 204] a comprehensive review of applications of ANN to renewable energy systems is performed. Within the scope of solar plants with distributed collectors, the application of the general identification methodology to obtain neural predictors for use in a non-linear predictive control scheme is shown in [16, 18]. Non-linear autoregressive models with exogenous input (NARX) models are used in this work, where several algorithms for selecting past signal values as inputs are developed for multilayer perceptron (MLP) and radial basis function (RBF) networks, while in [17] a comparison is made between different types of RBF neural network (NN) for the same plant. In [42], a static NN is used in an autoregressive configuration and a selection method is proposed based on the reduction of the estimated gradient for determining the past values that the network needs to construct the prediction. The works [289, 290] implied a neuro-fuzzy system based on a RBF network with support vector learning, while [179] used a recurrent network in combination with an on-line learning strategy to update both the weights of the network and the current state. In [339, 400], the identification of a DSCF is performed both by using ANN and by physical models. The non-linear identification problem is tackled by decomposing the complex system into two main components: an active part and a passive part. For the active part of the solar power plant, a model based on the parallel connection of ten ANN is built while for the passive part, a white-box model and a NN black-box model are developed. All models are identified and validated using measurement data. In [187] a model of the overall solar power plant is also developed using NN, to avoid overheads generated by training each of the networks presented in the work of [339, 400].

4.3.5.1 Linear Plant Models

When considering operation around a particular set point and after linearization most SISO plants, can be described in the discrete-time domain by linear models based on transfer functions $G(z^{-1}) = B(z^{-1})/A(z^{-1})$ (with z^{-1} being the backward shift operator), relating the sampled system output $y(k)$ to the system input $u(k)$ by $A(z^{-1})y(k) = B(z^{-1})u(k)$ with $A(z^{-1}) = 1 + a_1 z^{-1} + a_2 z^{-2} + \cdots + a_{na} z^{-na}$ and $B(z^{-1}) = b_0 + b_1 z^{-1} + b_2 z^{-2} + \cdots + b_{nb} z^{-nb}$, with $n_a \geq n_b$. This description is valid both for stable and unstable processes and has the advantage of needing few parameters to model the system, although it is fundamental to have an a priori knowledge of the system, mostly of the order of polynomials $A(z^{-1})$ and $B(z^{-1})$.

If noises and disturbances that can act on the system are taken into account, other types of linear model can be obtained, such as the following:

$$A(z^{-1})y(k) = z^{-d} B(z^{-1})u(k-1) + C(z^{-1})e(k) \tag{4.45}$$

where $u(k)$ and $y(k)$ are the control and output sequence of the plant, d is the dead time in discrete time of the system and $e(k)$ is a zero mean white noise with $C(z^{-1}) = 1 + c_1 z^{-1} + a_2 z^{-2} + \cdots + c_{nc} z^{-nc}$. This model is known as a Controller Auto-Regressive Moving-Average model (CARMA). It has been argued [113] that for many industrial applications in which disturbances are non-stationary, an integrated CARMA (CARIMA) model is more appropriate. A CARIMA model is given by

$$A(z^{-1})y(k) = B(z^{-1})u(k-1) + C(z^{-1})\frac{e(k)}{\Delta}; \quad \Delta = 1 - z^{-1} \tag{4.46}$$

The indicated structures of linear models (4.45) and (4.46) have been used for control design purposes. In the following paragraphs, several linear models obtained from input–output data of the plant are obtained.

Low-Order Linear Plant Models Low order models are adequate for many control structures, although their simplicity can produce unacceptable behavior if fast responses are required for the system, due to the influence of unmodeled dynamics (resonances).

When small changes around a particular set point are considered, most of the industrial processes can be described by a normally high-order linear model. The justification of this affirmation lies in the fact that the majority of the processes are composed of many dynamic elements, usually first order, so that the complete model has an order equal to the number of elements. In fact, each energy or mass storing element gives rise to a first order element in the model. For instance, a heat exchanger can be modeled by dividing it into tube segments each one being considered a first-order system. The resulting model will have an order equal to the number of segments the tube has been divided into. These high-order models are very difficult to cope with in control but, fortunately, the behavior of such processes can often be modeled by a system with a fundamental time constant and a delay.

Let us consider a process with n first-order elements in series, each one having a time constant τ/n. The system transfer function will be given by

$$G(s) = \frac{1}{(1+\frac{\tau}{n}s)^n}$$

If n is varied from 1 to ∞, the type of response changes from that of a first-order system to that of a pure delay (equal to τ). If, as happens in many processes, one of the time constants is greater than the rest, the smaller time constants join to produce a time lag which acts as a pure delay. In this situation, the dynamics are dominated by the major time constant. So, it is possible to approximate the high-order model of a complex dynamical system by a first-order process with a delay element.

For control design purposes, several linear models which relate changes in outlet HTF temperature to changes in the control variable (flow)[2] have been obtained. These models were obtained from the reaction curve after the injection of a step in the HTF flow signal (Fig. 4.3) at several operating points and can be approximated by a first-order transfer function with a pure delay:

$$G(s) = e^{-st_d}\frac{K}{(1+\tau s)} \tag{4.47}$$

Obviously, this is an approximate model, because the system is non-linear, but the approximation is made to obtain control schema in function of few parameters to simplify the identification mechanism in the case of adaptive controllers. From the transfer function of Eq. (4.47) the discrete-time transfer function can be obtained taking into account the zero-order hold. Two models that have been used in this kind of plant using high values of the sampling time (e.g. 39 s) [332] are given by

$$G_A\big(z^{-1}\big) = z^{-d}\frac{(bz^{-1})}{(1-az^{-1})}; \qquad G_B\big(z^{-1}\big) = z^{-2}\frac{(b_0+b_1z^{-1})}{(1-az^{-1})} \tag{4.48}$$

with $a = e^{-T/\tau}$ and $b = K(1-a)$. The values of d depend on the flow level conditions and the factor $(b_0+b_1z^{-1})$ acts as a discrete first-order Padé approximation to a delay term (useful for modeling a time delay non-integer multiple of the sampling period, e.g. $T_s < t_d < 2T_s$).

[2]If the feedforward analyzed in Sect. 4.4.1 is placed in series with the plant, the control signal will be the reference temperature to the feedforward controller.

These simplified linear models are based on the step response of the plant. Nevertheless, if the frequency response of one of the models mentioned is compared to that of the real plant or to the non-linear distributed parameter model at an operating point, it can be seen that the approximation made in this section is only valid for a narrow frequency range. The consequences of this will be commented on in following chapters.

High-Order Linear Plant Models A method complementary to that shown in Sect. 4.3.4.3 to account for resonances is to use high-order linear models. By using input–output data obtained with PRBS tests, degrees of polynomials $A(z^{-1})$ and $B(z^{-1})$ and that of the delay which best describe the system and minimize Akaike's Information Theoretic Criterion (AIC) [230], were found to be $n_a = 2$, $n_b = 8$, and $d = 0$. The value of the coefficients which define the previous polynomials in the backward shift operator z^{-1} were calculated using a LS algorithm. As an example, using the input–output data with which the frequency response shown in Fig. 4.5 was calculated (HTF flow around 6 l/s), the values of polynomials $A(z^{-1})$ and $B(z^{-1})$ given by the identification algorithm were: $A(z^{-1}) = 1 - 1.5681z^{-1} + 0.5934z^{-2}$ and $B(z^{-1}) = 0.0612 + 0.0018z^{-1} - 0.0171z^{-2} + 0.0046z^{-3} + 0.0005z^{-4} + 0.0101z^{-5} - 0.0064z^{-6} - 0.015z^{-7} - 0.0156z^{-8}$.

4.3.5.2 Non-linear Plant Models Based on Artificial Neural Networks

As has been mentioned in Sect. 4.3.5, ANN have demonstrated to be a good approximation of the system behavior and able to be used for output prediction purposes. This subsection presents a brief overview of an application of ANN identification to obtain models of DSCF, useful for use within MPC schemes [16, 18, 42]. Given a non-linear dynamic system, the identification problem will be posed here as the task of obtaining a mapping between past values of measured variables of the system and future values of some of those variables. In discrete time, the set of past measured variables of a multiple-input, single-output (MISO) system is a vector:

$$\varphi(k) = \left[y^T(k)\dots y^T(k-m_y)\mathbf{u}^T(k)\dots \mathbf{u}^T(k-m_u)\mathbf{p}^T(k)\dots \mathbf{p}^T(k-m_v)\right] \quad (4.49)$$

where $\mathbf{u}$ is the vector of input variables, y is the output variable, and $\mathbf{p}$ is a vector of measurable disturbances. The identification problem consists of obtaining a relationship f such that $y(k+1) \approx f(\varphi(k))$. The orders m_y, m_u and m_p in (4.49) are presumed to be known in many situations, but in the general case they have to be obtained from observations made about the system. To use the model for prediction purposes, the future values of the measurable disturbances and model inputs have to be known.

For non-linear system identification, several steps have to be followed [230]: data collection, selection of the model family, selection of the structural parameters of the model in the family (which is equivalent to finding the structure and size of the ANN), the selection of approximate values for the parameters of the model (training of the network) and the validation of the model obtained and its implementation in a control system.

Data Collection In the case of DSCF, the data collection problem involves measuring HTF flow, ambient, inlet and outlet temperatures and direct solar irradiance. To develop a model that is well suited for all operating regimes, care has to be taken in order to collect data that cover most of the working ranges of the variables. This will normally be the case when using ANN. In the case of DSCF, it is necessary to have data from clear and cloudy days to ensure that most operating regimes are covered. These data have to be filtered and normalized. The measurement of direct solar irradiance poses a problem, since the reflectivity of the mirrors is not constant, due to the accumulation of dust. For this reason, the most reliable data are those collected in experiments immediately after cleaning the mirrors, as there is less differences between the values of the measured solar irradiance and the effective irradiance. Normalization (usually within the $[-1,+1]$ range) is a very important step when using NN. Inputs with mainly dissimilar values cause neural learning to be very slow. Also, noisy data can be a source of problems during learning unless care is taken.

Non-linear Black-Box Models The problem in question is the identification of NARX models. The prediction of the output of the plant $\hat{y}(k+1 \mid k)$ is obtained as a non-linear function g acting over past values of the variables of interest. The TDL blocks are Tapped Delay Lines that provide an appropriate number of past values, n_y, n_u and n_p. It is easy to see that a NARX model can represent a non-linear system of the form given by Eq. (4.49).

Selection of Past Signal Values as Inputs When using static NN to predict future outputs of a system based on input–output information, the temporal domain is considered, treating past values of variables as different inputs and feeding them into a static network. When the order of the system is not known, most applications rely on one of its upper bounds. Such procedures can lead to inefficient models, due to the large number of inputs needed. The problem of finding the optimal number of past values has been widely investigated. In [313] the false nearest-neighbor method is developed. In respect to an input/output point (φ, y) consisting of a regression vector and the future output of the system, a false neighbor is another point (φ_f, y_f) that, although close to φ in the input space, is far away in the output space. This condition is tested using a threshold for the quotient $|y - y_f|/\|\varphi - \varphi_f\|$. The algorithm allows the model orders n_y, n_u and n_p to be determined by examining the percentage of false neighbors in the data. A similar method is used in [18], based on the descent in the gradient needed to explain the observed output when a new TDL is added. The method chooses the variable that provides the greatest descent at each stage. Pairs of points are considered neighbors only if the distance in the input space is less than a threshold.

In the case of the ACUREX plant, the number of past values has been denoted as n_u for the input variable q, n_y for the output variable T_{out}, n_{p1} for direct solar irradiance I and n_{p2} for the inlet HTF temperature T_{in} [18]. In Table 4.1(a), the approximation capabilities of several linear models are shown. In the first column, a short-hand notation is used to indicate the input variables used in the model. For

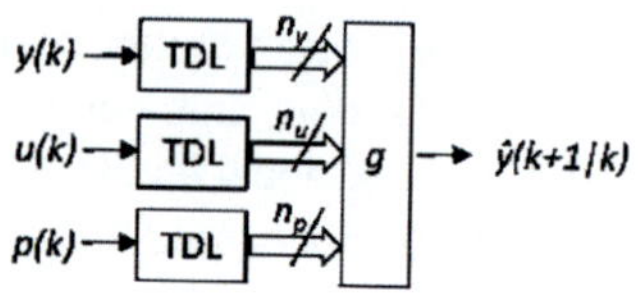

Fig. 4.11 Structure of the NARX model (courtesy of M.R. Arahal et al., [18])

example: 2111 is a model with $n_u = 2$, $n_y = 1$, $n_{p1} = 1$, $n_{p2} = 1$, so the regression vector is $\boldsymbol{\varphi}(t_k) = [q(k), q(k-1), T_{out}(k), I(k), T_{in}(k)]^T$. The second and third columns in the said table show the sum of the squared errors (SSE) in the one-step ahead prediction $e(i) = y(i+1) - \hat{y}(i+1 \mid i)$. This quantity is a figure of merit normally used to assess the quality of models and is expressed as $SSE = \sum_{i=1}^{n_P} e(i)^2$, n_P being the number of patterns. The SSE is measured by two sets of data coming from experiments performed at the plant on different days. The training set (TS) contains the input–output patterns $(\boldsymbol{\varphi}, y)$ used to fit the model while the validation set (VS) is the set used to validate the model and contains patterns not used for adjusting the model. The table has been constructed testing all possible combinations of inputs for a total of input variables $n_u + n_y + n_{p1} + n_{p2} \leq 9$ (only some of these combinations are shown) [18].

From Table 4.1(a), it is easy to select the inputs for a linear model of order up to 9. However, the optimality of the one-step ahead prediction does not guarantee good results when using the model recursively to obtain predictions over an horizon (as is needed in the model-based predictive control schema). A recursive predictor gives the N-step ahead prediction using the one-step predictor N times. For instance, the two-step ahead recursive prediction is $\hat{y}(k+2 \mid k) = \hat{f}(\boldsymbol{\varphi}(k), \hat{y}(k+1 \mid k), \mathbf{u}^T(k+1), \mathbf{p}^T(k+1))$, where $\hat{y}(k+1 \mid k) = \hat{f}(\boldsymbol{\varphi}(k))$.

In Table 4.1(b), the SSE for the one-step prediction and the recursive 25-step prediction are given for all linear models that use a total of eight input variables. The notation for the models is the same as before. It can be seen that the best one-step ahead predictor does not yield the best 25-step ahead predictions. The explanation is simple: obviously, for the one-step ahead prediction, the variable that best explains the output of the model $T_{out}(k+1)$ is $T_{out}(k)$. However, this value is affected by noise measurement that is propagated to successive predictions. This leads to less accurate results than if a variable less affected by noise, such as the HTF flow, is used. This observation points to the use of the input variables q and I as inputs for the model. Table 4.2 shows the errors for different prediction steps N.

It is good practice to start the search for input variables using linear models, because they serve as an indication of the non-linear case, reducing the number of trials. Care has to be taken when analyzing these latter results since, unlike the linear case, they depend not only on the input vector but also on the size of the neural network and on the training procedure. The results reported above were obtained by averaging over 10 runs.

Neural Architecture There are many kinds of neural network, most of them with a biological inspiration. Static nets perform a mapping from the input to the output space that does not depend upon past history. They can be used to represent dynamic behavior simply by being fed with delayed values of the variables, as shown in

Table 4.1 (**a**) SSE for linear (*left*) and non-linear (*right*) models using different input variables. The non-linear models are neural networks of size 8-6-3-1. (**b**) SSE in the VS for the one-step prediction $N=1$ and the recursive 25-step ahead prediction ($N=25$) for all models that use a total of eight input variables. The columns on the *left* correspond to linear models and the ones on the *right* to neural networks (courtesy of M.R. Arahal et al., [18])

Model input	Linear		Non-linear	
	TS	VS	TS	VS
1111	2.232	0.481	2.221	0.482
2111	1.963	0.412	1.801	0.410
1211	0.391	0.081	0.385	0.079
1121	2.191	0.474	1.967	0.472
1112	2.190	0.465	2.034	0.469
2211	0.232	0.051	0.221	0.050
1311	0.261	0.058	0.217	0.052
1221	0.362	0.086	0.331	0.078
1212	0.394	0.086	0.385	0.079
3211	0.231	0.051	0.196	0.044
2311	0.172	0.035	0.165	0.030
2221	0.202	0.046	0.184	0.033
2212	0.231	0.051	0.211	0.050
3311	0.155	0.032	0.136	0.029
2411	0.167	0.035	0.142	0.030
2321	0.166	0.035	0.138	0.029
2312	0.171	0.035	0.147	0.030
4311	0.154	0.032	0.131	0.028
3411	0.148	0.032	0.129	0.025
3321	0.154	0.032	0.136	0.028
3312	0.155	0.032	0.132	0.028

(a)

Model input	Linear		Non-linear	
	$N=1$	$N=25$	$N=1$	$N=25$
1151	0.531	10.1	0.345	8.01
1241	0.089	25.3	0.076	15.0
1331	0.057	13.2	0.045	14.1
1421	0.056	15.1	0.042	12.3
1511	0.055	14.4	0.039	10.1
2141	0.431	7.21	0.331	6.21
2231	0.049	9.44	0.028	9.02
2321	0.036	4.31	0.030	5.04
2411	0.036	6.12	0.031	5.23
3131	0.344	4.80	0.291	4.70
3221	0.049	9.20	0.026	8.91
3311	**0.032**	4.51	0.029	3.20
4121	0.291	**3.80**	**0.025**	2.96
4211	0.052	12.0	0.301	9.04
5111	0.270	4.03	0.092	**2.90**

(b)

Fig. 4.11. The block marked g can be carried out by a neural network. In this case, function g depends on a parameter vector W_g, which contains all the parameters in the neural network.

The two neural nets that have been most intensively used for identification are the MLP and the RBF network.

The MLP is a compound of several layers of neurons, as shown in Fig. 4.12(a). The first layer is called the input layer and serves only as a fan-out device propagating the input of the net $\mathbf{x} = [x_1, \ldots, x_{N1}]^T$. Hidden layers have non-linear activation functions, more precisely, each node performs a non-linear function of a weighted sum of its inputs from the previous layer and from a special node that provides a constant output. The connections to this node are called bias weights and they help in building the approximated function. Furthermore, in many cases the output

Table 4.2 SSE in VS for models that favor input variables. The *last* column is an average calculated for $N = 1, 5, 10, 15, 20$ (courtesy of M.R. Arahal et al., [18])

Linear models					Non-linear models				
Input	$N=1$	$N=10$	$N=20$	Average	Input	$N=1$	$N=10$	$N=20$	Average
5171	0.045	1.34	2.49	1.12	5171	0.035	1.11	1.79	0.722
6161	0.041	1.37	2.56	1.14	6161	0.032	0.99	1.86	0.781
5181	0.045	1.32	2.47	1.12	5181	0.040	1.05	1.27	0.531
6171	0.043	1.35	2.53	1.13	6171	0.033	1.15	1.21	0.529
5191	0.044	1.32	2.48	**1.11**	5191	0.027	0.89	1.28	0.512
6181	0.043	1.33	2.51	1.12	6181	0.030	0.93	1.11	**0.511**

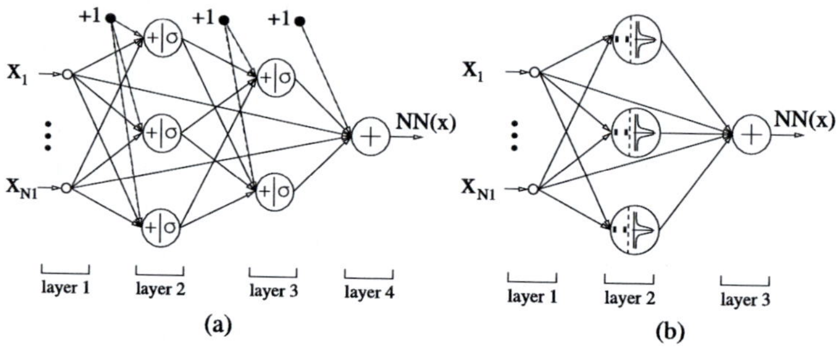

Fig. 4.12 **(a)** Multilayer perceptron with two hidden layers, bias weights and a linear output node with short-cut connections from the input layer. **(b)** A radial basis function network with three nodes in the hidden layer (courtesy of M.R. Arahal et al., [18])

nodes are linear and include short-cut connections from the input layers. This structure is capable of approximating any smooth non-linear mapping, provided that a sufficiently large number of nodes is used. This property, known as Universal Approximation, enables the use of the network by simply taking $\mathbf{x} = \boldsymbol{\varphi}$.

RBF nets are one-hidden layer feedforward networks with linear output nodes (see Fig. 4.12(b)). Each neuron in the hidden layer receives as input the whole input vector $\mathbf{x}$ and performs a non-linear function of this vector and an inner vector $\mathbf{c}$ referred to as center. The most used basis function is a Gaussian of the distance from the input vector to the center. The output of the network is the weighted sum of the outputs of all N nodes $NN(\mathbf{x})$, where σ_i is the width of the ith basis function and w_i the output weight. This neural scheme also holds the Universal Approximation property; hence, it can be used in the same way as MLP. However, since the approximation with RBF has local support, they are better suited to on-line identification, where fast adaptation is needed. On the other hand, the MLP structure usually needs fewer nodes to obtain the desired accuracy.

Network Training In black-box identification, the observed data are used to produce an estimation of the parameters of the model; this procedure is called training.

Selection of training examples among the available data reduces the computational time needed for estimating the parameters of a model. This is an important issue when using ANN, since its training typically involves a large number of cycles through the set of available examples. For this reason, it is convenient to have a small TS with the highest information content.

When the input–output patterns come from a system under control, the variables are highly correlated to each other and the information content is low. Passive learning consists of selecting patterns randomly. Active selection procedures use the experience already gained by a model to direct the selection of new training patterns [18].

To manage the problem in the way normal to the process industry, in the case of identification of the DSCF, the available input–output data were restricted to a number of sampled signals obtained from closed-loop operation [18]. The available data were divided into two sets in such a manner that most operating regimes are included in both sets. The main problem is the existence of redundant patterns in the training set. For the purpose of training an ANN to identify the plant, a procedure was devised [15] to extract a compact, yet sufficiently informative, data set. The main idea behind the method is that only input–output patterns with a great enough distance from other patterns already in the set should be included. The distance can be taken as the Euclidean distance in the input–output space but, since not all signals affect the output to the same extent, it is more convenient to use some weighted metric using a priori knowledge about the plant. This metric takes into account the fact that some signals (q, T_{out}, I) are more relevant than others (T_{in}) in the evolution of the plant. The application of the algorithm provided a TS with just 30% of the data, reducing the training time. The details can be found in [40].

In the case of MLP, there are two mainly types of training algorithm: those that select the neural structure during learning and those that do not. Algorithms in the first type are called constructive. Most applications of ANN, however, use a large, fully connected network and wait for the algorithm to set unnecessary weights to values close to zero. Some techniques exist to later prune those connections [18]. Non-constructive algorithms just select a value for the vector including all adjustable parameters of the network, including the popular backpropagation. For the present application the RPROP algorithm [314] has been selected because of its good performance and ease of implementation.

In Table 4.3(a), the SSE in the one-step ahead prediction in the TS and VS are shown for different networks. The input of the network is the regressor with $n_u = 5$, $n_y = 1$, $n_{p1} = 9$ and $n_{p2} = 1$. The network structure corresponds to an MLP with two hidden layers, one linear output and short-cut connections. The number of nodes in each layer is indicated in the first column by numbers separated by dashes. Thus, a 16-6-3-1 network has sixteen inputs, six and three nodes in the hidden layers and one output node. The neural structure refers to the number of nodes, the connections among them and the activation function used in ANN. Training of the network consists, in most cases, of selecting a set of weights that make the network behave in some desired way. In the case of the solar plant, a simple trial and error procedure has been used to determine the size of the MLP. The results obtained are shown in

Table 4.3 (**a**) SSE in the TS and VS for different networks; (**b**) SSE for the TS and VS for networks with a different number of nodes; (**c**) The same as (**b**) using the weighted metric (courtesy of M.R. Arahal et al., [18])

Size	Cycles $(\cdot 10^3)$	TS	VS
16-6-3-1	5	0.5729	0.2963
16-6-3-1	10	0.2668	0.1485
16-6-3-1	50	0.1339	0.0905
16-10-5-1	5	0.6312	0.2854
16-10-5-1	10	0.2701	0.1245
16-10-5-1	50	0.0986	**0.0172**
16-20-10-1	5	0.7821	0.3012
16-20-10-1	10	0.4565	0.1310
16-20-10-1	50	**0.0916**	0.0173

(a)

Nodes	$\frac{1}{2\sigma^2}$	TS	VS
0	–	0.155	0.032
1	15	0.150	0.032
2	15	0.143	0.032
5	15	0.129	0.030
8	20	0.123	0.030
10	50	0.121	0.030
15	15	0.116	0.029
20	25	0.109	0.028

(b)

Nodes	TS	VS
0	0.155	0.032
5	0.119	0.022
10	0.113	0.022
15	0.107	0.021
20	**0.101**	**0.021**

(c)

Table 4.3(a), where the third and fourth columns indicate the SSE in the TS and in the VS, respectively. The number of training cycles is shown in the second column.

The learning procedure for RBF networks can be data-driven for all parameters or just for some of them, the rest being selected by heuristics. In the application shown in this section, an intuitive technique has been used for the construction of RBF networks. The procedure begins with a linear model and progressively adds RBF nodes, selecting centers among input patterns that produce large errors in the linear model. The output weight of every new node is initially set to the value of the error, so that the RBF included helps to reduce the error of the linear model locally. The width is chosen so as to maximize the fall in the SSE. After the insertion of a node, the output weights can be fine-tuned. Table 4.3(b) presents the SSE of the one-step ahead prediction in the TS and VS for networks with a different number of nodes. The table begins with the results obtained by the linear approximation. The RBF network is used to approximate the residuals of said linear model as commented above. The second column indicates the value of $\frac{1}{2\sigma^2}$. It can be seen in the above table that the large descents in the SSE are given by nodes with larger σ. This observation can easily be explained: the greater the width, the greater the area covered by the RBF and the greater the improvement due to its addition. Hence, choosing the points where wider RBF can be placed produces a saving in the number of nodes. In other words, adding nodes with progressively narrower widths at points where the error exceed a threshold leads to networks that grow parsimoniously. In this way, the final network is close to optimality because of its small number of nodes [18].

Another issue to be considered is the choice of the distance for the RBF. In the previous results, the Euclidean distance has been used. A weighted metric allows a linear transformation of the input vector to be performed in order for its more relevant components to be better taken into account. In this way, instead of using the Euclidean distance: $\|z\|^2 = z^T z$, the following norm can be used: $\|z\|^2_{\mathbf{W}} = z^T \mathbf{W}^T \mathbf{W} z$. A diagonal matrix with components $[w_q, w_q, w_q, w_{T_{out}}, w_{T_{out}}, w_{T_{out}}, w_I, w_{T_{in}}]$ is

Table 4.4 Results obtained in the prediction with RBF networks over different horizons (courtesy of M.R. Arahal et al., [18])

Nodes	$N=1$	$N=5$	$N=10$	$N=20$	Average
0	0.0449	0.5912	1.3229	2.483	1.110
5	0.0456	0.5548	1.2300	2.494	1.081
10	0.0410	0.4931	1.0804	2.078	0.923
15	0.0392	0.4751	1.0684	2.146	0.932
20	0.0390	0.4735	1.0670	2.145	**0.931**

proposed. Physical considerations allow one to determine that I and T_{in} have less influence on the output than q and T_{out}. To diminish the importance of I and T_{in} in the norm, the following values were chosen after a process of trial and error: $w_q = 1$, $w_{T_{out}} = 1$, $w_I = 0.5$ and $w_{T_{in}} = 0.2$. Using this metric and the idea of only allocating nodes with large widths expounded above, the results shown in Table 4.3(c) were obtained using nodes with $\frac{1}{2\sigma^2} = 15$ showing the convenience of the scheme.

Model Validation The adequacy of a model to observed input/output data is normally assessed using the SSE as figure of merit. The error is calculated as the distance from the output pattern of the model $\hat{y}(i) = NN(\mathbf{x}(i))$ and the correct output $y(i)$. The tuning of the parameters of a parametric model is done, in many cases, guided by the SSE. Iterative procedures such as gradient descent adjust the value of the parameter vector, step by step, seeking a minimum of the SSE. Once at a local minimum, residual analysis can be used to decide if a given model can still improve its performance; that is, if the SSE can be further reduced. The residuals, or prediction errors given by a model in a set not used for training, should be independent of the signals used as model inputs. Any correlation is an indication that the model does not correctly explain the influence of some inputs on the system output.

Cross validation has been used to choose among different MLP sizes and to stop the training procedure. Table 4.3 shows that by stopping training before overtraining it is possible to obtain different neural structures as models for the DSCF.

Long-Range Predictors As stated previously, in model predictive control algorithms it is necessary to obtain a sequence of predictions $\{\hat{y}(k+i \mid k)\ i = 1, \ldots, N\}$ using information from the system up to time k. These predictions can be obtained using a one-step ahead predictor recursively or using a bank of predictors. In Table 4.4, the results obtained in the prediction with RBF networks over different horizons are presented. The values in the first column indicate the number of RBF. Columns 2 to 5 show the SSE on the VS for different prediction horizons using the model recursively and the last column is the average. The first row shows the results of a linear model corresponding to a network with short-cut connections and zero non-linear nodes.

Figure 4.13 has been set up to check the influence of the addition of new nodes to the RBF in the 10-step ahead prediction. The first value corresponds to a network

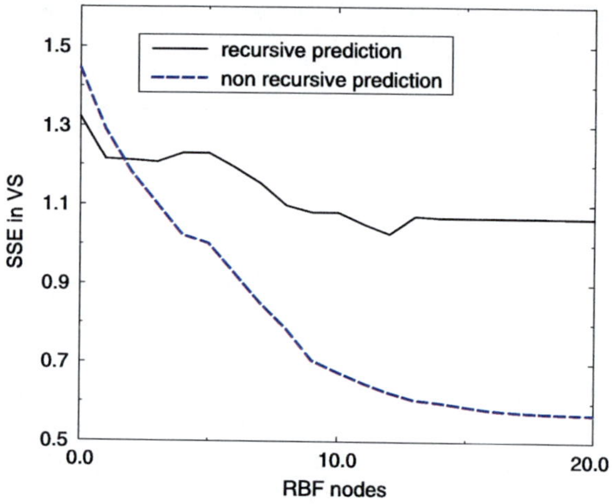

Fig. 4.13 SSE in the validation set for the 10-step ahead predictor realized by RBF networks with different number of nodes (courtesy of M.R. Arahal et al., [18])

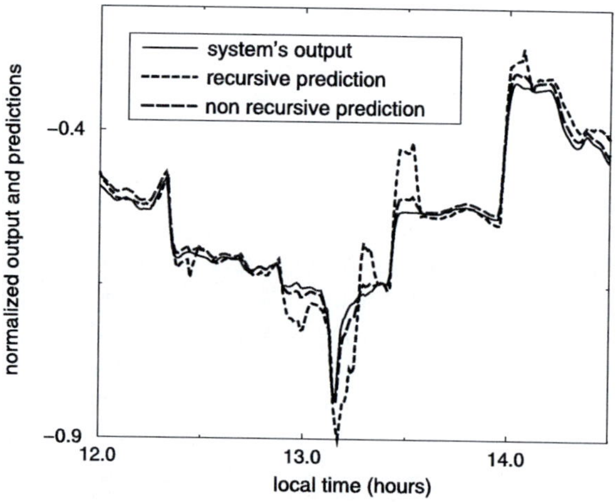

Fig. 4.14 Output of the plant and predicted output (horizon $N = 10$) in the VS using RBF networks with 20 nodes recursively and non-recursively (courtesy of M.R. Arahal et al., [18])

with zero basis functions; that is, just a linear model. It can be seen that the addition of certain nodes makes the 10-step ahead prediction worse because they were allocated and tuned using information from the plant just one step ahead in the TS. The quality of a 20 node RBF network as a 10-step ahead predictor in the TS is shown in Fig. 4.14.

The results of using the one-step predictor for long horizons recursively indicate the convenience of developing long-range predictors. As before, a linear model is first developed and then radial basis functions with progressively narrower widths are added. The model is able to produce the 10-step ahead prediction without iterations. In Figs. 4.13 and 4.14 it can be seen that the performance of the non-recursive predictor is better [18].

4.3.6 Object-Oriented Modeling

In [407, 408], a model of the thermohydraulic part of the system was developed, skipping the remaining subsystems (pneumatic, mechatronic, etc.) needed to maintain the proper instantaneous orientation of the PTC group and assuming a known input radiation power in the absorber pipe, as a consequence of the radiation reflected in the PTC mirrors. Due to the fact that the main phenomenon of interest is the thermofluid dynamics, the object-oriented Modelica language [257] was used to develop these models with the Dymola tool [127]. Within this modeling language the ThermoFluid library [128, 377] is a framework that helps to develop one's own libraries and final component models ready to be instantiated as components for simulations. The work analyzes each of the components of the thermohydraulic circuit and explains the modeling assumptions, trying to justify each one as they are oriented to get, by means of symbolic manipulations performed by the Dymola tool, a not too high index differential algebraic equation (DAE) system for the complete model in which the number of non-linear algebraic loops is minimized. For this purpose, all the components are classified following the modeling methodology derived from the Finite Volume Method (FVM) [287], in Control Volumes (CV in ThermoFluid nomenclature) and Flow Models (FM in ThermoFluid nomenclature). In some cases, information about the future control system architecture to be implemented is introduced in the modeling phase. This methodology, helps to simplify the design of the models and enhances the numerical behavior of the whole modeled system in the simulation execution phase without a significant loss of accuracy. Due to the existence of components with internal implementation which may vary according to the modeling hypotheses depending on the experimental framework, the polymorphism and the Modelica language constructs for classes and components parametrization has been extensively used and specifically applied in PTC models.

The main modeled components of a PTC are described in [408]: parabolic trough reflector surface, metal absorber pipe, energy loss rate to the environment by conduction, convection and radiation, HTF model, distributed CV, with a discretization level n in which mass, energy and momentum are conserved. Mass and energy conservation are stated in a dynamic formulation and the momentum (simplified) is in steady state with a staggered grid approach [287]. For modeling purposes, this component is considered to be a heat exchanger composed of one pipe with a HTF as the medium and a circular wall allowing thermal interaction with the fluid. This heat is fed by solar energy through the outer perimeter of the circular wall and, at the same time, some energy flow leaves through this external perimeter by conduction, convection and radiation. Models of other components (storage tank, pump, valves), are also developed in [407, 408].

4.4 Basic Control Algorithms

Although DSCF have all the characteristics needed for using advanced control strategies able to cope with changing dynamics (non-linearities and uncertainties) to allow the number of operational hours to be increased, most of them are controlled by detuned PID controllers producing sluggish responses or, if they are tightly tuned they may produce high oscillations when the dynamics of the process vary, due to changes in environmental and/or operating conditions. Thus, when the control specifications are very tight and the control system makes the process work at high frequencies where uncertainties are greater, more sophisticated or advanced control techniques are needed [340].

This chapter, describes the results obtained with standard control strategies that have been widely used for several decades (feedforward, proportional-integral-derivative and cascade control).

4.4.1 Feedforward Control (FF)

Feedforward controllers are extensively used in industry to correct the effect caused by external and measurable disturbances. The disturbances are sensed and used to calculate the value of the manipulated variable required to maintain control at the set point. The first step in designing a feedforward control system is to obtain a mathematical model of how the disturbances affect the process. Using the perturbation (or load) model and the process model, the manipulated variable is computed in order to cancel the effect of the disturbance in the process output. The offset resulting from modeling errors can be eliminated by adding feedback. DSCF suffer from changes in the received energy which can be slow, such as daily radiation variations, mirror reflectivity changes due to accumulation of dust, etc. or fast, mainly due to passing clouds and changes in the inlet HTF temperature at the starting phase of the power conversion system. These disturbances force the HTF flow to change, producing a variable residence time of the fluid within the field. Feedforward has been widely used in the control of DSCF [38, 82, 85, 100, 247, 325, 348, 382]. Both dynamic and static feedforward terms (and also white/black-box models) have been developed within this scope. The steady-state gain of the plant, although a function of the irradiance, ambient temperature, the inlet temperature, and the volumetric flow rate, can be predicted using simple static models of the plant [82, 98, 133]. The most extensively used feedforward compensation, both in parallel (Fig. 4.15(a)) and series (Fig. 4.15(b)) configurations, uses a steady-state energy balance from Eq. (4.5) and experimental data derived from a correlation for the HTF flow as function of the inlet and outlet HTF temperatures and direct solar irradiance [82, 97, 382],

$$(T_{out} - T_{in})q = 0.7869I - 0.485(T_{out} - 151.5) - 80.7 \tag{4.50}$$

where the constants that appear in the equation have been determined experimentally from the basic formulation [329]. Thus, the corrected direct solar irradiance and inlet HTF temperature serve to directly adjust the HTF flow to the values calculated to maintain the outlet temperature at the desired level. This restricts the outlet temperature excursions, which is desirable from the control viewpoint and ensures that the outlet temperature is predominantly a function of the HTF flow, which is the

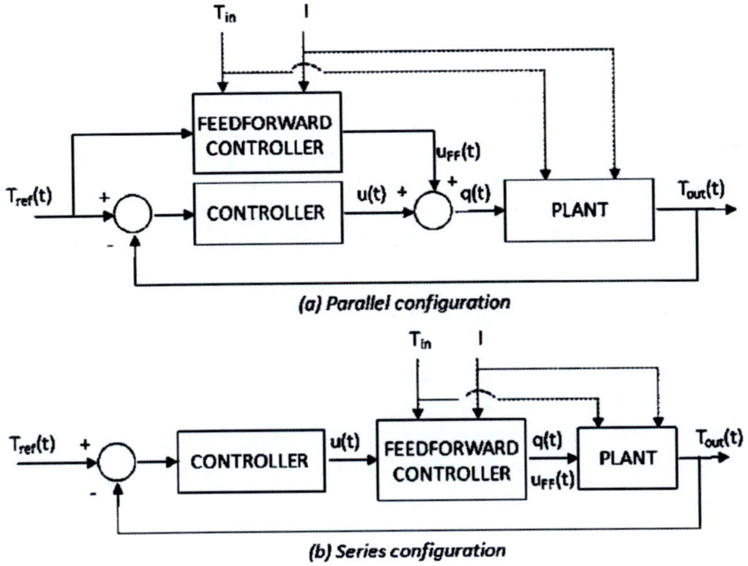

Fig. 4.15 Feedforward configurations

manipulated variable. These feedforward controllers have proved to be effective in many of the tests performed at the plant and have been used by many of the control algorithms tested at the plant [77, 83, 94, 195, 196, 208, 234, 333, 360].

4.4.1.1 Parallel Feedforward Compensation

The basic structure of a parallel feedforward controller is shown in Fig. 4.15(a), where variable u_{FF} is the flow calculated to provide the desired outlet temperature (T_{ref}) for the prevailing values of I and T_{in} and $q(t) = u(t) + u_{FF}(t)$; that is, the flow demanded to the pump is the sum of the contribution of the feedback and feedforward controllers. The calculation employed is

$$u_{FF} = \frac{0.7869I - 0.485(T_{ref} - 151.5) - 80.7}{T_{ref} - T_{in}} \tag{4.51}$$

When variations occur in I and T_{in}, signal u_{FF} is changed to a value which, in steady state, would maintain the desired outlet temperature at the desired value. This feedforward element serves to significantly reduce the dynamic variations in T_{out} due to changes in I and T_{in} and provides quick response to reference temperature changes.

4.4.1.2 Series Feedforward Compensation

An alternative approach to provide the essential compensation for variations in I and T_{in} is to introduce a serial element as shown in Fig. 4.15(b). The output of this serial element forms the desired HTF flow signal $q(t) = u_{FF}(t)$ and is calculated from the following expression:

$$u_{FF} = \frac{0.7869I - 0.485(u - 151.5) - 80.7}{u - T_{in}} \tag{4.52}$$

The variable $u(t)$ is the output of the feedback controller (reference temperature for the feedforward controller, also denoted as T_{rff}) and thus the input variable used for parameter estimation in the adaptive control schema, as shown in Chap. 5. If the controller incorporates integral action then, in steady state, the output temperature T_{out} is equal to the reference temperature T_{ref}. Incorporating a series compensator, the model employed in the control schema will always have a steady-state gain of approximate unity and dynamic information for estimation purposes can be provided simply by injecting appropriate variations in T_{ref}. The introduction of filters into the serial compensation can be employed to approximate dynamic characteristics.

4.4.1.3 General Comments About Feedforward Control

The feedforward signal provides control benefits when disturbances in solar irradiance and inlet temperature occur, but another reason for its inclusion is to preserve the validity of the assumed system models in those control schema that use a SISO description of the plant.

In order to illustrate the benefits of using feedforward compensation, three simulations of PID control are presented (using the distributed parameter model developed in Sect. 4.3.1: without feedforward, with parallel feedforward and with series feedforward compensation (the design of these controllers is presented in the next section trying in all cases to obtain the same closed-loop response). In the three cases, data from one typical operation day at the plant have been used (see Figs. 4.21 and 4.22). The reason for using data from this test to make the comparison between three PID-based control schemes is that it covers a wide range of HTF flow conditions (from 2 to 10 l/s), allowing the benefits achieved by using feedforward control to be demonstrated. Figure 4.22 shows direct solar irradiance corresponding to a real test that is also used for simulation purposes (data of the inlet temperature are not shown because it barely changed during the test).

Figure 4.16 shows in simulation how the fixed PID controller (without feedforward) works correctly at the middle operation point for which it was designed but, as can be seen, oscillations in the system response are present at high temperatures. This zone corresponds to low flow conditions where the plant is more difficult to control. Disturbances produced in the radiation level directly affect the outlet tem-

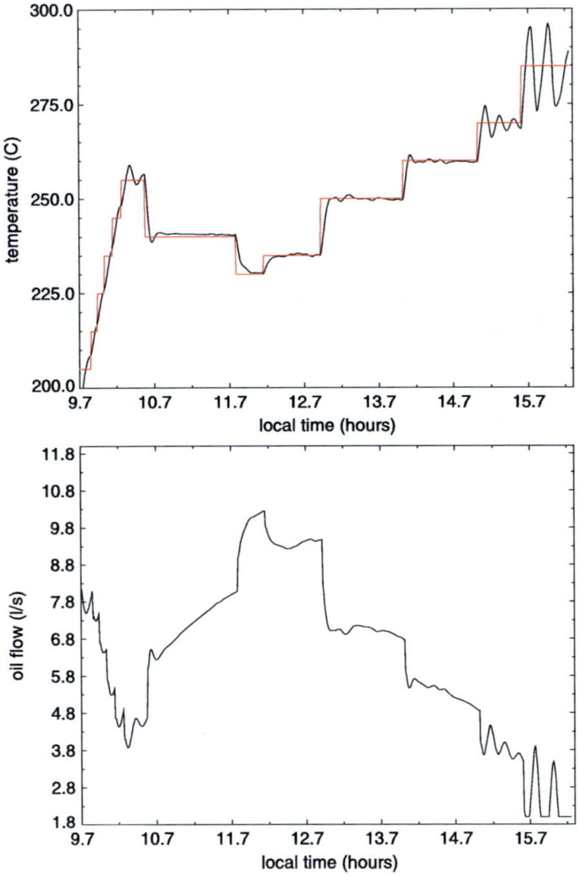

Fig. 4.16 Response without feedforward compensation, [85]

perature, therefore, the feedback controller reacts to these disturbances, although with a certain delay. Feedforward control provides corrective action before the disturbance is seen as an error in the controlled variable. Figures 4.17 and 4.18 show the same simulation with the incorporation of a parallel feedforward controller in one case and a series compensation in the other. As can be seen, good results have been obtained in both cases, the main design objective being achieved in spite of disturbances acting during operation, that is, to obtain a response with small overshoot.

The use of series feedforward compensation has been considered to be more advantageous since the whole plant and the controller become a system that approaches a linear one (at least in terms of small variations around an operation point) which provides large benefits when using an identification mechanism. In fact, the series feedforward controller acts as a static version of a feedback linearization controller. When variations occur in I or T_{in}, the series feedforward calculates the value of flow needed to maintain the desired outlet temperature.

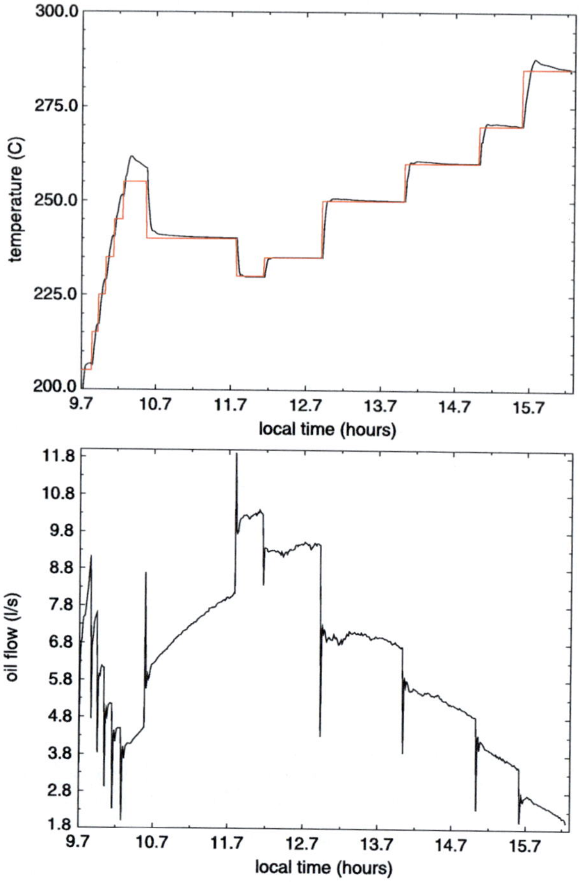

Fig. 4.17 Response with parallel feedforward compensation [85]

4.4.2 PID Control

Due to the significant variations in the dynamic characteristics of DSCF mentioned in Sect. 4.3.1, it is difficult to obtain a satisfactory performance over the total operation range with a fixed parameter controller, mainly if well damped responses are required, due to the existence of resonance dynamics. The use of PID controllers (Fig. 4.19) with fixed parameters has been restricted to safe operation conditions (backup controllers) [85, 98], but they cannot cope with nominal operation of the plant without including additional compensators in the control loop [28, 29, 87]. Even in these cases, performance is restricted by the excitation of resonance modes, but good results have been achieved in the reported literature both in terms of set point tracking and disturbance response when restricting the bandwidth of such controllers. Practically all the tested PID-based control schemes incorporate a feedforward term in the control loop to account for the effect of measurable disturbances [34, 82, 85, 334]. In [105, 106] a class of PID structure combined with a feedforward

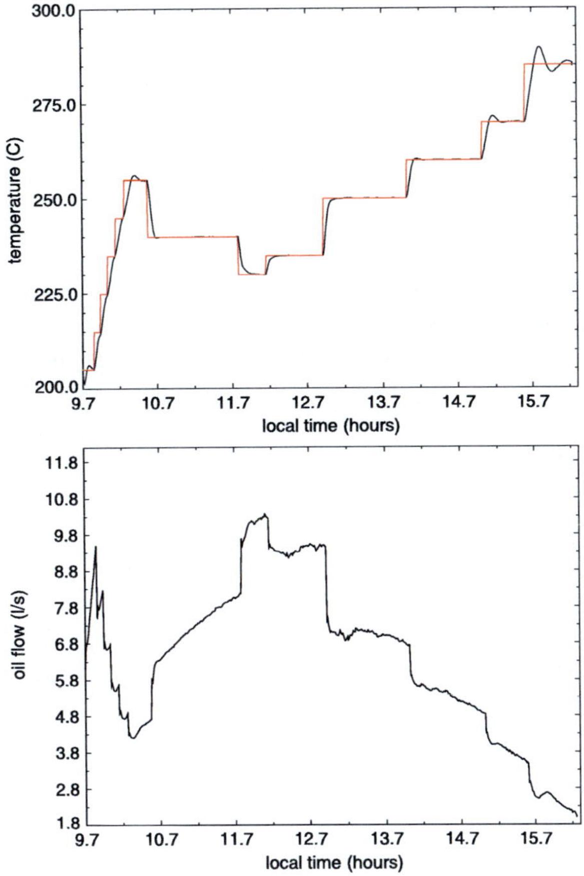

Fig. 4.18 Response with series feedforward compensation, [85]

term and a block for automatic generation of a set point has been satisfactory tested at the plant. Adaptive or gain-scheduling PI controllers [82, 85, 391], switching fuzzy logic or neural network based PID controllers [94, 178, 239], fuzzy logic PID controllers [41, 44, 360], and robust PID controllers [104, 111] are good examples of this philosophy of including a feedforward action and some kind of adaptation to plant dynamics when using PID controllers. In [391] a PID controller with gain interpolation is developed, while in [195, 196] a mixed feedback/feedforward energy based control using PID control is implemented in the form of a PID feedback with time-varying/non-linear gain. Some of these control schemes will be explained in the next chapter.

4.4.2.1 Fixed Ziegler–Nichols Rule Based PID Controllers

With this kind of fixed controller, good behavior can be achieved when operating near the design or nominal flow conditions, but when operating under other condi-

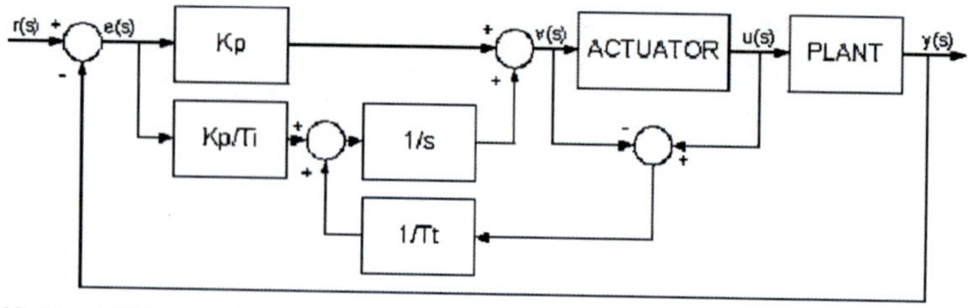

Fig. 4.19 Basic PID + anti-windup control scheme

tions behavior deteriorates. As an initial simple development, the Ziegler–Nichols (ZN) [423] rules have been used. The PID controller transfer function in the time domain can be written as

$$G_{PID}(s) = K_P\left(1 + \frac{1}{T_I s} + T_D s\right)$$

where K_P is the proportional gain of the controller, T_I is the integral time and T_D the derivative time. The derivative part can include a filter, with time constant T_D/N_D, as in [24]. The discrete-time transfer function can be obtained by applying the discretization Forward-Euler approximation for the integral term and the Backward-Euler approximation for the derivative [24], with a sampling time T_s and the value of the derivative filter design parameter $N_D = 5$.

$$u(k) = K_P\left(1 + \frac{T_s}{T_I(z-1)} + \left(\frac{T_D}{T_s + \frac{T_D}{N_D}}\right)\left(\frac{z-1}{z - \frac{T_D}{(N_D T_s + T_D)}}\right)\right)e(k)$$

The open loop Ziegler–Nichols rules have been used to obtain the controller parameters, due to the fact that the system (the DSCF plus the series feedforward controller) presents an open loop overdamped type of step response with delay. The calculation of the parameters is made by previously modeling the plant as a first-order system with a pure delay t_d, time constant τ and gain K (see Fig. 4.20) and then by applying the heuristic expressions contained in Table 4.5.

The controllers designed by this method usually provide fast responses with overshoots of about 25%. Thus, it is usual to reduce the value of the proportional gain in order to avoid highly underdamped responses.

The PID controller designed by the ZN rules can be improved by modifying the controller parameters after the first tests carried out at the plant. Several years of experience and the availability of the non-linear distributed parameter model has allowed a fixed PID controller to operate at medium flows (between 6 and 7 l/s) to be tuned. As will be seen, this controller has been tuned so that the responses obtained are as fast as many advanced control strategies based on low-order models around the nominal design point. At operation points other than the nominal one, the behavior deteriorates, producing high oscillations when operating at low flows.

As an illustrative example, Fig. 4.21 shows the results obtained when controlling the plant including the series feedforward controller with a PID controller with parameters ($K_P = 0.5$, $T_I = 75$ s, $T_D = (T_I/4)$ s) during an operation covering dif-

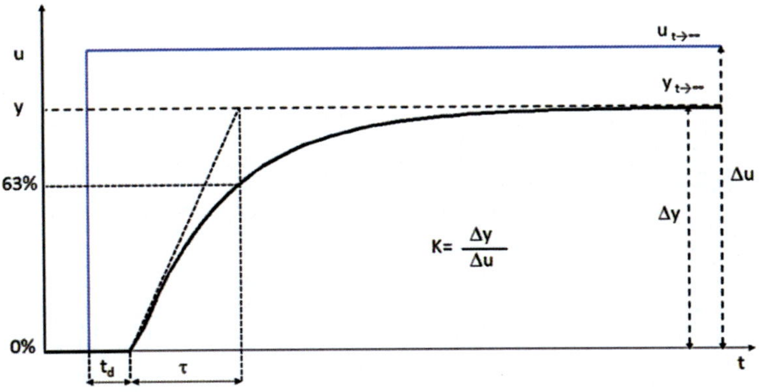

Fig. 4.20 First-order system response characterization

Table 4.5 Open loop Ziegler–Nichols heuristic rules

CONTROLLER	P	PI	PID
K_P	$\frac{\tau}{Kt_d}$	$\frac{0.9\tau}{Kt_d}$	$\frac{1.2\tau}{Kt_d}$
T_I	–	$3t_d$	$2t_d$
T_D	–	–	$0.5t_d$

ferent working points. Figure 4.22 shows the direct solar irradiance conditions and the HTF flow pumped through the field in this test.

As can be seen, the results obtained when operating around medium flows (6–7 l/s) are excellent, but the behavior of the system controlled deteriorates under other operation conditions (high or low flow). Notice that in conditions of low HTF flow (at the end of operation), a large overshoot and settling time appear in the response.

By looking at the responses obtained from the system under different operation conditions the following comments can be made:

- Under low flow conditions, the controlled systems tend to have an oscillatory response, mainly due to the following reasons: (i) As shown by the models, at low flow, the residence time of the fluid through the field is greater than with higher flow levels, the resonance modes lying at lower frequencies. This fact can be observed in the plant Bode plots, in which the first resonance mode lies at lower frequencies as the flow decreases. (ii) At low flow conditions, the system is more sensitive to pumped flow variations (the system gain increases as the flow decreases). The curves showing the response of different controllers show how, at low flow conditions, a variation of about 0.15 l/s produces a variation in the outlet HTF temperature of 5°C, while at medium and high flow conditions, variations of 0.6 and 1.6 l/s are required to obtain the same variation (5°C) in the outlet temperature due to the bilinear nature of the system. It is important to point out that this comment does not contradict the supposition made throughout the book that the system with series feedforward has an almost unitary steady-state gain

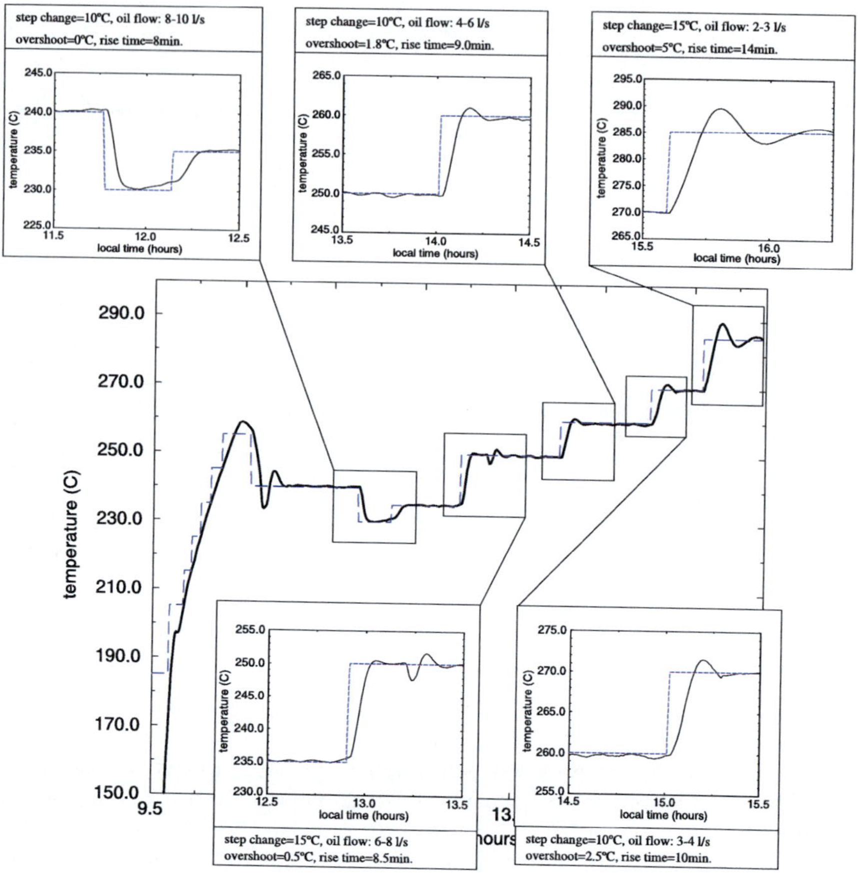

Fig. 4.21 Test with the fine-tuned fixed PID controller (27/03/96), [85]

and is mainly dependent on the steady-state conditions of direct solar irradiance and inlet HTF temperature. This is due to the formulation relating flow to given reference temperature to feedforward controller.

- At high flow level conditions the delay decreases. The resonance modes are not significant in this case as these modes lie at high frequencies.
- From the point of view of disturbance rejection capabilities, it can be seen that a great amount of the contribution needed to compensate for disturbances is provided by the series feedforward controller. The rest of the contribution is provided by the feedback controllers (especially when coping with large low-frequency-gain controllers).

The conclusion that can be drawn from this study is that a fixed PID controller (even a series feedforward controller) does not work appropriately at all operation points and more advanced controllers should be designed in order to take into account the variations in process dynamics and the resonance modes.

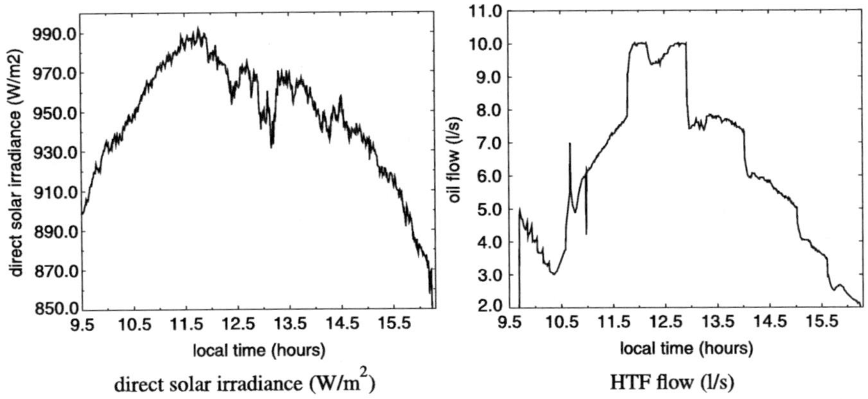

Fig. 4.22 Test with the fine-tuned fixed PID controller (27/03/96) [85]

4.4.3 Cascade Control (CC)

Cascade control is a traditional control technique aimed at canceling the effects of disturbances on the controlled output by splitting the control problem in two time scales and two control loops: an inner control loop (slave) devoted to compensating for disturbances and the outer control loop (master) controlling the process output. Few applications of cascade control are reported in the literature and are mainly developed in the scope of the cascade control of a DSCF for controlling the average of the temperatures at the outlet of the loops and the temperature of the HTF entering the storage tank. In [309, 347], the inner loop uses an adaptive model-based predictive controller exploiting the information conveyed by accessible disturbances (irradiance changes and inlet HTF temperature), while in the outer loop a PID is employed. The difference in the dominant time constants of the inner (faster) and outer (slower) control loops is explored by employing different sampling rates in each of them. Cascade control has recently been used in the scope of controlling solar plants with distributed collectors with direct steam generation [384], as dealt with in the following section.

4.5 New Trends: Direct Steam Generation (DSG)

As has been commented before, within the range of 200–400°C, present day PTC technology uses oil as a HTF in the absorber tubes, whereas a mixture of water and ethylene glycol can be used for lower temperatures. The working fluid is heated as it passes through the absorber tube of the solar collectors, thus converting direct solar irradiance into thermal energy. The hot working fluid is then sent to a heat exchanger, where its thermal energy is transferred.

A new PTC system prototype was implemented at the PSA to investigate the use of water as the working fluid in the solar field of a thermal power plant using a direct

Fig. 4.23 The DISS plant (courtesy of PSA)

steam generation (DSG) process [418]. Different operating strategies and configurations were evaluated taking several parameters (efficiency, cost, controllability, etc.) into consideration, and promising results were obtained for the commercial implementation of this new system, which at present constitutes the most advanced plant of this type. It is being industrialized mainly in Rankine-cycle electricity generation in which steam is delivered by the PTC. The DSG process increases overall system efficiency while reducing investment costs, since it eliminates the oil used at traditional plants as HTF between the solar field and the power block. The electricity generation cost will be reduced by 26% according to current available data [418]. Furthermore, a DSG solar field can be used to feed any other industrial process requiring thermal energy in the form of saturated or superheated steam at $T \leq 400°C$, $P \leq 100$ bar.

The main task of the control system for this type of plant is to provide a steady supply of live steam conditions at the outlet of the solar field under all operation conditions [384–386].

4.5.1 The PSA DISS Facility

The PSA DISS facility is a solar system that serves as a test-bed for investigating the DSG process in PTC, constituting the leading facility of this kind worldwide. Figure 4.23 shows two views of the facility; its main characteristics are listed in Table 4.6.

Although the solar field can be operated over a wide temperature/pressure range, the three main operation points investigated in the DISS project are listed in Table 4.7. The thermohydraulic behavior and system performance of three basic operation modes (once-through, recirculation and injection modes; see Fig. 4.24) were investigated under actual conditions to identify the specific advantages and disadvantages of each mode [384].

- In the once-through mode, feedwater is preheated, evaporated and converted into superheated steam as it circulates from the inlet to the outlet of the collector loop. The main disadvantage of this concept, which is the simplest of the three, is the

Table 4.6 Technical data for the PSA DISS test loop (courtesy of L. Valenzuela et al., [384])

Parameter	Value
Collectors row length	500 m
Collectors type	Modified LS-3
Collector aperture	5.76 m
Number of collectors	9,50-m-long collectors 2,25-m-long collectors
Orientation of the solar collectors	North–South
Absorber pipe outer diameter	70 mm
Absorber pipe inner diameter	50 mm
Optical efficiency of solar collectors	73%
Total mirror surface	2760 m^2
Maximum pressure at the field outlet	100 bar
Maximum outlet temperature	400°C
Maximum steam production	0.85 kg/s

controllability of the superheated steam parameters at the collector field outlet. A water injector is placed in front of the last collector to control the outlet steam temperature. The selective coating of the absorber pipes would be degraded if the metal piping reached temperatures of around 450°C, which is possible when the system is working in operation mode 3 (Table 4.7).

- In the injection mode, water is injected at several points along the row of collectors. The measurement system necessary to assist the control scheme designed for this mode did not work properly during experiments [129, 418]. The complexity and cost of this operating mode make it advisable to discard it in favor of new developments.
- In the recirculation mode, the most conservative of the three, a water-steam separator is placed at the end of the evaporation section of the row of solar collectors. The amount of water fed in at the inlet of the evaporator is greater than the amount that can be evaporated. In the intermediate separator, the excess water is recirculated to the collector loop inlet where it is mixed with the preheated water. The excess water in the evaporation section guarantees good wetting of the absorber tubes and makes stratification impossible. The steam produced is separated from

Table 4.7 Operating points studied in the DISS solar field (courtesy of L. Valenzuela et al., [384])

	Solar field conditions			
	Inlet conditions		Outlet conditions	
	Pressure [bar]	Temperature [°C]	Pressure [bar]	Temperature [°C]
Mode 1	40	210	30	300
Mode 2	68	270	60	350
Mode 3	108	300	100	375

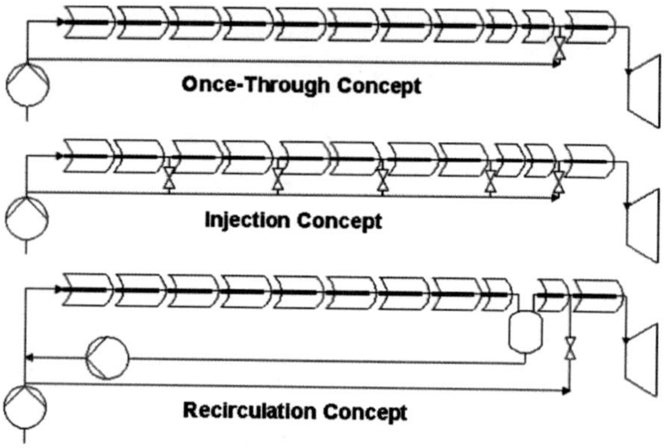

Fig. 4.24 Basic concepts for direct solar steam generation in parabolic trough collectors (courtesy of L. Valenzuela et al., [384])

the water by the separator and fed into the inlet of the superheater section. This type of DSG system is highly controllable [384], but the excess water that must be recirculated, the middle water-steam separator and the water recirculation pump all increase the parasitic load of the system.

The preheating, evaporation and superheating sections are not precisely defined in the once-through and injection modes. The length of these zones depends on the inlet water flow rate and temperature, the pressure in the solar field and the radiation available. In the recirculation mode, the superheating process starts in the next to last collector but the length of the preheating section and, consequently, the starting point of the evaporation section are not exactly defined. This also depends on the operation conditions. All three modes present advantages and disadvantages that have been studied during the DISS project. One of the objectives of the DISS project has been to demonstrate that it is possible to operate the plant under the once-through operating mode guaranteeing flow stability and acceptable controllability [384].

4.5.2 Simulation Models

As in the case of DSCF using oil as HTF, both first principles-based models for simulation and simple models for control purposes (as those shown in the next subsection) were developed for this kind of plant. Regarding models based on energy, mass and moment balances, the DSG process poses new challenges as discontinuities due to two-phase flow may lead to numerical integration problems.

This facility has been modeled using the object-oriented Modelica language with the Dymola tool [406]. The main features of this approach are:

- The base component is the CV, modeled as a class in which mass, energy and momentum conservation equations are taken into account. There are two different CVs in which both mass and energy are conserved. In addition, there are predefined usage rules for connecting them. All the classes constituting the DISS row are generated from the interconnection of these components which are developed based on the ThermoFluid framework [377].

- The FVM [287] is used for the spatial independent variable discretization. It is the discretization level that has been used for the major longitudinal independent variable. This discretization is fixed and does not depend on either the space or time independent variables.
- The thermodynamic properties of the medium are those of the IAPWS-IF97 standard [393] which is currently the most precise reference for their calculation.
- Several experimental correlations developed at PSA were used for calculating energy flux to the ambient.

The final purpose of the model in [406] is to predict the transient behavior of the thermodynamic variables associated to the thermohydraulic output power of the evaporator (temperature, pressure, specific enthalpy, etc.), when external disturbances (mainly concentrated solar irradiance, ambient temperature, subcooled water inlet temperature and subcooled injector inlet water temperature) and controllable input (subcooled inlet mass-flow rate, final injector inlet mass-flow rate and outlet superheated steam pressure) change [406].

After compiling the model using the Dymola tool, a set of parameterized nonlinear differential equations are obtained:

$$\frac{d\mathbf{x}}{dt} = \mathbf{F}(\mathbf{x}, \mathbf{u}, \mathbf{p}); \qquad \mathbf{y} = \mathbf{G}(\mathbf{x}, \mathbf{u}, \mathbf{p}) \tag{4.53}$$

where $\mathbf{p}$ is a vector with 26 parameters that are not completely determined from first principles and are subjected to uncertainty, $\mathbf{x}$ is the vector of state variables of dimension $3N_{CV}$, where N_{CV} is the number of CV in the complete model of the DISS row. It is constituted by the pressure and specific enthalpies of each volume and the temperature of each section of absorber tube in thermo contact with each CV. $\mathbf{y}$ is the vector of output variables of the models. The vector of boundary conditions $\mathbf{u}$ is constituted by row input pressure, temperature of the water at the row inlet, direct solar irradiance normal to the collector, ambient temperature, pressure and temperature at the entrance of the injector, mass flow at the field inlet, mass flow at the injector input and outlet steam pressure. From the control viewpoint, the mass flows are manipulated variables to control the outlet temperature of the last collector and the outlet pressure is a controlled variable.

The model includes several components for pumps and injectors, but the main components are those of the PTC: (i) PTC mirror surface: reflects the direct solar irradiance incident on the focal line of the mirror, (ii) metal absorber pipe: absorbs most of the energy reflected by the mirror, (iii) energy loss to the environment by conduction-convection and radiation, (iv) HTF model medium: in the case at hand, this medium is water-steam, (v) distributed CV, with discretization level n in which mass, energy and momentum are conserved.

For modeling purposes, this component could be considered a heat exchanger with only one pipe with water and/or steam as the media fluid and a circular wall for thermal exchange with the fluid. This heat exchanger is fed by solar energy entering through the outer perimeter of the wall and, at the same time, some energy flow leaves through this outer perimeter by conduction-convection and radiation. The water/steam pipe is 50 m long and under normal operation conditions, the inlet/outlet flow may be in any of the three states of water, two-phase mix of saturated liquid and vapor, or superheated steam. This depends on the position of the PTC in the row as well as the incident solar irradiance on it.

Thus, the dynamic behavior of each PTC varies along the DISS row depending on the thermodynamics and transport state of the water/steam in each PTC. Most of the length of the PTC is fully discretized in n CVs, in which mass, energy and momentum balances are given. Momentum conservation is stated in CVs staggered half spatial grid with regard to mass and energy balance CVs. To solve the PDE system stated from balance equations, ThermoFluid provides partial classes [377] in which the discretization with the FMV is applied. To close the equation system, the heat transfer coefficient for the water-steam flow and the solid media must be entered. This coefficient depends on heat transfer correlations using adimensional fluid numbers (Reynolds, Prandtl, Pecklet, ...), geometry of the contact surface and thermodynamic and transport properties of the fluid (i.e. water–steam). Some of the correlation parameters depend strongly on the experimentation and parameter adjustment stage of modeling. In developing experimental correlation classes for the heat transfer coefficients, sliding modes have appeared with some frequency around the water/steam-CV phase boundaries.

These phenomena are more frequent when CVs go from subcooled (Region 1 in IAPWS-IF97 standard for water/steam properties) to saturated (Region 4 in IAPWS-IF97), for two reasons: firstly, the existence of discontinuities in the heat transfer coefficients on the boundary between water and walls and secondly, the opposite gradients in the state velocity vectors present around the phase-change boundaries. To avoid chattering in the simulation, another polymorphic evaporator model has been developed in which the subcooled and saturated regions of the water/steam pipe are replaced by an equivalent Moving Boundary Model (MBM) [404]. Although the mixed model reduces the likelihood of finding chattering during integration, it is theoretically less accurate, it is harder to find consistent initial DAE conditions experimentally and the model's range of validity is more limited than that of the fully discretized one [60].

The boundary conditions are defined by reservoir components for pressure, specific enthalpy and temperature boundary conditions. For one-phase flow, the (pressure, temperature) pair is selected and in two-phase flow the (pressure, specific enthalpy) pair is selected. Representative results of this model are shown in [406].

4.5.3 Control Problem

Added to the control problems associated to solar plants with distributed solar collectors studied in this chapter (using synthetic oil as HTF), the control in DSG system is still more complex because of the two-phase flow which complicates not only the engineering of the system but also the control system that must be designed for the solar field. In the DISS test loop, both the temperature and pressure of the fluid must be controlled to maintain the desired steam conditions at the outlet (i.e., turbine specifications). The main objective of the control system is to obtain steam at a constant temperature and pressure at the outlet of the solar field in such a way that changes produced in the inlet water conditions and in the solar radiation will only affect the amount of steam produced by the system and not its quality.

4.5.3.1 Control Scheme for the Once-Through Mode

As the first step, the main dynamics were approximated by linear models [384, 386]. After studying the control schemes and analyzing possible loop interactions, SISO transfer functions of all relevant control loops (Table 4.8) were experimentally investigated for three different operation points defined in Table 4.7. The identification method followed was to find in open loop the process parameters (gains, dead times and time constants) that experimentally fit step-response data. Based on the low-order models obtained, the various PI controller parameters were chosen using the reaction curve method, studying the closed-loop responses by simulation and modifying the parameters when necessary to provide safe stability margins. A final optimization of the parameter values was made in subsequent tests at the plant. Additionally, the control scheme designed for the once-through mode includes mixed feedforward-cascade control schemes to control the outlet steam temperature. The PI controllers were implemented using a classical interactive formulation including anti-windup, bumpless proportional band tuning.

The process diagram including the most important feedback loops for the once-through operating mode is shown in Fig. 4.25. The main control loops for the solar field in the once-through operating mode are as follows:

- *Feed pump control loop*: The rotational speed of the feed pump is adjusted to maintain a specific pressure drop in the feed valve. With a constant pressure drop in a feed valve, its flow in steady state is directly proportional to its valve opening. The feed pump control loop, therefore, provides a linearized flow relationship between valve position and flow (PI control).
- *Outlet steam pressure control*: The steam produced by the collector row feeds a steam separator and the outlet steam pressure is kept constant by adjusting a steam control valve (PI control).
- *Outlet steam temperature control loops*: The outlet temperature control is achieved by both inlet feed flow control and by water injection in the superheater. The former control ensures that the steady-state inlet flow matches radiation conditions, whereas the latter control provides the means for rapid response to sudden disturbances (PI-feedforward control-based loops) via a water injection point at the inlet of the last collector of the solar field.

PI functions for the first two control loops have been implemented based on the reaction curve method and study of the stability margins using simplified linearized models (see Table 4.7). Outlet steam temperature control required a more detailed

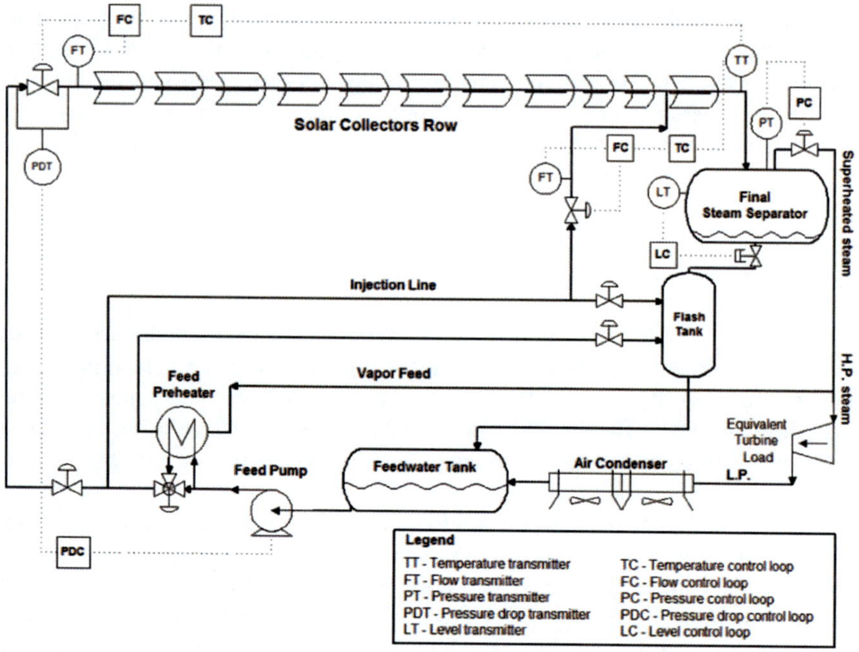

Fig. 4.25 Schematic diagram of the DISS test loop configured in the once-through mode (courtesy of L. Valenzuela et al., [384])

design because the process is strongly affected by disturbances at the inlet and by disturbance variables and acceptable control cannot be achieved with conventional PI or PID schemes. Contrary to the recirculation mode (the process diagram with the most important feedback loops is shown in Fig. 4.26 [385]), in the once-through mode there is no intermediate separator in the field that muffles disturbances occurring in the preheating and evaporation sections and the starting point of the superheating section is not precisely defined.

This reduces the controllability of this operating mode when compared to the recirculation one in which the control loops are based on simple PI controllers:

- *Recirculation pump control loop*: recirculation flow is controlled by PI control of the rotational speed of the recirculation pump.
- *Feed pump control loop*: the rotational speed of the feed pump is adjusted by a PI controller to maintain a specific pressure drop across the feed valve.
- *Middle steam separator liquid level control loop*: to maintain the level around a nominal value, the feed flow is adjusted to control the aperture of the feed valve whose pressure drop is being controlled by the feed pump, PI control.
- *Outlet steam pressure control loop*: by adjusting using a PI a steam control valve in the steam separator.
- *Outlet steam temperature control loop*: by water injection in the inlet of the last collector using PI control of the injector valve.

Table 4.8 Once-through mode: Models and PI control loops parameters (courtesy of L. Valenzuela et al., [384])

Control loop	Model	PI parameters	
		K_P	T_I [s]
Feed pump	$G(s)=\frac{0.1375}{s^2+1.332s+0.3329}$	1.1 %/bar	10
Outlet steam pressure	$G(s)=\frac{-4.543\cdot10^{-3}s-5.05\cdot10^{-5}}{s^2+4.976\cdot10^{-2}s+9.693\cdot10^{-5}}$	−5.3 %/bar	184
Outlet steam temperature control via injector valve (G_1 master loop, G_2 slave loop)	$G_1(s)=\frac{-a}{s+b}e^{-cs}$ where $a\in[3.12, 8.13]$, $b\in[5\cdot10^{-3}, 6.6\cdot10^{-3}]$ and $c\in[70, 100]$	−0.0015 kg/s/°C	600
	$G_2(s)=\frac{-3.2\cdot10^{-4}}{s+0.2}$	−500 %/kg/s	12
Outlet steam temperature control via feed valve (G_1 master loop, G_2 slave loop)	$G_1(s)=\frac{-a}{s+b}e^{-cs}$ where $a\in[1.365, 2.526]$, $b\in[1.8\cdot10^{-3}, 9.6\cdot10^{-4}]$ and $c\in[395, 750]$	$-8\cdot10^{-5}$ kg/s/°C	250
	$G_2(s)=\frac{3\cdot10^{-2}}{s+0.1}$	20 %/kg/s	12

The last two control loops are the main controllers of the system in recirculation mode for guaranteeing the steam quality at each time. The rest of the controllers are required to improve the behavior of the whole control system and for operational feasibility. The method for obtaining the PI parameters has been the same as that explained for the case of the PI controllers for the once-through mode.

Regarding the once-through mode, the solution adopted in the DISS project has been to control the outlet steam temperature with control schemes based on forward action. The parameters of the PI functions appearing in these schemes were also chosen using the reaction curve method and by studying the closed-loop system stability margins. Parameters a, b, and c in Table 4.8 are related to the uncertainty of the models obtained. Depending on the operation conditions (outlet steam flow production, temperature, pressure and solar radiation available) the gain, time constants and time delays vary. The different model parameters influence the PI control design. Therefore, once a set of PI parameters was chosen, closed-loop simulations were performed varying the model parameters to guarantee wide stability margins for the whole range of model parameters. In this way, the selection of the PI parameters was conservative.

The detailed schemes are discussed in the two following subsections. Although interactions among loops do exist, they are small (as can be seen in the results) because the two slave loops explained in the following sections are very fast (compared to the other loops) and able to rapidly reject slow disturbances due to the interactions caused by other loops. The outlet steam pressure loop is also faster (faster time constant and no dead time) and is also able to reject the somewhat slow disturbances coming from the slower temperature loops. The temperature loops also have an inherent interaction reduction mechanism (see following sections). As a result, the interactions are canceled by the control strategy designed.

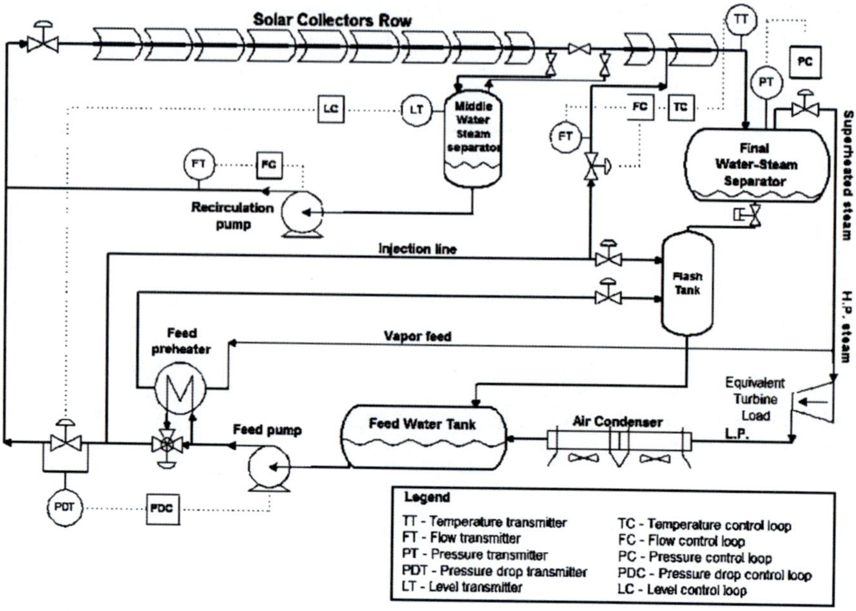

Fig. 4.26 Schematic diagram of the DISS test loop configured in the recirculation mode (courtesy of L. Valenzuela et al., [384])

4.5.3.2 Feedforward Control of Outlet Steam Temperature via Feed Valve Adjustment

The great variations in solar radiation and the long residence time of the fluid in the field call for the use of forward action to anticipate the effect of load changes on the controlled outputs; that is, the control system should calculate the adequate value of the inlet mass flow in advance so that the outlet steam temperature remain within the range of desirable reference values. The performance of the system in the once-through mode is very dependent on the inlet flow control. Changes in solar irradiance, inlet fluid temperature, and so on, require the flow rate to change in order to maintain the desirable output. If changes involve wide oscillations, the solar field performance is strongly affected. Not only are thermal and pumping losses increased but the relatively narrow margin between the design maximum outlet temperature and the actual temperature, which triggers the alarm signal, may be bridged by wide oscillations [384].

To manage these instabilities, the designed and tested outlet steam temperature control loop is a mixed cascade-feedforward control loop aimed at guaranteeing a desired flow in the face of valve non-linearities and changes in disturbances affecting the loop (see Fig. 4.27 and nomenclature). The feedforward term uses a model of the process to make changes in the controller output in response to measured changes in a major load variable without waiting for the error to occur. The outer loop is composed of a feedforward function, FF_{FV}, in parallel to a PI controller

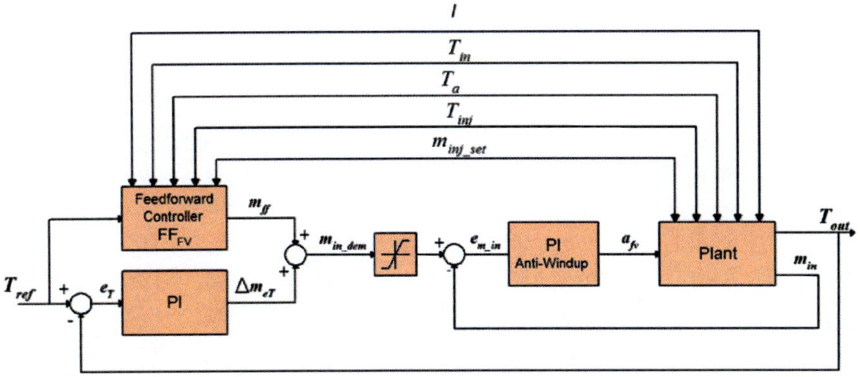

Fig. 4.27 Outlet steam temperature control scheme based on forward action (feed valve) (courtesy of L. Valenzuela et al., [384])

of fixed parameters. Block FF_{FV} calculates a nominal mass flow $\dot{m}_{ff}$ and the parallel PI controller corrects this value according to the current output T_{out}. In a set point change, this PI controller only uses the integral part because the new temperature reference also passes to the FF_{FV} block that calculates the nominal flow $\dot{m}_{ff}$. The flow calculated by this master loop, $\dot{m}_{in\_dem}$, is the input for the inner slave PI control loop which calculates a new aperture of the feed valve, a_{fv}. The saturation included in front of the PI inner control loop limits the inlet water flow to a minimum value of 0.3 kg/s; this guarantees a turbulent flow in the absorber pipes and consequently limits the temperature gradients in the cross-sectional area of the pipes (below 50°C, which is a temperature gradient limit from the point of view of the pipe thermal stress).

As has been previously mentioned, the PI parameters of the outer and inner loops were calculated from open loop responses using the reaction curve method as for feed pump control and pressure control loops. Very conservative parameters were chosen for the PI controller of the master loop, reducing interactions with the temperature control loop using the injector. The feedforward action is obtained from a simplified steady-state energy balance formulation for the collector row. The energy collected is corrected using an efficiency factor estimated for the collector row that implicitly considers the optical efficiency and consequently the optical losses. Then, the simplified energy balance equation can be written as

$$(\dot{m}_{in} + \dot{m}_{inj})h_{out} - (\dot{m}_{in}h_{in} + \dot{m}_{inj}h_{inj}) = \eta_{col}A_f LI - H_l S_{abs}(\bar{T}_f - T_a) \quad (4.54)$$

In this equation, the specific enthalpy at the outlet, h_{out}, is substituted by the outlet enthalpy reference, h_{ref}, and the water flow rate injected in the last collector, $\dot{m}_{inj}$, by the nominal injection flow established in the temperature control loop via the injector valve, $\dot{m}_{inj\_set}$, to avoid a feedback of the variations (which could be oscillatory) dictated by the temperature control loop via the injector valve in block FF_{FV}. Such feedback would deteriorate the temperature response obtained. Then, taking these substitutions into account, the feedforward control equation used to calculate

Table 4.9 Specific enthalpy of the fluid: Parameters of the linear regressions (courtesy of L. Valenzuela et al., [384])

Pressure [bar]	Phase	Temperature [°C]	a_1	a_2	R^2	Standard deviation [kJ/kg]
30	Water	$100 < T_f < 234$	−21	4.38	0.99989	3
	Steam	$234 < T_f < 400$	+2225	2.54	0.99868	7
60	Water	$100 < T_f < 276$	−34	4.48	0.99965	7
	Steam	$276 < T_f < 400$	+1940	3.12	0.99686	11
100	Water	$100 < T_f < 312$	−54	4.61	0.99905	12
	Steam	$312 < T_f < 400$	+1480	4.10	0.99415	15

the nominal feed water flow, $\dot{m}_{ff}$ in order to achieve the desired outlet temperature, T_{ref}, is the following

$$\dot{m}_{ff} = \frac{\eta_{col} A_f L I - H_l S_{abs}(\bar{T}_f - T_a) - \dot{m}_{inj\_set}(h_{ref} - h_{inj})}{h_{ref} - h_{in}} \tag{4.55}$$

where η_{col} has been estimated from experimental data as 0.53; A_f, L, S_{abs} are geometric parameters (see the table of variables and parameters defined at the beginning of the book); I is a corrected value of the measured direct solar irradiance; h_{ref}, h_{inj}, and h_{in} are specific enthalpy values calculated from the outlet pressure P and the corresponding temperatures T_{ref}, T_{inj} and T_{in} as follows: $h_f|_P = a_1 + a_2 T_f$ [kJ/kg], where a_1 and a_2 are coefficients estimated performing linear regressions using the enthalpies and temperature values in the thermodynamic tables (Table 4.9). H_l is a factor related to the thermal losses which for an LS-3 type collector can be approximated by [384] $H_l = b_1 + b_2(\bar{T}_f - T_a) + b_3(\bar{T}_f - T_a)^2$, where b_1, b_2, and b_3 depend on the average temperature of the fluid in the absorber pipes (Table 4.10). To simplify the control loop structure, the average temperature, $\bar{T}_f$, of the fluid in the field is approximated by a constant value for the three different operation points. Values calculated considering inlet and outlet conditions and the conditions in the preheating, evaporation and superheating sections are listed in Table 4.11.

4.5.3.3 Feedforward Control of Outlet Steam Temperature via Injector Valve Adjustment

The outlet steam temperature can also be controlled by injecting preheated water into the last collector providing another degree of freedom to allow a fast reaction

Table 4.10 Thermal loss factor H_l in LS-3 collectors: b_1, b_2 and b_3 values (courtesy of L. Valenzuela et al., [384])

Fluid average temperature [°C]	b_1	b_2	b_3
$\bar{T}_f < 200$	0.687257	0.00194	0.000026
$200 < \bar{T}_f < 300$	1.433242	−0.00566	0.000046
$300 < \bar{T}_f$	2.895474	−0.01640	0.000065

Table 4.11 Average temperature values used in the FF_{FV} controller for the three operation points of the DISS test loop (courtesy of L. Valenzuela et al., [384])

Pressure [bar]	Inlet temperature [°C]	Outlet temperature [°C]	$\bar{T}_f$ [°C]
30	210	300	237
60	240	350	277
100	280	400	316

in the outlet temperature; however, there are regular changes in the outlet steam temperature of the previous collector, in the steam flow rate, or in the injection water temperature influencing the behavior of this single loop. These changes are more frequent and stronger in the once-through mode than in the recirculation mode due to the lack of an intermediate separator, as previously mentioned. A controller based on forward action was also designed for this loop. The feedforward block corrects the injection water flow rate at the inlet of the last collector taking into account changes in the collector inlet temperature and mass flow, the injection water temperature and the outlet temperature reference. The mixed cascade-feedforward control scheme designed is shown in Fig. 4.28 (again, the cascade structure is used to compensate for actuator non-linearities).

The outer loop is composed of a feedforward function, FF_{IV}, in parallel to a PI controller with fixed parameters. The output of the block FF_{IV}, $\Delta\dot{m}_{ff\_iv}$, corrects the PI controller output, $\Delta\dot{m}_{eT}$. When the controller is set in automatic mode, a nominal injection water flow $\dot{m}_{inj\_set}$ is established (around 10% of the expected steam mass production). The outer control loop corrects this nominal value and a new injection water flow value, $\dot{m}_{inj\_dem}$, is calculated and dictated from the injector. This value is the input for the inner loop, a PI control loop which calculates a new aperture of the injection valve, a_{iv}. Injection valve non-linearity detected during experiments is compensated by the cascade structure. The saturation included in the master loop avoids the flow rate dictated by the injector valve being zero, as this would deteriorate the control action due to the non-linearity of the injector when this actuator is nearly closed, as was observed in real tests. The PI parameters of the outer

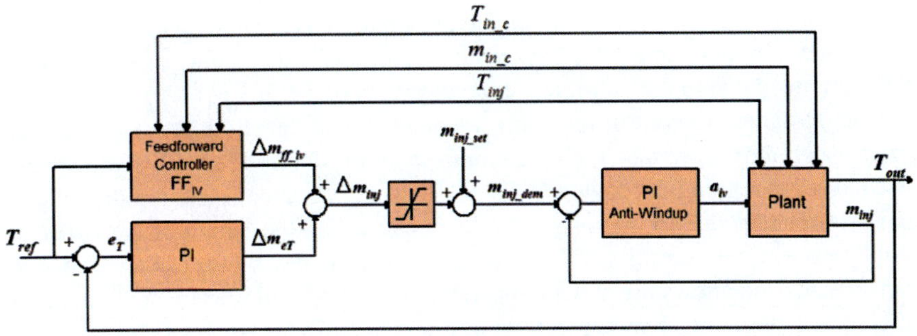

Fig. 4.28 Outlet steam temperature control scheme based on forward action (injector) (courtesy of L. Valenzuela et al., [384])

and inner loops were also calculated from open loop responses using the reaction curve method. Linearized models detailed in Table 4.8 were used to simulate the closed-loop responses and study the stability margins in the worst cases (considering the uncertainty of the models). The final selection of values for the PI parameters was made in a conservative way to avoid instabilities in the system and to diminish interactions with the rest of controllers [384].

As in the case of the feedforward function formulated for the feed valve control loop, this feedforward action is obtained from a simplified steady-state energy balance formulation for the collector. In this case, the energy collected is also corrected by the global efficiency of the collector η_{col} that implicitly considers the thermal and optical efficiency and consequently the thermal and optical losses. The simplified energy balance equation can then be written as

$$(\dot{m}_{in\_c} + \dot{m}_{inj})h_{out} - (\dot{m}_{in\_c}h_{in\_c} + \dot{m}_{inj}h_{inj}) = \eta_{col}A_f LI \tag{4.56}$$

where the right term of the equation includes the energy collected and losses.

Substituting the specific enthalpy at outlet h_{out} with the outlet enthalpy reference h_{ref} that is directly calculated from the temperature and pressure references, the corresponding injection flow in steady state should be

$$\dot{m}_{inj} = \frac{\eta_{col}A_f LI - \dot{m}_{in\_c}(h_{ref} - h_{in\_c})}{h_{ref} - h_{inj}} \tag{4.57}$$

Using this equation, the thermodynamic tables for calculating the enthalpies corresponding to each temperature, pressure and data obtained using the data series detailed in Table 4.12, a regression analysis has been performed to obtain the feedforward function implemented to calculate the injection flow rate correction. The multiple regression model stated is

$$\Delta\dot{m}_{ff\_iv} = c_1 T_{in\_c} + c_2 \dot{m}_{in\_c} + c_3 T_{ref} + c_4 T_{inj} + c_5 \tag{4.58}$$

The values obtained for the parameters c_1, c_2, c_3, c_4 and c_5 at the various operation points are listed in Table 4.13. These parameters depend not only on the thermodynamic properties of the fluid, but also on the geometry and global efficiency of the collector. Table 4.13 includes the correlation coefficient and standard deviation of the residual errors. A good approximation is obtained within the operation ranges listed in Table 4.12, but outside these ranges the quality of the model is not guaranteed due to the non-linear characteristics of the process.

Both temperature control loops are working in parallel. Both are necessary, as pointed out above, because the temperature control via the feed valve adjustment calculates a nominal inlet flow rate for the field but it cannot react rapidly to sudden disturbances due to the long time delay caused by the length of the collector loop. Temperature control via the injector valve provides a faster control to sudden changes and allows the outlet temperature to be adjusted more accurately to the reference. Interactions between both controllers are avoided by the inclusion of parameter $\dot{m}_{inj\_set}$ in block FF_{FV} and by choosing conservative PI parameters in the case of the temperature control via the feed valve.

Table 4.12 Design of the FF_{IV}: Input data sets (courtesy of L. Valenzuela et al., [384])

Input	30 bar	60 bar	100 bar
Direct solar irradiance[1]	[650, 1000] W/m^2	[650, 1000] W/m^2	[650, 1000] W/m^2
Global collector efficiency[1]	[0.40, 0.60] –	[0.40, 0.60] –	[0.40, 0.60] –
Collector inlet mass flow	[0.35, 0.70] kg/s	[0.35–0.70] kg/s	[0.35, 0.70] kg/s
Outlet temperature reference	[280, 320] °C	[320,370] °C	[340, 400] °C
Collector inlet fluid temperature	[250, 310] °C	[290, 370] °C	[330, 390] °C
Injection water temperature	[180, 215] °C	[220, 260] °C	[260, 300] °C

[1]Changes in this parameter do not have significant influence on the adjusted model.

Table 4.13 Outlet steam temperature control with injector valve: FF_{IV} parameters (courtesy of L. Valenzuela et al., [384])

Outlet steam pressure [bar]	c_1	c_2	c_3	c_4	c_5	R-square	Standard deviation [kg/s]
30	$6.212 \cdot 10^{-4}$	0.00313	$-1.7 \cdot 10^{-6}$	$3.0 \cdot 10^{-6}$	−0.00171	0.95035	0.00112
60	$8.457 \cdot 10^{-4}$	0.00505	$-4.4 \cdot 10^{-6}$	$7.1 \cdot 10^{-6}$	−0.00279	0.95219	0.00146
100	$8.942 \cdot 10^{-4}$	0.00494	$-4.2 \cdot 10^{-6}$	$5.8 \cdot 10^{-6}$	−0.00261	0.95167	0.00117

4.5.4 Representative Experimental Results

Figure 4.29 shows the results obtained during an experiment in the once-through mode with 30 bar of outlet steam pressure. The objective of the experiment was to evaluate the response of the control system to a defocusing of one collector of the evaporation section that is equivalent to producing a decreasing 10% step change in the inlet energy to the field. Collector number 6 was defocused at 13:15 and stayed out of focus for 5 min. The resulting outlet temperature deviation was 21°C (7% of the reference). When the temperature came close to the reference again, the irradiance dropped 300 W/m^2, taking the temperature close to the saturation temperature value and changing the outlet steam pressure about 0.8 bar (2.6% of the reference). The nominal conditions of the system later recovered in only 15 min. Prior to defocusing, the irradiance had dropped around 150 W/m^2 at 12:30, which mainly affected steam flow, but the outlet steam pressure was maintained constant and the maximum outlet steam temperature deviation was 4.5°C.

The dead time in the response of the steam temperature observed in Fig. 4.29 is due to the fact that the temperature control via the feed valve was in manual mode during the start-up and the feed flow established by the operator was too high according to solar radiation available.

In [384] more results are shown, where all set points could be maintained during steady-state conditions and even with short transients in solar radiation. During longer solar radiation transients it is more difficult to maintain the steam temperature as a minimum flow must be guaranteed to avoid high temperature gradients in pipe

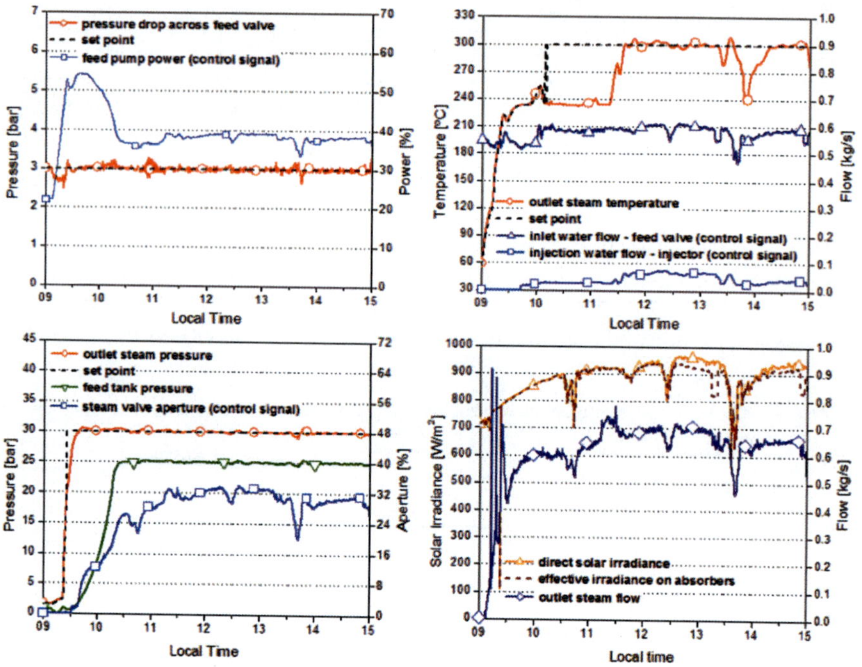

Fig. 4.29 Actual control loop responses during operation at 30 bar (April 26th, 2002) (courtesy of L. Valenzuela et al., [384])

cross-sections when solar radiation recovers (in any case, null values of inlet mass flow lead to zero production, which is commercially undesirable). In this situation, a mixed steam/water flow feeds the separator tank where it condenses, returning to the feed water via the separator drain valve. This increases the parasitic load of the system but also increases security. A control system configured to operate the system with zero inlet flow would have to satisfy very stringent specifications (mostly under actuator saturation) and it would probably be necessary to defocus all the collectors to avoid dangerous conditions in the solar field.

4.6 Summary

The main features of the different modeling and basic control approaches used during the last twenty five years to control DCS have been outlined. The DCS may be described by a distributed parameter model of the temperature. It is widely recognized that the performance of PI and PID type controllers will be inferior to model-based approaches [85, 245]. Even when the plant is linearized about some operation point and approximated by a finite dimensional model, the frequency response contains resonance modes near the bandwidth that must be taken into consideration in the controller in order to achieve high performance [243, 245]. Thus, the "ideal" controller should be high-order and non-linear. The simplest control techniques are outlined in this chapter, others, with high complexity, are studied in the following one, looking for a trade-off between commissioning time and performance [87].

As the main example of the new generation of PTC, the DISS project has demonstrated that it is possible to directly produce high-pressure high-temperature steam in parabolic trough solar collectors. A leading plant using this type of technology has been operated in two different modes. Using a scheme based on PI and feedforward controllers, the controllability of the plant is guaranteed on clear days and even during short transients in the solar radiation. Longer transients in solar radiation make it difficult to maintain the steam temperature in favor of guaranteeing a minimum flow in the field to avoid high temperature gradients in the cross-sectional area of the pipes when the solar radiation level recovers. A structure partially based on classical controllers was chosen because the plant operators are familiar with this type of controller and are able to adapt the controller parameters in the face of situations affecting plant dynamics and controller performance, such as modifications in plant layout or system changes over time. The control structure developed has demonstrated the technical controllability of the system.

Chapter 5
Advanced Control of Parabolic Troughs

5.1 Introduction

In the previous chapter, the main features of solar plants with a distributed collector system (DCS) have been studied from the viewpoints of modeling, simulation and basic control. A classification of control strategies introduced by [340], has been considered to explain the different control approaches used successfully to control this kind of plant [88]. This chapter is devoted to overviewing advanced control techniques aimed at taking into account the special dynamic features of distributed solar collector fields (DSCF). Many of the applications included in this chapter have been tested at the ACUREX field of the PSA (see the previous chapter for a complete description).

5.2 Adaptive Control (AC)

The main idea behind AC is to modify the controller when the process dynamics changes. It can be said that adaptive control is a special kind of non-linear control where the state variables can be separated into two groups moving in two different time scales. The state variables which change faster correspond to the process variables (internal loop), while the state components which change more slowly correspond to the estimated process (or controller) parameters. Adaptive controllers have traditionally been classified into one of the following families: model reference adaptive controllers (MRAC, Fig. 5.1(a)) and self-tuning controllers (STC, Fig. 5.1(b)) [25]. As has been mentioned, when controlling DSCF, the control schemes that vary the HTF flow rate applied to the collector field generally use a combination of feedback and feedforward control, incorporating a non-linear mechanism to operate effectively. Initial studies conducted on the ACUREX field attributed the oscillatory behavior obtained with classical proportional-integral-derivative (PID) controllers to the variability of plant dynamics with operating point [82, 329, 331, 332, 334]. Different adaptive-control schemes were thus developed

E.F. Camacho et al., *Control of Solar Energy Systems*, Advances in Industrial Control, DOI 10.1007/978-0-85729-916-1_5,

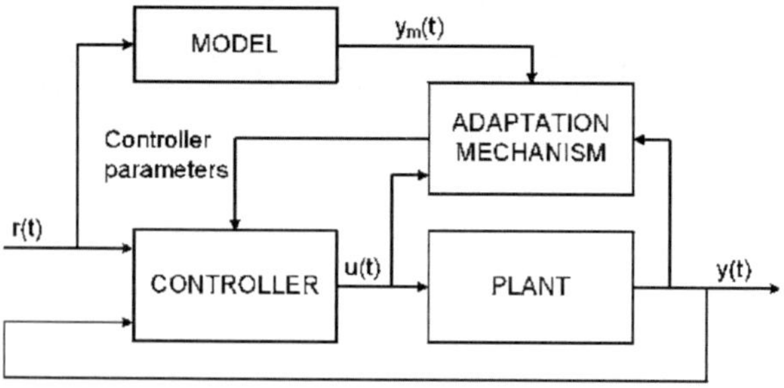

(a) Model reference adaptive control (MRAC)

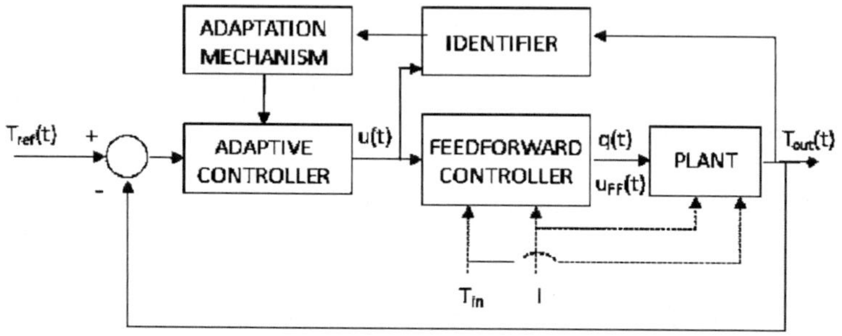

(b) Self-tuning controller (STC) used to control DSCF

Fig. 5.1 Block diagrams of adaptive-control (AC) schemes

to cope with this problem of changing dynamics. The major role played by changes of solar radiation and plant uncertainty lead to the approach of [82] where a pole-placement self-tuning PI controller with the series feedforward compensator given in Sect. 4.4.1 is used, also using a modified recursive least squares (LS) identification mechanism and low-order discrete transfer functions representing plant dynamics, as on-line estimation based upon high-order transfer function models tends to perform poorly with slow parameter convergence being the norm. Thus, lower-order transfer function models have been used when desirable control bandwidths are not too stringent, but decreasing the commissioning time of the controller and producing a simple control law. Simulation results prove the advantage of the adaptive controller above a fixed PI controller, as will be treated in this section. The same feedforward term and identification mechanism were then embedded within an adaptive predictive-control scheme providing very simple control laws [83]. In [77] an adaptive robust predictive controller is developed based on a simplified transfer function model of the plant, where the pole location is fixed and the parameters of the numerator are on-line robustly identified, in such a way that the uncertainty of the closed-loop system decreases quickly.

Different forms of the multivariable self-tuning multipredictor adaptive regulator MUSMAR [162] were also demonstrated with success [116, 117, 308, 309, 347]. In [310, 344, 348], a new dual version of the MUSMAR algorithm is presented based on a bicriterial optimization, in order to improve start-up transients. The adaptive MUSMAR controller proposed in [116, 117], which includes the accessible disturbances from the irradiance and the HTF inlet temperature, shows good results especially for experiments at fast changing values for the irradiance. However, the complexity of this controller is also high. Many of the proposed adaptive-control schemes show oscillatory behavior when requiring fast responses to set point changes and disturbances. In [242] it is established that the cause of these problems was the existence of resonance dynamics lying at a low frequency, as commented in the previous chapter. A simple linear transfer function model of these characteristics was developed by [243, 244] from a system representation derived from its basic thermodynamic equations. Using this model, a series cancelation controller was developed, which in simulation studies achieved faster control than a PI controller, while maintaining a similar level of damping. However, the controller was seen to oscillate the input signal vigorously. This controller was tuned using experimental frequency response data and implemented as a prescheduled scheme [241, 245]. This controller, when combined with feedforward, was shown to be capable of effectively regulating the outlet temperature during both irradiance and inlet temperature disturbances. In [36, 37] an application of adaptive frequency-based internal model control (IMC) for accounting the resonance characteristics is developed. The key idea of the method is to implement both the model of the plant and the controller with frequency-based interpolation models, in such a way that a determined frequency response of the system controlled is imposed at the chosen interpolating points. This can be obtained by using banks of band filters for the input and output of the plant. For estimating the parameters describing these simple models of the plant at each interpolating frequency, a common recursive LS algorithm was used. With this approach, the problem of identifying high-order polynomials is decomposed into a few simple problems in which only low-order models need to be identified.

The rest of this section deals with describing of the parameter estimation algorithm used in all the developed adaptive-control schema and the application of self-tuning PID control, which mainly consists of calculating controller parameters taking the system model parameters to be those obtained by the identification algorithm. The conventional adaptive-control scheme can be seen in Fig. 5.1(b). At each sampling time, the adaptive-control strategy consists of:

1. Estimating the linear model parameters using input–output data from the process.
2. Adjusting controller parameters (this does not necessarily have to be done at each sampling time).
3. Calculating the control signal.
4. Supervising the correct behavior of the controlled system.

5.2.1 Parameter Identification Algorithm

System identification can be defined as the process of obtaining a model for the behavior of a plant based on plant input and output data. If a particular model structure is assumed, the identification problem is reduced to obtaining the parameters of the model. The usual way of doing this is by optimizing a function that measures how well the model fits the existing input–output data with a particular set of parameters. When process variables are perturbed by noise of a stochastic nature, the identification problem is usually interpreted as a parameter estimation problem. This problem has been extensively studied in literature for the case of processes which are linear on the parameters to be estimated and perturbed by a white noise.

If we suppose that the system can be modeled as a stable, single-input single-output (SISO) linear time invariant process (LTI), it can be described by an auto-regressive moving average (ARMA) process given by

$$y(k) = \frac{B(z^{-1})}{A(z^{-1})} z^{-d} u(k) + \frac{C(z^{-1})}{A(z^{-1})} e(k) \tag{5.1}$$

where $A(z^{-1}) = 1 + a_1 z^{-1} + a_2 z^{-2} + \cdots + a_n z^{-n}$, $B(z^{-1}) = b_1 z^{-1} + b_2 z^{-2} + \cdots + b_n z^{-n}$ and $C(z^{-1}) = 1$ without loss of generality; $u(k) = U(k) - U_\infty$, $y(k) = Y(k) - Y_\infty$, $U(k)$ and $Y(k)$ being the system input and output signals at the kth discrete-time instant, U_∞ the mean value of the input signal, Y_∞ is the mean value of the output signal and $e(k)$ mean signal statistically independent and stationary is a noise with normal distribution and zero mean. Equation (5.1) can be represented as

$$y(k) = \boldsymbol{\varphi}^T(k)\boldsymbol{\theta}(k) + e(k) \tag{5.2}$$

where $\boldsymbol{\theta}^T = [a_1, a_2, \ldots, a_n, b_1, b_2, \ldots, b_n]$ and

$$\begin{aligned}\boldsymbol{\varphi}^T(k) = \big[&-y(k-1), -y(k-2), \ldots, -y(k-n), u(k-d-1), u(k-d-2),\\ &\ldots, u(k-d-n)\big]\end{aligned}$$

with d being the pure discrete-time input–output delay and the residual error is supposed to be uncorrelated to the elements of $\boldsymbol{\varphi}(k)$.

The first term in the right hand of Eq. (5.2) can be interpreted as the one-step prediction $\hat{y}(k/k-1)$ of the output $y(k)$ with data available at time $k-1$, thus, the error is the difference between the real output and its prediction: $e(k) = y(k) - \hat{y}(k/k-1)$. Suppose we have a set of measurements $y(k), y(k+1), \ldots, y(k+N)$. The problem is solved by finding the set of parameters $\boldsymbol{\theta}$ which minimizes the sum of the weighted square errors:

$$J(\boldsymbol{\theta}) = \sum_{j=0}^{N} w(j) e(k+j)^2$$

which can be written as

$$J(\boldsymbol{\theta}) = (\mathbf{Y} - \boldsymbol{\Phi}\boldsymbol{\theta})^T \mathbf{W} (\mathbf{Y} - \boldsymbol{\Phi}\boldsymbol{\theta})$$

where

$$\mathbf{W} = \begin{bmatrix} w(0) & 0 & \ldots & 0 \\ 0 & w(1) & \ldots & \vdots \\ \vdots & \ldots & \ddots & 0 \\ 0 & \ldots & 0 & w(N) \end{bmatrix}; \qquad \mathbf{Y} = \begin{bmatrix} y(k) \\ y(k+1) \\ \ldots \\ y(k+N) \end{bmatrix}$$

$$\boldsymbol{\Phi} = \begin{bmatrix} \varphi^T(k) \\ \varphi^T(k+1) \\ \cdots \\ \varphi^T(k+N) \end{bmatrix}$$

The weight factors $w(j)$ are used to penalize the errors of the different measurements taken. If the process is time variant, these factors can be chosen in order to give more weight to the latest measurements. Usually $w(j)$ is chosen to be $w(j) = f^{N-k}$ where f is a constant smaller than one known as the forgetting factor.

The minimum can be found by solving $dJ(\boldsymbol{\theta})/d\boldsymbol{\theta} = 0$, which proves to be

$$\boldsymbol{\theta} = \underbrace{\left(\boldsymbol{\Phi}^T \mathbf{W} \boldsymbol{\Phi}\right)^{-1}}_{\mathbf{P}} \boldsymbol{\Phi}^T \mathbf{W} \mathbf{Y} \tag{5.3}$$

As the plant is a non-linear system which is going to be identified as a linear one, parameters of the linear model will change with the plant operating point and the linear model will be time variant. Equation (5.3) can be solved recursively by the well known recursive least squares identification (RLS) algorithm [25, 145]. In the RLS algorithm, the estimation of parameters ($\boldsymbol{\theta}$) is given by the values of the previous estimation corrected by a linear term in the error between the output and the prediction, $\mathbf{L}$ being the gain of the correction. The RLS method has been chosen with a variable forgetting factor in order to reduce the identifier memory and to avoid the identifier gain reaching zero. This factor is made equal to 1 if the trace of the covariance matrix $\mathbf{P}$ exceeds a certain value. Furthermore, the trace must not be less than a priori selected value. The algorithm is well known and consists of the following steps in the recursive version [330]:

1. Select initial values of $\mathbf{P}(k)$, $\hat{\boldsymbol{\theta}}(k)$ and $f(k)$ (covariance matrix, estimated parameters vector and forgetting factor, respectively).
2. Read the new values of $y(k+1)$ and $u(k+1)$ (identifier inputs).
3. Calculate the a priori residual error:

$$e(k+1) = y(k+1) - \boldsymbol{\varphi}^T(k+1)\hat{\boldsymbol{\theta}}(k)$$

4. Calculate $\mathbf{L}(k+1)$ given by expression

$$\mathbf{L}(k+1) = \frac{\mathbf{P}(k)\boldsymbol{\varphi}(k+1)}{f(k) + \boldsymbol{\varphi}^T(k+1)\mathbf{P}(k)\boldsymbol{\varphi}(k+1)}$$

5. Calculate the new estimated parameters given by

$$\hat{\boldsymbol{\theta}}(k+1) = \hat{\boldsymbol{\theta}}(k) + \mathbf{L}(k+1)e(k+1)$$

6. Calculate the new forgetting factor $f(k+1)$:

$$f(k+1) = 1 - \left(1 - \boldsymbol{\varphi}^T(k+1)\mathbf{L}(k+1)\right)\frac{e(k+1)^2}{S_o}$$

If $f(k+1) < f_{min}$ then $f(k+1) = f_{min}$.
If $f(k+1) > 1$ then $f(k+1) = 1$.
Parameter S_o must be a priori known [143] and is related to the sum of the square errors.

7. Actualize the covariance matrix:

$$\mathbf{P}'(k+1) = \left(\mathbf{I} - \mathbf{L}(k+1)\boldsymbol{\varphi}^T(k+1)\right)\frac{\mathbf{P}(k)}{f(k)}$$

If $trace\left(\frac{\mathbf{P}'(k+1)}{f(k+1)}\right) > trace_{max}$ then $f(k+1) = 1 \rightarrow \mathbf{P}(k+1) = \frac{\mathbf{P}'(k+1)}{f(k+1)}$.

8. Make $k = k+1$ and return to step 2.

Values of $trace_{max}$ and the initial forgetting factor will depend on the system and the desired convergence speed and stability. Notice that for a value of the forgetting factor equal to unity, the covariance matrix decreases monotonously, allowing the identifier gain to be zero. Moreover, if the operating point is fixed, the product $\mathbf{P}(k)\boldsymbol{\varphi}(k)$ can be zero and therefore $\mathbf{P}(k+1) = \mathbf{P}(k)/f(k)$, which for $f(k) < 1$ will make $\mathbf{P}(k)$ increase excessively, making the identifier very sensitive to any change. These reasons justify the selection of a variable forgetting factor in the identification algorithm so that, if the trace of the covariance matrix exceeds a determined value, the forgetting factor is fixed $f(k) = 1$. Similarly, the trace of the covariance matrix is not allowed to decrease below a certain prefixed value. This is done by a supervisory level of the identification process and by adding some heuristic considerations [85]. UDU factorization [49] can also be used to avoid ill conditioning of the calculation of $\mathbf{P}(k+1)$ [85].

In order to obtain stability increases, several modifications have been carried out in the algorithm originally designed for the ideal case. An increase in the robustness of the adaptive-control schema can be achieved by using data prefiltering for identification purposes to obtain better behavior of the adaptive controller and to ensure the validity of the assumed low-order model. An excellent development and overview of all these techniques can be found in [262]. In [85] a summary of simple modifications that can be performed to increase the robustness of the RLS identification mechanism is included.

Moreover, some improvements can be achieved in the performance of the closed-loop controlled system if, parallel to the identification algorithm, a certain supervisory mechanism is implemented to check the evolution of fundamental process parameters. Several methods to characterize DSCF have been studied in Chap. 4. This characterization, when extended to the whole range of possible operating conditions, provides relevant information not only for controller design purposes but also for supervising the correct behavior of the identification mechanism in the case of adaptive-control strategies. The following supervisory mechanisms have been implemented when controlling the distributed solar collector field [85]:

- Supervision of the estimated parameters as they have to be within a pre-specified range provided by known dynamical responses.
- Supervision of the evolution of the identifier, that is, of the trace of the covariance matrix and low-pass filtering of the signals used for identification purposes.
- Stopping identification during the starting phase of the operation or when a considerable decrease in solar irradiance occurs due to passing clouds. Under these circumstances, the disturbances cannot be completely compensated for by the feedforward controller and may cause the identification of dynamics not corresponding to those modeled by the simplified linear models.

- Stopping identification when the input and output signal do not contain enough dynamical information; the basic idea is to identify only at those moments in which input signal changes are produced (changes in the reference temperature to the feedforward controller).

5.2.2 Adaptive PID Controllers

As an initial development of adaptive controllers, the Ziegler–Nichols (ZN) [423] rules have been used within the control scheme shown in Fig. 5.1(b). The PID controller transfer function is the same as that explained in Sect. 4.4.2. In the adaptive version of the algorithm, two different adaptation mechanisms have been used:

1. In the first approach, both the integral and derivative constants are fixed (that is, the time delay is considered to be fixed) and only a proportional gain adaptation is performed from the estimation of the pole and gain of the system linear model (Type A model including the series feedforward controller as part of the plant, G_A, given in Eq. (4.48) with parameters $a = e^{-T_s/\tau}$, $b = K(1-a)$). In the implementation, parameters a and b are estimated on-line using the RLS algorithm. The pure delay t_d is considered to be fixed (between 39 and 78 s), causing the integral and derivative constants of the PID controller to be fixed and equal to $T_I = 2t_d$ and $T_D = T_I/4$. The proportional gain is calculated using the ZN rules with a scale factor α_{PID} to avoid oscillatory behavior: $K_P = \frac{\alpha_{PID}1.2\tau}{Kt_d}$.
2. In the second approach, integral and derivative constants are also estimated by adding a zero to the transfer function of the system (Type B model including the series feedforward controller as part of the plant, G_B, given in Eq. (4.48) and represented by three parameters a, b_0 and b_1) in order to estimate a variable delay by using a first order Padé approximation to a delay. The estimation of a, b_0 and b_1 allows for the adaptation of all the controller parameters. The estimated plant parameters are $\tau = -T_s/\log(a)$, $K = (b_0 + b_1)/(1-a)$ and the dead-time is approximated by $t_d = T[1 + |b_1/(b_0 + b_1)|]$, which can be used to calculate the controller parameters (K_P, T_I and T_D) on-line using ZN rules.

Another kind of PID design strategy is the pole-placement approach, developed in [397] and [3] and used by some of the authors in [332]. By using a Type B model, the following controller design procedure can be developed. The control structure used with this controller is shown in Fig. 5.1(b). The discrete-time transfer function of a PI controller is given by

$$G_{PI}(z) = \frac{g_0(1 + (g_1/g_0)z^{-1})}{(1 - z^{-1})} \tag{5.4}$$

If the zero of the controller is chosen such that it cancels the plant pole, that is, $a = -g_1/g_0$, the system closed-loop characteristic polynomial $P_{cl}(z)$ is given by $P(z) = z^3 - z^2 + g_0b_0z + g_0b_1$. If the closed-loop system is required to have its dominant pole at $z = A$, then the desired closed-loop characteristic polynomial can be expressed as $P_d(z) = (z - A)p(z)$, where $p(z)$ is a second-order polynomial in z. Equating $P(z)$ and $P_d(z)$ gives

$$p(z) = \frac{z^3 - z^2 + g_0 b_0 z + g_0 b_1}{(z - A)}$$

For $(z - A)$ to be a factor of $P_d(z)$, the remainder term in the previous equation must be zero. Hence the closed-loop system will have a pole at $z = A$, if g_0 is chosen such that

$$g_0 = \frac{A^2(1 - A)}{(b_0 A + b_1)} \tag{5.5}$$

and $g_1 = a g_0$. Hence, knowing a, b_0 and b_1 from the model, the controller parameters g_0 and g_1 can be calculated to provide a specified closed-loop dominant pole at $z = A$. Analysis of $p(z)$ provides the location of the other two system poles:

$$p(z) = \left[z^2 - (1 - A)z + g_0 b_0 - A(1 - A)\right]$$

Root locus analysis of the system indicates that apart from the pole at $z = A$, the system will have two other real poles. Solving $p(z)$ enables the locations of these poles to be determined to ensure that they correspond to rapidly decaying modes and that the specified pole at $z = A$ is the dominant pole.

The previous development is valid for both Type A and Type B models (the Type A model being a particularization of the Type B model by making $b_0 = 0$ or $b_1 = 0$).

In order to show the behavior of the adaptive-control schema described in the previous paragraphs, the results of simulation tests showing the evolution of outlet temperature, flow and estimated parameters are shown in Fig. 5.2 using data from a reference test where the main disturbances are shown in Fig. 5.3. This simulation serves to draw two logical conclusions about the behavior which can be expected from any adaptive-control scheme using a reduced linear model of the system:

- When the HTF flow increases, the delay decreases and the response speed increases. Thus, when direct solar irradiance I is high (usually corresponding to a high flow), the delay is reduced and also the effective time constant associated to the plant model. This corresponds to a small value of the estimated plant pole a.
- The parameter variations show that the expected trends in the values are obtained.

At the highest flow condition (the central part of the operation), the estimated system pole decreases, corresponding to a faster system. In the operation around 255°C, less flow is needed to achieve a higher outlet temperature and the evolution of this variable follows the daily solar cycle. The gain is mainly dependent on the relation between the irradiance level and the flow. Theoretically, the existence of the feedforward controller in series with the plant will provide a system with near unitary gain as, in fact, happens. Changes around this value are due to the fact that the adjustment of the feedforward controller is not perfect. With the HTF flow near 10 l/s, the gain is slightly less than that corresponding to medium flows, because of the value of the ratio effective irradiance/flow.

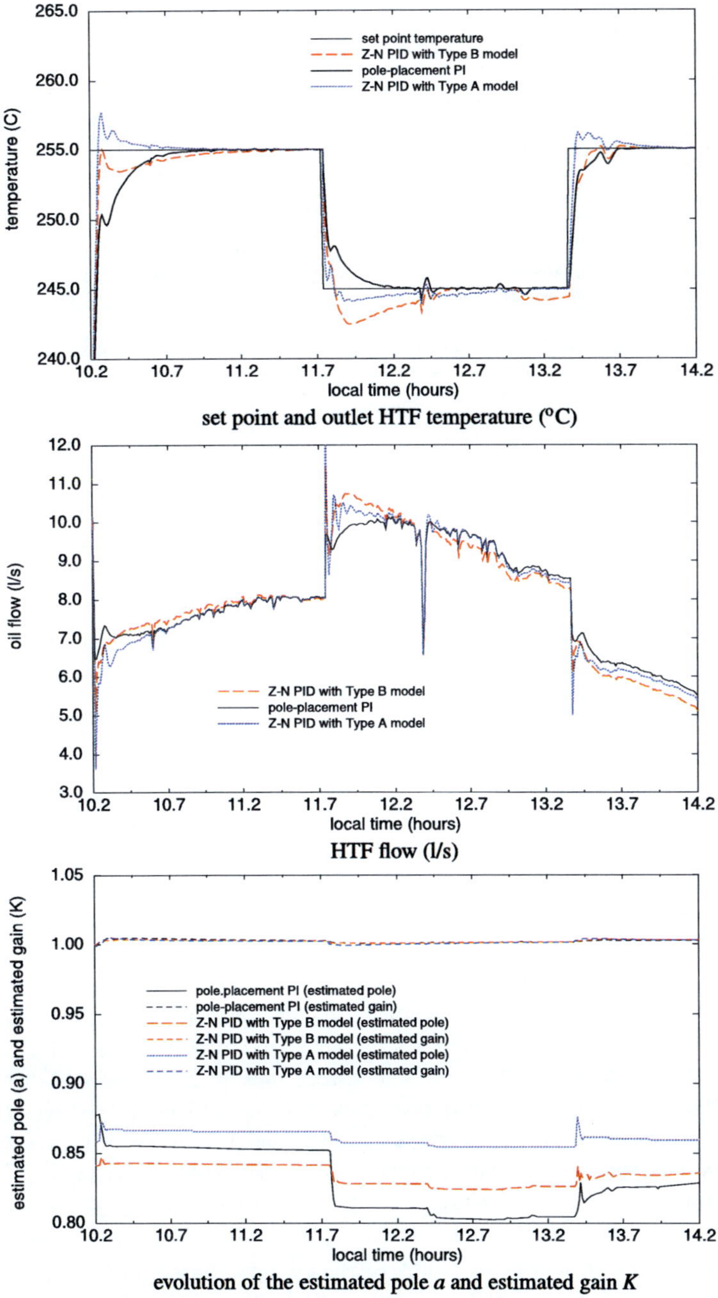

Fig. 5.2 Simulation with adaptive PID controllers [85]

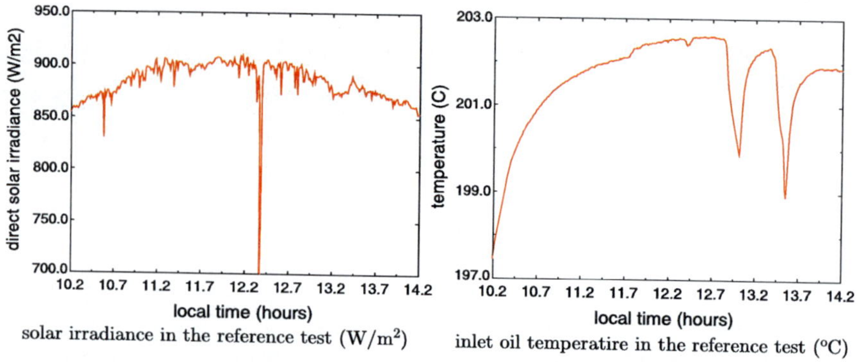

Fig. 5.3 Disturbances used in the simulation of the adaptive PID controllers [85]

As can be observed, at the first two set point changes, the output temperature follows the set point but a peak appears in the response after which the output follows the previous evolution. In most control texts, this *kickback* effect is attributed to a wrong estimation of the gain. However, it has been observed that this effect may not only be due to this but also to the existence of resonance characteristics of the plant controlled. This effect is manifested when a fast response is required from the controlled system. In this case, the first resonance mode is excited, producing peaks and oscillations in the time response.

Figure 5.4 shows plant results using the pole-placement adaptive controller incorporating the series feedforward element. These results are significant since they show performance over a 3-hours test period when the irradiance level is changing substantially, due to scattered cloud conditions. Injected disturbances take the form of step changes in the reference temperature. These generate dynamic information for estimation purposes and the resulting HTF outlet temperature responses also serve to demonstrate the quality of the control performance achieved.

The period from 11:39 am to 12:46 pm (local time) covers the starting-up phase of the plant when HTF is recirculated. After this initial period well-damped step responses are obtained and the HTF outlet temperature is maintained close to the reference values in spite of the significant changes experienced in the irradiance level. This latter feature demonstrates the beneficial effects of the series feedforward element, which not only serves to reduce unwanted variations in the outlet temperature, but also reduces erroneous excursions in the estimated parameters.

This can be seen in the estimated parameter response of Fig. 5.4, in which the estimated parameter (a) is shown to be smoothly adjusted during the test period. This follows the expected variation needed to match the changing system response characteristics. The levels of irradiance and the required outlet temperature determine the HTF flow rate, which, in turn, influences both the speed of response of the system and its time delay. The normalized term $|b_0/(b_0+b_1)|$ will increase with an increase in the direct irradiance level I (the HTF flow increases and the delay time decreases). The response obtained at the plant can be seen to follow the variation

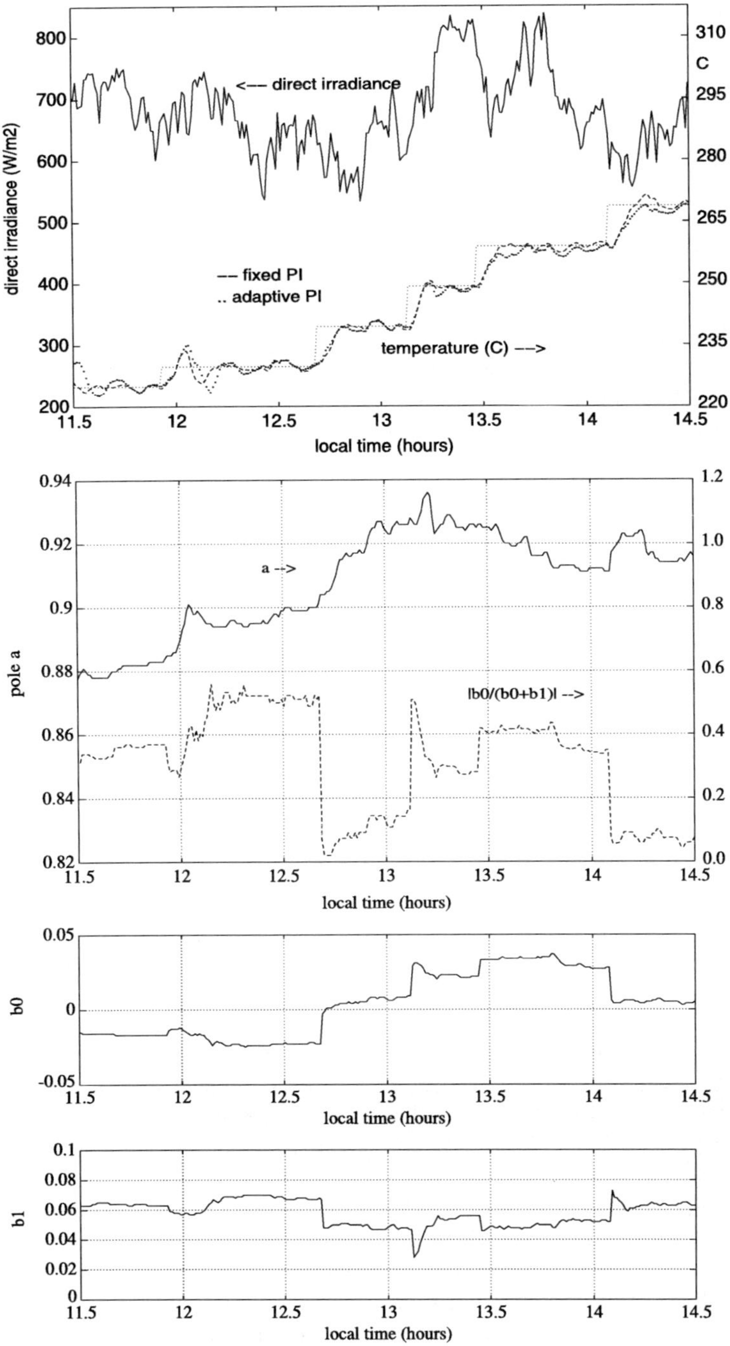

Fig. 5.4 Test with the adaptive PI controller (04/08/91) [82]

at I and to give an indication of how the model adapts to the variation in the time delay.

5.3 Gain Scheduling (GS)

Some controllers have the ability of adapting to changes in process dynamics but are not considered to be proper adaptive controllers. This is the case of gain-scheduling (GS) controllers (Fig. 5.5), where process dynamics can be associated to the value of some process variables that can be measured related to the operating point or to environmental conditions. If the dynamic characteristics of the process can be inferred from measurable variables, the controller parameters can be computed from these variables. Notice that only the inner loop appears in this control structure and the parameter updating can be considered as a sort of feedforward term which changes controller gains. When dealing with DSCF control, due to the presence of resonance modes, conventional discrete transfer functions must be approximately tenth order to represent the resonance dynamics. Scheduling avoids the problems that arise when using AC with high-order models. First steps in this direction were made by employing gain-scheduling control using high-order models of the plant [74, 75, 84, 197, 244, 245, 296] and switched multiple model supervisory controllers [225, 285].

As will be shown in Sect. 5.9.3, where the main results in [39, 84] are summarized, high-order discrete plant auto-regressive models with exogenous inputs (ARX) were used for four operating conditions defined by fluid flow values (because the feedforward in series was used) and the parameters of a generalized predictive controller (GPC) are determined based on this model. The experimental results show quite good reference tracking without an offset of the output temperature. An adaptive gain-scheduled linear quadratic design approach, also based on local linear ARX models depending on HTF flow conditions, was investigated by [295].

A control strategy based on switching between multiple local linear models and controllers was suggested and tested by [308], using the MUSMAR adaptive algorithm in the design of the controllers. The control structure consists of a bank of candidate controllers and a supervisor. Each of the candidate controllers is tuned in order to match a region in the plant operating conditions. The MUSMAR adaptive controller is applied off-line to a plant model corresponding to the operating condition considered. The candidate controller gains are obtained as the MUSMAR convergence gains. The supervisor consists of a shared-state estimator, a performance weight generator and a time switching logic scheme.

In [264], supervised linear quadratic Gaussian (LQG) multicontrollers were developed, where third-order ARX models were identified for six HTF flow conditions. Simpler approaches such as that presented in [391] can also be found, where a PID controller with gain interpolation is developed. In [197] the authors elegantly showed that gain-scheduling can effectively handle plant non-linearities using high-order local linear ARX models that form the basis for the design of local linear controllers using pole placement and flow rate and direct solar irradiance as scheduling

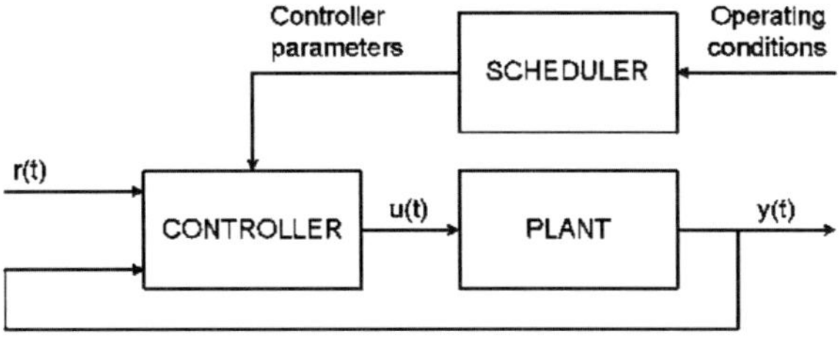

Fig. 5.5 Block diagram of gain-scheduling (GS) controllers

variables. They also pointed out that slow variation of the scheduling variables is a sufficient condition for stability of gain-scheduled control, but the input used for scheduling in the solar plant varies at approximately the same rate as the output, thus stability is proven by experiments. Results in this work are similar to those in [84] and better than those in [295, 308] as more accurate higher-order local models are used while the lower-order models used in [295] and [308] cannot be expected to capture the resonance modes. Another reason is that the local controllers are scheduled on the actual HTF flow rate rather than the predicted steady-state HTF flow rate, as in [295], or local model performance measures, as in [308], both of which correspond to lower bandwidth in the scheduler.

In [296], the application of indirect adaptive control (LQG controller) is shown. The region of operation is split up into five operating points which are represented by five different dynamical linear third-order auto-regressive moving average with exogenous inputs (ARMAX) models to describe the plant characteristics. The actual operating point of the plant is determined by a characteristic value combining several measurements. The algorithm contains an on-line identification procedure to determine and to update the respective model of the operating point. If the operating conditions change slowly, a soft transition between the different operating points is carried out.

In [178] a hierarchical control strategy consisting of the supervisory switching of PID controllers, simplified using the c-Means clustering technique is developed, providing real results. To guarantee good performance at all operating points, a local PID controller is tuned to each operating point and a supervisory strategy is proposed and applied to switch among these controllers according to the actual measured conditions. Each PID controller is tuned off-line by combining dynamic recurrent non-linear artificial neural network (ANN) model with a pole-placement control design.

A control scheme that employs a fuzzy PI controller with feedforward is developed in [360] for the highly non-linear part of the operating regime and the gain-scheduled control over the more linear part of the operating envelope, only showing simulation results based on the model developed in [38]. In [154, 179] a hybrid scheme is presented combining the potentialities of ANN for approximation purposes with PID control. As in other gain-scheduling control schemes, the HTF flow is considered to be the main variable governing the switching of the controller, but to account for the other variables affecting this value, the scheduling variable is obtained from an ANN having as inputs the values of direct solar irradiance, inlet HTF temperature and reference (or outlet) temperatures. Thus, the scheduler implements

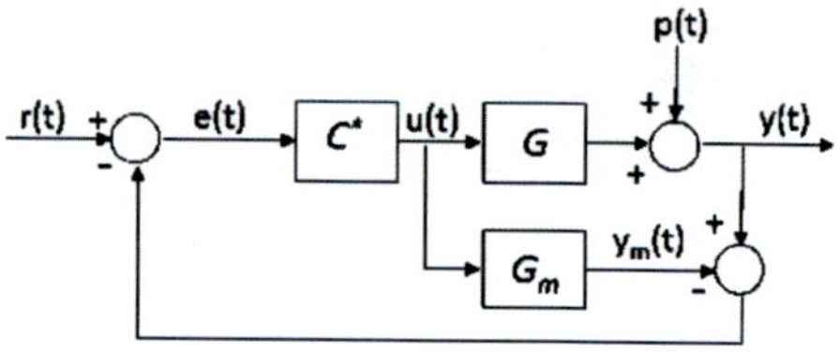

Fig. 5.6 Block diagram of internal model controllers (IMC)

an inverse of the plant at steady state and uses this signal to select the adequate PID controller.

5.4 Internal Model Control (IMC)

The generic structure of an IMC controller is shown in Fig. 5.6. If there are no modeling errors and there are no external disturbances ($p = 0$), the output (y) of the plant coincides with the output of the model (y_m) and as there is no feedback signal, then the controller can be designed in an open-loop manner and the resulting control structure is stable if, and only if, the process is open-loop stable and the controller is also stable. The feedback signal is included to account for uncertainties and disturbances. IMC [260] has the advantages of open-loop (the controller is easy to design) and closed-loop design methods (the feedback signal in this structure represents the uncertainty about the process and the disturbances). In [135, 136] a model based on non-linear partial differential equations (PDE) was developed and used as a part of the control design along with its realizable inverse (static version of the model) as the controller.

5.4.1 An IMC-Based Repetitive Control of DSCF

This section describes the design, based on the models of the resonances described in Chap. 4 of a repetitive-control scheme able to improve control system performance in distributed systems like solar collectors characterized by the presence of resonances [8, 11].

First of all, an IMC PI controller is designed based on an approximate first-order model to show that performance deteriorates due to resonance when fast responses are required and resonance has not been taken into account. Secondly, a repetitive controller structure is used to counteract system resonance dynamics. Repetitive control arises from the model of resonances developed in Chap. 4, Sect. 4.3.4.3 (Eqs. (4.41)–(4.44)), not from the approximate first order model used with IMC PI, as a logical way to cancel out the resonance dynamics. The fact of using a repetitive controller avoids the need for a main controller with an integrator term, as will be seen later. For this reason, the control system includes, as main controller, an IMC PD controller in the feedback loop for tracking step changes at set point without steady-state error. The control system is supplemented by several parallel feedforward controllers which compensate for disturbances, based on Eqs. (4.42) to (4.44).

5.4.1.1 IMC PI Design

As mentioned above, a first-order approximation of the low-frequency system dynamics was used to design an IMC PI controller according to the rules in [260, 315]. Following Fig. 5.6, it is possible to set a relation between a classic feedback controller, C and an IMC controller, C^*:

$$C = \frac{C^*}{1 - C^* G_m}, \qquad C^* = \frac{C}{1 - C G_m} \tag{5.6}$$

Usually, the IMC controller is specified as

$$C^* = \frac{1}{G_m{}^-} F \tag{5.7}$$

where $G_m{}^-$ is the model invertible part and F is a low-pass filter usually of the form:

$$F = \frac{1}{(\varepsilon s + 1)^n} \tag{5.8}$$

The fact of augmenting the IMC controller by a low-pass filter ensures closed-loop robustness if the filter parameter, ε, is adequately chosen based on the model uncertainty (mainly at high frequencies). Moreover, the filter parameter can be adjusted on-line if necessary. The complexity of C^* and, similarly, the complexity of the equivalent classic controller C is determined by the complexity of the model. So, simple models give rise to simple controllers. In particular, for several simple models the IMC design procedure yields classic PI and PID controllers [315].

Two issues should be highlighted when applying this technique to control DSCF and, in particular, the ACUREX plant, (i) the manipulated variable is not the fluid velocity but the volumetric flow. However, it is not difficult to change Eq. (4.41) to another are relating output fluid temperature to volumetric flow by dividing by the tube cross area, A_f. (ii) There is a small delay between output fluid temperature and volumetric flow in the experiments due to the position of the temperature sensor in the field, which introduces a transport delay, although this is relatively small compared to the fundamental time constant. To design an approximate model, Eq. (4.41) is rewritten in standard form as

$$G(s) = \frac{\tilde{T}_f(L,s)}{\tilde{q}(s)} = \underbrace{K_a \frac{-\beta_a s + 1}{s(\tau_a s + 1)} e^{-t_d s}}_{P(s)} \underbrace{\left[1 - e^{-t_r s} \underbrace{\left(\frac{-\beta s + 1}{\tau s + 1}\right)}_{R'(s)}\right]}_{R^*(s)} \tag{5.9}$$

where it is easy to see from Eq. (4.41) that

$$K_a = -\frac{1}{A_f} \frac{T_{a_s} - T_{f_{so}} + I_s \gamma \tau_2}{c} e^{\frac{-L}{c}} \frac{\tau_1 (\tau_{12} + \tau_2)^2}{\tau_1 (\tau_{12} + \tau_2)^2 + \tau_{12} \tau_2^2} \tag{5.10}$$

$$\beta_a = -\frac{\tau_{12} \tau_2}{\tau_{12} + \tau_2} \tag{5.11}$$

$$\tau_a = \frac{\tau_{12} \tau_2 \tau_1 (\tau_{12} + \tau_2)}{\tau_1 (\tau_{12} + \tau_2)^2 + \tau_{12} \tau_2^2} \tag{5.12}$$

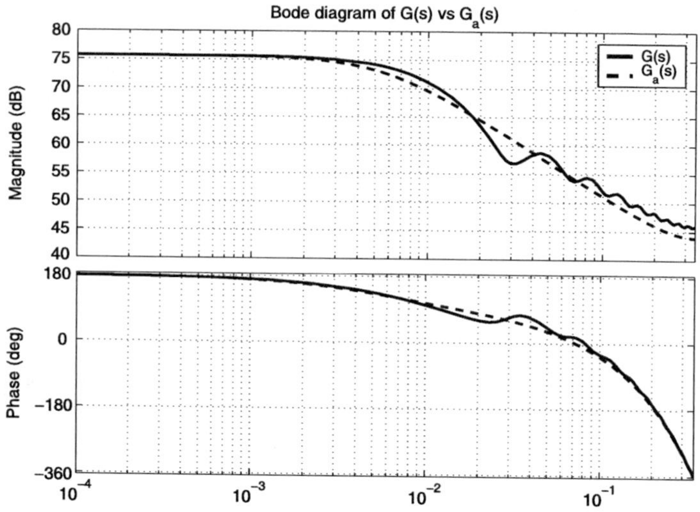

Fig. 5.7 Comparison of the Bode diagram of $G(s)$ model versus its approximated first-order model $G_a(s)$ (courtesy of J.D. Álvarez et al., [11])

$t_r = L/v_s$, $t_d \approx 30s$, $P(s)$ is a low-order transfer function without including resonances and $R^*(s)$ is a transfer function which models the resonance modes.

The variable t_d is not really a fixed delay, but inversely proportional to fluid velocity, although its variations are very small when compared to the fundamental time constant. The value chosen is the mean calculated from model steady-state conditions. Therefore, the approximate first-order model must have the same gain as the gray-box model, given by

$$\lim_{s\to 0} G(s) = K_a[t_r + \beta + \tau] \tag{5.13}$$

From these results, it can be deduced that the gray-box model gain depends on fluid residence time t_r. Moreover, the response time of the gray-box model also depends on t_r. In other words, as $t_r = \frac{L}{v_s}$, the HTF flow value establishes a "natural" time scale for the system, this fact was previously pointed out in [351]. Consequently, the approximate first-order model chosen for control in the ACUREX plant taking the volumetric flow rate as control input is

$$G_a(s) = \frac{K}{\tau s + 1} e^{-t_d s} = \frac{-6.55}{174s + 1} e^{-30s} \tag{5.14}$$

Figure 5.7 compares Bode plots for $G(s)$ and $G_a(s)$, in which it may be observed that the approximate function $G_a(s)$ can model the low-frequency dynamics of $G(s)$, but cannot capture resonance dynamics at medium or high frequencies.

In [260, 315] there are several design rules to calculate IMC PI and IMC PID controllers based on model parameters. The IMC PI controller parameters were calculated using the model in Eq. (5.14) and row "*A*" in [315], this row establishes that $K_P = \tau/K\varepsilon$ and $T_I = \tau$, where K and τ are first-order system characteristic

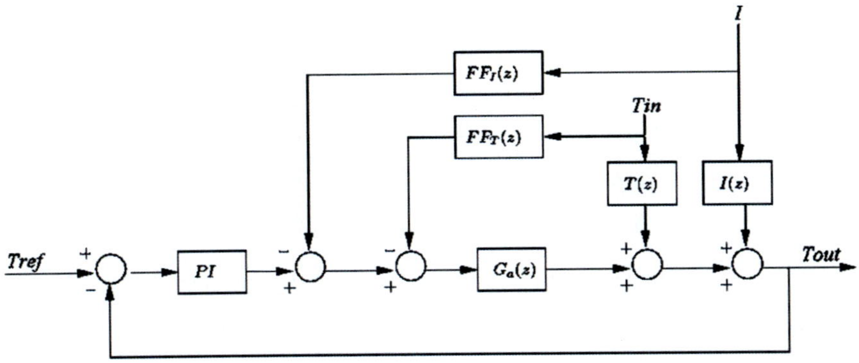

Fig. 5.8 Control system structure with IMC PI plus feedforward controllers (courtesy of J.D. Álvarez et al., [11])

Table 5.1 IMC PI controller and model parameters (courtesy of J.D. Álvarez et al., [11])

IMC-PI	K [°C s/l]	τ [s]	ε	K_P [l/(s °C)]	T_I [s]
1st controller	−6.55	174	1330	$-2 \cdot 10^{-2}$	174
2nd controller	−6.55	194	1480	$-2 \cdot 10^{-2}$	194
3th controller	−6.55	174	2655	$-1 \cdot 10^{-2}$	174
4th controller	−6.55	174	1770	$-1.5 \cdot 10^{-2}$	174

parameters and K_P and T_I are the controller gain and integral time parameters of a classic PI controller. Several IMC PI controllers were developed by changing the filter parameter ε and performing two ±10°C step changes in the reference. Notice in Eqs. (4.42) to (4.44) that there are several disturbance variables which influence the outlet HTF temperature and, once more, fast changes in solar energy excite resonances causing oscillation. For this reason, classical parallel feedforward controllers were designed based on transfer functions from Eqs. (4.42) to (4.44), in the form of $D(s)/G(s)$ where $D(s)$ is the transfer function relating outlet temperature changes to disturbance variable and $G(s)$ is the transfer function relating outlet temperature changes to control variable changes. Due to the non-causal nature of the feedforward controllers obtained, a causal version of the resulting controllers was implemented. These parallel feedforward controllers were used along with the IMC PI controller to counteract possible changes in the disturbance variables, mainly solar irradiance and fluid inlet temperature, during operation; disturbances generated by ambient temperature are insignificant and can be neglected. This control system structure is shown in Fig. 5.8.

Four different IMC PI controllers were developed based on Eq. (5.14) and different ε and τ (model parameters) were used to achieve fast responses. Table 5.1 summarizes IMC PI controller and model parameters used during operation.

Figure 5.9 shows real results from the ACUREX plant. The IMC PI controllers designed stabilize the system, but with oscillation and up to 20% overshoot in the

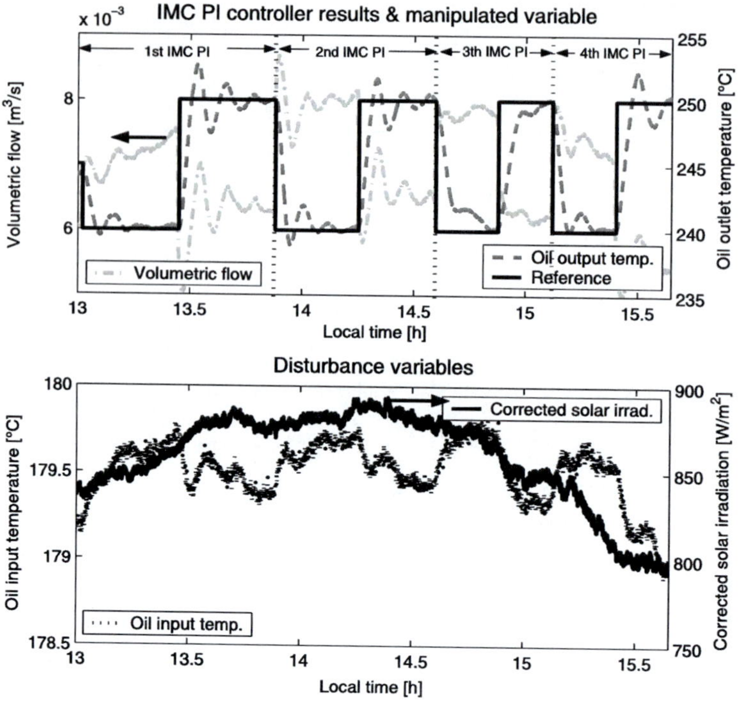

Fig. 5.9 IMC PI controllers representative results (courtesy of J.D. Álvarez et al., [11])

1st and 4th controllers and a settling time of 15 to 20 min depending on the case. If overshoot is not desirable, ε must be increased as in the 3rd controller, which reduces the settling time by up to 15 min, but even in this case, the closed-loop system has an oscillatory behavior and kickback typical of resonance.

From the results, it can be seen how the models developed capture system features. If the reference is increased, the controller must decrease the volumetric flow and, therefore, the fluid velocity is diminished to reach the new set point. This fact causes residence time t_r to increase and, as Eq. (5.13) shows, the system gain increases as well. At the same time, the increment in the residence time causes resonance dynamics at lower frequencies, amplifying oscillatory system behavior.

Obviously, to avoid the oscillatory behavior, the IMC PI controllers could be detuned, but the aim of this section is to show how the model developed can be used to achieve well-damped, closed-loop responses with broad bandwidths.

5.4.1.2 Repetitive Control-Based Scheme

Repetitive control has demonstrated its ability to counteract resonance dynamics by treating them as internal system disturbances [8, 307]. Figure 5.10 shows the scheme for a typical discrete repetitive controller, where z^{-d} is a discrete-time delay of d

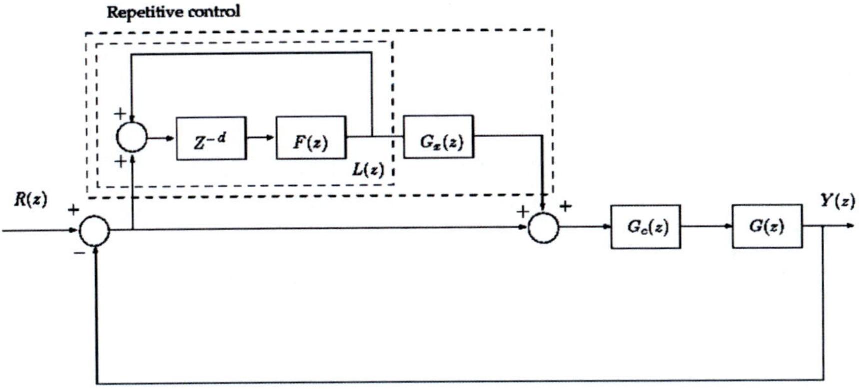

Fig. 5.10 Repetitive-control scheme (courtesy of J.D. Álvarez et al., [11])

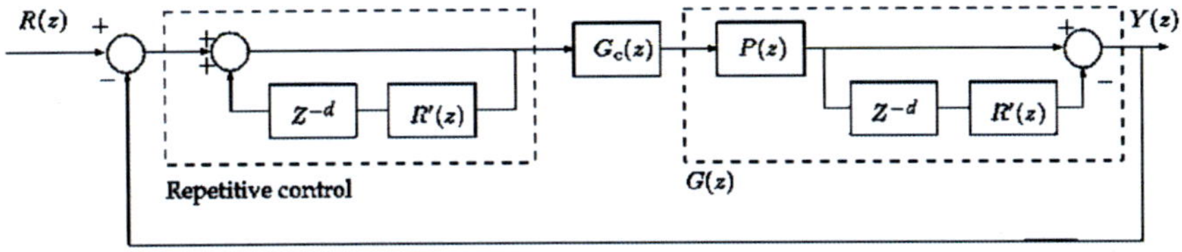

Fig. 5.11 Simplified repetitive-control scheme (courtesy of J.D. Álvarez et al., [11])

sampling periods, $F(z)$ is a low-pass finite impulse response (FIR) filter with a null period, which attempts to reduce uncertain frequency gains, $L(z)$ is the internal repetitive-control loop transfer function, $G_x(z)$ is a linear filter to ensure system stability, $G_c(z)$ is the main controller and $G(z)$ is the system to be controlled.

A more detailed description of such control systems is provided in [122]. Suitable choices for filters $F(z)$ and $G_x(z)$ can be found in [8], such that the repetitive-control transfer function is equal to the discrete form of the inverse transfer function in brackets in Eq. (5.9). In a perfect model, the repetitive controller is able to cancel the term in brackets in Eq. (5.9) completely and, thereby, the resonance dynamics of the system. Figure 5.11 shows the scheme resulting from a suitable choice of filter parameters for canceling out the resonance dynamics.

Although a perfect model is impossible and total resonance cancelation cannot be achieved, it is possible to damp it; the main controller needs only to concentrate on $P(s)$, in Eq. (5.9), which is a low-order transfer function with an integrator plus a small dead-time. For this reason, an IMC PD was chosen as the main controller, unlike Sect. 5.4.1.1 in which an IMC PI was the main controller because the simplified plant model $G_a(s)$ in Eq. (5.14), was used. In this case row "P" in [315] was used, this row establishes that $K_P = 1/K_a(\beta + \varepsilon)$ and $T_D = \tau_a$, where K_a and τ_a are characteristic parameters of $P(s)$.

The repetitive-controller uncertainty, is taken into account when filter parameter, ε, of the IMC PD is adjusted. In this way no additional filter is necessary. Moreover,

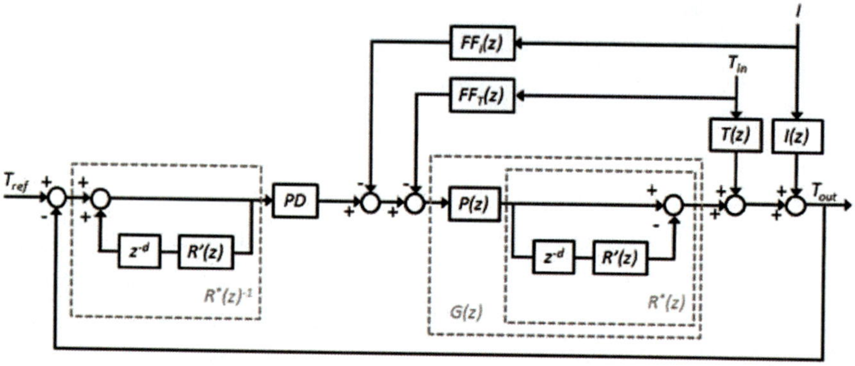

Fig. 5.12 Control system structure with IMC PD plus repetitive controller (courtesy of J.D. Álvarez et al., [11])

in [260] it is stated that the correct choice of the filter parameter ensures closed-loop robust stability against model uncertainty; at the same time, closed-loop performance is increased by means of a repetitive controller. As has been pointed out previously, the absence of an integral term in the main controller is justified by the integral term in the transfer function $P(s)$ in Eq. (5.9) if step changes are performed at the set point.

Furthermore, to avoid problems due to actuator saturation, the repetitive controller incorporates anti-windup action, consisting of adjusting repetitive-controller output by means of 'actuator error' (the difference between the actual actuator output and the desired control action) modulated by a parameter N_r/K_o, K_o being the static gain of $G(z)$ transfer function and N_r is a parameter that modulates the velocity at which the repetitive controller is reset (when saturated, the repetitive controller acts like an integrator). As in the IMC PI controller, the control system was supplemented with classical feedforward controllers in order to compensate for both solar irradiance and fluid inlet temperature, based on Eqs. (4.42), (4.43) and (4.44), as explained in [8, 11, 12]. Figure 5.12 shows the control system structure consisting of a repetitive-control block, an IMC PD controller in the forward path and the parallel feedforward controllers, whereas Fig. 5.13 shows the results achieved when the ACUREX plant is controlled by this control system configuration.

The control system was tested using ±10°C step changes in the reference. Results shown in Fig. 5.13 are better than those with an IMC PI controller. There is only one overshoot of less than 15% and the settling time is reduced to 7–9 min. Notice that the control signal oscillates as it attempts to avoid overshoots in the controlled signal. Theoretically, the frequencies of these oscillations should match those of the resonance modes. However, although results are better, they are not as good as expected. Plant behavior is slightly oscillatory and non-linear, as can be observed in the plant response to the two step changes. This is because the manipulated variable dictates the resonance dynamics frequencies and so they are not totally canceled out.

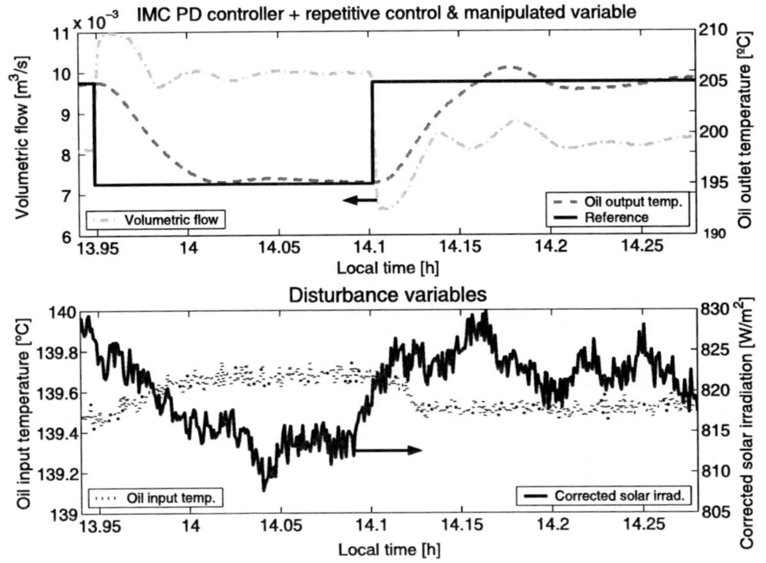

Fig. 5.13 Representative repetitive-controller results (courtesy of J.D. Álvarez et al., [11])

5.4.1.3 Adaptive and Robust Versions of the Basic Control Scheme

Further improvements of the basic algorithm presented can be found in [11] where an adaptive version of the basic algorithm is developed and implemented. In the above paragraphs, resonances are dealt with using a linear repetitive controller with a fixed frequency, but the resonance frequency modes change at the same time as the volumetric flow or velocity does because they are associated with fluid residence time inside the collector. In [242] the physical mechanism behind the resonance phenomena is explained. Moreover, an estimation of the frequencies in which the resonance modes are located is provided. So, as treated in Sect. 4.3.3.2, the minimum resonance modes are located around frequencies $\omega_{min} = 2n\pi/t_r$, whereas the maximum resonance modes are located around frequencies $\omega_{max} = 2(n+1/2)\pi/t_r$, where n is a positive integer and t_r is the residence time of the fluid within the tube.

In solar collectors, the fluid output temperature can only be controlled by fluid velocity. This makes the resonance mode frequencies change with the control signal values. One way to cope with this problem is to implement an adaptive repetitive-control scheme which changes its working frequency according to fluid residence time. However, the fluid residence time is a parameter which is not easy to calculate, under the assumptions that the fluid is incompressible and its specific heat capacity and density are constant and the thermal resistance of the tube wall is neglected, it is possible to estimate a residence time value from the volumetric flow which is measured. So $t_r = L/v$ where L is the pipe or collector length and v is the fluid velocity, which, at the same time, is estimated by $v = q/A_f$ where q is the volumetric flow and A_f the cross-sectional area for flow inside the pipe.

Two main repetitive-control techniques have been reported to deal with variable frequency disturbances [57, 93, 358]. In general, controller design is based on multirate concepts but, while the main controller works with an invariable sample time (T_s), the repetitive controller can use either the same sample (T_s) or a variable sample time. In both cases, the general adaptive repetitive-controller development methodology is:

1. Estimate the frequency of either the repetitive disturbance signal to reject or repetitive reference to track.
2. Decimate and interpolate the samples necessary for repetitive control with either a variable or fixed sample time.
3. Calculate the control signal.
4. Interpolate and decimate the control signal to the fixed main controller frequency.

Following this methodology, an adaptive repetitive control with a fixed sample time was designed in [11], providing real results when tested at the ACUREX field of the PSA.

A robust approach was developed and also tested at the ACUREX DSCF in [12], where an alternative structure of the repetitive controller is proposed which also works with a fixed frequency but is tuned taking into account the uncertainty in the resonance mode frequencies in order to achieve fast, well-damped system responses.

5.4.2 *Adaptive Frequency-Domain IMC*

Another approach to cope with resonances within an IMC framework are those developed in [36, 37, 85], where an adaptive frequency-domain IMC controller is developed to cope with resonances, both the plant and the controller using frequency-domain interpolating models. In the frequency-based IMC control structure, the error signal is fed to a bank of comb filters (CF) where it is separated into its spectral components. Each of the components is then multiplied by a complex gain and fed to the process and model. The calculation of the complex gain can be made by imposing a determined frequency response of the closed-loop system at the interpolating points. As the frequency response of the plant is needed at all interpolating points, banks of band filters are used for the input and the output of the plant, such that simple models of the plant can be used in each interpolating frequency, using a common recursive least squares (RLS) identification algorithm to estimate only two parameters describing these models (using a regressor in each band).

Figure 5.14 outlines the method that is fully developed in [85] and not included here to save space. For each frequency ω_i, the frequency response of the compensated system at the interpolating points (G_d) is imposed to fulfill the closed-loop specifications and, thus, the controller frequency response is obtained from $C_i(\omega_i) = G_d(\omega_i)/G_{mi}(\omega_i)$, $i = 0, \ldots, L$, where $G_{mi}(\omega_i)$ is obtained from an interpolation model with two parameters c_i and d_i that are identified using a standard RLS algorithm.

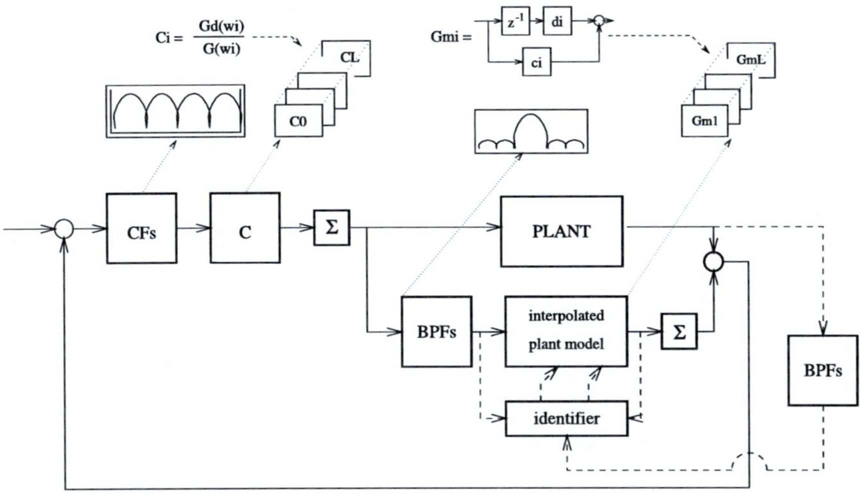

Fig. 5.14 Adaptive-control scheme in the frequency domain, [37]

The scheme includes the design of the controller in the frequency domain, frequency response estimation, banks of bandpass filters and time-varying controller. All the development and results achieved at the ACUREX field can be found in [85].

The main advantages are that the estimation is concentrated in frequencies used for design purposes, the disturbances with frequency components out of these bands are filtered, it is not necessary to over excite in cases with processes of high-order because only a few points of the Nyquist curve are estimated, low-order models can be used to control high-order processes, the order of the process need not be known, and the control scheme can respond to changes in plant dynamics.

The main drawbacks that have been encountered in the control scheme are that the existence of resonance characteristics causes the identification of the frequency response of the plant close to these modes to reach very low values. Due to the fact that the control scheme obtains the control signal using the inverse of the estimated plant, very high values of the control signal can be obtained, thus requiring a supervising scheme. On the other hand, the choice of the transfer function which provides the desired behavior of the controlled system and, particularly, the choice of the desired magnitude and phase values of the system near the resonance modes is critical, leading to a different kind of behavior in function of the chosen values.

5.5 Time Delay Compensation (TDC)

Time delay compensation (TDC) schemes (Fig. 5.15) aim at designing controllers without taking the pure delay of the plant into account and obtaining a closed-loop response after the delay to be that expected from the design without considering the delay in the dynamics of the process. Dead-times appear in many industrial processes, usually associated with mass or energy transport, or due to the accumulation

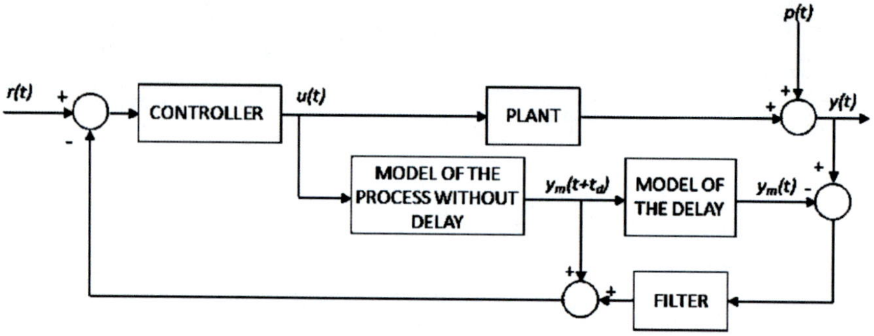

Fig. 5.15 Block diagram of time delay compensators (TDC)

of a great number of low-order systems. Dead-times produce an increase in the system phase, therefore decreasing the phase and gain margins. When using classical methods to control processes where the effect of dead-time is dominant, the controller must be detuned so as to achieve stability, providing a slow behavior. In these cases, it is convenient to use a dead-time compensator (DTC). The Smith Predictor (SP) is perhaps the best known and most widely used algorithm for dead-time compensation. With this structure (Fig. 5.15 without the filter block), if there are no modeling errors or disturbances, the error between the current process output and the model output will be null and the controller can be tuned as if the plant had no dead-time.

Several TDC schemes have been developed for controlling DSCF. In [246], an alternative-control scheme using a simplified transfer function model including the resonance characteristics has been developed. This controller adopts a parallel control structure similar to that of a SP. This parallel structure is shown to effectively counter the resonance dynamics of the system while avoiding the excitable control signal of the controller developed by [243, 244].

In [274], an easy-to-use PI controller with dead-time compensation that presents robust behavior and that can be applied to plants with variable dead-time is applied to control the temperature entering the storage tank of the ACUREX field. The formulation is based on an adaptive SP structure plus the addition of a filter acting on the error between the output and its prediction in order to improve robustness. The implementation of the control law is straightforward as the controller has only three tuning parameters that can be tuned using a classical step identification test and the filter needs no adjustment. In this approach, the system is described by a first-order plus dead-time (FOPDT) transfer function $G(s) = Ke^{-t_d s}/(1+\tau s)$. With the addition of the PI, the resulting controller has therefore five tuning parameters: the three plant coefficients K, τ and t_d and the two PI parameters K_P and T_I. This can constitute an increase in operational complexity compared with a PID controller, so in [171] a predictive PI controller (PPI) is proposed with only three adjustable parameters, choosing the controller gain equal to the inverse of the process gain and the integral time equal to the process time constant. This results in a simple controller that has the same advantages as the PID controllers since it can easily be tuned manually and it is also suitable for the control of processes with long dead-times. However, this formulation does not take the robustness of the closed-loop system into account and presents stability problems when uncertainties exist in the plant parameters, mainly in the dead-time.

A robustness analysis of the PPI can be found in [273], where the authors have shown that the PPI controller has undesirable closed-loop behavior when dead-time uncertainties (high-frequency modeling errors) are considered. Some modifications to the PPI controller were introduced to improve robustness: a filter $F(s)$ with unitary static gain ($F(0)=1$) was introduced (filter block in Fig. 5.15), maintaining the structure in [171]. Filter F can be used to improve the robustness of the system at the desired frequency region. If it is desirable to maintain the simple structure of the PPI in order to use the filtered PPI (FPPI) controller in industrial applications, then filter F will be defined as a first-order filter with only one parameter (the time constant τ_f) related to t_d. The relation between τ_f and t_d was computed in [273] as $\tau_f = t_d/2$. This choice of τ_f gives good results for error in the dead-time of up to 30%. However, if the changes in the dead-time are greater than 50% of the nominal value and robust closed-loop performance specifications are defined, the FPPI will not give a satisfactory solution. As in most industrial cases, high varying dead-times are associated to the transport of mass or flow and so an adaptive scheme can be used to improve the performance of the FPPI.

This is the case of DSCF in some applications. As was mentioned in Chap. 4, in order to ensure safe operation, the controlled variable in this kind of plant is usually the outlet temperature of the loop with the highest value at each sampling time. Nevertheless, situations may exist in which hot HTF should be used for directly feeding the heat exchanger of the electricity generation system or a desalination plant. In these situations, it is preferable to control the temperature of the HTF entering the top of the thermal storage tank. In this way, the considerable length of the tube joining the output of the collector field to the top of the tank introduces a large variable delay within the control loop which depends on the value of the flow. If the system is modeled as a FOPDT process, part of the dead-time can be used to model the effect of high-order dynamics and the other part correspond to the actual transport of fluid. Thus, the estimated dead-time t_d can be computed as $t_d = t_{d0} + g(q)$. It is assumed that t_{d0} has only small variations around its nominal value (less than 30%) and the FPPI can cope with the effects of these model uncertainties. The relation between the flow and the dead-time, $g(q)$, is given by Eq. (4.14), where $L \approx \frac{T_s}{A_f}\sum_{i=0}^{i=n-1} q(k-i)$, $q(k-i)$ being the flow rate demanded by the ith previous sampling time, so that $t_r \approx nT_s$ and the dead-time of the process can be estimated as $t_d = t_{d0} + nT_s$.

Notice that: (i) the digital implementation of the previous relationship is very simple and does not need any tuning parameter; (ii) the complete tuning procedure for the controller is the same as in the FPPI and only the values of L and A_f must be given for the adaptive part of the control law; (iii) in general, the estimated value of t_d will be different from the real dead-time because of approximation errors, but the filter can cope with these modeling errors.

In order to use the distributed-parameter model given by Eqs. (4.1) and (4.2) to account for longer passive tube lengths for comparing different DTC control

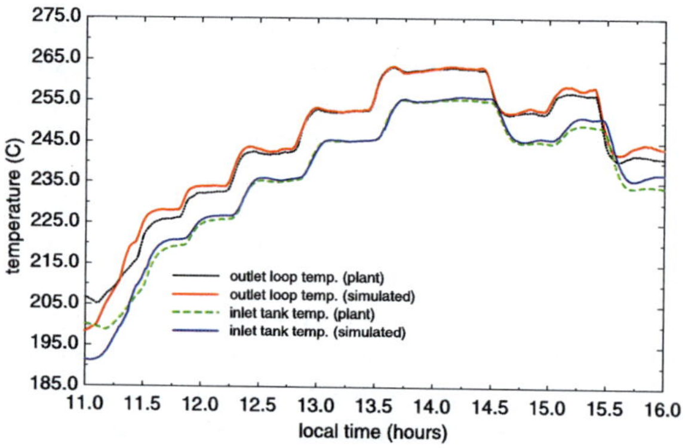

Fig. 5.16 Validation of the non-linear model accounting for transport delays (courtesy of J.E. Normey-Rico et al., [274])

schema, it has been taken into account [309] that the dynamic characteristics of the tube joining the output of the loops to the top of the storage tank are given by a gain of less than one, a time constant (of about 2 or 3 min) and a variable delay. This approximation was adopted in order to modify the basic formulation of the non-linear model to account for dynamic characteristics introduced by the tube. The modified model was validated with data obtained at the plant in closed-loop operation [309]. Figure 5.16 shows representative results obtained in closed-loop operation at the ACUREX field. Both the outlet temperature of the collector field and the inlet temperature at the top of the storage tank are shown and compared with those obtained from the non-linear modified model. As can be seen, results are quite approximate (mostly around solar midday) and so the model has been used as an appropriate test-bed for dead-time compensation control schema.

To illustrate the performance of the FPPI controller some simulation results are shown. When the controlled variable is the temperature of the HTF entering the top of the storage tank (T_{st}), classical PID control schema performance deteriorates, as the dead-time dynamics become dominant. One way to cope with this problem is to detune PID controllers to obtain stable operation in the whole range of operating conditions and with a bounded set of possible values of the varying delay, providing sluggish responses. In [274] results using a detuned PID controlling T_{st} are provided, where performance deteriorates at low flow conditions.

The performance of classical PID controllers can be improved by including some kind of dead-time compensation. The complete control structure proposed for the adaptive FPPI is shown in Fig. 5.17 for a digital implementation, where $a = e^{-T_s/\tau}$, $b = K(1-a)$; $a_f = e^{-T_s/\tau_f}$ and d is the discrete-time representation of the dead-time t_d. The identification of step tests has been used to obtain different models of the plant (including the series feedforward controller) without the varying dead-time for different operating points. In the application shown in this section, the model for the predictor without the varying dead-time has been chosen as $G_n(s) = \frac{1.2e^{-78s}}{1+320s}$ and so the control parameters were: $K_P = 0.833$, $T_I = 320$ s and $t_{d0} = 78$ s. The sampling time was chosen as $T_s = 39$ s and the measurement of the equivalent length and section of the tube (which at the solar plant has different sections at different parts of the tube) gives the following values: $L = 200$ m

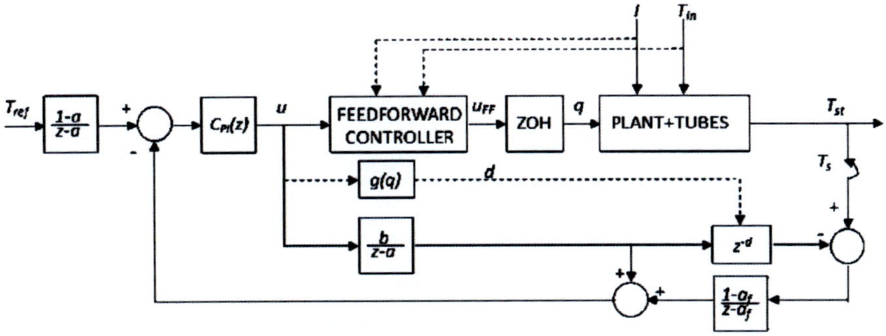

Fig. 5.17 Structure of the filtered PPI (courtesy of J.E. Normey-Rico et al., [274])

and $A_f = 5.30929 \cdot 10^{-4}$ m^2. The time constant of the filter (τ_f) was chosen according to the value of the estimated delay. In Fig. 5.18 the results obtained with the basic design of the FPPI are presented. As can be seen, the controlled system accounts for the (estimated) varying dead-time (which varies according to the HTF flow changes) and ensures acceptable, good operation under the whole range of working conditions. It must be noted that the tuning of the adaptive FPPI is simpler than the procedure used to obtain the detuned PID and gives better performance and robustness.

To analyze the effect on performance when dead-time estimation errors occur, a non-adaptive FPPI controller was implemented by using two extreme fixed values of the dead-time: one near the maximum possible value and other near the minimum one, obtaining the results shown in Fig. 5.19. As can be seen, the performance deteriorates but the control scheme can cope with the operation covering a wide range of operating conditions without leading the system to instability, proving the robustness inherent in the design of the controller.

5.6 Optimal Control (LQG)

In the framework of controlling DSCF, [325] and [280] suggested an optimal control formulation where the objective is to maximize net produced power when the pumping power is taken into consideration. An alternative approach is taken by [99, 100], where a quadratic control Lyapunov function is formulated for the distributed-parameter model and a stabilizing control law is derived. In [330], a LQG/LTR controller using the series feedforward controller explained in Chap. 4 was developed to obtain a robust fixed-parameter controller, able to acceptably control the distributed solar collector field under a wide range of operating conditions. A similar approach to that of [100] was developed by [195, 196], but relies on using a storage function with a physical interpretation leading to a conceptually simpler stabilizing control law with more transparent tuning parameters and less involved analysis. In [388, 389] a finite dimensional approxima-

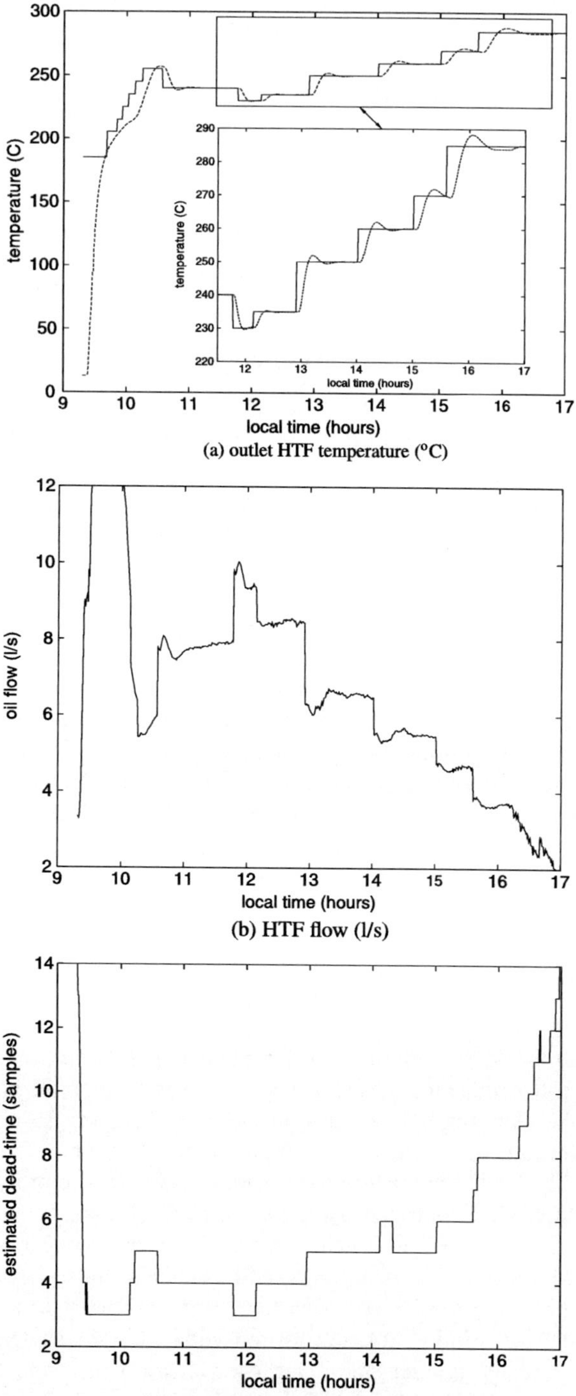

Fig. 5.18 Results obtained with the nominal FPPI controller (courtesy of J.E. Normey-Rico et al., [274])

(a) outlet HTF temperature (°C)

(b) HTF flow (l/s)

(c) estimated dead-time [sampling times]

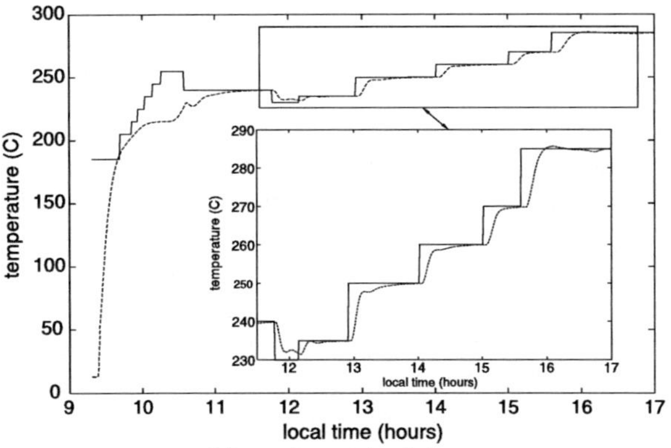

(a) maximum dead-time

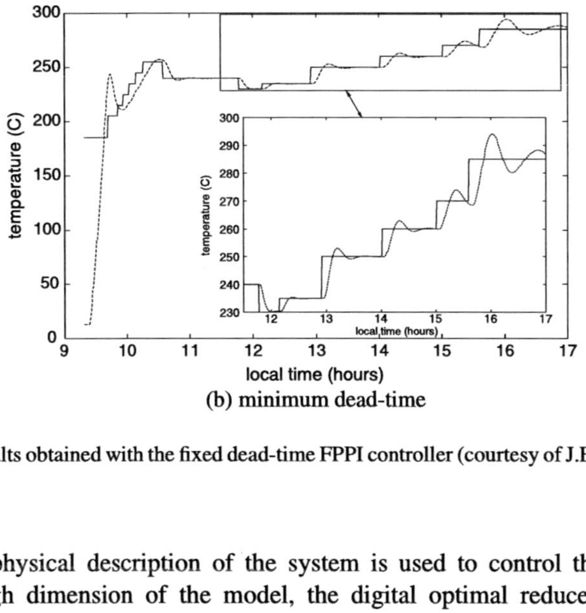

(b) minimum dead-time

Fig. 5.19 Results obtained with the fixed dead-time FPPI controller (courtesy of J.E. Normey-Rico et al., [274])

tion of the physical description of the system is used to control the plant. Despite the high dimension of the model, the digital optimal reduced-order controller design procedure enables an optimal controller with just one state variable to be synthesized. The synthesis of the digital optimal reduced-order controller takes place at two levels. At the first level, based on the full non-linear model representing the finite-difference approximation, the associated initial state and the cost function to be minimized, a digital optimal control is computed as well as the associated state trajectory and the output trajectory, this computation being performed off-line. At the second level, based on the linearized dynamics about the digital optimal control and state trajectory computed at the first level, a quadratic cost function and a description of the model and measurement uncertainties by small additive white noise, a digital optimal reduced-order LQG compensator is computed. This compensator is used to attenuate on-line errors.

5.7 Robust Control (RC)

In the previous section, some aspects of the application of a robust optimal LQG/LTR control scheme [330] have been commented on. Robust control tries to apply principles and methods that allow the discrepancies between the model and the real process to be explicitly considered. There are many techniques for designing feedback systems with a high degree of robustness, some of which are commented on in [85] in the scope of the control of solar plants. In [282], a controller based on the H_∞ theory was developed and successfully tested at the ACUREX field. The approach used in this work takes advantage of the assurance of high stability margins in the face of a norm bounded perturbation and uses the feedforward controller in series with the developed controller. One particular approach to robust control design is the so-called sliding mode control methodology, which is a particular type of variable structure control system (VSCS), which is characterized by a suite of feedback control laws and a decision rule (termed the switching function) and can be regarded as a combination of subsystems where each subsystem has a fixed control structure and is valid for specified regions of system behavior. In sliding mode control, the VSCS is designed to drive and then constrain the system state to lie within a neighborhood of the switching function. In [292], the application of three predictive sliding mode controllers is presented (reaching law approach, equivalent control approach and a non-linear GPC law that forces the reachability of a differential predicted surface) using a first-order plus dead-time model for controller tuning purposes. Thanks to a predictive strategy, these control laws provide optimal performance, even in presence of constraints that cannot be considered in the classical sliding modes theory.

5.7.1 *QFT Control (QFT)*

Based on the quantitative feedback theory (QFT), in [104] a robust controller was developed for the ACUREX field incorporating the series feedforward controller to solve a simplified problem in which a non-linear plant subjected to disturbances is treated in the design as an uncertain linear plant with only one input (the reference temperature to the feedforward controller). The frequency response of the plant was analyzed for different operating points using dynamic tests performed on the simulator of the field [38]. In this way, in the process design of the QFT controller, the disturbances affecting the plant are not explicitly taken into account, as the feedforward term partially compensates them. A set of plant models was obtained and used to design a robust controller to maintain the desired response within specific frequency-domain bounds and taking into account the uncertainties of the system in the process design. This work was improved in [111] by incorporating an automatic set point generator plus anti-windup to avoid the system entering into saturation. For each operation point and using the series feedforward, a model of the plant is estimated, obtaining a set of plant models with the same structure where all the operating points are included as the parameter values vary from one model to another

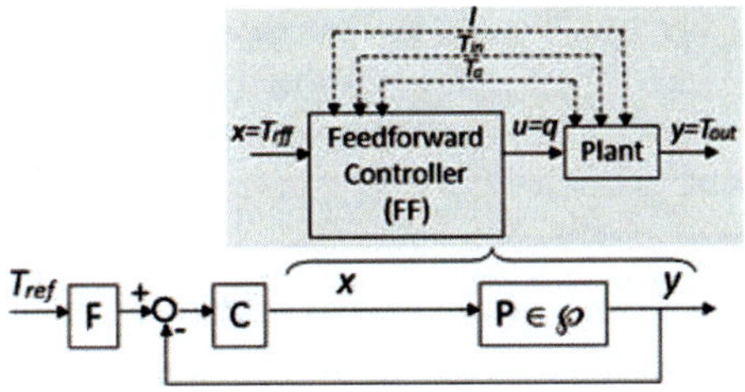

Fig. 5.20 A 2DoF control system with uncertain plant including the series feedforward controller [111]

so that the ACUREX field model can be expressed as a model with uncertain parameters (as they depend on the flow rate). Therefore, in order to control the system with a fixed-parameter controller, the following model has been used:

$$\wp(s) = \left\{ P(s) = \frac{K\omega_n^2}{s^2 + 2\zeta\omega_n s + \omega_n^2} e^{-t_d s} : \zeta = 0.8, \right.$$

$$\left. t_d = 39s,\ \omega_n \in [0.0038, 0.014] \text{ rad/s},\ K \in [0.7, 1.05] \right\} \quad (5.15)$$

where the chosen nominal plant is $P_0(s)$ with $\omega_n = 0.014$ [rad/s] and $K = 0.7$.

The robust control technique QFT [182] was used to obtain a controller fulfilling a set of performance and stability specifications, so that all the uncertainty in the process is taken into account in the design process. This technique allows robust controllers to be designed which fulfill some minimum quantitative specifications taking into account the presence of uncertainties in the plant model. The Nichols chart is used to achieve the desired robust design over the specified region of plant uncertainty where the aim is to design a compensator $C(s)$ and a prefilter $F(s)$ (if necessary) according to Fig. 5.20, so that performance and stability specifications are achieved for the family of plants.

Thus, once the uncertain model has been obtained, the specifications must be determined in time domain and translated into the frequency domain for the QFT design. In this case, the tracking and stability specifications were established [182]. By tracking specification the effect of the uncertainties will be reduced. It is only necessary to impose the minimum and maximum values for the magnitude of the closed-loop system in all frequencies. With respect to the stability specification, the desired gain (GM) and phase (PM) margins are set. The tracking specifications were required to fulfill a settling rise time of between 5 and 35 min and an overshoot of less than 30% after 10–20°C set point changes for all operating conditions (realistic specifications, see [85]). These specifications are translated to the following frequency conditions using the QFT framework obtaining [111]

$$\left|B_l(j\omega)\right| \le \left|F(j\omega)\frac{P(j\omega)C(j\omega)}{1+P(j\omega)C(j\omega)}\right| \le \left|B_u(j\omega)\right|, \quad \forall\omega > 0,\ \forall P \in \wp \quad (5.16)$$

where $B_l(j\omega)$ and $B_u(j\omega)$ are the minimum and maximum values for the magnitude of the closed-loop system in all frequencies to fulfill tracking specifications.

Table 5.2 Tracking specifications for the C compensator design (courtesy of C.M. Cirre et al., [111])

ω [rad/s]	0.0006	0.001	0.003	0.01
$\varrho(\omega)$	0.55	1.50	9.01	19.25

For stability specification, the following condition must be fulfilled:

$$\left|\frac{P(j\omega)C(j\omega)}{1+P(j\omega)C(j\omega)}\right|_{dB} \leq \varrho(\omega) = 3.77, \quad \forall\omega > 0,\ \forall P \in \wp \qquad (5.17)$$

in order to guarantee a phase margin of 35° for all operating conditions.

In order to design the compensator $C(s)$ the tracking specifications in (5.18) are transformed into

$$\left|\frac{P(j\omega)C(j\omega)}{1+P(j\omega)C(j\omega)}\right| \leq \left|B_u(j\omega)\right| - \left|B_l(j\omega)\right| = \Delta(\omega), \quad \forall\omega > 0,\ \forall P \in \wp \quad (5.18)$$

Table 5.2 shows the specifications in (5.18) for each frequency in the set of design frequencies W. Using these specifications, the stability specification in (5.17), system uncertainties, tracking and stability boundaries are computed using the algorithm in [269]. The computed boundaries are shown in Fig. 5.21, as well as stability and tracking boundaries with the shaped $L_0(j\omega) = C(j\omega)P_0(j\omega)$ satisfying all of them for all frequencies in W.

The resulting compensator $C(s)$ synthesized in order to achieve the stability specifications in (5.17) and the tracking specifications in (5.18) is the following PID-type controller $C(s) = 0.75(1 + \frac{1}{180s} + 40s)$. In order to satisfy the specifications in (5.16), the prefilter $F(s)$ must be designed, where the synthesized prefilter is given by $F(s) = \frac{0.1}{s+0.1}$.

Figure 5.22 shows that the tracking specifications (5.16) are fulfilled for all uncertain cases. Note that the different appearance of Bode diagrams in closed-loop for five operating conditions is due to the changing root locus of $L(s)$ when the PID is introduced. The time response for these operation conditions is shown in Fig. 5.23.

Assuming the 2DoF controller given by compensator $C(s)$ and prefilter $F(s)$, stability problems can appear when HTF flow rate saturates at $u_{min} = 2$ l/s or $u_{max} = 12$ l/s. Moreover, the absolute stability cannot be ensured using QFT due to the flow-outlet temperature dependent time delay present in the process [182] (this is small

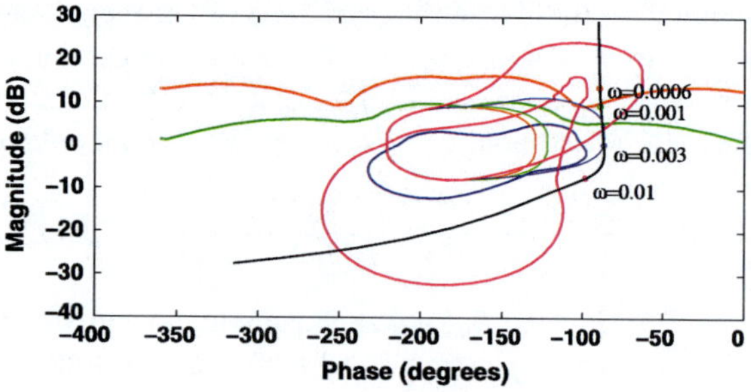

Fig. 5.21 Tracking and stability boundaries with the designed $L_0(j\omega)$ (courtesy of C.M. Cirre et al., [111])

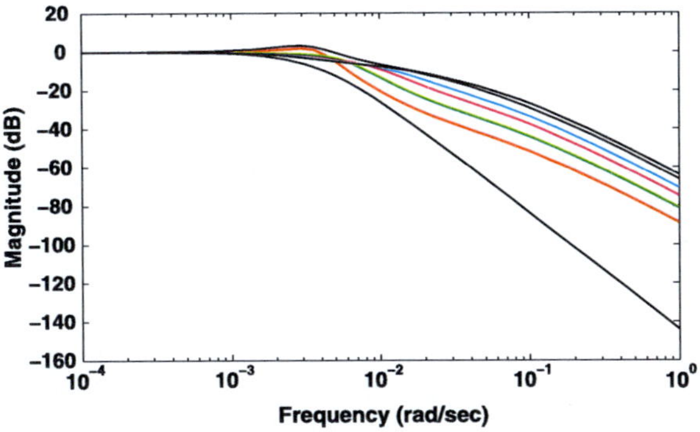

Fig. 5.22 Tracking specifications (*dashed-dotted*) and magnitude Bode diagram of some closed-loop transfer functions (courtesy of C.M. Cirre et al., [111])

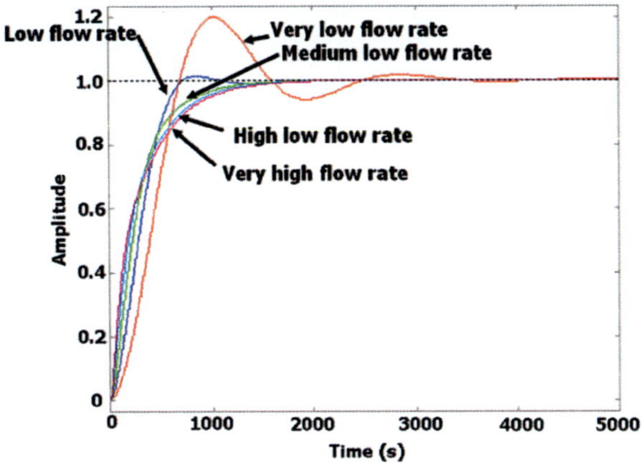

Fig. 5.23 Time step responses for some operation conditions (courtesy of C.M. Cirre et al., [111])

when compared to the main time constant of the system). Therefore, two strategies have been included to deal with this problem: the use of an anti-windup mechanism and the development of a reference governor to avoid flow saturation to be used as a support decision tool by the operator.

Figure 5.24 shows the anti-windup scheme used (where T_r is the tracking constant of the anti-windup term). This mechanism is a soft modification of the classical anti-windup [23]. In this case, saturation is located between the feedforward term and the plant and the compensator is placed in front of the feedforward term. Thus,

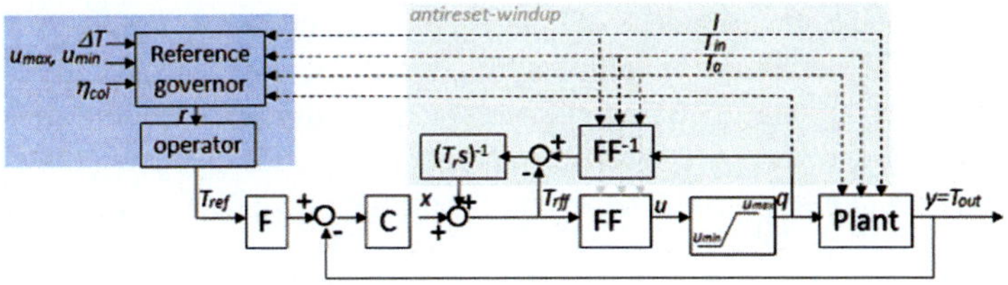

Fig. 5.24 Control scheme with anti-windup and reference governor, [111]

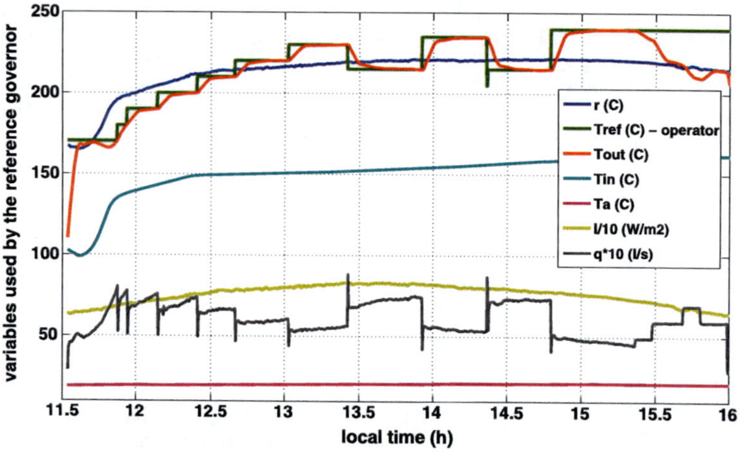

Fig. 5.25 Results provided by the reference governor to avoid input and output constraints violation (courtesy of C.M. Cirre et al., [111])

when the actuator saturates to u_{max} (or u_{min}) value the corresponding feedforward term input, T_{rff}, which would provide an output u_{max} (or u_{min}) taking into account the measurable instantaneous solar irradiance, is computed using the term FF^{-1}, which implements the inverse of the feedforward compensator.

The second element used to try to avoid control signal saturation and to maximize outlet power taking into account the security constraints is the use of a reference governor to support the decision of the operator on the actual set point (also shown in Fig. 5.24). The reference governor supplies the reference r taking into account the actual operating conditions (outlet temperature, inlet temperature, direct solar irradiance, ambient temperature and both the maximum allowed temperature gradient $\Delta T = (T_{out} - T_{in})$ and the PTC global efficiency η_{col}). This reference is used by the operator in order to take decisions about set point changes. This element uses static models of the DSCF based on mass and energy balances as those developed in Sects. 4.3.4.1 and 4.3.4.2, see [111] for a detailed description. As will be seen in Sect. 5.12, another advantage when using the reference governor is that it is a very useful tool for giving adequate set points during one of the most difficult parts of the operation, the start-up, where control signals are usually saturated during long intervals and the inlet temperature suffers from the greatest changes during operation due to the existence of cold HTF inside the tubes. Figure 5.25 shows an example of the reference obtained by the reference governor and the set point used by the operator, also including the outlet temperature and HTF flow demanded when operating with the QFT-based control scheme. Notice that during start-up the operator took a more conservative set point control policy than the reference governor.

In order to prove the fulfillment of the tracking and stability specifications of the control structure, experiments in the ACUREX plant were performed at several op-

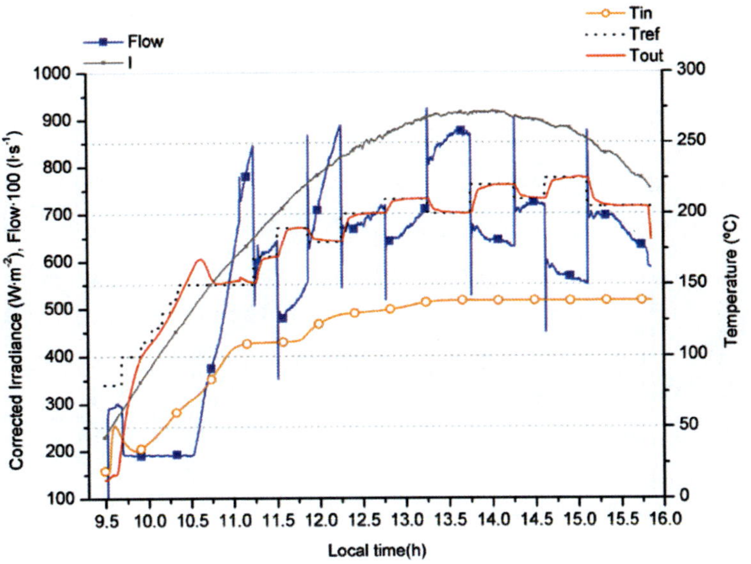

Fig. 5.26 Test with PID without anti-windup (24/03/2009) (courtesy of C.M. Cirre et al., [111])

erating points and under different conditions of disturbances. The outlet controlled was the loop with the maximum temperature for all the experiments. The sampling time was chosen equal to 15 s. Figure 5.26 shows an experiment with the robust controller but without anti-windup. At the beginning of the experiment, the flow is saturated until the outlet temperature is higher than the inlet one (the normal situation during the operation). This situation always appears due to the fact that the HTF inside the pipes is cooler than the HTF from the tank. Once the HTF is mixed in the pipes, the outlet temperature reaches a higher temperature than the inlet one. During the start-up, steps in the reference temperature are made until reaching the nominal operating point. The overshoot at the end of this phase is 18°C approximately and thus the specifications are fulfilled. Analyzing the time responses, a settling time between 11 and 15 min is observed at the different operating points. Therefore, time specifications, overshoot and settling time are properly fulfilled. Disturbances in the inlet temperature (from the beginning until $t = 12.0$ h), due to the temperature variation of the stratified HTF inside the tank, are observed during this experiment and correctly rejected by the feedforward action.

Finally, in order to check the behavior of the control structure in the most complicated situation, changing radiation due to clouds passing, a new test under these conditions is presented in Fig. 5.27. When the step responses are analyzed, it can be seen that the settling time is between 12 and 30 min in this case and the overshoot is under 15% in all cases. Hence, the tracking and stability specifications are fulfilled for a cloudy day despite the system uncertainty in presence of different operating points. At $t = 12.39$ h, a change in the inlet temperature is observed due to a variation in the three-way valve, causing an increment of 6°C in the outlet temperature.

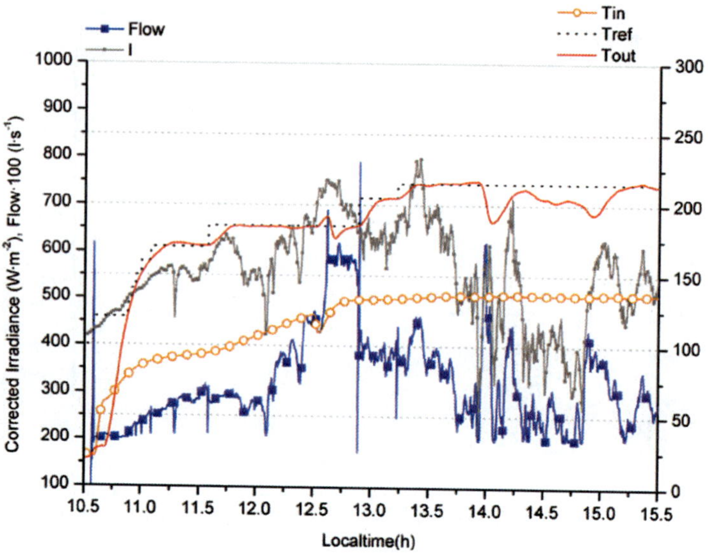

Fig. 5.27 Test with PID with anti-windup (07/10/2009) (courtesy of C.M. Cirre et al., [111])

However, the error was reduced in 17 min thanks to the feedforward action. From 14 h to the end of the test, the direct solar irradiance was too low and it was impossible for the controller to track the proposed reference (the operator did not change the reference according to that suggested by the reference governor).

5.8 Non-linear Control (NC)

As has been mentioned, explicit recognition of plant non-linearities and their exploitation could lead to performance and robust stability improvements but at the cost of increasing the controller complexity. Steps in this direction were made by employing traditional non-linear control strategies where non-linear transformations of input or output variables take place. In [28, 29], a feedback linearization (FL) scheme is proposed including Lyapunov-based adaptation and using a simplified plant model. For dealing with plant non-linearities and external disturbances, a non-linear transformation is performed on the accessible variables such that the transformed system behaves as an integrator, to which linear control techniques are then applied. In [195, 196], a control design is proposed based on a distributed model, using ideas from passivity theory, that is using internal energy as a storage function and then using energy considerations and Lyapunov-like arguments to derive stable and robust control laws relying on feedback from the distributed collector field's internal energy. It is shown that if the internal energy is controlled, the outlet temperature is under control as well.

The design also incorporates feedforward from the measured disturbances in order to achieve passivity and high disturbance rejection performance. In [349], a PDE physical plant model is used for non-linear control purposes. Two situations are considered: in the first, a constant space discretization is imposed, this causes the sampling period to vary according to the flow; in the second, constant sampling in both space and time causes temperature to be a function of flow and direct solar irradiance, linearizing this function by a convenient variable transformation suggests the use of FL techniques. The first situation considered is exploited in [351, 352], where the key point is the observation that the value of HTF flow establishes a *natural* time scale for the system. This is achieved in discrete time by indexing the sampling rate to the value of flow. As a consequence, the model equations become linear and it is possible to achieve good performance on step responses of big amplitude. In the approach followed by [186], the PDE describing the field is approximated by a lumped parameter bilinear model whose states are the temperature values along the field. By using feedback exact linearization together with a Lyapunov approach, an AC is designed. This paper improves the previous work of [29] by using a better approximated model, which takes into account that, in the field considered, temperature measures are only made at the input and at the output and not along the pipe.

In [107] an automatic control approach using a simple FL method and a lumped parameter model of a DSCF is also developed. The control scheme resembles that of a feedforward controller in combination with a classical feedback controller as those presented in [82], but with the difference that an embedded feedback from the output is used both for linearization and feedforward purposes, as shown in the next subsection. All the control schemes based on FL have shown excellent results when tested at the real plant and are very adequate for the starting-up phase of the operation. Model predictive-control (MPC) extensions of the basic algorithm will be treated in Sect. 5.9.5, while non-linear ANN-based controllers based on output regulation (OR) theory are commented on in Sect. 4.3.5.2.

5.8.1 Feedback Linearization Control of DSCF

Using the simplified bilinear model accounting for varying delays given by Eq. (4.15) developed in Sect. 4.3.4.2, input–output linearization can be used to control the DSCF when the design of the controller is performed on a linear representation of the system as the non-linear dynamics are embedded in the definition of a virtual control signal. Conditions for the existence of an FL controller are met if the inlet and outlet HTF temperatures are different, this is always true during nominal operation and it is ensured during the start-up stage by replacing the real outlet HTF temperature by the reference temperature, thus helping to improve the dynamical response of the system during this phase. Any linear controller could be used in the design of the linear part of the control system.

The basic idea of the feedback linearization technique [189, 354] is to treat non-linear systems as linear ones by means of algebraical transformations and feedback.

Considering that the coordinate transformation (or mapping) is represented by a function Ω depending on the new virtual control signal u' and the system state x, the control signal is obtained as $u = \Omega(x, u')$. Moreover, if the non-linear system includes disturbances ($\mathbf{p}$) with a linear dependence (such as in the case of the DSCF), this algebraical transformation, Ω, will also depend on these disturbances and thus, $u = \Omega(x, u', \mathbf{p})$. Following this interpretation, FL technique may be applied to track a desired outlet reference temperature using the HTF volumetric flow as the control signal. For this purpose, the outlet temperature model (4.15) is transformed into the canonical form, where the functions g and b also depend on the system disturbances, $\mathbf{p} = (I, T_{in}, T_a)$:

$$\frac{dx}{dt} = g(x, \mathbf{p}) + b(x, \mathbf{p})u \tag{5.19}$$

$$y = h(x) \tag{5.20}$$

where $x = T_{out}$, $u = q(t - t_d)$, $h(x) = x$ and

$$g(x, \mathbf{p}) = \frac{\eta_{col} G I(t)}{\rho_f(\bar{T}_f) c_f(\bar{T}_f) A_f} - \frac{\widetilde{H}_l(\bar{T}_f, T_a)}{\rho_f(\bar{T}_f) c_f(\bar{T}_f) A_f L_2}$$

$$b(x, \mathbf{p}) = \frac{T_{out}(t) - T_{in}(t - t_r)}{A_f n_{ope} L}$$

This yields a simple input–output relationship:

$$\frac{dT_{out}(t)}{dt} = u'(t - t_d) \tag{5.21}$$

which represents a pure integrator with time delay where the input is the virtual signal u'. Then, in this case, the equivalent linear dynamics takes the following form:

$$\begin{aligned} \dot{\psi}(t) &= u'(t - t_d) \\ \eta(t) &= \psi(t) \end{aligned} \tag{5.22}$$

with $\psi = x = T_{out}$. Thus, it is possible to use any linear control over the system (5.22), which represents the evolution of the outlet solar field temperature. The virtual signal u' must be mapped into the real control signal q. From Eq. (5.19), the control input needed to compensate for non-linearities would be

$$\begin{aligned} q(t - t_d) = {} & \frac{A_f n_{ope} L}{(T_{out}(t) - T_{in}(t - t_r))} \left(\frac{\eta_{col} G I(t)}{\rho_f(\bar{T}_f) c_f(\bar{T}_f) A_f} \right. \\ & \left. - \frac{\widetilde{H}_l(\bar{T}_f, T_a)}{\rho_f(\bar{T}_f) c_f(\bar{T}_f) A_f L_2} - u'(t - t_d) \right) \end{aligned} \tag{5.23}$$

with no thermal inversion in the solar field, $T_{out}(t) - T_{in}(t - t_r) \neq 0$, so that the control law is valid in the state space defined by the operation range. Only plant start-up has to be supervised to avoid numerical problems. Therefore, $q(t)$ does not only depend on the actual value of $u'(t)$, but also on both the future values of the disturbances $\mathbf{p}(t+t_d)$ and the output, $T_{out}(t+t_d)$. For disturbances, it is assumed that $\mathbf{p}(t) \approx \mathbf{p}(t+t_d)$. On the other hand, the prediction of $T_{out}(t+t_d)$ may be estimated by using a discrete Euler approximation (as the system is going to be implemented in discrete time) such as

$$\hat{T}_{out}(k+d) = T_{out}(k) + T_s \sum_{i=1}^{d} u'(k-d+i) \tag{5.24}$$

in such a way that the actual control signal can be computed as

$$q(k) = \frac{A_f n_{ope} L}{(\hat{T}_{out}(k+d) - T_{in}(k-n))} \left(\frac{\eta_{col} G I(k)}{\rho_f(\bar{T}_f) c_f(\bar{T}_f) A_f} - \frac{\widetilde{H}_l(\bar{T}_f, T_a)}{\rho_f(\bar{T}_f) c_f(\bar{T}_f) A_f L_2} - u'(k) \right) \tag{5.25}$$

with $nT_s \approx t_r - t_d$.

A basic I-PD control structure was employed to control the linearized system. This structure is a modification of a classical PID, but only the integral term is in the direct trajectory between the input and the output. The proportional and derivative terms are in the feedback trajectory. When the reference input is a step function, PID control involves a step function in the manipulated signal which may not be desirable. Therefore, I-PD moves the proportional and derivative actions to the feedback path so that these actions affect the feedback signal only. In the absence of the reference input and noise signals, the closed-loop transfer function between the disturbance input and the output is the same as for PID control. A first set of parameters of the I-PD controller [276] was obtained by using the plant simulator [85] and tuning the parameters using the minimum integral of time-weighted absolute error (ITAE) rules [277], that were subsequently refined at the real plant to improve the response of the system. The I-PD parameters were fixed to $K_P = 0.015$ [1/s], $T_I = 300$ [s], $T_D = 50$ [s] [109]. A diagram of the final control structure is shown in Fig. 5.28. A sampling time of $T_s = 39$ s was chosen.

As mentioned above, the inlet temperature may be higher than in the outlet during start-up. This could cause a negative temperature difference in (5.23). Therefore, a variation in the control structure used during normal plant operation, Fig. 5.28 (top), has been developed to control T_{out} during start-up. This structure, shown in Fig. 5.28 (bottom), is very similar to the one used for normal operation except for the inputs to the non-linear block change. Notice that T_{ref} is a new input in this block and also that this makes the difference in enthalpy positive, which is normal during operation. Once the outlet temperature reaches the set point, the control is switched to routine control as shown in Fig. 5.28 (top). The transition between the two controllers is made safely by bumpless transfer.

Since the limits of the virtual signal depend on the actual conditions affecting non-linear mapping, the method's performance decreases if they are not properly set (see [219]). So the physical limits imposed by the pump can be mapped into limits in the virtual control signal u', again using the non-linear mapping represented by Eq. (5.23). The physical limits in the real control signal are transformed into

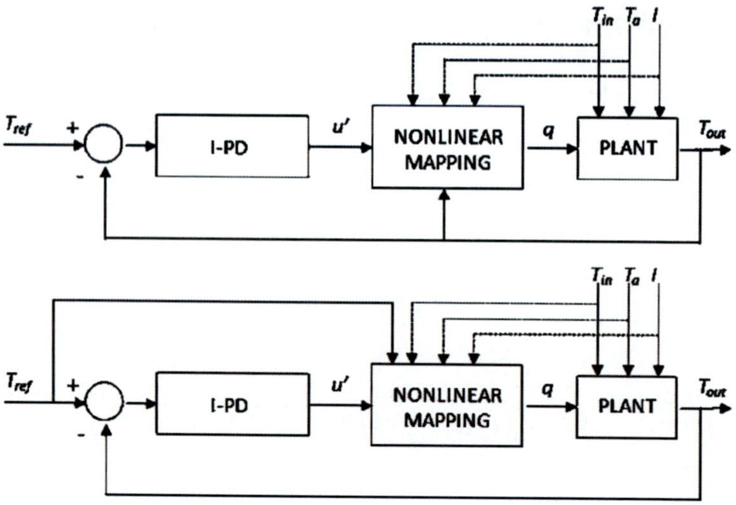

Fig. 5.28 Feedback linearization controller (*top*) and modification for start-up stage (*bottom*) (courtesy of C.M. Cirre et al., [109])

variable constraints in the virtual control signal which depends on the operating conditions. Equations (5.26) and (5.27) represent the corresponding limits of the virtual signal u' (u'_{max}, u'_{min}) depending on the lower and upper physical limits of the pump (q_{max}, q_{min}). This mechanism can be considered an anti-windup strategy, so it has therefore been included to account for possible saturation in the control signal. Virtual signal limits estimated at each sample time have provided very useful results, mainly during operation start-up. Notice that a global efficiency modifier can also, easily be included to modulate the scheme's disturbance rejection capabilities.

$$u'_{min} = \min_{q_{min} \leq q \leq q_{max}} u' \tag{5.26}$$

$$u'_{max} = \max_{q_{min} \leq q \leq q_{max}} u' \tag{5.27}$$

The control scheme in Fig. 5.28 resembles feedforward controllers explained in Chap. 4, which were also based on a simplified physical system model but considered steady-state conditions in such a way that an HTF flow correlation could be derived as a function of the inlet and outlet HTF temperatures, corrected for direct solar irradiance and ambient temperature. Thus, the main difference in the scheme presented here is that internal feedback is now included both for linearizing and disturbance cancelation, providing smoother control.

Figure 5.29 shows the controller responses under real conditions used as plant simulator input. The figure illustrates the part of the test where the behavior of three control schemes (see Table 5.3) is compared. The set point step responses of the three controllers seem to be very similar, but, the IPDFL (see Fig. 5.28) controller output overshoots the reference temperature less and so would seem to provide better inlet disturbance rejection. In [109] a complete comparison is carried out using

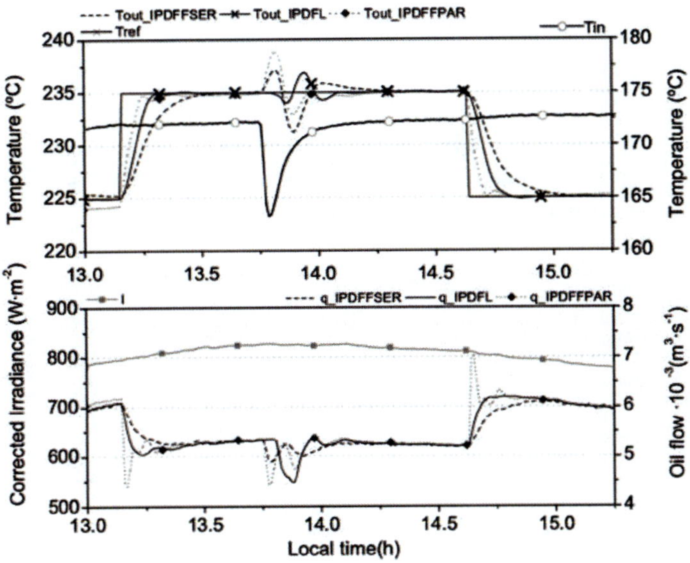

Fig. 5.29 Simulation with test on July 15th, 2004, of I-PD with a series and parallel feedforward and the feedback linearization controller (courtesy of C.M. Cirre et al., [109])

Table 5.3 Parameters tuned for the simulated control schemes (courtesy of C.M. Cirre et al., [109])

Controller	K_P	T_I [s]	T_D [s]
I-PD series feedforward (IPDFFSER, Eq. (4.52))	1	140	30
I-PD parallel feedforward (IPDFFPAR, Eq. (4.51))	−0.15 [l/(s °C)]	400	110
Feedback linearization (IPDFL)	0.012 [1/s]	300	50

as indices the integral of the square error (ISE), the integral of time-weighted square error (ITSE), the integral of the absolute error (IAE) and the ITAE [277] for the same period of time when there is a step in the reference temperature and a variation in the inlet temperature is considered a disturbance.

The FL controller was tested at the ACUREX solar field. Both the set point tracking features and the disturbance rejection capabilities were tested under operating conditions varying due to changes in set point, direct solar irradiance and inlet HTF temperature variables. Start-up and full-day operation are also shown. The operating procedure is the same in all experiments using the scheme in Fig. 5.28. First, the set point temperature is used as the actual temperature in the non-linear mapping block (arrow from T_{ref} to that block in Fig. 5.28) to improve the starting phase and avoid problems that could arise if the conditions for the applicability of the FL technique were not fulfilled. Once T_{ref} is reached, the actual temperature is then used in the non-linear mapping block and during this transition period, bumpless transfer guarantees good current controller tracking. Parameters T_I and T_D of both controllers

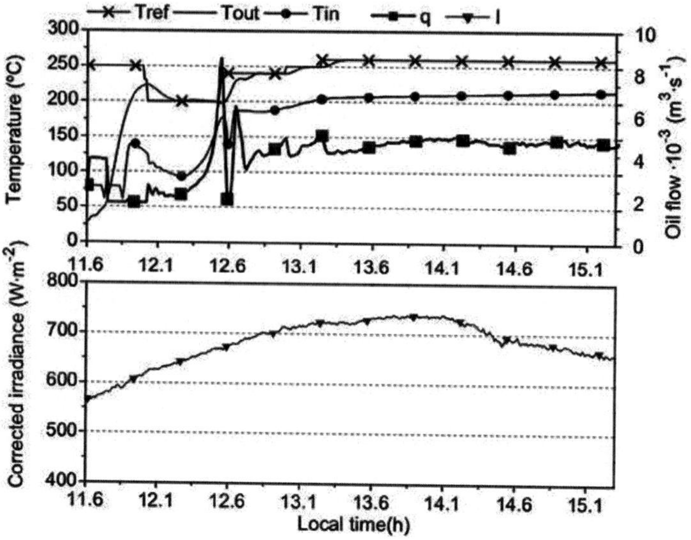

Fig. 5.30 Test on July 21th, 2004 with strong inlet HTF temperature disturbance (courtesy of C.M. Cirre et al., [109])

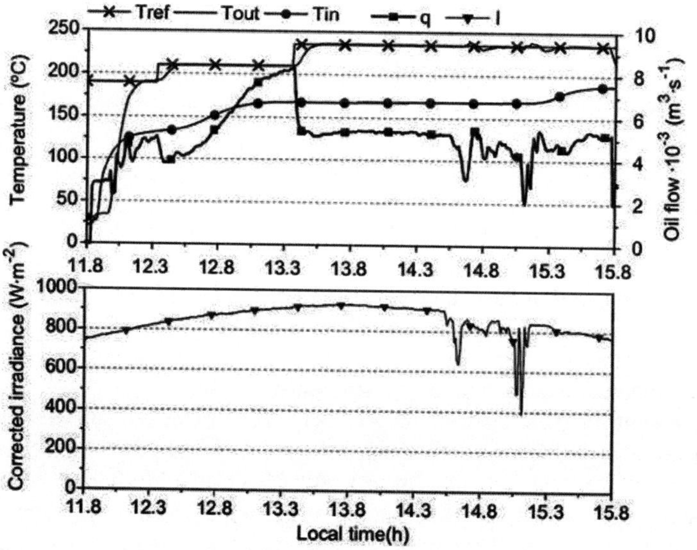

Fig. 5.31 Test on October 1st, 2004 with solar irradiance momentary variations and inlet HTF temperature disturbances (courtesy of C.M. Cirre et al., [109])

are the same, but controller parameter K_P, used for start-up, was tuned to provide a more conservative response. Figure 5.30 shows experiment results from start-up to the end of the test where T_{out} is controlled automatically by the control schemes described above. At first, the cold HTF in the pipes in the field is sent to the bottom of the storage tank instead of the top, which prolongs start-up. Therefore, the first reference temperature set was diminished to avoid long controller output saturation at minimum flow, which is normal when operating under these circumstances (even with manual operation). As may be observed, the system then reached steady state without overshooting. To reject the fast increase in the inlet HTF temperature, the controller had to perform fast changes in the required HTF flow rate over the whole operating range. The performance of the rest of the operation can be described by settling times of around 10 min without overshooting with set point changes of 10°C. Smooth changes in direct solar irradiance of about 70 W/m^2 and inlet HTF temperature increases were compensated in such a way that no tracking errors (more than ±1°C) were found.

The test in Fig. 5.31 shows the behavior of the controller when there are slow inlet HTF increases and fast variations in solar irradiance. Two inlet HTF temperature disturbances occurred during the test. The first one started at 12:05 and the second one at 15:03 h. In both cases, the tracking error was always less than ±1°C. The rejection of solar radiation disturbances for a period of approximately 50 min with variations of 390 W/m^2 is good, as the maximum tracking error is less than 4°C. The output settling time (±5%) is around 10 min after a set point step of 25°C. Throughout the test, the controller provided an HTF flow that produced a smooth output response (in spite of disturbances), which is desirable in this kind of system. The next test in Fig. 5.32 shows the controller behavior with continuous solar radiation disturbances along with a steep rise in inlet HTF temperature. Test start-up with this controller was short (around 30 min). The output response shows a small overshoot of less than 5°C. At approximately 13:05 h, the inlet temperature started to rise very quickly so the operator had to make a step change in the reference temperature (45°C) due to the safety conditions of the operation and not to help the operation of the automatic controller. The controller regulated the HTF flow correctly during the whole test, but at about 13:05 h, there was a fast and big increase in the inlet temperature because the HTF inside the tank is stratified depending on its density. To track the same reference temperature with a very small $\Delta T = (T_{out} - T_{in})$ and with large values of irradiance, a very high flow is required, irrespective of whether it is demanded from an automatic controller or from manual mode and the pump will saturate to its physical limits, causing the system to operate in open loop; thus, to prevent this situation, the set point was increased; in fact, automatic set point governors should avoid this problem [46, 105, 106, 108, 110].

In spite of these disturbances, including solar irradiance, the tracking error was less than 1.5°C. The settling time in this step was 9.1 min. This is a good result, taking into account the disturbances affecting the plant and the wide set point change, as steps are not usually more than 20°C. The importance of calculating the virtual signal limits, depending on the conditions affecting the plant such as solar irradiance and inlet HTF temperature, can be checked in the test shown in Fig. 5.32. Due to

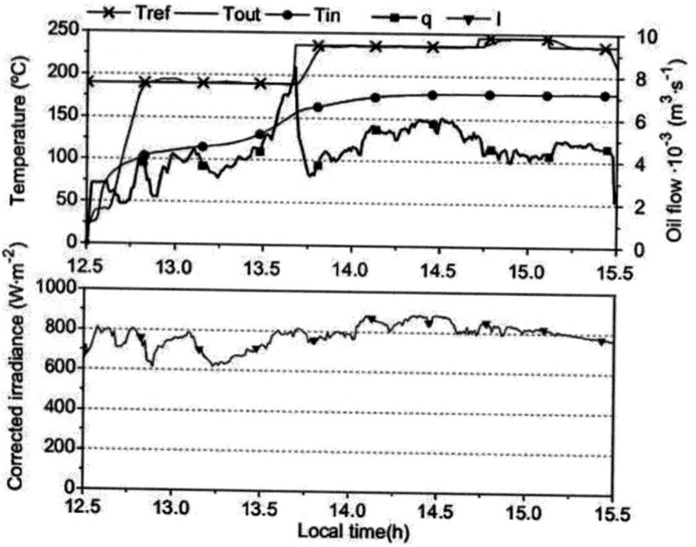

Fig. 5.32 Test on October 7th, 2004 with solar irradiance continuous disturbances added to inlet HTF temperature disturbances (courtesy of C.M. Cirre et al., [109])

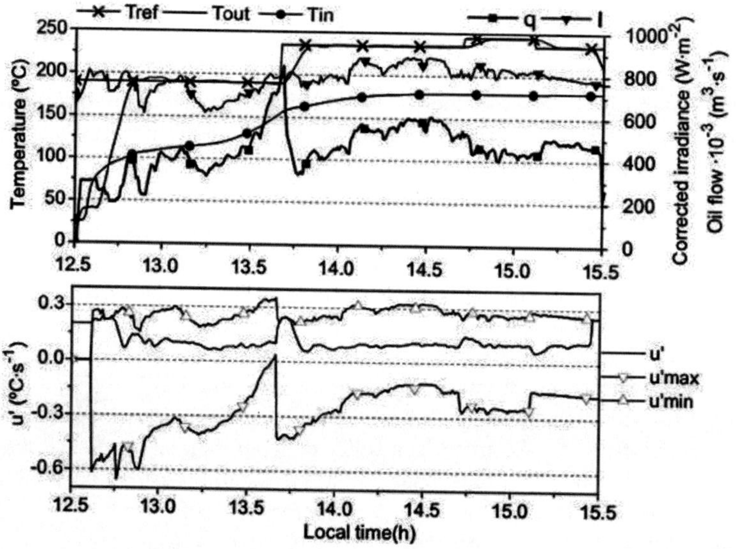

Fig. 5.33 Test on October 7th, 2004 the respective virtual signal and its limits (courtesy of C.M. Cirre et al., [109])

the small difference between the reference temperature set and the inlet HTF temperature (at 13:50), the controller had to increase the HTF flow to track properly. This situation involved varying the virtual signal limits as the external conditions had also changed. Figure 5.33 shows the test shown in Fig. 5.32 with the virtual signal calculated at each sample time. Note how the virtual signal is displaced from zero, however, it increases or decreases correctly as the derivative of T_{out} increases or decreases. This is because the non-linear mapping parameter η_{col}, which cannot be measured and depends on how clean the mirrors are, is not fine-tuned. This fact is studied in depth in [109], where more experimental results can be found.

5.9 Model-Based Predictive Control (MPC)

Many different MPC strategies (Fig. 5.34) have been applied to control DSCF. The ideas appearing in greater or lesser degree in all the predictive-control family are basically [80]: explicit use of a model to predict the process output at future time instants (horizon), calculation of a control sequence minimizing a certain objective function subject to constraints and receding strategy so that at each instant the horizon is shifted toward the future, which involves the application of the first control signal of the sequence calculated at each step. Most of the MPC strategies applied to the control of DCS are in adaptive, robust, or non-linear fields and include a feedforward term as part of the controller [34, 85]. Few implementations of MPC controllers with fixed parameters have been reported in the literature (e.g. [74, 75, 85]). In [366], a linear model predictive controller is developed to cope with the control of the 30 MWe SEGS VI parabolic trough collector plant. A non-linear distributed-parameter model is discretized and linearized obtaining a set of ordinary differential equations (ODE) that are then used within an MPC framework including a state estimator. The differences between the collector outlet temperatures as predicted by the linear model and the detailed model are multiplied by an observer gain and fed back to the linear model to minimize the difference. The observer gain is calculated as the discrete steady-state Kalman filter gain with the intention of minimizing the mean-square error of the state estimate. For offset-free control, the set point used in the receding horizon regulator has to be updated with respect to the measured disturbance and the estimated difference between the collector outlet temperature prediction and the measurement. The latter represents the second part of the integral action implementation. The target calculation is formulated as a mathematical program to determine the new set point.

The most important applications to DSCF are: MPC adaptive control [74, 75, 83, 116, 117, 246, 308–310, 347, 366], MPC gain-scheduling control [34, 39, 84, 225], MPC robust control [77, 228, 291] and MPC non-linear control, including NN and FL approaches [16–18, 40, 42, 76, 192, 297, 299, 344–346, 351, 352], all these techniques we briefly summarized in the following section, based on the Generalized Predictive Control (GPC) approach.

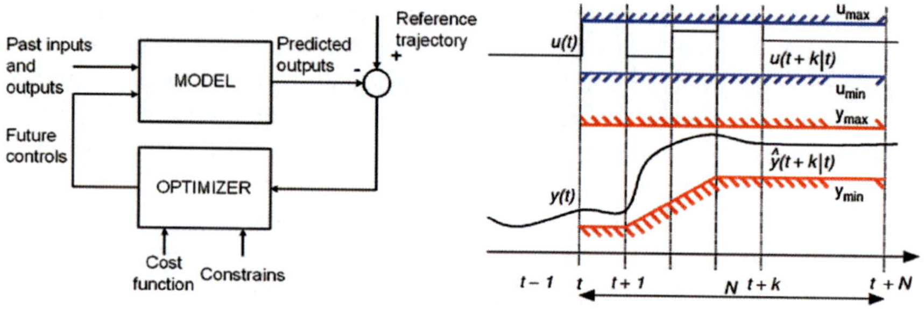

Fig. 5.34 Block diagram of a model-based predictive controller (MPC) and MPC strategy

5.9.1 Generalized Predictive Control (GPC)

5.9.1.1 Internal Model

Most SISO plants, when considering operation around a particular set point and after linearization, can be described by a controlled auto-regressive integrated moving average model with an exogenous inputs (CARIMA) model accounting for non-stationary disturbances:

$$A\left(z^{-1}\right)y(k) = z^{-d}B\left(z^{-1}\right)u(k-1) + T\left(z^{-1}\right)\frac{e(k)}{\Delta} \tag{5.28}$$

where $u(k)$ and $y(k)$ are the control and output sequence of the plant, d is the dead-time of the system and $e(k)$ is a zero mean white noise. A, B and T are polynomials in the backward shift operator z^{-1} (see Chap. 4) and $\Delta = 1 - z^{-1}$. The polynomial $T(z^{-1})$ usually equals one in GPC formulations. To enhance robustness, the T-polynomial is used as a design element, acting as a filter to attenuate the prediction error caused by high-frequency unmodeled dynamics and unmeasured load disturbances. If there are not modeling errors, the T-polynomial affects only the disturbance rejection; it does not affect the algorithm's tracking ability. T can thus be used as a design parameter to increment robust stability [80, 326].

5.9.1.2 Cost Function

The quadratic cost function in the GPC algorithm accounts for the error between the predicted trajectory and a prescribed reference, as well as for the control effort. This cost function has the form

$$J = E\left\{\sum_{j=N_1}^{N_2}\delta(j)\left[\hat{y}(k+j\mid k) - r(k+j)\right]^2 + \sum_{j=1}^{N_u}\lambda(j)\left[\Delta u(k+j-1)\right]^2\right\} \tag{5.29}$$

where $E\{\cdot\}$ denotes expectation, $\hat{y}(k+j \mid k)$ is an optimal system output prediction sequence performed with data known up to discrete time k, $\Delta u(k+j-1)$ is a future control increment obtained from cost function minimization, N_1 and N_2 are the minimum and maximum prediction horizons, N_u is the control horizon and $\delta(j)$ and $\lambda(j)$ are weighting sequences that penalize the future tracking errors and control efforts along the horizons, respectively. The horizon and weighting sequences are design parameters used as tuning knobs. The reference trajectory, $r(k+j)$, can be the set point or a smooth approximation from the current value of the system output $y(k)$ toward the known reference by means of a first-order system [80].

5.9.1.3 System and Performance Constraints and GPC Solution

MPC techniques enable constraints to be included in the algorithm during the design stage. As stated above, the control increments calculated by the GPC strategy are obtained by minimizing a quadratic function of the form

$$J = \delta(\mathbf{y}-\mathbf{r})^T(\mathbf{y}-\mathbf{r}) + \lambda \Delta\mathbf{u}^T\Delta\mathbf{u}$$

where $\delta(j) = \delta$ and $\lambda(j) = \lambda$ for all j in the GPC cost function. The sequence of future predictions $\mathbf{y} = [\hat{y}(k+1 \mid k) \dots \hat{y}(k+N_2 \mid k)]^T$ consists of the free and forced response $\mathbf{y} = \mathbf{G}\Delta\mathbf{u} + \mathbf{f}$, where $\Delta\mathbf{u} = [\Delta u(k) \dots \Delta u(k+N_u-1)]^T$, matrix $\mathbf{G}$ contains the system open-loop step response coefficients and $\mathbf{f}$ contains terms that depend on past and present plant outputs and past inputs. By substituting the sequence of future outputs in

$$J = \frac{1}{2}\Delta\mathbf{u}^T\mathbf{H}\Delta\mathbf{u} + \mathbf{b}^T\Delta\mathbf{u} + \mathbf{f}_0 \tag{5.30}$$

with $\mathbf{H} = 2(\delta\mathbf{G}^T\mathbf{G} + \lambda\mathbf{I})$, $\mathbf{b}^T = 2\delta(\mathbf{f}-\mathbf{r})^T\mathbf{G}$ and $\mathbf{f}_0 = \delta(\mathbf{f}-\mathbf{r})^T(\mathbf{f}-\mathbf{r})$, the optimal solution with no constraints is linear and given by $\Delta\mathbf{u} = -\mathbf{H}^{-1}\mathbf{b}$. When constraints are taken into account, there is no explicit solution; a quadratic programming (QP) problem must be solved using a quadratic cost function with linear inequality and equality constraints of the form $\mathbf{R}\Delta\mathbf{u} \le \mathbf{c}$ and $\mathbf{A}\Delta\mathbf{u} = \mathbf{a}$ in the control increment $\Delta\mathbf{u}$. The different constraints considered are shown in Table 5.4 [80, 169], where Γ is a N-dimensional vector (N being the receding horizon) the elements of which are all equal to one and Υ is an $N \times N$ lower triangular matrix in which all elements equal one. The GPC treatment of constraints is conceptually simple and can be included systematically during the design process. An interactive tool, GPCIT, was developed in [169] where all the GPC-related concepts can easily be understood, also including closed-loop relationships.

5.9.2 Adaptive Generalized Predictive Control

One of the main reasons of the success of traditional PID controllers in the industry is the easiness of their implementation and syntonization using in most cases certain heuristic rules, Ziegler–Nichols rules being the most commonly used in control engineering practice. The development of suitable controllers to be implemented

Table 5.4 Constraints in GPC [169]

Variable	Linear constraint
Control signal amplitude $u_{min} \le u(k) \le u_{max}$	$\Gamma u_{min} \le \Upsilon \Delta \mathbf{u} + u(k-1)\Gamma \le \Gamma u_{max}$
Control signal increment $\Delta u_{min} \le u(k) - u(k-1) \le \Delta u_{max}$	$\Gamma \Delta u_{min} \le \Delta \mathbf{u} \le \Gamma \Delta u_{max}$
Output signal amplitude $y_{min} \le y(k) \le y_{max}$	$\Gamma y_{min} \le \mathbf{G}\Delta \mathbf{u} + \mathbf{f} \le \Gamma y_{max}$
Envelope constraints $y_{min}(k) \le y(k) \le y_{max}(k)$	$\mathbf{G}\Delta \mathbf{u} \le \mathbf{y}_{max} - \mathbf{f}$ $\mathbf{y}_{max} = [y_{max}(k+1)\dots y_{max}(k+N)]$ $\mathbf{G}\Delta \mathbf{u} \ge \mathbf{y}_{min} - \mathbf{f}$ $\mathbf{y}_{min} = [y_{min}(k+1)\dots y_{min}(k+N)]$
Output overshoot $y(k+j) \le \gamma r(k)$; $j = N_{o1}\dots N_{o2}$	$\mathbf{G}\Delta \mathbf{u} \le \Gamma \gamma r(k) - \mathbf{f}$
Output monotone behavior $y(k+j) \le y(k+j+1)$ if $y(k) < r(k)$ $y(k+j) \ge y(k+j+1)$ if $y(k) > r(k)$	$\mathbf{G}\Delta \mathbf{u} + \mathbf{f} \le [\mathbf{0}^T, \mathbf{G}']\Delta \mathbf{u} + [y(k), \mathbf{f}']$, $\mathbf{G}'$ and $\mathbf{f}'$ are the result of eliminating the last row of $\mathbf{G}$ and $\mathbf{f}$
Limit inverse response (NMP) $y(k+j) \le y(k)$ if $y(k) > r(k)$ $y(k+j) \ge y(k)$ if $y(k) < r(k)$	$\mathbf{G}\Delta \mathbf{u} \ge \Gamma y(k) - \mathbf{f}$
Final state $y(k+N+1)\dots y(k+N+m) = r$	$\mathbf{y}_m = [y(k+N+1)\dots y(k+N+m)]^T$ $\mathbf{y}_m = \mathbf{G}_m \Delta \mathbf{u} + \mathbf{f}_m$, $\mathbf{G}_m \Delta \mathbf{u} = \mathbf{r}_m \mathbf{f}_m$

in the process industry is becoming an important motivation in a great amount of research work [79]. In this section an adaptive generalized predictive controller is presented. The adaptive controller has been developed by a method that makes use of the fact that a generalized predictive controller using a quadratic function results in a linear control law that can be described by a few parameters. These parameters can be computed over the range of interest of the process parameters [75, 83]. A Ziegler–Nichols-type function which relates the generalized predictive-controller parameters to the process parameters is used to obtain an approximation of the real controller parameters. The method avoids the heavy computation requirement of this type of controller. The method can be applied to processes that can be modeled by the reaction curve method, that is, a wide range of processes in industry.

5.9.2.1 Application to the ACUREX DSCF

According to the method described in [79] and the requirements for self-tuning control, a first-order CARIMA model is used:

$$(1 - az^{-1})y(k) = bz^{-1}u(k-1) + \frac{e(k)}{\Delta} \tag{5.31}$$

This expression can easily be transformed into

$$y(k+1) = (1+a)y(k) - ay(k-1) + b\Delta u(k-1) + e(k+1) \tag{5.32}$$

For simplicity, fixed values of the weighting sequences have been chosen ($\lambda(j)=\lambda$, $j=1,\dots,N_2-d$ and $\delta(j)=1$, $j=N_1,\dots,N_2$). Taking into account the values of the fundamental time constant and sampling period used for control purposes and that the system considered has a dead-time of $d=1$ sampling periods, values of $N_1=2$, $N_2=16$ and $N_u=15$ are chosen (notice that a lesser value of N_u could be chosen, but in this case, the value of the sampling time allows high control horizons to be used). The range of possible values of λ has been obtained via simulation ($3\le\lambda\le7$). If λ diminishes faster controllers are obtained. Equation (5.32) can be applied to obtain the expected value of $y(k+j+1)$:

$$\begin{aligned}\hat{y}(k+j+1\mid k)=(1+a)\hat{y}(k+j\mid k)-a\hat{y}(k+j-1\mid k)\\+b\Delta u(k+j-1)\end{aligned}\tag{5.33}$$

If Eq. (5.33) is applied recursively for $j=1,2,\dots,i$ we get

$$\hat{y}(k+i+1\mid k)=G_i\big(z^{-1}\big)\hat{y}(k+1\mid k)+D_i\big(z^{-1}\big)\Delta u(k+i-1)\tag{5.34}$$

where $G_i(z^{-1})$ is of degree 1 and $D_i(z^{-1})$ is of degree $i-1$.

If $\hat{y}(k+i+1\mid k)$ is introduced in Eq. (5.29), J is a function of $\hat{y}(k+1\mid k)$, $y(k)$, $\Delta u(k+14)$, $\Delta u(k+13),\dots,\Delta u(k)$ and the reference sequence. Minimizing J with respect to $\Delta u(k)$, $\Delta u(k+1),\dots,\Delta u(k+14)$ leads to

$$\mathbf{M}\begin{bmatrix}\Delta u(k)\\ \Delta u(k+1)\\ \vdots\\ \Delta u(k+14)\end{bmatrix}=\mathbf{P}\begin{bmatrix}\hat{y}(k+1\mid t)\\ y(k)\end{bmatrix}+\mathbf{R}\begin{bmatrix}r(k+2)\\ r(k+3)\\ \vdots\\ r(k+16)\end{bmatrix}\tag{5.35}$$

where $\mathbf{M}$ and $\mathbf{R}$ are of dimension 15×15 and $\mathbf{P}$ of dimension 15×2. Let us call $\mathbf{q}$ the first row of matrix $\mathbf{M}^{-1}$. Then $\Delta u(k)$ is given by

$$\Delta u(k)=\mathbf{qP}\begin{bmatrix}\hat{y}(k+1\mid t)\\ y(k)\end{bmatrix}+\mathbf{qR}\begin{bmatrix}r(k+2)\\ r(k+3)\\ \vdots\\ r(k+16)\end{bmatrix}\tag{5.36}$$

If the future references $r(k+j)$ are unknown and are considered to be equal to the current reference $r(k)$, the control increment $\Delta u(k)$ can be written as

$$\Delta u(k)=l_1\hat{y}(k+1\mid k)+l_2y(k)+l_3r(k)\tag{5.37}$$

where $\mathbf{qP}=[l_1l_2]$ and $l_3=\sum_{i=1}^{15}q_i\sum_{j=1}^{15}r_{ij}$.

Notice that if future references were known, $l_3=\mathbf{qR}$ would be a gain vector multiplying future references. The coefficients l_1, l_2, l_3, are functions of a,b, δ and λ. If the GPC is designed considering the system to have a unit static gain, the coefficients in (5.37) will only depend on δ and λ (which are supposed to be fixed) and on the pole of the plant which will change for the adaptive-control case. Notice that by doing this, a normalized weighting factor λ is used which should be corrected accordingly for systems with different static gains. Most of all, for systems with a static gain different from unity, the GPC can be obtained in this way by changing coefficients in Eq. (5.37) accordingly.

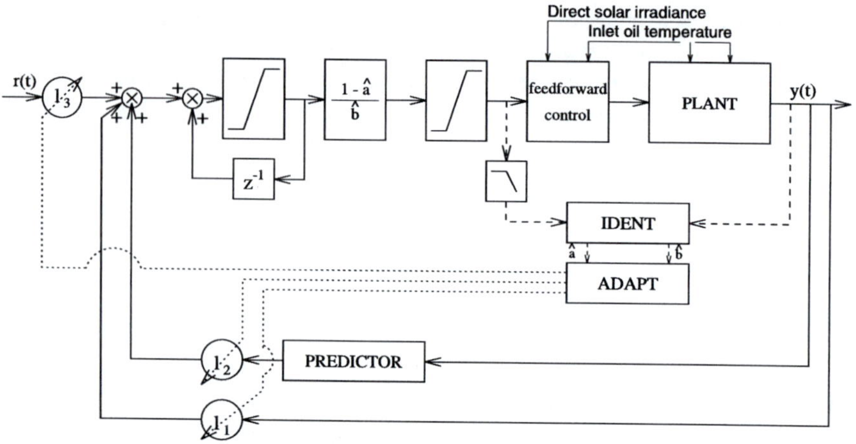

Fig. 5.35 Adaptive GPC control scheme [83]

The value $\hat{y}(k+1 \mid k)$ is obtained by the use of the predictor described previously (5.33). The control scheme proposed is shown in Fig. 5.35. The plant's estimated parameters $(\hat{a}, \hat{b})$ are used to compute the controller coefficients (l_1, l_2, l_3) via the adaptation mechanism. Notice that in this scheme, the series feedforward term developed in Chap. 4 is considered as part of the plant (the control signal is the set point temperature for the feedforward controller instead of the HTF flow). This signal is saturated and filtered before its use in the estimation algorithm. The control signal is divided by the estimated gain because the design has been performed presuming a system with unitary static gain.

As suggested in [78, 83], the controller coefficients can be obtained by interpolating in a set of previously computed values. The number of points of the set used depends on the variability of the process parameters and on the accuracy needed. The set does not need to be uniform and more points can be computed in regions where the controller parameters vary substantially in order to obtain a better approximation or to reduce the computer memory needed.

A set of GPC parameters (l_1, l_2, l_3) were obtained for $\delta = 1$, $\lambda = 5$ and $N = 15$. The pole of the system has been changed with a 0.0005 step from 0.8 to 0.95, which are values that guarantee the system stability if the parameter set estimation is not accurate enough (the same procedure has been performed for different values of λ between 3 and 7). Notice that owing to the fact that the closed-loop static gain must equal the value unity, the sum of the three parameters equals zero. This result implies that only two of the three parameters need to be known.

The curves shown in Fig. 5.36 correspond to the controller parameters l_1, l_2, l_3 for the values of the pole mentioned above. A set of simple Ziegler–Nichols-type functions which approximate the computed values of l_1, l_2 and l_3 has been obtained as follows. By looking at Fig. 5.36, it can be seen that the functions relating the

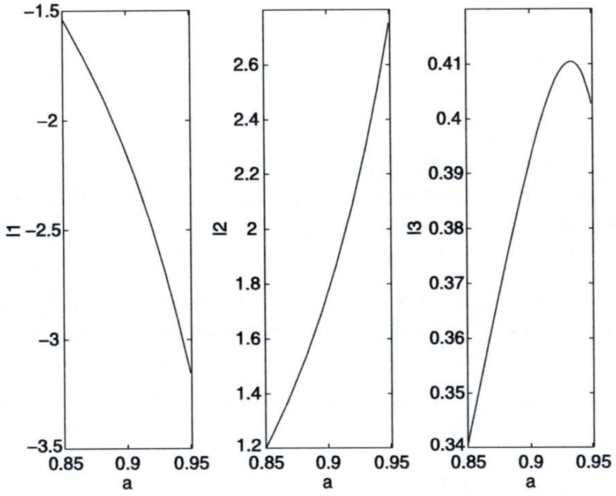

Fig. 5.36 GPC controller parameters for $\lambda = 5$, [83]

controller parameters to the process pole can be approximated by functions of the form

$$l_i = k_{1i} + k_{2i}\frac{a}{(k_{3i} - a)}, \quad i = 1,2 \tag{5.38}$$

The coefficients k_{ji} can be calculated by a least squares adjustment using the set of known values of l_i for different values of a.

For the selected design parameters, the controller coefficients are given by

$$\begin{aligned} l_1 &= 0.4338 - 0.6041\hat{a}/(1.11 - \hat{a}) \\ l_2 &= -0.4063 + 0.4386\hat{a}/(1.082 - \hat{a}) \\ l_3 &= -l_1 - l_2 \end{aligned} \tag{5.39}$$

These expressions give a very good approximation to the true controller parameters and fit the set of computed data with a maximum error of less than 0.6% of the nominal values for the range of interest of the open-loop pole.

In each sampling period k, the adaptive controller consists of the following steps:

1. An estimation of the parameters of a linear model by measuring the inlet and outlet values of the process.
2. The adjustment of the parameters of the controller using expressions obtained for l_i (5.39).
3. The computation of $\hat{y}(k+d \mid k)$ using the predictor (5.33).
4. The calculation of the control signal using (5.37).
5. The supervision of the correct working of the control.

5.9.2.2 Plant Results

When operating with the real plant, the objective is to obtain a response as quickly as possible, trying to avoid oscillations due to the excitation of the resonance modes or to a wrong controller parameter tuning. Due to the fact that simplified linear models have been used at the design stage of the control strategy, when a fast response is required, oscillations may appear in the outlet temperature due to unmodeled dynamics.

The disturbance rejection capabilities of this controller are similar to those of controllers analyzed in Chap. 4 (notice that in all these control schema the feedforward controller in series with the plant has been used).

As an example, Fig. 5.37 shows the outlet HTF temperature and reference temperature when controlling the plant with an adaptive GPC with a value of $\lambda = 5$. As can be seen, quite a fast response is obtained (rise time of about 6 min). The evolution of the direct solar irradiance in this test can be seen in the same figure. It corresponds to a day with scattered clouds. The HTF flow changed from 4.5 l/s to

Fig. 5.37 Test with the adaptive GPC controller (10/03/92) [83]

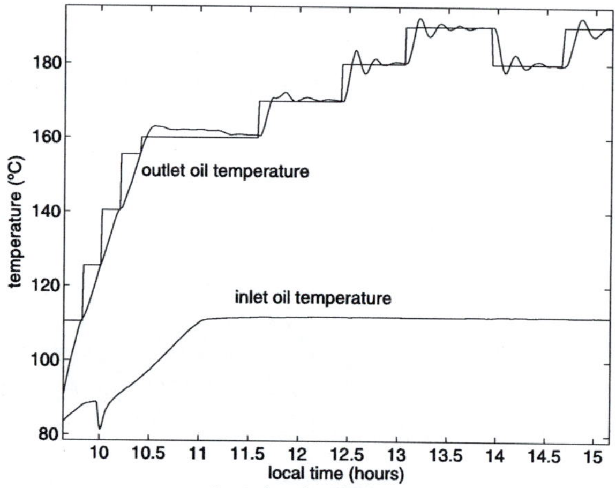

set point and outlet HTF temperature (ºC)

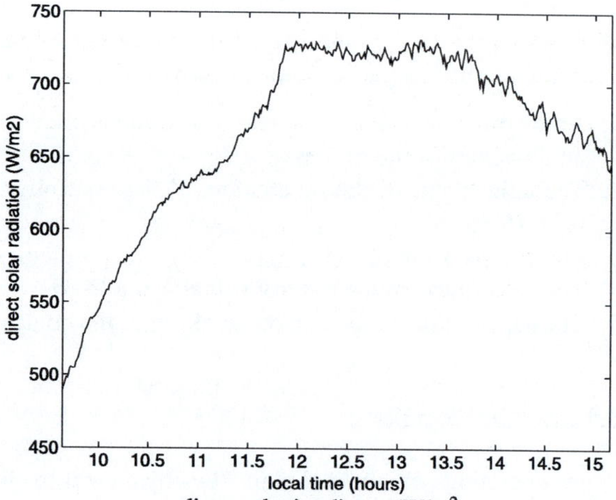

direct solar irradiance (W/m^2)

Fig. 5.37 (continued)

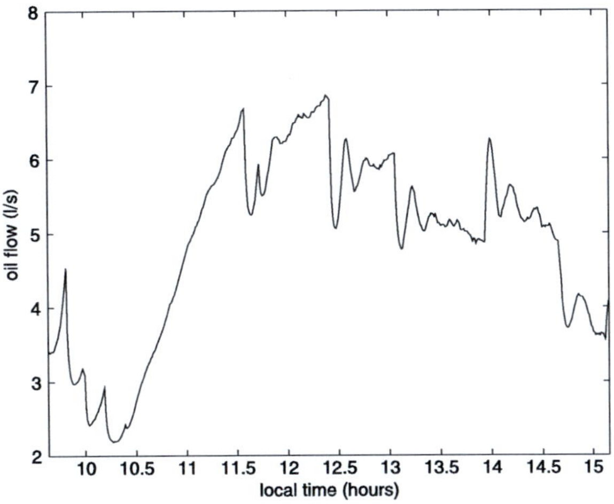

HTF flow (l/s)

7 l/s and the controller was able to maintain good behavior of the system controlled in spite of changes in the process dynamics.

Figure 5.38 shows the evolution of the controller parameters l_1 and l_2 (l_3 is linearly dependent on l_1 and l_2). As can be observed, the controller parameters follow the evolution of the HTF flow, which is the variable that mainly dictates the behavior of the system. More simulations and real results including constraints can be found in [85].

5.9.3 Gain Scheduling Generalized Predictive Control

The working principle of basic gain-scheduling controllers is simple and has been studied in Sect. 5.3. It is based on the possibility of finding auxiliary variables which guarantee a good correlation with process changing dynamics. In this way, it is possible to reduce the effects of variations in the plant dynamics by adequately modifying the controller parameters as functions of auxiliary variables. In the case of DSCF, the dynamics mainly depend on the HTF flow if the series feedforward controller is considered as a part of the plant, so that this variable can be used as the gain-scheduling auxiliary variable. Different sets of model parameters can be found for several operating conditions dictated by the volumetric flow so that a table of controller parameters can be computed for the defined operating points (in this case using GPC methodology). When coping with gain-scheduling control schema, stability and performance of the controlled system is usually evaluated by simulation studies [262]. A crucial point here is the transition between different operating points. In those cases in which non-satisfactory behavior is obtained, the number of inputs to the table of controller parameters must be augmented. As has been mentioned, it is important to point out that there is no feedback from the behavior of

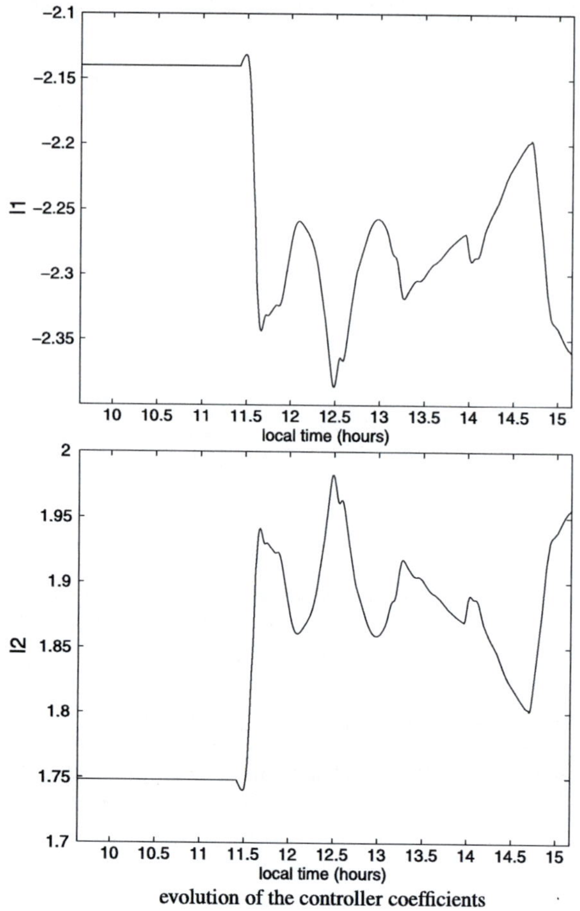

Fig. 5.38 Test with the adaptive GPC controller (10/03/92), [83]

the system controlled to the controller parameters. Thus, this control scheme is not considered as an adaptive one but as a special case of a non-linear controller.

The control structure proposed is shown in Fig. 5.39. As can be seen, the output of the generalized predictive controller is the input (T_{rff}) of the series compensation controller which also uses the solar irradiance and inlet HTF temperature to compute the value of the HTF flow which is sent to the pump controller.

The controller parameters were obtained from a linear model of the plant. Data for identification purposes was obtained from pseudo random binary sequence (PRBS) tests (see Sect. 4.3.3.2), so that the degrees of the polynomials A and B and the delay (of a CARIMA plant model) that minimizes Akaike's Information Theoretic Criterion (AIC) were found to be $n_a = 2$, $n_b = 8$ and $d = 0$. By a least squares estimation algorithm, the following polynomials were obtained using input–output data of one test with HTF flow around 6 l/s (which Bode diagram is shown in Fig. 4.5):

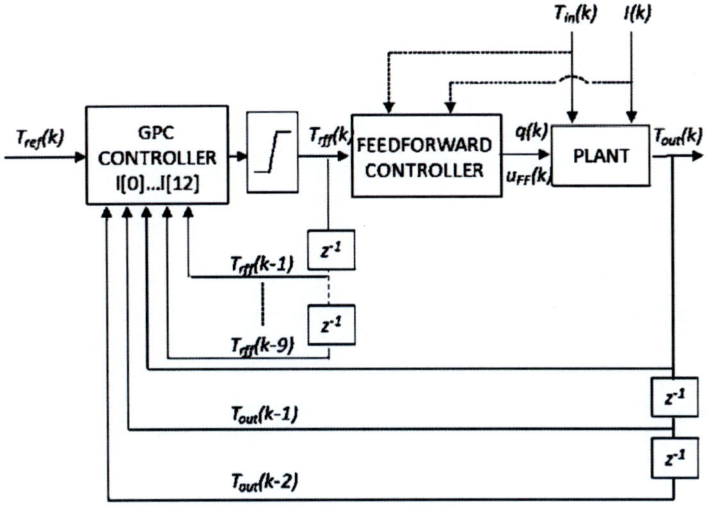

Fig. 5.39 Control scheme using high-order models [84]

$$A(z^{-1}) = 1 - 1.5681z^{-1} + 0.5934z^{-2}$$
$$B(z^{-1}) = 0.0612 + 0.0018z^{-1} - 0.0171z^{-2} + 0.0046z^{-3} + 0.0005z^{-4} + 0.0101z^{-5} - 0.0064z^{-6} - 0.015z^{-7} - 0.0156z^{-8}$$

The tuning knobs selected in the GPC algorithm were $N_1 = 1$, $N_2 = N_u = 15$, $\lambda \in [6, 7]$. Following the design procedure of the linear GPC methodology, the controller parameters corresponding to $\lambda = 7$ were obtained (Table 5.5).

The control law can be written by

$$T_{rff}(k) = \sum_{i=0}^{2} l[i]T_{out}(k-i) + \sum_{j=1}^{9} l[j+2]T_{rff}(k-j) + l[12]T_{ref}(k) \quad (5.40)$$

The behavior of this fixed-parameter controller was analyzed with these values, in the operation with the distributed solar collector field [85], showing good results around the design operating point but when operating conditions in the field change, the dynamics of the plant also change and the controller should be redesigned to cope with control objectives.

Table 5.5 Fixed GPC controller coefficients

$l[0]$	$l[1]$	$l[2]$	$l[3]$	$l[4]$	$l[5]$	$l[6]$
−4.7091	6.8215	−2.4483	1.0553	0.0231	−0.0631	0.0311

$l[7]$	$l[8]$	$l[9]$	$l[10]$	$l[11]$	$l[12]$
0.0161	0.0629	−0.0084	−0.0526	−0.0644	0.3358

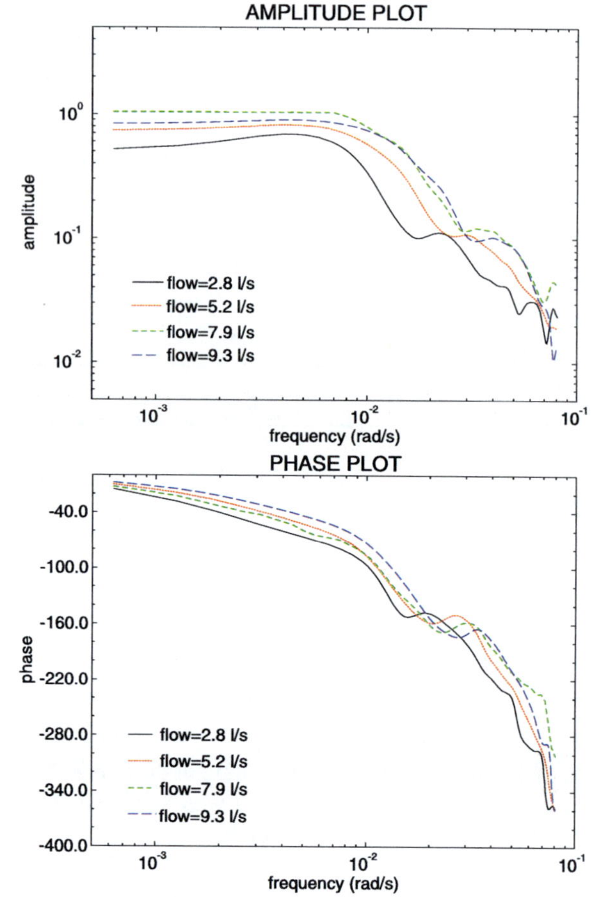

Fig. 5.40 Frequency response of the field in different operating conditions [84]

The dynamics of the field are mainly dictated by HTF flow which depends on the general field operating conditions: solar irradiance, reflectivity, HTF inlet temperature, ambient temperature and outlet HTF temperature. Figure 5.40 shows the frequency response of the non-linear distributed-parameter dynamic model of the field in series with the feedforward controller, obtained by a spectral analysis of the input–output signals of the model at different operating points (PRBS signals were used for the input). As can be seen, the frequency response changes significantly for different operating conditions. The steady-state gain changes for different operating points, as well as the location of the resonance modes.

Taking into account the frequency response of the plant and the different linear models obtained from it, it is clear that a self-tuning controller based on this type of model is very difficult to implement [85].

With the input–output data used to obtain the frequency responses shown in Fig. 5.40 and using the method and type of model previously described for the case of a high-order fixed-parameter controller, process ($a[i]$ and $b[i]$) and controller

Table 5.6 Coefficients of polynomials $A(z^{-1})$ and $B(z^{-1})$

HTF flow [l/s] / Model parameters	q_1	q_2	q_3	q_4
$a[1]$	−1.7820	−1.438	−1.414	−1.524
$a[2]$	0.81090	0.5526	0.5074	0.7270
$b[0]$	0.00140	0.0313	0.0687	0.0820
$b[1]$	0.03990	0.0660	0.0767	0.0719
$b[2]$	−0.0182	−0.0272	−0.0392	−0.0474
$b[3]$	−0.0083	0.0071	0.0127	0.0349
$b[4]$	0.00060	0.0118	0.0060	0.0098
$b[5]$	−0.00001	0.0138	−0.0133	−0.0031
$b[6]$	0.00130	0.0098	−0.0156	0.0111
$b[7]$	0.00160	0.0027	−0.0073	0.0171
$b[8]$	0.00450	−0.0054	0.0037	0.0200

($l[i]$) parameters were obtained for several HTF flow conditions ($q_1 \approx 2.8$ [l/s], $q_2 \approx 5.2$ [l/s], $q_3 \approx 7.9$ [l/s] and $q_4 \approx 9.3$ [l/s]), using different values of the weighting factor λ. Tables 5.6 and 5.7 contain model and control parameters, respectively, for a weighting factor $\lambda = 6$. A value of $\lambda = 7$ has also been used to obtain responses with less overshoot.

The controller parameters which are applied in the real operation are obtained by using a linear interpolation with the data given in Table 5.7. It is important to point out that to avoid the injection of disturbances during the controller gain adjustment,

Table 5.7 GPC controller coefficients in several operating points ($\lambda = 6$)

HTF flow [l/s] / Controller coefficients	q_1	q_2	q_3	q_4
$l[0]$	−9.5455	−2.7794	−2.6527	−2.0142
$l[1]$	16.2223	3.84390	3.48440	3.02280
$l[2]$	−7.0481	−1.4224	−1.1840	−1.3603
$l[3]$	0.82620	0.83010	0.89360	0.87390
$l[4]$	0.36470	0.16410	0.19600	0.12480
$l[5]$	−0.1575	−0.0822	−0.0869	−0.1098
$l[6]$	−0.0793	0.00880	0.03980	0.05070
$l[7]$	−0.0016	0.02480	0.02630	0.00460
$l[8]$	−0.0070	0.03390	−0.0239	−0.0197
$l[9]$	0.00560	0.02610	−0.0352	0.01080
$l[10]$	0.00980	0.00830	−0.0184	0.02730
$l[11]$	0.03910	−0.0139	0.00860	0.03740
$l[12]$	0.37130	0.35800	0.35230	0.35170

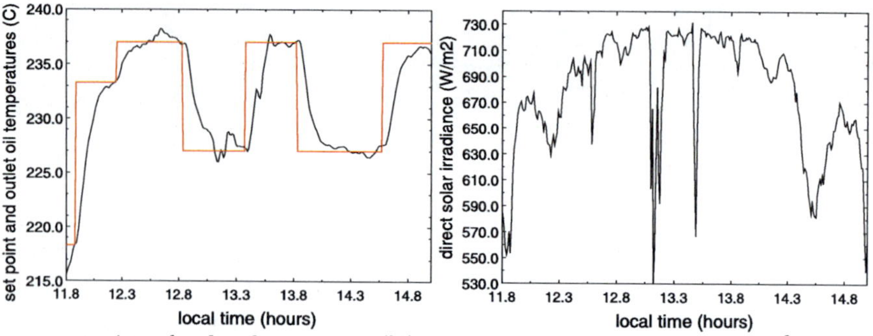

set point and outlet oil temperatures (°C) direct solar irradiance (W/m^2)

Fig. 5.41 Test with the gain scheduling GPC controller; $\lambda = 7$ (21/03/93) [84]

it is necessary to use a smoothing mechanism of the transition surfaces of the controller gains. In this case, a linear interpolation in combination with a first-order filter has been used, given a modified flow $q_f(k) = 0.95q_f(k-1) + 0.05q(k)$, being $q_f(k)$ the filtered value of the volumetric flow in discrete time k used for controller parameter selection. The linear interpolation has also been successfully applied by [194]. Another kind of gain-scheduling approach can be obtained by switching from one controller parameter to another depending on the flow conditions, without interpolating between controller parameters. The set of controller parameters c can be obtained by choosing between one of the sets c_i in Table 5.7, related to flow conditions q_i $(i = 1, 2, 3, 4)$:

$$\text{if} \quad \frac{q_{i-1} + q_i}{2} < q \leq \frac{q_i + q_{i+1}}{2} \quad \text{then} \quad c = c_i,\ i = 2, 3$$
$$\text{if} \quad q \leq q_1 \quad \text{then} \quad c = c_1$$
$$\text{if} \quad q \geq q_4 \quad \text{then} \quad c = c_4$$

The optimal realization of the gain-scheduling controller consists of calculating the controller parameters in a number of operating conditions and taking the values of the controller coefficients to be constant among different operating conditions, generating a control surface based on an optimization criterion which takes into account the tracking error and control effort. It is evident that if the procedure is applied at many working points, an optimum controller will be achieved for these operating conditions if there is a high correlation between the process dynamics and the auxiliary variable. The problem of this solution is that the design process becomes tedious. This is one of the main reasons for including a linear interpolation between the controller parameters.

As examples of the application of this technique to the ACUREX DSCF, Fig. 5.41 shows the results of a test corresponding to a day with sudden changes in the solar irradiance caused by clouds. As can be seen the controller (designed with $\lambda = 7$) is able to handle different operating conditions and the sudden perturbations caused by the clouds.

Fig. 5.42 Test with the gain-scheduling GPC controller; $\lambda = 6$ (29/05/95) [84]

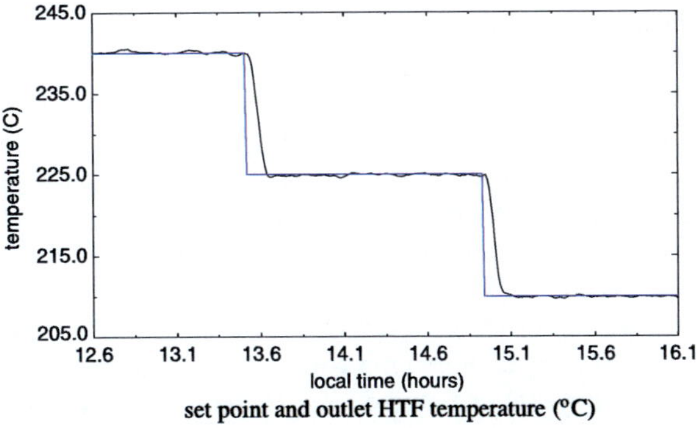

Figures 5.42 and 5.43 show the results obtained in the operation on a day with normal levels of solar irradiance but in which a wide range of operating conditions is covered (HTF flow changing between 2 and 8.8 l/s) by performing several set point changes. At the start of the operation there is an overshoot of 6°C, due to the irregular conditions of the HTF flow through the tubes because the operation starts with a high temperature level at the bottom of the storage tank. After the initial

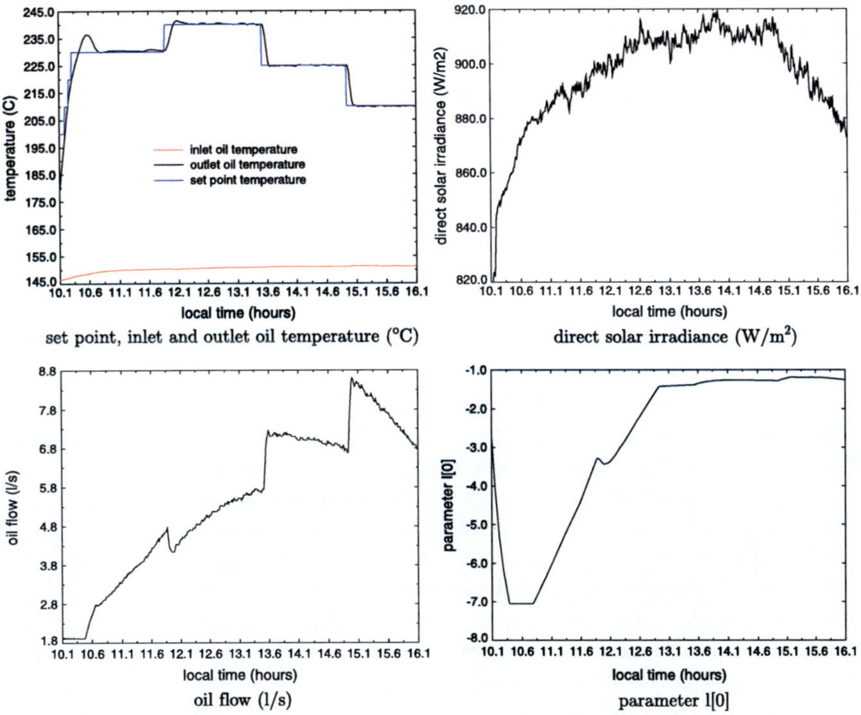

Fig. 5.43 Test with the gain scheduling GPC controller; $\lambda = 6$ (29/05/95) [84]

transient, it can be observed that the controlled system responds quickly to set point changes in the whole range of operating conditions with a negligible overshoot. The rise time is about 6 min with a set point change of 15°C, as can be seen in Fig. 5.42 (in which a zoom of the response is presented), with smooth changes in the control signal, constituting one of the best controllers implemented at the plant. It is important to note that the controller behaves well even with great set point changes. In Fig. 5.43, the evolution of one of the thirteen controller parameters ($l[0]$) is presented to show its variation in function of the operating conditions.

Figure 5.44 shows a test in which the solar desalination plant was connected to the field. The HTF coming from the desalination plant produced a decrease in the values of the temperature of the HTF entering the field. In this case, a decrease of 60°C was produced and the controller was able to cope with it. The outlet HTF temperature suffered an oscillation of less than 10°C. Moreover, the solar irradiance changes considerably and this disturbance was rejected by the control scheme. An operation with a highly constant temperature level (adequate for operating with the desalination plant) can be achieved. A change in the reflectivity value with which the feedforward controller calculates the adequate HTF flow was produced in order to analyze the controller robustness. At 10:50 am, the value of programmed reflectivity changed from 0.8 to 0.72. The transients generated were very small because of the corrections performed by the gain scheduling of the controller. More tests and simulations including disturbances and constraints can be found in [85].

5.9.4 *Robust Adaptive Model Predictive Control with Bounded Uncertainties*

In Sect. 5.7, classical robust control techniques have been commented on within the framework of control of DSCF, aimed at design controllers, which preserve stability and performance in spite of model inaccuracies or uncertainties. There are different approaches for modeling uncertainties, mainly depending on the type of technique used for designing the controllers. The most widespread types are frequency response uncertainties and transfer function parametric uncertainties. Most of the approaches assume that there is a family of models and that the plant can be exactly described by one of the models belonging to the family. That is, if the family of models is composed of linear models, the plant is also linear.

In [228] a robust MPC is developed for tracking piece-wise constant references applied to the ACUREX DSCF. The real plant is assumed to be modeled as a linear system with additive bounded uncertainties on the states. Under mild assumptions, the proposed controller can steer the uncertain system in an admissible evolution to any admissible steady state, that is, under any change of the set point. This allows constant disturbances to be rejected, compensating for their effect by changing the set point. The feasibility, stability and asymptotical convergence of the proposed controller for any admissible set point is achieved by adding an artificial steady state as decision variable, penalizing the deviation between this artificial steady state and

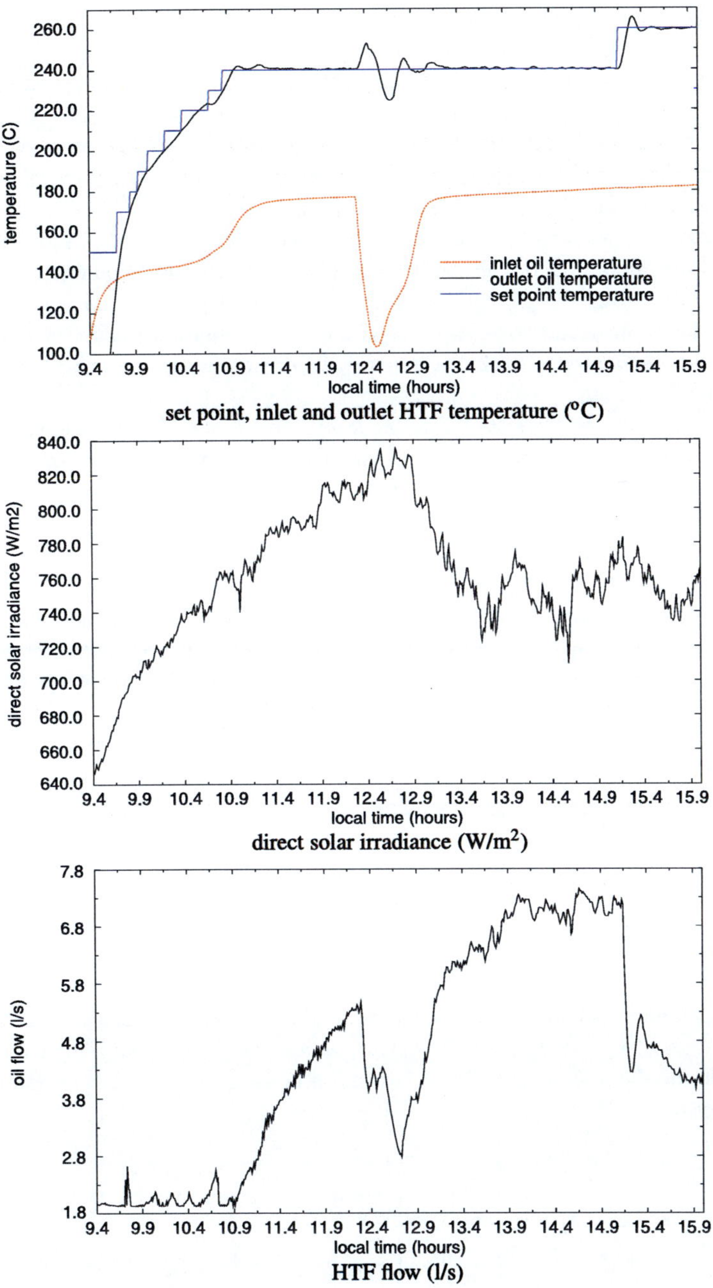

Fig. 5.44 Test with the gain-scheduling GPC controller, $\lambda = 6$ (12/06/95), [84, 85]

the real one in the cost function and using an invariant set for tracking as a terminal constraint. Robust constraint satisfaction is guaranteed by the tube-based approach; a nominal control problem is defined whose solution (a trajectory) defines the center of a tube; and where the *cross-section* of the tube is an invariant set. The state trajectory of the controlled system will be forced to lie in this tube by the control. At each instance, a new tube is determined by solving a control optimization in which, the decision variables are the initial state of the nominal system, the control sequence over a finite horizon and the artificial steady state.

This section presents the application of a robust adaptive-control scheme which uses a robust identification mechanism combined with a finite receding horizon controller to cope with the process dynamics with bounded uncertainties [77]. Uncertainties about the prediction capability of the model can be defined. The uncertainties will be considered to affect the transfer function parameters and the 1-step ahead prediction equation.

The robust controller is used in connection with a robust identification technique, which consists of determining membership sets for the parameter of the plant. Several robust identification methods have been proposed in the literature. These methods can be classified into three kinds, depending on the form of the membership sets: polyhedric, ellipsoidal and hypercubic methods. An estimation method based on a hypercubic parameter uncertainty set will be used here [252].

The key idea of the controller is that the identification algorithm determines (and progressively reduces) the uncertainty level about plant parameters. These uncertainty levels are used by a robust model predictive controller which optimizes the objective function for the worst possible case of the uncertainties.

5.9.4.1 Robust Identification Mechanism

Consider a SISO process whose behavior is dictated by the following equation:

$$y(k) = \sum_{i=1}^{n_a} a_i y(k-i) + \sum_{i=0}^{n_b} b_i u(k-d-i) + e(k) \tag{5.41}$$

where $y(k)$ and $u(k)$ are the output and input of the system, respectively, at discrete-time instant k; $e(k)$ is the modeling error, which is unknown but bounded (UBB); a_i and b_i are the parameters of the plant and d is the delay time. Let us define $\boldsymbol{\varphi}(k)$ and $\boldsymbol{\theta}$ as

$$\boldsymbol{\varphi}(k) = \begin{bmatrix} y(k-1) \\ \vdots \\ y(k-n_a) \\ u(k-d) \\ u(k-d-1) \\ \vdots \\ u(k-d-n_b) \end{bmatrix}; \quad \mathbf{a} = \begin{bmatrix} a_1 \\ a_2 \\ \vdots \\ a_{n_a} \end{bmatrix}; \quad \mathbf{b} = \begin{bmatrix} b_0 \\ b_1 \\ \vdots \\ b_{n_b} \end{bmatrix}; \quad \boldsymbol{\theta}^T = [\mathbf{a}^T \ \ \mathbf{b}^T]$$

with $n = n_a + n_b + 1$; $\boldsymbol{\varphi}(k), \boldsymbol{\theta} \in \mathbb{R}^n$; $\mathbf{a} \in \mathbb{R}^{n_a}$; $\mathbf{b} \in \mathbb{R}^{n_b+1}$ so that Eq. (5.41) can be rewritten as

$$y(k) = \boldsymbol{\varphi}^T(k)\boldsymbol{\theta} + e(k) \tag{5.42}$$

The modeling UBB errors are defined by

$$e_{min} \leq e(k) \leq e_{max} \tag{5.43}$$

and taking into account Eq. (5.41) the fundamental inequality of the systems with bounded uncertainties can be written:

$$y(k) - e_{max} \leq \boldsymbol{\varphi}^T(k)\boldsymbol{\theta} \leq y(k) - e_{min} \tag{5.44}$$

This relation implies that at each sampling instant k and after a new output is known, two hyperplanes are generated in $\mathbb{R}^n$ and the parameter vector $\boldsymbol{\theta}$ belongs to the region included between both hyperplanes. In this way, after ι sampling times, a set $\mathbf{S}(\iota)$, which is delimited by the ι pairs of hyperplanes, will be generated. This set $\mathbf{S}(\iota)$ must be consistent with the known error bounds (5.43) and the measurements (5.41).

$$\mathbf{S}(\iota) = \left\{\boldsymbol{\theta}/y(k) - e_{max} \leq \boldsymbol{\varphi}^T(k)\boldsymbol{\theta} \leq y(k) - e_{min},\ k = 1, \ldots, \iota\right\}$$

When the number of measurements is too large, $\mathbf{S}(\iota)$ becomes too complex to be used. In this case, $\mathbf{S}(\iota)$ can be approximated by a hypercube. The determination of this hypercube, which must include the set $\mathbf{S}(\iota)$, is carried out by applying the method proposed in [256]. This method has three steps:

1. Make $\mathbf{P}(0)$ the initial hypercube, defining its vertices $\mathbf{v}_i(0)$, $i = 1, \ldots, 2^n$, which must be large enough to include $\mathbf{S}(\iota)$, $\forall \iota \geq 0$.
2. Acquire $y(\iota)$ and calculate $\boldsymbol{\varphi}(\iota)$.
3. Determine the components of the vertices of the geometrical form resulting from the intersection of the hypercube $\mathbf{P}(\iota - 1)$ and the two new hyperplanes. Then the hypercube $\mathbf{P}(\iota)$ is determined by the maxima and minima of the components of the previous vertices. Let $\mathbf{V}(\iota)$ be the matrix that includes the components of all vertices of the geometrical form resulting from the intersection of the hypercube $\mathbf{P}(\iota - 1)$ and the two new hyperplanes generated. This matrix has n columns but its number of rows is variable at each sampling instant. Then, if r_v is the number of rows of $\mathbf{V}(\iota)$:

$$\mathbf{P}(\iota) = \left\{\boldsymbol{\theta}/\theta^j_{min}(\iota) \leq \theta^j \leq \theta^j_{max}(\iota),\ j = 1, \ldots, n\right\} \tag{5.45}$$

with $\theta^j_{min}(\iota) = \min_i v^j_i(\iota)$ and $\theta^j_{max}(\iota) = \max_i v^j_i(\iota)$, $i = 1, \ldots, r_v$ where $v^j_i(\iota)$ is the jth component of the vertex $\mathbf{v}_i(\iota)$; $\theta^j_{min}(\iota)$ is the minimum of the jth component of $\boldsymbol{\theta}$ and $\theta^j_{max}(\iota)$ is the maximum of the jth component of $\boldsymbol{\theta}$.

By using this robust identification mechanism the uncertainty level about plant parameters can be progressively reduced.

5.9.4.2 Robust Adaptive Model Predictive Control

In MPC techniques, the usual way of operating when considering a stochastic type of uncertainty is to minimize a function J for the most expected situation; that is, supposing that the future trajectories are going to be the future expected trajectories (as when using Eq. (5.29)). When bounded uncertainties are considered explicitly, bounds on the predictive trajectories can be calculated and it would seem that a more robust control would be obtained if the controller were to try to minimize the objective function for the worst situation. That is, by solving the following min–max problem:

$$\min_{\mathbf{u}} \max_{\boldsymbol{\theta},\mathbf{e}} J(\boldsymbol{\theta}, \mathbf{e}, \mathbf{u}) \tag{5.46}$$

subject to:

(1) $e_{min} \leq e(k+j) \leq e_{max},\ j = d, \dots, d+N-1$
(2) $\boldsymbol{\theta} \in \mathbf{P}(k)$
(3) $u_{min} \leq u(k+j) \leq u_{max},\ j = 0, \dots, N-1$

Define

$$\mathbf{u} = \begin{bmatrix} u(k) \\ u(k+1) \\ \vdots \\ u(k+N-1) \end{bmatrix}, \qquad \mathbf{e} = \begin{bmatrix} e(k+d) \\ e(k+d+1) \\ \vdots \\ e(k+d+N-1) \end{bmatrix}, \qquad \boldsymbol{\theta}^* = \begin{bmatrix} \boldsymbol{\theta} \\ \mathbf{e} \end{bmatrix}$$

$$\mathbf{u}, \mathbf{e} \in \mathbb{R}^N, \qquad \boldsymbol{\theta}^* \in \mathbb{R}^{N+n}$$

where $\boldsymbol{\theta}^*$ is the vector of generalized uncertain parameters and N is the prediction horizon. If we take into account that the error vector $\mathbf{e}$ must be included in the error hypercube $\boldsymbol{\Xi}$, which is defined by the bounds e_{min} and e_{max}, then we can define the generalized hypercube $\mathbf{T}(k)$ as

$$\mathbf{T}(k) = \{\boldsymbol{\theta}, \mathbf{e}\colon\ \boldsymbol{\theta} \in \mathbf{P}(k),\ \mathbf{e} \in \boldsymbol{\Xi}\} \tag{5.47}$$

Then the min–max problem (5.46) can be rewritten as

$$\min_{\mathbf{u}\in\mathbf{U}} \max_{\boldsymbol{\theta}^*\in\mathbf{T}(k)} J(\boldsymbol{\theta}^*, \mathbf{u}) = \min_{\mathbf{u}\in\mathbf{U}} J^*(\mathbf{u}) \quad \text{with } J^*(\mathbf{u}) = \max_{\boldsymbol{\theta}^*\in\mathbf{T}(k)} J(\boldsymbol{\theta}^*, \mathbf{u}) \tag{5.48}$$

The function to be minimized $J^*(\mathbf{u})$ is the maximum of a quadratic norm that measures how well the process output follows the reference trajectories. Let us consider a finite horizon quadratic criterion

$$J(\boldsymbol{\theta}^*, \mathbf{u}) = \sum_{j=N_1}^{N_2} \big(y(k+j \mid k) - r(k+j)\big)^2 + \lambda \sum_{j=1}^{N_u} \big(\Delta u(k+j-1)\big)^2 \tag{5.49}$$

where $\mathbf{r} = [r(k+d), \dots, r(k+d+N-1)]^T$ is a vector containing the future reference sequences and $y(k+j \mid k)$ is the worst case prediction output, taking into account that the future sequence error $\{e(k+d), \dots, e(k+d+N-1)\}$ is evaluated in the vertices of the hypercube Ξ. Other types of objective function have been used in literature. In [90] a ∞–∞ norm is used while in [4] a 1–∞ norm is proposed.

On the other hand, we define the following matrices:

$$\mathbf{G_a} = \begin{bmatrix} g(1) & 0 & \cdots & 0 \\ g(2) & g(1) & \cdots & 0 \\ \vdots & \vdots & \ddots & \vdots \\ g(N) & g(N-1) & \cdots & g(1) \end{bmatrix}, \qquad \mathbf{G_a} \in \mathbb{R}^{(N,N)}$$

$$g(i) = \sum_{j=1}^{\min(i-1,n_a)} a(j)g(i-j)$$

$$g(1) = 1; \qquad \boldsymbol{\vartheta} = [\boldsymbol{\vartheta}_\mathbf{y} \quad \boldsymbol{\vartheta}_\mathbf{u} \quad \mathbf{I}_N]; \qquad \boldsymbol{\vartheta} \in \mathbb{R}^{(N,n+N)}$$

Matrix $\boldsymbol{\vartheta}_\mathbf{y}$ depends on past output values $\{y(k+d-1), \ldots, y(k+d-n_a)\}$, $\boldsymbol{\vartheta}_\mathbf{u}$ is a matrix depending on past and future input values $\{u(k+N-1), \ldots, u(k), \ldots, u(k-n_b)\}$ and $\mathbf{I}_N$ is the identity matrix of order N. In this way, if the prediction Eq. (5.41) is used, then

$$\mathbf{y} = \begin{bmatrix} y(k+d) \\ y(k+d+1) \\ \vdots \\ y(k+d+N-1) \end{bmatrix} = \mathbf{G_a}\boldsymbol{\vartheta}\boldsymbol{\theta}^* \tag{5.50}$$

Equation (5.49) can now be written as

$$J(\boldsymbol{\theta}^*, \mathbf{u}) = (\mathbf{G_a}\boldsymbol{\vartheta}\boldsymbol{\theta}^* - \mathbf{r})^T(\mathbf{G_a}\boldsymbol{\vartheta}\boldsymbol{\theta}^* - \mathbf{r}) + J_u \quad \text{with } J_u = \lambda\sum_{j=1}^{N}(\Delta u(k+j-1))^2 \tag{5.51}$$

The function $J(\boldsymbol{\theta}^*, \mathbf{u})$ can be expressed as a quadratic function of the parameters b_i, $i = 0, \ldots, n_b$ and the errors $e(k+d+j)$, $j = 0, \ldots, N-1$, for each value of the vector $\mathbf{u}$:

$$J(\boldsymbol{\theta}^*, \mathbf{u}) = \frac{1}{2}\boldsymbol{\theta}^{*T}\mathbf{H}_\theta(\mathbf{a})\boldsymbol{\theta}^* + \mathbf{q}_\theta^T(\mathbf{a})\boldsymbol{\theta}^* + p_\theta \tag{5.52}$$

where the matrix $\mathbf{H}_\theta(\mathbf{a})$ and the vector $\mathbf{q}_\theta(\mathbf{a})$ depend on the parameters a_i, $i = 1, \ldots, n_a$. The Hessian matrix of function $J(\boldsymbol{\theta}^*, \mathbf{u})$ with respect to parameters $\mathbf{b}$ and $\mathbf{e}$ is $\mathbf{H}'_\theta(\mathbf{a}) = 2(\mathbf{G_a}\boldsymbol{\vartheta}')^T(\mathbf{G_a}\boldsymbol{\vartheta}')$, which is a positive semidefinite matrix, and where $\boldsymbol{\vartheta}' = [\boldsymbol{\vartheta}_\mathbf{u}\mathbf{I}_N]$. Since matrix $\mathbf{H}'_\theta(\mathbf{a})$ is positive semidefinite, function $J(\boldsymbol{\theta}^*, \mathbf{u})$ is convex in the hypercube of the parameters $\mathbf{b}$ and $\mathbf{e}$ [30] and we can assume the maximum of $J(\boldsymbol{\theta}^*, \mathbf{u})$ will be reached on one of the vertices of the hypercube of the parameters $\mathbf{b}$ and $\mathbf{e}$.

For a given $\mathbf{u}$ the maximization problem is solved by determining which of the $2^{(N+n_b+1)}$ vertices of the hypercube of $\mathbf{b}$ and $\mathbf{e}$ produces the maximum value of $J(\boldsymbol{\theta}^*, \mathbf{u})$. This operation is carried out by evaluating $J(\boldsymbol{\theta}^*, \mathbf{u})$ at all the $2^{(N+n_b+1)}$ vertices of the hypercube of $\mathbf{b}$ and $\mathbf{e}$ and by looking for the maximum of $J(\boldsymbol{\theta}^*, \mathbf{u})$ at an interior point of the hypercube of $\mathbf{a}$, due to the convexity of $J(\boldsymbol{\theta}^*, \mathbf{u})$ with respect to parameters $\mathbf{a}$ is not guaranteed. Also the maximum of $J(\boldsymbol{\theta}^*, \mathbf{u})$, at an interior point of the hypercube of $\mathbf{a}$, can be found by applying a grid of precision p^{n_a} because of the application of a robust identification algorithm, which yields a very small hypercube at a. If a finite truncated impulse model is used the convexity of $J(\boldsymbol{\theta}^*, \mathbf{u})$ is guaranteed with respect to all the parameters [4, 90].

A method was presented in [392], which looks for the maximum of J at the vertices of the hypercube of the predictor parameters. However, there are $N(N+n+1)$ predictor parameters and therefore the hypercube of the predictor parameters has $2^{N(N+n+1)}$ vertices. On one hand, this algorithm can be too conservative because there are vertices of the hypercube of the predictor parameters that do not correspond to the vertices of the hypercube of the real parameters and on the other hand, when N is big the amount of computation required becomes too large. When n_a is small and N big, the algorithm explained here can be faster than the above method.

The prediction vector can be written as

$$\mathbf{y} = \mathbf{G_u u} + \mathbf{f} \tag{5.53}$$

where matrix $\mathbf{G_u}$ depends on parameter vector $\boldsymbol{\theta}$ and vector $\mathbf{f}$ depends on $\boldsymbol{\theta}$ and past values. In this way, Eq. (5.49) can now be expressed as

$$J(\boldsymbol{\theta}^*, \mathbf{u}) = (\mathbf{G_u u} + \mathbf{f} - \mathbf{r})^T (\mathbf{G_u u} + \mathbf{f} - \mathbf{r}) + J_u \tag{5.54}$$

where $J_u = \lambda(\mathbf{Mu} - \mathbf{m})^T (\mathbf{Mu} - \mathbf{m})$ with

$$\mathbf{M} = \begin{bmatrix} 1 & 0 & 0 & \cdots & 0 \\ -1 & 1 & 0 & \cdots & 0 \\ 0 & -1 & 1 & \cdots & 0 \\ \vdots & & \ddots & \ddots & \vdots \\ 0 & 0 & \cdots & -1 & 1 \end{bmatrix}, \quad \mathbf{m} = \begin{bmatrix} u(k-1) \\ 0 \\ \vdots \\ 0 \end{bmatrix}, \quad \mathbf{M} \in \mathbb{R}^{(N,N)}, \quad \mathbf{m} \in \mathbb{R}^N$$

Thus, Eq. (5.54) can now be rewritten as

$$J(\boldsymbol{\theta}^*, \mathbf{u}) = \frac{1}{2}\mathbf{u}^T \mathbf{H_u u} + \mathbf{q_u}^T \mathbf{u} + p_\mathbf{u} \tag{5.55}$$

with

$$\begin{aligned} \mathbf{H_u} &= 2(\mathbf{G_u}^T \mathbf{G_u} + \lambda \mathbf{M}^T \mathbf{M}) \\ \mathbf{q_u} &= 2[\mathbf{G_u}^T (\mathbf{f} - \mathbf{r}) - \lambda \mathbf{M}^T \mathbf{m}] \\ p_\mathbf{u} &= (\mathbf{f} - \mathbf{r})^T (\mathbf{f} - \mathbf{r}) + \lambda \mathbf{m}^T \mathbf{m} \end{aligned}$$

where $\mathbf{H_u} = \mathbf{H_u}(\boldsymbol{\theta}^*)$, $\mathbf{q_u} = \mathbf{q_u}(\boldsymbol{\theta}^*)$ and $p_\mathbf{u} = p_\mathbf{u}(\boldsymbol{\theta}^*)$. Note that $J(\boldsymbol{\theta}^*, \mathbf{u})$ is a quadratic function of $\mathbf{u}$.

It can easily be seen that function $J^*(\mathbf{u})$ is a piece-wise quadratic function of $\mathbf{u}$. Let us divide the $\mathbf{u}$ domain $\mathbf{U}$ into different regions $\mathbf{U_p}$ so that $\mathbf{u} \in \mathbf{U_p}$ if the maximum of $J(\boldsymbol{\theta}^*, \mathbf{u})$ is attained for the vertex $\boldsymbol{\theta}_p^*$ (note that $\boldsymbol{\theta}_p^*$ is only a vertex with respect to vectors $\mathbf{b}$ and $\mathbf{e}$). For region $\mathbf{U_p}$ the function $J^*(\mathbf{u})$ is defined by

$$J^*(\mathbf{u}) = \frac{1}{2}\mathbf{u}^T \mathbf{H_u}(\boldsymbol{\theta}_p^*)\mathbf{u} + \mathbf{q_u}(\boldsymbol{\theta}_p^*)^T \mathbf{u} + p_\mathbf{u}(\boldsymbol{\theta}_p^*) \tag{5.56}$$

The Hessian matrix of function $J^*(\mathbf{u})$, $\mathbf{H_u}(\boldsymbol{\theta}_p^*)$ can be ensured to be positive definite by choosing a value of $\lambda > 0$. This implies that function J is strictly convex and that there are no local optimal solutions other than the global optimal solution [30]. These results imply that the maximum function has no local optimal minimum other than the global optimal minimum. In this way, it can be ensured that the solution of the min–max problem is *unique* and it is the minimum of function $J^*(\mathbf{u})$. One of the main problems of numerical optimization, the existence of local minima, is avoided and any non-linear optimization method can be used to solve the problem.

5.9.4.3 Application to the ACUREX DSCF

This method was applied in the ACUREX DSCF. The model chosen to implement the algorithm has been a simplified one obtained from the characterization of the step response of the plant (Type B Model including the series feedforward controller, Eq. (4.48)), given by $G_B(z^{-1}) = z^{-1}\frac{b_0 z^{-1}+b_1 z^{-2}}{1-az^{-1}}$. In [85], simulation studies were carried out to analyze the performance of the technique, varying the static gain within the range [0.9 1.2] and the pole (a) within [0.8 0.95]. An analysis of a large number of tests at the real plant indicated that the maximum level of uncertainty for the UBB errors is about $e_{min} = -0.5$ and $e_{max} = 0.5$. The design values chosen for the robust adaptive GPC controller were: weighting factor $\lambda = 15$, prediction and control horizons $N_1 = 1$, $N_2 = N_u = 7$. With these values of the tuning knobs, the computational effort was quite high (the number of vertices to be analyzed was $2^{N(N+n+1)} = 2^{70}$ and applying a grid of precision $p = 5$ results in 1280 vertices) to allow a real time implementation of the algorithm [85] (at the time of the tests, nowadays, there are more efficient algorithms and computing facilities to implement the min–max approach).

In the real tests, in order to diminish the computational effort required to implement the robust adaptive controller, parameter a was fixed at a value of 0.85. Parameters b_0 and b_1 may change between limits chosen to take into account the maximum and minimum value of the static gain. These values were $0.14 \le b_0 \le 0.16$, $-0.005 \le b_1 \le 0.005$, which produce $0.9 \le K \le 1.06$. Changes in the gain of the process and delay time (the zero acts as a Padé approximation of a non-integer delay time [82]) are therefore taken into account. In this case, the design values chosen were: $\lambda = 5$, $N_1 = 1$, $N = N_2 = 4$, $e_{min} = -0.5$, and $e_{max} = 0.5$. A gradient method was used to find the optimal constrained control sequence. The problem could be solved in real time because only two parameters (b_0, b_1) were identified and a small control horizon ($N = 4$) was used. Notice that to calculate $J^*(\mathbf{u})$ requires the computation of a quadratic function at $p \cdot 2^{N+n_b+1}$ points.

Figure 5.45 shows the evolution of the outlet temperature when controlling the plant with the proposed control scheme. The starting phase has been carried out using a fixed-parameter GPC ($a = 0.85$, $b_0 = 0.15$ and $b_1 = 0$) to avoid a large overshoot. As can be seen, the response is slow and takes a long time to reach the first set point. The inlet HTF temperature changes substantially during this phase, because the field had been out of operation for a long period before this test. After this phase, a change in the set point is performed, freezing the evolution of the

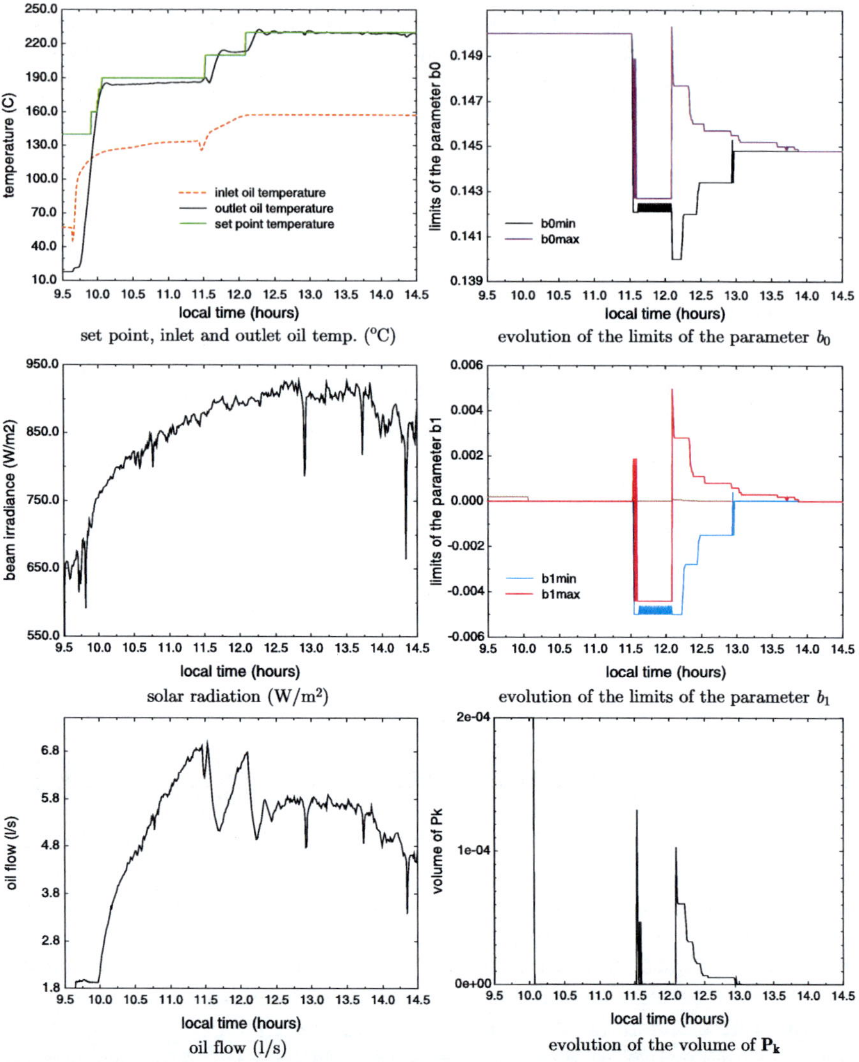

set point, inlet and outlet oil temp. (°C)

evolution of the limits of the parameter b_0

solar radiation (W/m^2)

evolution of the limits of the parameter b_1

oil flow (l/s)

evolution of the volume of $\mathbf{P_k}$

Fig. 5.45 Test carried out with the robust adaptive GPC controller (20/10/95) [77]

limits of b_0 and b_1, as can be seen in the same figure. The inlet HTF temperature changes at this point due to non-uniformities in the storage tank. The outlet HTF temperature shows a considerable offset because of the increase of the inlet HTF temperature and the increase of solar irradiance according to its daily cycle. After this intermediate phase of the test, the set point value is augmented 20°C to reach the final operating point. The limits of the polytope are opened up to allow the robust identification mechanism to choose the correct parameters corresponding to the dynamics of the new operating point. As can be seen, the evolution of the limits of b_0 and b_1 is such that the volume of the hypercube decreases with time and the level of uncertainty of the closed-loop system decreases quickly. Some disturbances due to changes in solar irradiance are produced between half past twelve and half past two. The feedforward controller diminishes the effect of these disturbances, but some large changes of about 200 W/m^2 affect the outlet HTF temperature. The controller reaches the set point again before the next big disturbances are produced.

5.9.5 Non-linear MPC Techniques (NMPC)

Because the majority of controlled processes have inherently non-linear behavior, there are incentives to develop MPC control strategies based on non-linear process models, both obtained from physical principles or data, in this last case mainly black-box models based on ANN (see [203, 204] for a comprehensive review of applications of ANN in renewable energy systems). In these cases, a non-linear programming problem must be solved in real time at every sampling period instead of the quadratic linear problem typical of standard MPC. The main difficulties of these methods are that the theoretical analysis of properties of the closed-loop such as stability and robustness is very complicated because of the appearance of non-linear models in the formulation (application or simulation examples are, therefore, usually used which do not guarantee the generality of the results or are not representative of them) and that if the solution of solving a non-linear programming problem at each sampling period is adopted, it is difficult to guarantee the convergence of the algorithm in an adequate lapse of time.

In [76] a MPC control scheme is presented where the response of the plant is divided into the forced and free terms. This division allows the use of a linear model for the forced response, from which the optimal sequence of control actions is obtained without a need for numerical methods. Also, the effect of disturbances is taken into account, thanks to a non-linear model of the free response. In [34] a non-linear model based on first principles is used to obtain the free response of the plant and in [18] an ANN. In [297–299] a non-linear model-based predictive controller (NMPC) is developed using a simplified mathematical model of the plant and a search strategy minimizing a cost function for a given prediction horizon. The parameters of the non-linear model are estimated on-line in order to compensate for time-varying effects and modeling errors. In [319], an improved FL strategy is proposed based on the simplified bilinear models developed in Sect. 4.3.4.2 and used in a basic FL approach in Sect. 5.8.1. The benefits of input–output FL are improved using a FSP-MPC algorithm [275] with embedded variable constraints mapping in order to take advantage of: (i) using a linear control without loosing the intrinsic non-linearities typical of thermal power plants; (ii) including input amplitude constraint handling capabilities due to control signal saturations induced, for example, by hard irradiance disturbances or plant start-up; and (iii) avoiding unstable or highly oscillatory responses caused by plant–model mismatch.

In [71] a non-linear optimal feedforward control of the solar collector field of an air conditioning plant is developed. The controller operates in an MPC framework, without feedback and constraints, to highlight feedforward controller performance gains. To reduce the computational complexity, state extension is avoided when modeling the time delays. The motivation for this is provided by a proof of the equivalence between state extension and direct prediction in optimal control for non-linear multiple-input multiple-output (MIMO) systems with time delays in disturbances and controls. The solar collector controller also uses flow-dependent sampling to reduce the time variation of the delays. To obtain an accurate model for feedforward controller design, a black-box recursive prediction error identification algorithm was used for modeling the non-linear plant using measured data [70]. Experimentally, accurate feedforward control was obtained when the controller design was tested by simulation, using measured disturbances from the plant. About two-third of the control effort needed appears to be available by feedforward control only. In order to evaluate the algorithms under difficult conditions, the evaluation was performed on data obtained during a day with partly cloudy weather, this caused large flow variations and consequently large time delay variations.

In [146] an NMPC extended with a DTC is proposed to control a DSCF using data of the ACUREX field while in [372] the approach is applied to the collector field of a solar desalination plant. This non-linear controller uses the non-linear extended prediction self-adaptive-controller (NEPSAC) algorithm. A non-linear gray-box model of the plant, based on first principles (Eqs. (4.1) and (4.2)) and tuned according to real measurements, is used in the simulation tests, including a model of the transport delay given by Eq. (4.14) [274]. The resulting controller is compared to other architectures based on DTC, showing very good performance for reference tracking and for disturbance rejection.

In [301], the temperature control of DSCF which employs molten salt as HTF is developed, based on a hybrid adaptive-control scheme and a time-warped predictive controller, where the HTF temperature can be effectively controlled within prescribed constraints and also in presence of uncertainty in model parameters and faults on collectors. This work illustrates how the uncertainty on the local flow rate can be incorporated into a distributed state-space model of the loop. Then a novel adaptive-control scheme is formulated in order to accomplish the requirements of the distributed individual control for molten salt solar plants. This scheme relies on the coupling of a non-linear filter with a time-warped controller. The state estimation and the parameter identification problems are addressed by a Dual Unscented Kalman Filter, designed on the basis of a constant sampling time distributed-parameter bilinear model. A variable sampling time Lagrangian model can be constructed with the same parametrization of the Eulerian one. The Lagrangian approach provides an exact linearization of the plant dynamics, so it is suitable for the design of a predictive controller. The prediction of the optimal control action is performed in a time-warped time scale, whereas the control moves can be calculated whenever a new estimate of system state and parameters is available. In this way the cycle time of the controller is made equal to the arbitrary sampling period of the filter. In [155] a non-linear adaptive constrained MPC scheme is presented where the methodology exploits the intrinsic non-linear modeling capabilities of non-linear state-space ANN and their on-line training by

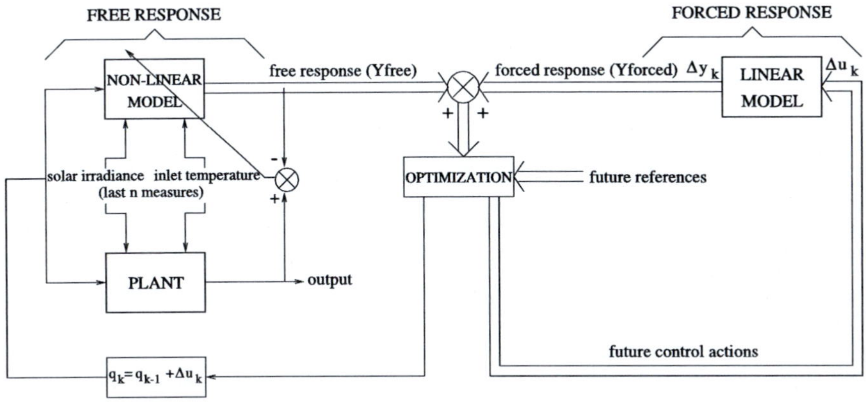

Fig. 5.46 Non-linear prediction control scheme [76]

means of an unscented Kalman filter. In [154, 155] another non-linear adaptive constrained MPC scheme with steady-state offset compensation is developed and implemented.

5.9.5.1 GPC Scheme with Non-linear Prediction of the Free Response

MPC of non-linear processes leads to non-linear and usually non-convex optimization problems which are computationally very demanding. This section describes a method which makes use of the fact that although the plant characteristics change from operating point to operating point, linearity can be assumed in the neighborhood of a particular operating point. The GPC control algorithm is modified to include a kind of free response obtained from a non-linear model at the optimization stage of the algorithm. This control strategy allows for including the effect of measurable disturbances within a GPC framework [76]. This scheme does not use the series feedforward controller developed in Chap. 4, but includes the disturbance dynamics within a non-linear process model used for prediction purposes.

The control structure can be seen in Fig. 5.46. A non-linear model of the plant such as that given by Eqs. (4.1) and (4.2) is used to generate the free response of the plant due to past control actions and past and future disturbances (calculated by using a convenient predictor) and considering the control signal as being at the last value. Notice that the predictive nature of the GPC and the availability of a model of the process make it possible to incorporate the dynamic disturbances caused by changes in solar irradiance and inlet HTF temperature into a GPC framework. A linear incremental model is used to generate the forced response (in this case a first-order model with a delay of one sampling period). The incremental control actions are obtained from the linear model by minimizing:

$$J = E\left\{ \sum_{j=N_1}^{N_2} \left[\hat{y}_{free}(k+j \mid k) + \sum_{i=N_1}^{i=j} \Delta\hat{y}_{forced}(k+i \mid k) - r(k+j) \right]^2 + \sum_{j=1}^{N_u} \lambda(j)\big[\Delta u(k+j-1)\big]^2 \right\} \tag{5.57}$$

where $\hat{y}_{free}(k+j \mid k)$ is a j-step ahead prediction of the free response on data up to time k and $\Delta\hat{y}_{forced}(k+i \mid k)$ is a i-step ahead prediction of the forced response on data up to time k. The output of the plant is used to update the non-linear model state vector.

A good load disturbance prediction model is also necessary. As justified in [85], for prediction purposes, the inlet HTF temperature has been considered constant at the prediction horizon in the optimization part of the algorithm, while two approaches for solar irradiance prediction were adopted: (i) considering a constant value during the prediction horizon equal to the last measured value, (ii) using the beam solar irradiance prediction model explained in Sect. 2.4.2.1.

The completely free response prediction scheme consists mainly of the following steps:

1. Calculate off-line the clear day direct solar irradiance prediction. Values obtained from this calculation are called $I_m(k)$.
2. Calculate direct solar irradiance prediction increments $\Delta I_m(k) = I_m(k) - I_m(k-1)$.
3. During the operation, at each sampling time k:
   a. Measure the inlet HTF temperature and consider it constant along the prediction horizon.
   b. Measure the real direct solar irradiance $I(k)$.
   c. Perform the direct solar irradiance predictions $\hat{I}(k+j)$, $j = 1, \ldots, N$ in the following way:
      i. $\hat{I}(k+1) = I(k) + \Delta I_m(k+1)$.
      ii. $\hat{I}(k+j) = \hat{I}(k+j-1) + \Delta I_m(k+j)$ with $j = 2, \ldots, N$.

      In this way, an implicit feedback of the disturbance values is obtained. Notice that these computations are not necessary if the approach $\hat{I}(k+j) = I(k)$, $j = 1, \ldots, N$ is adopted.
   d. Use the predicted values in the non-linear model to calculate the free response (calculated considering $u(k) = u(k+1) = \cdots = u(k+N-1)$).

When applying this technique to the ACUREX DSCF, a forced response CARIMA model for medium flows was obtained using polynomials $A(z^{-1}) = 1 - 0.89973z^{-1}$, $B(z^{-1}) = -0.629035z^{-2}$. Model polynomials have to be changed to take into account other different operating points. In this particular case, due to the strong influence of solar irradiance on the absolute value of the control signal, the contribution of the linear part in the control signal is smaller than that of the free response. In this way, even in cases of erroneous estimation of A and B polynomials, good results can be achieved. As can be seen, the model has negative steady-state gain because of the relation between HTF flow changes and outlet temperature variations.

The minimization of the cost function (5.57) provides the vector of control increments $\Delta\mathbf{u} = (\mathbf{G}^T\mathbf{G}+\lambda\mathbf{I})^{-1}\mathbf{G}^T(\mathbf{y}_{free}-\mathbf{r})$, where matrix $\mathbf{G}$ contains the step response coefficients of the forced response model.

The control algorithm presented in this section has been applied to control the DSCF ACUREX. Figure 5.47 presents the results obtained with the proposed control scheme considering the CARIMA model in the forced response calculation. This figure shows the evolution of set point temperature, inlet HTF temperature, outlet HTF temperature, HTF flow, HTF flow increments, direct solar irradiance and clear day prediction of this variable. The operation corresponds to that of an almost clear day. The rain of previous days had cleaned the collectors, originating high values of reflectivity (not known when the test was performed). The inclusion of the integral effect in the controller contributed to obtaining offset-free control. The rise time obtained was quite fast (less than 5 min in changes of 10°C in the set point temperature), with overshoot of less than 3°C and small settling time. The greatest overshoot occurs after the last set point change, because of fall in the solar irradiance levels. The HTF flow evolution is such that sudden changes are avoided. It is very important to mention that with these control algorithms, fairly fast responses are obtained with little control effort (small oscillations in the HTF flow), which is a very desirable property.

Figure 5.48 shows the evolution of the outlet HTF temperature and its predictions (5th and 10th). In this case, the predictions are very similar to the actual values obtained.

The same approach was used in [16] using as non-linear model of the free response an ANN.

5.9.5.2 Robust Constrained Predictive Feedback Linearization Controller

The FL technique developed and tested in Sect. 5.8.1 can be combined with GPC to control the outlet temperature of the solar field, where the constraints can be computed using the predicted future outputs and the state and disturbances at each sampling time using similar ideas to those in [218, 219]. This combination of GPC with FL allows constraint management but, as in all MPC techniques, it is necessary to bear in mind that unmodeled dynamics may lead to aggressive control actions or even unstable responses. Hence, a modification of the GPC algorithm to cope with uncertain dead-time processes was used, namely, the DTC-GPC [270]. In this algorithm, the GPC is modified so that a FSP [270] is used for the predictions up to the dead-time. The algorithm provides the following benefits [319]:

- By using a typical non-linear mapping of the FL technique, a linear control may be used over the non-linear plant without loosing the intrinsic non-linearities of the process.
- An on-line constraint-mapping treatment with state and disturbance dependences is embedded in the controller algorithm.
- Dead-time errors are handled using a robust solution.

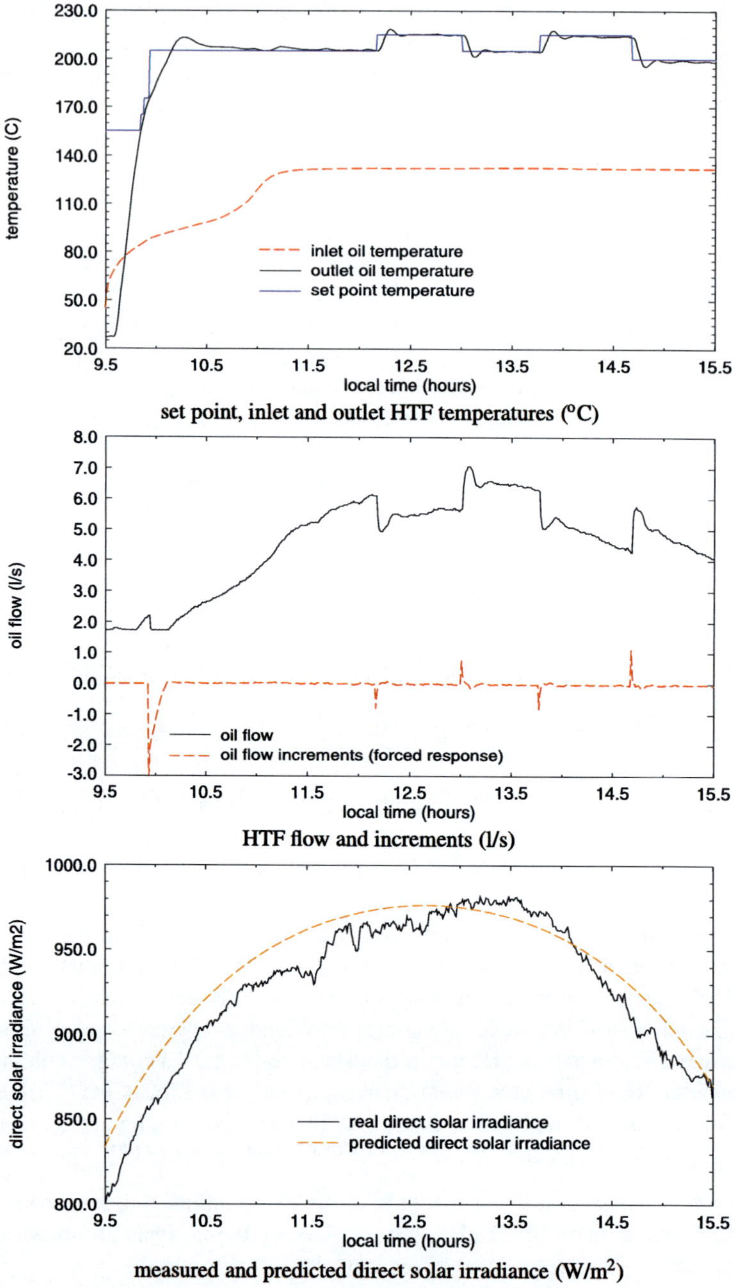

Fig. 5.47 Test with the non-linear GPC controller (03/11/95) [76, 85]

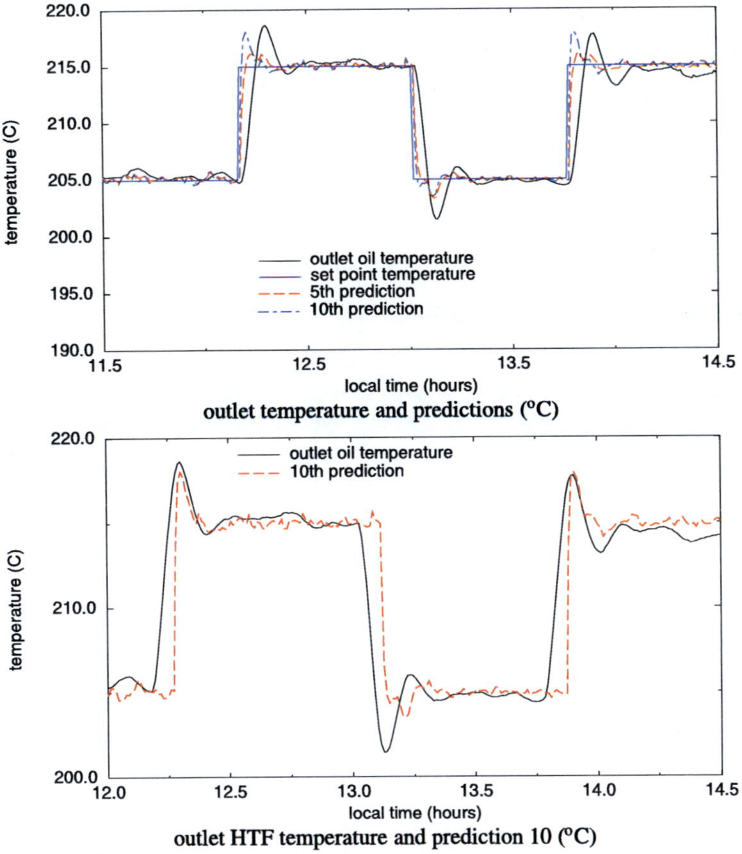

Fig. 5.48 Test with the non-linear GPC controller (03/11/95) [85]

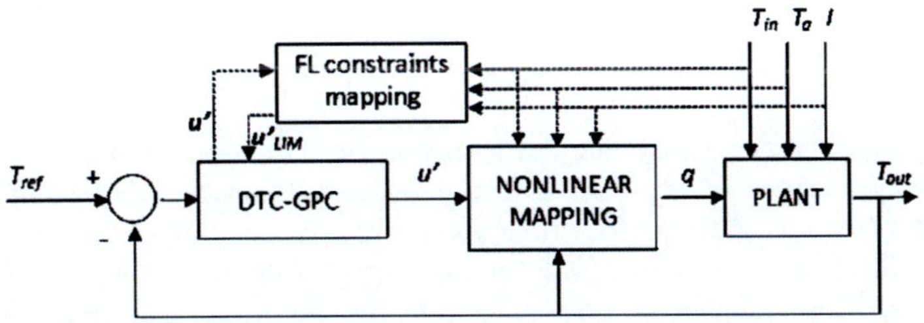

Fig. 5.49 DTC-GPC-FL control diagram (courtesy of L. Roca et al., [319])

The control algorithm is depicted in Fig. 5.49. The DSCF model used is the same as that in Sect. 5.8.1, including the estimation of variable transport delay (see [319] for more details when applied to the DSCF of a solar desalination plant). A DTC-GPC algorithm acts as the linear control for the outlet solar field temperature

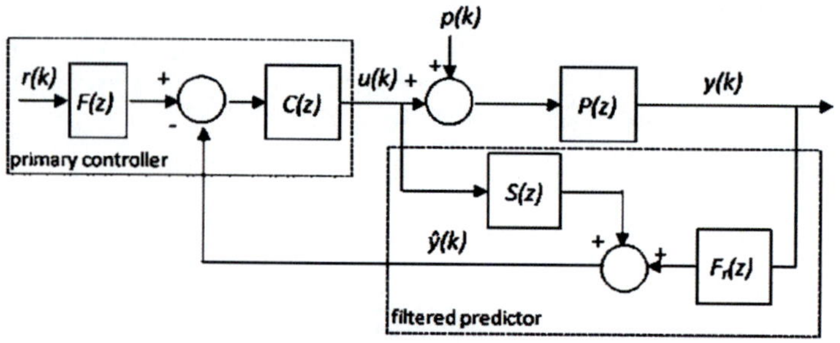

Fig. 5.50 DTC-GPC structure for the unconstrained case, [270]

using the virtual signal, u', as the control signal. The non-linearities of the plant are included in the non-linear mapping block in which the state and disturbances must be added as inputs to obtain the real control signal of the plant (solar field flow, q). On the other hand, the DTC-GPC algorithm provides a vector, $\mathbf{u}'$, with the future predictions of the virtual signal that, together with the actual state and disturbances, are used to recalculate at each sampling time the suitable constraints over the virtual signal, $\mathbf{u}'^{LIM}$. The calculation of the FL control law (involving both the real q and virtual u' control signal) is performed following the same approach explained in Sect. 5.8.1, providing the actual control signal $q(k)$ given by Eq. (5.25).

The DTC-GPC algorithm is used in combination with the FL technique to control the outlet temperature of the solar field despite disturbances and system uncertainties. As the traditional GPC control technique, the DTC-GPC algorithm consists of applying a control sequence that minimizes the multistage cost function given by Eq. (5.29) using an incremental model of the form $\Delta A(z^{-1})y(k) = z^{-d}B(z^{-1})\Delta u(k-1)$.

The prediction of the output of the plant up to $k+d$ is computed using an FSP structure such as the one shown in Fig. 5.50, where $S(z^{-1}) = G_n(z^{-1})[1 - F_r(z^{-1})z^{-d}]$ is the filter tuning parameter and $G_n(z^{-1}) = z^{-1}B(z^{-1})/A(z^{-1})$ is the dead-time free process model. The prediction can be written in a compact form as

$$\hat{y}(k+d \mid k) = S(z^{-1})u(k) + F_r(z^{-1})y(k) \tag{5.58}$$

In the unconstrained case the final structure of the controller can be drawn as in the block diagram of Fig. 5.50. This scheme is a linear dead-time compensator equivalent to the FSP where the primary controller is tuned using an optimization procedure. As has been shown in [270], the tuning of the filter can be used to improve closed-loop robustness and also to allow the use of the controller with unstable open-loop dead-time processes. Note that when the process model is unstable, $S(z^{-1})$ must be implemented with a stable transfer function [270]. The previous structure is useful for analysis and in practice, the control law can be computed using the step and free response of the system, as in the GPC algorithm [112]. In this case, the prediction is considered as $\mathbf{y} = \mathbf{G}\mathbf{u} + \mathbf{f}$, where the free response, $\mathbf{f}$, is computed as follows. From $k+1$ to $k+d$, $\mathbf{f}$ is computed using the FSP structure and from $k+d+1$ to $k+N_2$ it is calculated using the normal procedure in the GPC.

In the unconstrained case, the following two-step procedure can be used for tuning the controller, based on ideas of dead-time compensation control [272]:

- Compute the horizons and weighting factors in order to obtain the desired set point performance for the nominal plant.
- Estimate the uncertainties of the plant and tune the filter $F_r(z)$ in order to obtain robust stability and the highest bandwidth for the disturbance rejection performance.

Following the ideas in [271, 272, 319], for the particular case of an integrative process resulting from the FL approach (using the virtual control signal u', $G_n(z) = T_s/(z-1)$ as the input and considering step disturbances) a second-order filter is selected

$$F_r(z) = \frac{f_{b1}z + f_{b0}}{(z - a_f)^2} \tag{5.59}$$

where $f_{b0} = (1 - a_f)^2 d + 2(1 - a_f)$, $f_{b1} = (1 - a_f)^2 - f_{b0}$ and a_f is a free parameter.

$S(z) = G_n(z)z^{-d}[z^d - F_r(z)]$ must be implemented without canceling of the root at $z = 1$.

$$\begin{aligned} S(z) &= T_s \frac{z^{-d}}{z-1}\left[z^d - \frac{f_{b1}z + f_{b0}}{(z - a_f)^2}\right] \\ S(z) &= \frac{T_s}{(z - a_f)^2} z^{-d}\big[z^{d+1} + (1 - 2a_f)z^d + \big(1 - a_f^2\big)z^{d-1} + \cdots \\ &\quad + \big(1 - a_f^2\big)z + f_{b0}\big] \end{aligned} \tag{5.60}$$

Therefore, the predictor output $\hat{y}(k) = F_r(z)y(k) + S(z)u(k)$ is computed as

$$\begin{aligned} \hat{y}(k) &= F_r(z)y(k) + \big[z^{-1} + (1 - 2a_f)z^{-2} + (1 - a_f)^2 z^{-3} + \cdots + (1 - a_f)^2 z^{-d-1} \\ &\quad + f_{b0}z^{-d-2}\big] \frac{T_s}{(1 - a_f z^{-1})^2} u(k) \end{aligned} \tag{5.61}$$

Thus, Eq. (5.61) is used to compute the predictions up to the dead-time.

In the constrained case, the effect of the predictor structure on the controller can be interpreted using the scheme in Fig. 5.51. As can be observed from this figure, the predictor on the DTC-GPC only affects the computation of the free response. Thus, qualitatively, the influence of the internal DTC structure of the controller is the same as in the unconstrained case. Notice that if the DTC-GPC solution does not breach any constraints in the vicinity of the process operating point, the properties and tuning of the predictor can be analyzed using the closed-loop transfer functions of the system.

As usual in GPC, constraints can be handled easily. Three main kinds of constraint have to be taken into account when controlling DSCF: (i) the outlet temperature must be below a maximum value, (ii) the temperature difference between the outlet and inlet HTF temperature should be upper limited to avoid stress on the absorber tube material and lower limited to allow for applying FL, as mentioned in

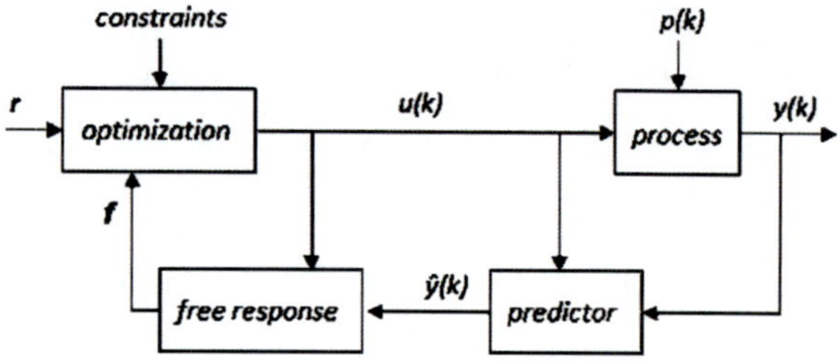

Fig. 5.51 MPC structure for the constrained case [270]

Sect. 5.8.1, (iii) a minimum flow is required in order to maintain, as far as possible, hydraulic equilibrium in all the collectors and a maximum water flow is limited because of the pump characteristics. All these constraints are amplitude constraints included in Table 5.4.

Although the feedback linearization technique produces quite good responses with a low computational cost in respect to other control strategies, the constraint treatment is very limited [319]. The control design is carried out without taking into account different constraints such as actuators or limits in the process variables. This can mean a final control far removed from the desired ideal one. In [109, 318], the dynamics of the outlet temperature, T_{out}, is regulated using a PI linear control to obtain the virtual control signal u' of the FL controller. In this case, constraints related to the physical limits of the water flow pump are included using an anti-windup mechanism over this variable. Although saturation is applied to the control signal, input and output limits are transformed into virtual signal saturation values using Eqs. (5.26) and (5.27). In this case, the constraints obtained following this procedure are just for the control signal, whereas the output variable constraints are taken into account in the reference choice.

As commented above, predictive control provides a clear advantage over the FL technique, being possible to include constraint handling during the design phase in a systematic way. Therefore, the DTC-GPC algorithm described in the previous paragraph is used for this purpose. Nevertheless, as the FL technique is used to obtain a linear model equivalent to the system, the predictive algorithm uses the virtual signal control, u' as control input, which will be mapped using the transformation $\Omega(x)$ to compensate for process non-linearities and transform into the real control signal (the inlet solar field flow). This method implies that although the constraints over the output variable do not suffer from variations, the constraints over the control signal, u, must be mapped into constraints over the virtual signal, u'.

In this section, the solution proposed by [218] has been used. Considering the virtual signal such as

$$u'(k) = \alpha_1 x(k+d) + \alpha_2 x(k+d)u(k) + \alpha_3 p(k+d) \tag{5.62}$$

with α_1, α_2, and α_3 constants, the constraints over u' may be calculated every sampling time along the whole control horizon using the future control predictions obtained with the predictive controller.

Assuming that the DTC-GPC algorithm is developed over a linear system written such as that in Eq. (5.22), the objective is to obtain the future constraints in u' over the entire control horizon. Let us consider the control signal sequence generated at time instant $k-1$ by the DTC-GPC:

$$\mathbf{u}'(k-1\mid k-1)=\big[u'(k-1\mid k-1)\quad u'(k\mid k-1)\quad \dots\ u'(k+N_u-2\mid k-1)\big] \tag{5.63}$$

Whereas the first element of the sequence $\mathbf{u}'$, $u'(k-1\mid k-1)$, is used to evaluate the signal control $u(k-1)$ by means of the Ω transformation, the following elements may be used to make estimations about the future values of the state variables, ψ and about the output, η, from the discretized version of the linear system (5.22):

$$\begin{aligned}\Psi(k+d\mid k+d-1)&=\big[\psi(k+d\mid k+d-1)\quad \psi(k+d+1\mid k+d-1)\\&\quad\dots\ \psi(k+d+N_u-1\mid k+d-1)\big]^T\\H(k+d\mid k+d-1)&=\big[\eta(k+d\mid k+d-1)\quad \eta(k+d+1\mid k+d-1)\\&\quad\dots\ \eta(k+d+N_u-1\mid k+d-1)\big]^T\end{aligned}$$

From these future state variable values, Ψ and the inverse coordinate transformation Ω^{-1}, it is possible to calculate the future state variables x of the non-linear system:

$$\begin{aligned}\mathbf{x}(k+d\mid k+d-1)&=\big[x(k+d\mid k+d-1)\quad x(k+d+1\mid k+d-1)\\&\quad\dots\ x(k+d+N_u-1\mid k+d-1)\big]^T\end{aligned}$$

Therefore, the constraint values over u':

$$\begin{aligned}\mathbf{u}'_{min}(k\mid k-1)&=\big[u'_{min}(k\mid k-1)\quad u'_{min}(k+1\mid k-1)\\&\quad\dots\ u'_{min}(k+N_u-1\mid k-1)\big]^T\\\mathbf{u}'_{max}(k\mid k-1)&=\big[u'_{max}(k\mid k-1)\quad u'_{max}(k+1\mid k-1)\\&\quad\dots\ u'_{max}(k+N_u-1\mid k-1)\big]^T\end{aligned}$$

are computed, every sample time, solving the optimization problem in the variable u as follows [319]:

$$\begin{aligned}u'_{min}(k+d+j\mid k+d-1)&=\min_u\big[\alpha_1 x(k+d+j\mid k+d-1)+\cdots\\&\quad+\alpha_2 x(k+d+j\mid k+d-1)u+\alpha_3 p(k+d)\big]\\u'_{max}(k+d+j\mid k+d-1)&=\max_u\big[\alpha_1 x(k+d+j\mid k+d-1)+\cdots\\&\quad+\alpha_2 x(k+d+j\mid k+d-1)u+\alpha_3 p(k+d)\big]\end{aligned}$$

with $0\le j\le N_u-1$, subject to the constraint $u_{min}\le u\le u_{max}$ and where the disturbances are assumed to be constant and equal to the last measured value along the prediction horizon.

Note that in the case of the solar field, the sequences Ψ, H and X are similar due to the chosen coordinate transformation (5.22). Moreover, the input–output relationship has the following discretized form:

$$\hat{T}_{out}(k+d+1)=T_{out}(k+d)+T_s u'(k) \tag{5.64}$$

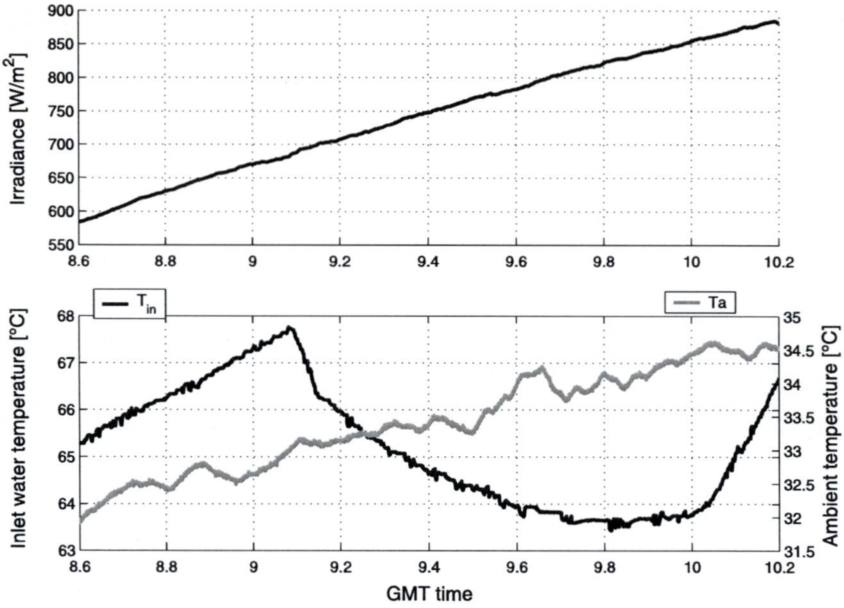

Fig. 5.52 Experimental disturbances. August 12, 2008 (courtesy of L. Roca et al., [319])

facilitating evaluation of the virtual constraints for the solar field case using the procedure above. Note also that the variable $\hat{T}_{out}(k+d)$ in Eq. (5.64) is the same prediction as that obtained for the control signal mapping in Eq. (5.24).

More details of the technique can be found in [319], including simulation results, robustness analysis and real tests. As a representative example, the DTC-GPC strategy with dynamic mapping constraints has been tested at the DSCF of a solar desalination plant (described in [318, 319] with similar bilinear models as those used in the ACUREX field using water as HTF) using $T_s = 5$ s, $\lambda = 100$, $\delta = 1$, $d = 8$, $N_2 = d + 50$, $N_u = 8$ as tuning knobs and $a_f = 0.925$ as the free parameter of the FSP filter. For the real experiments, it is important to mention that the control target is to reach an outlet–inlet temperature gradient in order to maximize solar field efficiency. The storage system configuration in the field produces a continuous increment in inlet temperature when the distillation unit is off and thus, this outlet–inlet temperature gradient requirement involves ramp-reference tracking rather than constant or step-type references.

The disturbances for the experiment on August 12th, 2008 are depicted in Fig. 5.52 between the 8.6 h and 10.2 h, whereas the output system and the control signal are shown in Fig. 5.53. At the beginning of the day, the ramp reference was chosen to maintain a difference of 5°C with the inlet water temperature, but the good level of irradiance and the high input temperature (around 65°C) produces the saturation of both virtual and control signals. After 8.77 h, the reference is changed to obtain a gradient temperature of 6°C. Although the reference is reached with a low overshoot of 0.2°C, the control signal is again super-saturated so that a new

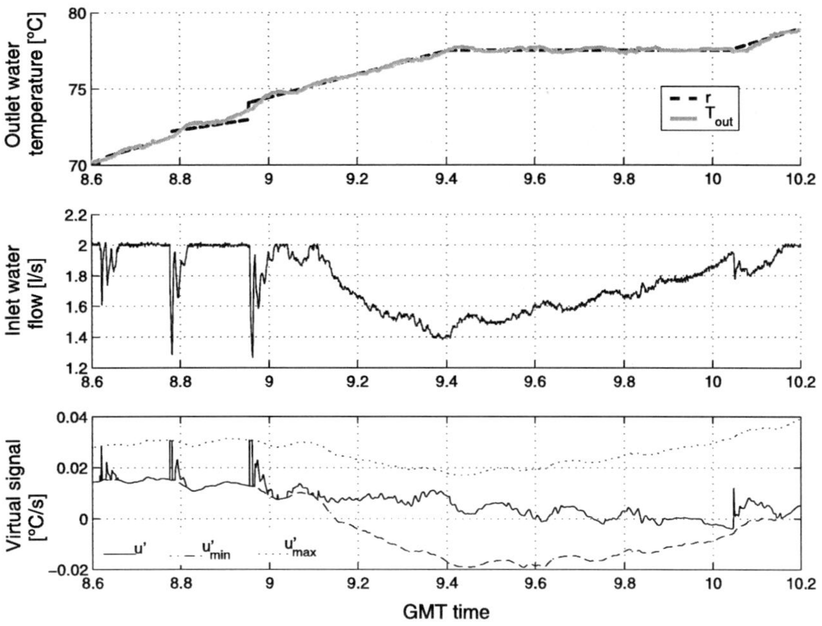

Fig. 5.53 Experimental results with DTC-GPC-FL, August 12, 2008 (courtesy of L. Roca et al., [319])

ramp reference of 7°C is selected. Notice how this reference is perfectly tracked despite the small changes in irradiance (due to the solar cycle) and the hard inlet temperature fall at 9.01 h.

It is important to mention that the first three reference changes produce some oscillations of the control signal. Notice that, in these cases, the system is super-saturated when the reference change takes place and the controller tries to move the system out of saturation using a hard control action. In addition, FL technique is based on a model of the plant and it may show aggressive behavior under reference changes due to plant–model mismatches. In the middle of the experiment, when the outlet temperature reaches 77.5°C, the decision is to maintain that reference in order to avoid a higher temperature difference. At 10 h GMT, a new ramp reference is chosen with good tracking performance. Notice that the virtual constraints vary at each sample time so that the inlet solar field flow is always within the defined limits.

5.10 Fuzzy Logic Control (FLC)

Fuzzy logic provides a conceptual base for practical problems where the process variables are represented as linguistic variables which can only present a certain limited number of possible values and then be processed using a series of rules. FLC seems to be appropriate when working with a certain level of imprecision, uncertainty and partial knowledge and also in cases where the knowledge of operating

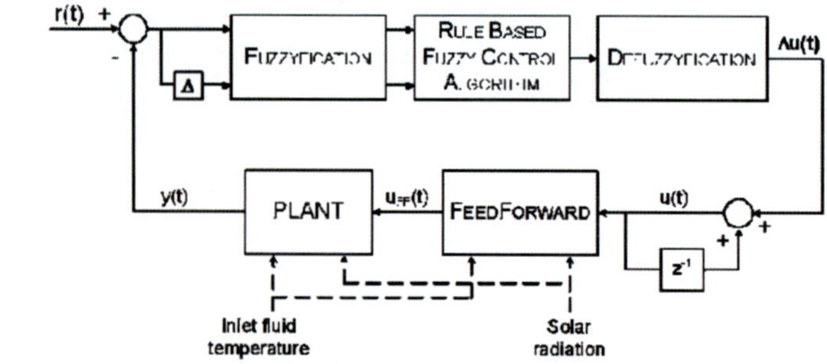

(a) Direct fuzzy logic control

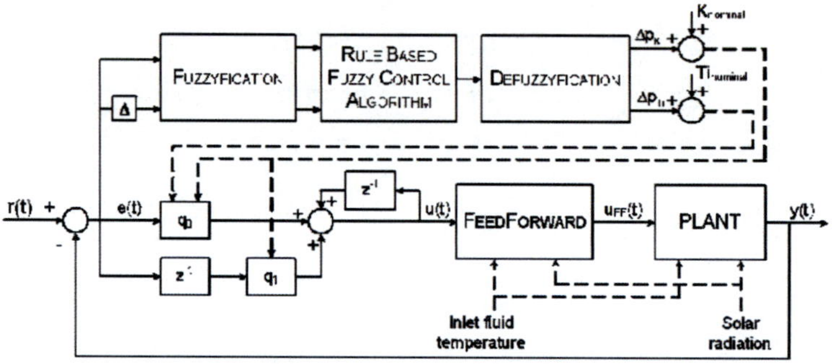

(b) Incremental fuzzy PI control

Fig. 5.54 Block diagrams of typical fuzzy logic controllers (FLC)

with the process can be translated into a control strategy that improves the results reached by other classical strategies. In the framework of this type of system, FLC has attracted much attention since it was first applied to the control of DSCF [333]. In this seminal work, a direct FLC was developed incorporating a feedforward term [82], aimed at finding a non-linear control surface to control the output temperature of the field using previous knowledge about the system. In this case, a special subclass of fuzzy inference systems, the triangular partition and triangular partition with evenly spaced midpoints, was used to obtain adequate control signals within the whole range of possible operating conditions.

An FLC is commonly described by a set of fuzzy rules that constitute the control protocol. With these rules, the interconnected relationships between measurable variables and control variables can be expressed. In Fig. 5.54(a), three parts can be seen which constitute the design parameters of the FLC: the block fuzzifier, the control block (fuzzy rule base and inference procedure) and the block defuzzifier. The fuzzification interface converts the numerical values of the input variables into linguistic variables (fuzzy sets). The conversion requires scale mapping that transforms the range of values of input variables into corresponding universes of discourse. The rule-based fuzzy control algorithm provides definitions of linguistic control rules which characterize the control policy. The block includes decision making logic, which infers fuzzy control actions employing fuzzy implication and the inference rules. The defuzzification block converts the inferred control action, which interpolates between rules that are fired simultaneously, to a continuous signal. The control scheme uses the error between the output of the plant and the set point signal and its increment as inputs for the FLC. The output variable of the FLC is the increment in the control signal (reference temperature for the feedforward controller). The implementation of the controller was made by means of a fuzzy associative memory (FAM) the centers of which were heuristically obtained.

After this first application of FLC, the same idea was used in [233] to obtain direct FLC and in [41] to obtain fuzzy logic based PID controllers, where the parameters of PID controllers were modified according to a fuzzy logic inference mechanism (Fig. 5.54(b)). In [160], the application of genetic algorithms (GA) was first introduced to automatically tune the FLC developed by [333], by obtaining the centers of the FAM describing the controller using an optimization algorithm with intensive use of data obtained in closed-loop under manual and automatic operation. A comprehensive overview of these developments can be found in [44, 86, 334]. In [239], PID-type fuzzy controllers similar to that of [41] were used in series with a feedforward controller developed by [382] in order to maintain a reference temperature in the ACUREX DSCF. The controller was tuned and tested on the non-linear computer model of the plant [38] and then tested on the actual plant. Reference [208] used a unique hierarchical GA (HGA) for the design and optimization of an FLC similar to that developed by [333], minimizing the number of fuzzy membership functions and rules applied. HGA is used to optimize the fuzzy membership functions, while the fuzzy rules also undergo an evolution process for the realization of a set of fuzzy rules that can be obtained optimally. In [234], GA are again used to develop an FLC which rule base encompasses an empirical set of *if-then* rules. The authors of [360] and [232] developed a control scheme that employs a fuzzy PI controller for the highly non-linear part of the operating regime and gain scheduled control over the more linear part of the operating envelope. In order to satisfy performance characteristics for the plant at different points in the operating regime, a multiobjective GA is used to design the parameters of the fuzzy controller. The resulting controller is shown to both satisfy the desired performance criteria and have a reduced number of terms compared to a conventional design approach. All these works use the feedforward controller and the simulator developed by [38]. In [343] a DSCF is modeled using GA. The relation among temperature, pressure and flow rate, time, day and Sun heat flux is modeled and the heat flux is expressed versus other variables through genetic optimization. To control the outlet temperature, adaptive network-based fuzzy inference system (ANFIS) was employed including fuzzy switching control to prevent chattering phenomena in the multi-loop plant where switches between loops take place complicating HTF cycle control. Simulation results of the solar power plant and the control system show that the applied control system can manage the HTF cycle in different situations with safe operating conditions and with better performances.

For the linguistic equation approach presented in [198, 200–202, 278], the fuzzy rules are replaced by linguistic equations. In fuzzy linguistic equations, fuzziness is taken into account by membership functions—the linguistic equation approach does not necessarily need any uncertainty or fuzziness. Real valued linguistic equations provide a basis for sophisticated non-linear systems where fuzzy set systems are used as a diagnostic tool. The linguistic equation controller applied in [198, 200–202, 278] is based on the PI-type fuzzy controller. Non-linearities are introduced to the system by membership definitions that correspond to membership functions used in FLC; this is the main difference to the FLC presented in [333], where the non-linearities are handled through the rule base. Operation of the controller is modified by variables describing operating conditions (temperature difference between inlet and outlet temperatures and solar irradiance); the implemented controller consists of a three level cascade controller. The linguistic equation controller reacts very efficiently to variable irradiance conditions since the control surface is only gradually changed. In [199], new results of the multilevel linguistic equation controller are shown. The controller smoothly combines various control strategies into a compact single controller. Control strategies ranging from smooth to fast are chosen by setting the working point of the controller. The controller takes care of the actual set points of the temperature. The operation is very robust under difficult conditions: start-up and set point tracking are fast and accurate under variable radiation conditions. Reference [94] developed a fuzzy switching supervisor PID control approach using a feedforward compensator. The use of a supervisor is easy to implement because it needs very little knowledge about the process, it can lead to a highly non-linear control law increasing the robustness of the control system and it can provide a user interface for precisely expressing the specifications in terms of closed-loop performance. The supervisor was implemented using a Takagi–Sugeno fuzzy strategy to implement an on-line switching between each PID controller according to real time conditions. The local PID controllers were previously tuned off-line using an ANN approach that combines a dynamic recurrent non-linear ANN model with a pole-placement control design. The number of local controllers to be employed by the supervisor is reduced using the c-Means clustering technique. The feedforward controller proposed by [82] was used in parallel configuration. The same approach was used by [178]. In [289, 290], a neuro-fuzzy system based on a radial basis function (RBF) network and using support vector learning is considered for non-linear modeling and applied to the output regulation problem. In [152], the application of a neuro-fuzzy identification for predictive control is performed. The same idea is exploited in [191, 192], where an intelligent predictive controller is proposed. However, in both cases, only simulation results are provided using the simulator in [38]. In [142] a fuzzy predictive-control scheme is developed. The proposed predictive controller uses fuzzy characterization of goals and constraints based on the fuzzy optimization framework for multiobjective satisfaction problems. This approach enhances MPC allowing the specification of more complex requirements.

5.10.1 Heuristic Fuzzy Logic Controllers

Since Mamdani [237] published his experiences using a FLC on a test-bed plant of a laboratory, many different control approaches have appeared based on this theory

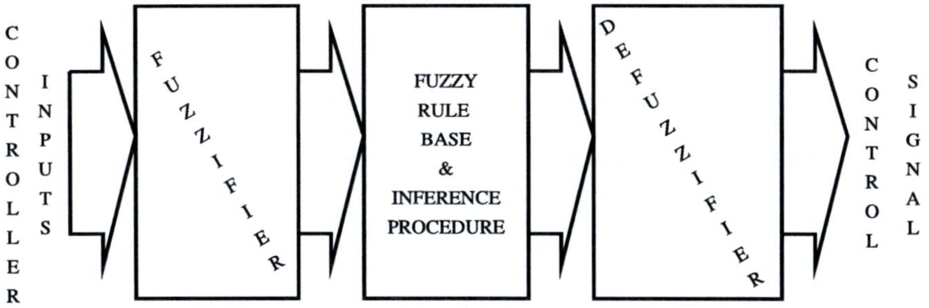

Fig. 5.55 Fuzzy logic inference scheme

and also many applications of this type of controller to a diversity of processes. An extensive introduction to the historical development, state and concepts involving fuzzy control systems can be found in [223, 224].

Many advantages of FLC have been mentioned in literature. One relevant advantage is the possibility of converting a linguistic control strategy based on experience and expert knowledge into an automatic control strategy. Another advantage is that FLC can easily be implemented. Moreover, it has been demonstrated [367] that by using fuzzy inference systems with triangular partition rule bases with evenly spaced midpoints (TPE systems) like those used in this chapter, the amount of computation required for processing input is independent of the number of rules.

The methodology of fuzzy controllers seems to be the most adequate in cases where the process is difficult to control and there is wide experience in operating the plant, due to the possibility of incorporating such knowledge in terms of qualitative rules. This is the case of the control schema presented in this chapter.

A special subclass of fuzzy inference systems, the TP and TPE systems, are used both in the development of an incremental fuzzy PI controller (IFPIC) and in the implementation of a FLC strategy which provides an adequate control signal over the entire range of possible operating conditions. Both strategies have been implemented in connection with the simple series feedforward controller analyzed in Sect. 4.4.1.2, which allows for compensating effects of the disturbances acting on the system.

5.10.1.1 Fuzzy Logic Inference Scheme

A fuzzy logic inference scheme is commonly described by a set of fuzzy rules that constitute the control protocol. With these rules, the interconnected relationships between measurable variables and control variables can be expressed.

In Fig. 5.55 three parts constituting the design parameters of a FLC can be seen: the block fuzzifier, the control block (fuzzy rule base and inference procedure) and the block defuzzifier.

- The fuzzification interface converts the numerical values of the input variables into linguistic variables (fuzzy sets). The conversion requires scale mapping that transforms the range of values of input variables into corresponding universes of discourse.

- The rule-based fuzzy control algorithm provides definitions of linguistic control rules which characterize the control policy. In this case, the block includes decision making logic which infers fuzzy control actions employing fuzzy implication and the inference rules mentioned.
- The defuzzification block converts the inferred control action which interpolates between rules that are fired simultaneously into a continuous signal.

The expert knowledge usually takes the form:

> **if** (*a set of conditions are satisfied*) **then** (*a set of consequences can be inferred*)

The set of conditions (fuzzy sets) belongs to the input domain and the set of consequences to the output domain. The fact that various rules can be fired simultaneously is due to the conversion of values obtained from sensors to linguistic terms, assigning a membership function μ_A to each one. If triangular partition rule bases are used [367], the input and output universes are subdivided using triangular membership functions of the form

$$\mu_{A_i}(x) = \begin{cases} (x - a_{i-1})/(a_i - a_{i-1}) & \text{if } a_{i-1} \le x \le a_i \\ (-x + a_{i+1})/(a_{i+1} - a_i) & \text{if } a_i \le x \le a_{i+1} \\ 0 & \text{otherwise} \end{cases} \tag{5.65}$$

the membership values of $A_i \neq 0$ being only at interval (a_{i-1}, a_{i+1}). Point a_i is the unique element that has membership value 1 at A_i (midpoint of A_i). A triangular decomposition of a universe consists of a sequence of triangular fuzzy subsets $A_1, \ldots, A_n$ such that the leftmost and rightmost fuzzy regions satisfy $\mu_{A_1} = 1$ and $\mu_{A_n} = 1$ at its midpoints (we are assuming that the triangular decomposition forms a fuzzy partition of the underlying universe, that is: $\sum_{i=1}^{n} \mu_{A_i}(u) = 1$ for every u belonging to the universe). A decomposition of the input and output domains that satisfies these requirements is called a TP (triangular partition) system. The inference process can be made more efficient by requiring the membership functions to be isosceles triangles with bases of the same width. These fuzzy inference systems with evenly spaced midpoints are called TPE systems.

The inference mechanism in a TP or TPE system is very straightforward and efficient. As is demonstrated in [367], considering SISO systems (the extension to MIMO system is straightforward), if A_i and A_{i+1} are two fuzzy sets (with midpoints a_i and a_{i+1}, respectively) providing non zero membership for x, the appropriate action for any input at an interval $[a_i, a_{i+1}]$ using weighted averaging defuzzification is given by

$$z = \frac{\mu_{A_i}(x)c_r + \mu_{A_{i+1}}(x)c_s}{\mu_{A_i}(x) + \mu_{A_{i+1}}(x)} = \frac{x(c_s - c_r) + a_{i+1}c_r - a_i c_s}{a_{i+1} - a_i} \tag{5.66}$$

with c_s and c_r being the midpoints of C_s and C_r (fuzzy regions in the output space).

Equation (5.66) shows that only two completely determined constants $(c_s - c_r)/(a_{i+1} - a_i)$ and $(a_{i+1}c_r - a_i c_s)/(a_{i+1} - a_i)$, are required to obtain the control action, allowing a fuzzy inference system to be represented in tabular form. The amount of computation required for processing input is independent of the number of rules with a TPE system. The tabular information can be stored in a manner that permits direct addressing, avoiding searching procedures to find the appropriate rules in the inference table.

A fuzzy associative memory (FAM) can be used instead of the table. A FAM is a k-dimensional table where each dimension corresponds to one of the input universes of the rules. The ith dimension of the table is indexed by the fuzzy sets that comprise the decomposition of the ith input domain. The FAM representation may be modified to produce a numeric inference (NI) table. In this way, the indices will represent the corresponding midpoints of the set. The entries in the table are the midpoints of the consequent of the associated rule.

5.10.2 Incremental Fuzzy PI Control (IFPIC)

In this section, a control scheme to improve performance of classical PI controllers is used. The approach assumes that nominal controller parameters are available (obtained from classical tuning methods, as in [380]). By using an appropriate fuzzy matrix, small changes in each controller parameter are performed in order to improve transients and steady-state performance of the closed-loop system. From experience of operating the field, a fuzzy matrix that contains this knowledge (in condensed rule form) is obtained for each parameter of a PI controller. These matrices describe the changes that the nominal controller parameters experiment.

In [380], three different classical PID tuning techniques were used (Ziegler–Nichols, analytical and Kalman PID tuning techniques). Here, another approximation based on pole-placement techniques is used. Empirical rules are used to build a 7×7 fuzzy matrix for each parameter, which gives an acceptable quantization. The basis for the control rules is the error $r - y = T_{ref} - T_{out}$ and its increment. In order to obtain good performance, two main rules have to be taken into account:

- To decrease overshoot, the integral term has to be decreased when the output is approaching the set point. To diminish the rise time, the integral term has to be increased during the transient.
- To decrease rise time, the proportional term has to be increased during the transient and decreased when the output is approaching the set point.

These rules and previous knowledge of the behavior of the controlled system produces the control matrix of each parameter. Each matrix can be described by the midpoint element of each region (if triangular partitions are chosen for the input and output universes of discourse). The inputs chosen are the error e and its increment Δe. In this way, the proportional and integral terms (controller parameters) can be defined:

$$
\begin{aligned}
K_P &= K_{P_{nominal}} + \Delta K_P\{e, \Delta e\} \\
T_I &= T_{I_{nominal}} + \Delta T_I\{e, \Delta e\}
\end{aligned}
\tag{5.67}
$$

where K_P denotes the proportional term, T_I denotes the integral time, the subindex *nominal* refers to nominal parameters, ΔK_P, ΔT_I are the values obtained from the matrices.

5.10.2.1 Application to the ACUREX DSCF

The fuzzy logic inference scheme described in the previous section has been used to control the distributed solar collector field, ACUREX. In order to define input and output signals of the fuzzy logic inference scheme, the series feedforward controller developed in Sect. 4.4.1.2 has also been used here as part of the plant.

From step response observations and stability analysis, it was found that appropriate parameters that describe the system model within the range of operating points (including the series feedforward term) are $0.8 \le a \le 0.95$, $0.9 \le K \le 1.2$. In order to obtain nominal PI control parameters, the type A model given in Eq. (4.48) was used to represent the system dynamics in discrete time and a PI transfer function given by $C(z) = (g_0 + g_1 z^{-1})/(1 - z^{-1})$, where $g_0 = K_P(T_I + T_s)/T_I$ and $g_1 = -K_P$. If the controller zero is chosen to cancel the plant pole, i.e., choosing $g_1/g_0 = -a$, the system closed-loop transfer function is given by

$$G_{cl}(z) = \frac{g_0 b}{z^2 - z + g_0 b}$$

The closed-loop characteristic polynomial roots are $z_{1,2} = 0.5 \pm 0.5\sqrt{1 - 4g_0 b}$. By making the dominant pole of the closed-loop system faster than the open-loop one ($z_{cl} = \gamma_{cl} a$), with $0.9 \le \gamma_{cl} \le 1$ in order to avoid high oscillations), one obtains

$$g_0 = \frac{1 - (2(\gamma_{cl} a - 0.5))^2}{4b} \tag{5.68}$$

which gives another relation ship of the controller parameters limited to the system ones (remember that $g_1/g_0 = -a$). With these relationships and taking into account the possible values achievable by a, b and γ_{cl}, the range of possible values of controller parameters can be obtained:

$$\frac{g_1}{g_0} = -a = -\frac{T_I}{T_s + T_I}$$

Then $T_I = \frac{aT_s}{(1-a)}$, obtaining from the relationship $0.8 \le a \le 0.95$ the following inequalities: $156 \le T_I \le 741$ [s]. With Eq. (5.68) and the relationship $g_1/g_0 = -a$, $K_P = \frac{\gamma_{cl} a^2}{b}(1 - \gamma_{cl} a)$ can be obtained, which gives $0.188 \le K_P \le 3.65$. These theoretical ranges have been slightly modified, using experience in operating the plant to avoid extremely low or fast responses, to $0.5 \le K_P \le 3.5$ and $150 \le T_I \le 675$ [s].

As has been mentioned, nominal controller parameters are needed to implement an IFPIC. A conservative fixed PI controller used as a backup controller at the starting phase of the operation has been used as the nominal one. This conservative controller gives the response shown in Fig. 5.56 (dotted-line) and is included in the family of possible controllers, being designed with $a = 0.95$, $b = 0.05$, and $\gamma_{cl} = 0.95$, which gives values of $g_0 = 1.65$ and $g_1 = -1.56$ ($K_P = 1.56$ and $T_I = 675$ s).

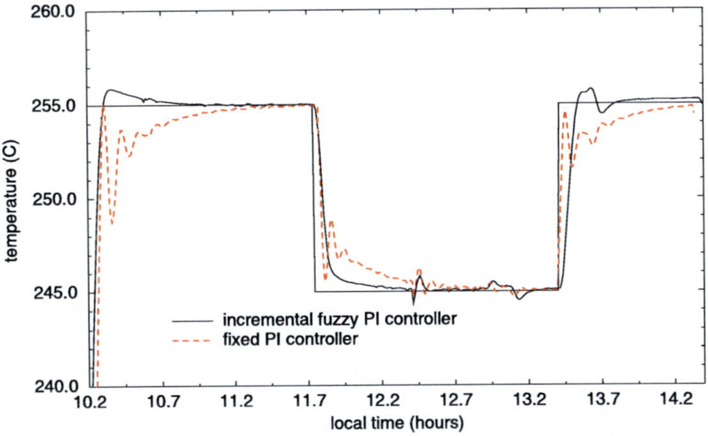

Fig. 5.56 Comparison between fixed PI control and incremental fuzzy PI control [41]

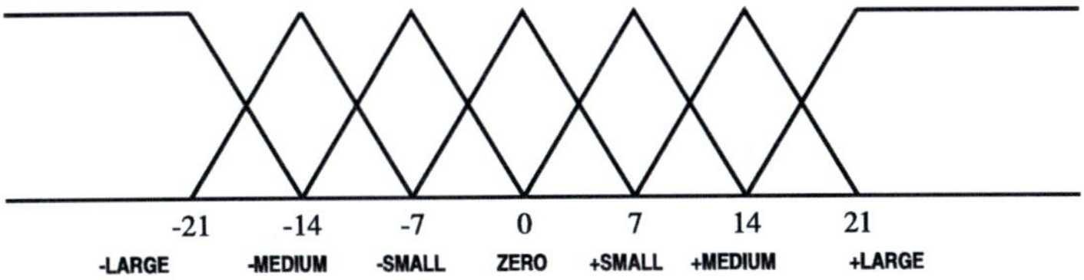

Fig. 5.57 Membership functions of the antecedents (e and Δe) [41]

The control scheme is shown in Fig. 5.54(b). The error between the output of the plant and the set point signal (e) and its increment (Δe) are considered to be the inputs for the fuzzy inference mechanism. The output variables of the fuzzy controller are the increments in the controller parameters.

TP and TPE systems have been chosen for the inference mechanism. A triangular decomposition of the input universes of discourse can be seen in Fig. 5.57. As can be seen, the extremes of the input universes of discourse (corresponding to e and Δe) are $[-21, 21]$. Those of the output have been chosen to be $[-1, 1]$ (corresponding to ΔK_P) with centers $[1.95, 1.40, 1.20, 1.0, -0.1, -1.0]$ and $[-525, 0]$ (corresponding to ΔT_I), with centers $[0, -220, -250, -320, -450, -525]$. In this case (two-input single-output system), the rules take the form (e.g. with K_P):

> **if** *error is positive large* **and** *the change in error is negative small* **then** *make the K_P change positive large*

where terms small, large, medium, etc. are used to describe the fuzzy variables and the control action. As mentioned previously, because of the partial matching attribute of fuzzy control rules and the fact that the preconditions of rules do overlap, four rules can fire at the same time.

Table 5.8 FAM of the controller parameters [41]

Kp

Δe \ e	-L	-M	-S	Z	+S	+M	+L
-L	+EL	+EL	+L	+M	+S	+L	+EL
-M	+EL	+L	+M	+S	-S	+M	+EL
-S	+L	+M	+S	-ES	-ES	+M	+L
Z	+M	+S	-ES	-S	-ES	+S	+M
+S	+L	+M	-ES	-ES	+S	+M	+L
+M	+EL	+L	-S	+S	+M	+L	+EL
+L	+EL	+L	+S	+M	+L	+EL	+EL

Ti

Δe \ e	-L	-M	-S	Z	+S	+M	+L
-L	-EL	-EL	-L	-M	-S	-L	-EL
-M	-EL	-L	-M	-S	Z	-M	-EL
-S	-L	-M	-S	-ES	-ES	-M	-L
Z	-M	-S	-ES	Z	-ES	-S	-M
+S	-L	-S	-ES	-ES	-S	-M	-L
+M	-EL	-L	Z	-S	-M	-L	-EL
+L	-EL	-L	-S	-M	-L	-EL	-EL

EL: extra-large; L: large; M: medium; S: small; ES: extra-small; Z: zero

Assuming that the *algebraic product* operator is used as the conjunction operator, each rule recommends an action ($\Delta P^{PI}_{i,j}$, P^{PI} denoting a PI parameter) with a membership function $\mu_{\Delta P^{PI}_{i,j}} = \mu_{e_i}\mu_{\Delta e_j}$, where μ_{e_i} and $\mu_{\Delta e_j}$ are calculated by Eq. (5.65). The methodology used in deciding what control action should be taken results in the firing of four rules. The combination of these rules produces a non-fuzzy action ΔP^{PI}, which is calculated using the weighted averaging defuzzification method. Then, the resulting action can be calculated by

$$\Delta P^{PI} = \frac{\sum_i^{i+1}\sum_j^{j+1}(\mu_{\Delta P^{PI}_{i,j}}\Delta P^{PI}_{i,j})}{\sum_i^{i+1}\sum_j^{j+1}\mu_{\Delta P^{PI}_{i,j}}} \tag{5.69}$$

and, as in this case, $\sum_i^{i+1}\sum_j^{j+1}\mu_{\Delta P^{PI}_{i,j}} = 1$, we have

$$\Delta P^{PI} = \sum_i^{i+1}\sum_j^{j+1}(\mu_{\Delta P^{PI}_{i,j}}\Delta P^{PI}_{i,j})$$

Taking into account these maximum and minimum values and using the exposed inference mechanism (using triangular partitions in the input and output spaces and the weighted averaging defuzzification method), a description of the fuzzy control system can be obtained. A two-dimensional FAM for each one of the controller parameters can be used (Table 5.8).

Here, we also show gain and integral time constant increment surfaces, obtained after a few simulations (Fig. 5.58). The midpoints were slightly changed from the initial chosen values in order to achieve good results in the controller performance.

Results showing the behavior obtained with this approach have been given in Fig. 5.56, comparing them to the behavior obtained with the fixed nominal PI controller. As can be seen, the overall performance is greatly improved, with faster rise times and with less than 3°C of overshoot when great changes in the set point signal occur.

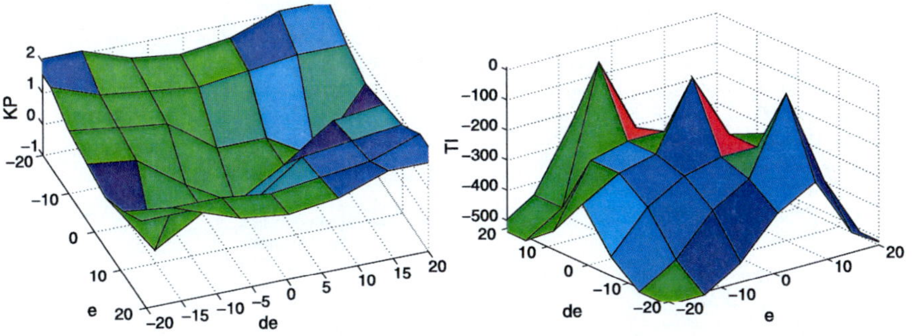

Fig. 5.58 Gain (K_P) and integral time constant (T_I) increment surfaces [41]

5.10.2.2 Plant Results

The proposed control algorithm has been tested at the ACUREX DSCF with very acceptable results. Figures 5.59 and 5.60 show the characteristic variables in a test in the presence of passing clouds. As can be seen, the outlet HTF temperature follows the reference temperature without great changes in the HTF flow that could deteriorate this actuator. The estimated controller gain K_P and integral time T_I evolution are shown in Fig. 5.60, following the tendency pointed out by the rules used to design the IFPIC.

Figures 5.61 and 5.62 show operation with the IFPIC on a day with strong disturbances caused by large passing clouds which produce drastic changes in the direct solar irradiance level. Under these conditions, the effect of the feedforward controller in series with the IFPIC allows for very acceptable behavior even under such extreme conditions (with deviations from the set point of less than 10°C with long time direct solar irradiance changes greater than 300 W/m^2). Figure 5.62 shows the evolution of the direct solar irradiance, HTF flow and controller parameters.

5.10.3 Fuzzy Logic Controller (FLC)

This section introduces a FLC which directly calculates the control signal. In this case, the FLC has been applied in an incremental form in series with the feedforward controller, that is, the signal obtained from the FLC is the increment needed in the control signal (reference temperature to the feedforward controller) to provide a desired behavior (the output universe is related to control increments).

The control scheme is shown in Fig. 5.54(a). As in the previous section, the error between the output of the plant and the set point signal (e) and its increment (Δe) are considered to be the inputs for the fuzzy controller. The output variable of the fuzzy controller is the increment in the control signal (reference temperature for the feedforward controller). The fuzzy controller could give the control signal directly, but it has been chosen in incremental form in order to introduce an integral effect

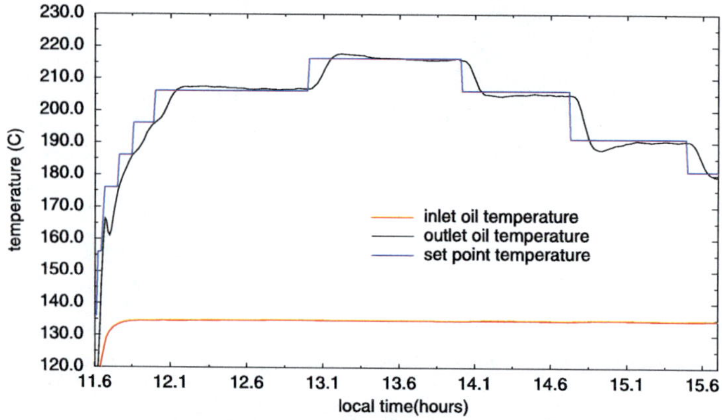

Set point, inlet and outlet HTF temperatures (°C)

Fig. 5.59 Test with the IFPIC controller (17/01/95) [41]

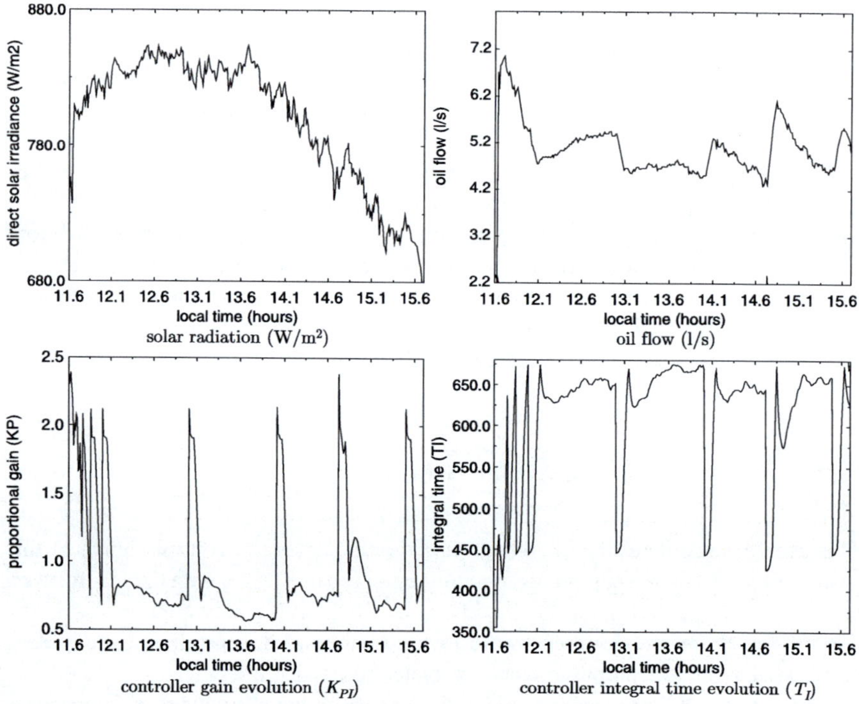

Fig. 5.60 Test with the IFPIC controller (17/01/95), [41]

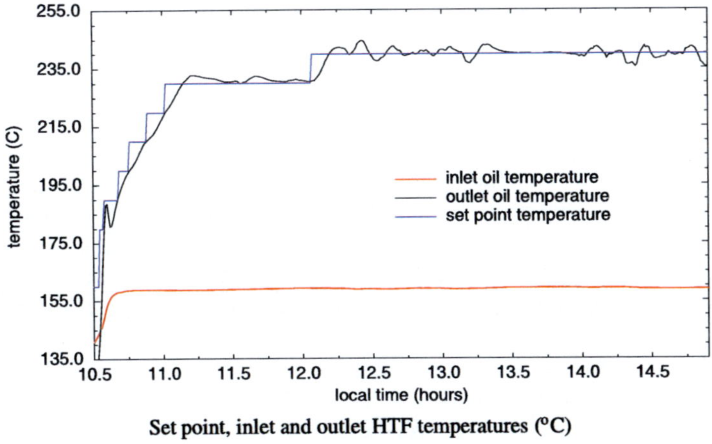

Fig. 5.61 Test with the IFPIC controller (03/03/95), [41]

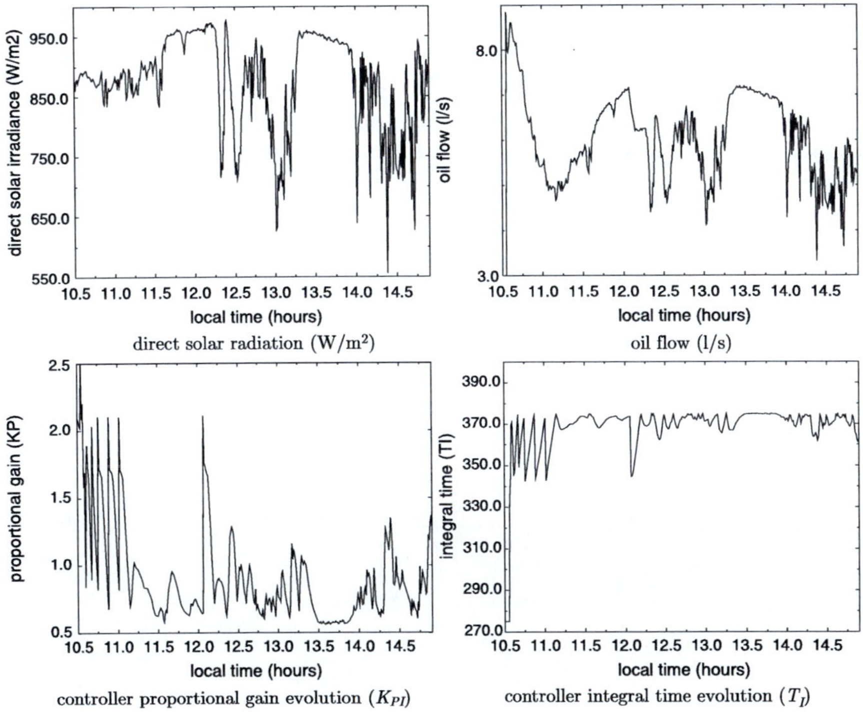

Fig. 5.62 Test with the IFPIC controller (03/03/95), [41]

in the control action and to reduce the fuzzy partition of the output domain. TP and TPE systems have also been chosen for the inference mechanism.

The triangular decomposition of the input universes of discourse has been shown in Fig. 5.57. Again, the extremes of the input universes of discourse (corresponding to e and Δe) are $[-21, 21]$. Those of the output have been chosen to be $[-10, 10]$ (corresponding to $\Delta u = \Delta T_{rff}$), with centers $[\pm 10, \pm 6, \pm 4, \pm 2, \pm 0.7, 0]$. The domains under consideration are not normalized. The discretization of the universes of discourse plays an important role in the final behavior of the controlled system. In this case, the discretization has been selected taking into account the operating range of possible HTF flows (between 2 and 12 l/s), the outlet temperature (with a maximum of 300°C) and previous knowledge of the plant dynamics. As is commented below, there is great experience in controlling of the plant with other control approaches. This fact and the existence of the non-linear distributed-parameter model for simulation purposes, allows for good discretization of the universes of discourse and testing the controller behavior before its actual implementation at the plant. In this way, few changes were made when implementing the controller at the plant.

In this case (two-input single-output system), the rules take the form:

if *the error is positive large* **and** *the change in error is negative small* **then** *make the control signal change positive large*

Terms small, large, medium, etc. are used to describe the fuzzy variables and the control action. As mentioned previously, due to the partial matching attribute of fuzzy control rules and the fact that the preconditions of rules do overlap, four rules can fire at the same time.

Again, the *algebraic product* operator is also used as the conjunction operator, each rule recommends a control action ($\Delta u_{i,j}$) with a membership function $\mu_{\Delta u_{i,j}} = \mu_{e_i} \mu_{\Delta e_j}$, where μ_{e_i} and $\mu_{\Delta e_j}$ are calculated by Eq. (5.65). The methodology used in deciding what control action should be taken results in the firing of four rules. The combination of these rules produces a non-fuzzy control action Δu, which is calculated using the weighted averaging defuzzification method, as mentioned before. Then the resulting control action can be calculated by

$$\Delta u = \frac{\sum_i^{i+1} \sum_j^{j+1} (\mu_{\Delta u_{i,j}} \Delta u_{i,j})}{\sum_i^{i+1} \sum_j^{j+1} \mu_{\Delta u_{i,j}}} \tag{5.70}$$

and, as in this case, $\sum_i^{i+1} \sum_j^{j+1} \mu_{\Delta u_{i,j}} = 1$, we have

$$\Delta u = \sum_i^{i+1} \sum_j^{j+1} (\mu_{\Delta u_{i,j}} \Delta u_{i,j})$$

A 2-dimensional FAM can also be used in this case (Table 5.9). The controller surface obtained from the resulting NI table can be seen in Fig. 5.63.

The first step in the design procedure was to produce the FAM table shown in Table 5.9 and the centers of Δu as $[\pm 20, \pm 15, \pm 10, \pm 6, \pm 3, 0]$. These values were chosen based on previous experience in controlling the plant. The resulting control

Table 5.9 FAM of the controller [333]

Δe \ e	-L	-M	-S	Z	+S	+M	+L
-L	-EL	-EL	-L	-M	-S	+L	+EL
-M	-EL	-L	-M	-S	Z	+M	+EL
-S	-L	-M	-S	-ES	+ES	+M	+L
Z	-M	-S	-ES	Z	+ES	+S	+M
+S	-L	-M	-ES	+ES	+S	+M	+L
+M	-EL	-L	Z	+S	+M	+L	+EL
+L	-EL	-L	+S	+M	+L	+EL	+EL

EL: extra-large; L: large; M: medium; S: small; ES: extra-small; Z: zero

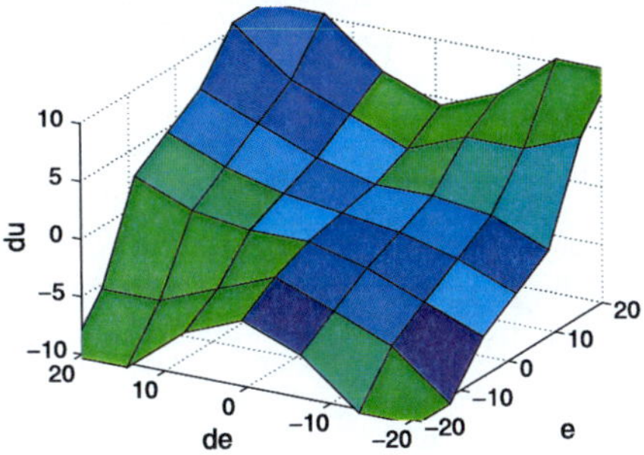

Fig. 5.63 Fuzzy control surface [333]

was simulated by the non-linear computer model and the results shown in Fig. 5.64 (dotted line) were obtained. The simulations of the first FLC designed showed an undesirably high overshoot. Thus, the centers of the FAM were reduced, taking the following new values in the intermediate design: $[\pm 13, \pm 10, \pm 8, \pm 5, \pm 3, \pm 1.2, 0]$. With these values and the help of the non-linear computer model, results shown in Fig. 5.64 (dashed line) were obtained. This result can be considered to be good, but, due to the characteristics of the solar plant, the main design objective is to achieve a response with low overshoot. Thus, the centers of the FAM were again reduced, obtaining the definitive ones mentioned above $[\pm 10, \pm 6, \pm 4, \pm 2, \pm 0.7, 0]$. The simulation which corresponds to these values can also be seen in Fig. 5.64 (solid line). This figure includes the values of the HTF flow in order to show the control effort obtained by each of the designs. These values have been used in the control of the ACUREX DSCF, as shown in the next paragraph.

5.10.3.1 Plant Results

Figure 5.65 (outlet HTF temperature, inlet HTF temperature, set point temperature, direct solar irradiance and HTF flow, respectively) corresponds to a step response test (27/04/94), covering a wide range of HTF flow conditions (from 5 to 7.5 l/s).

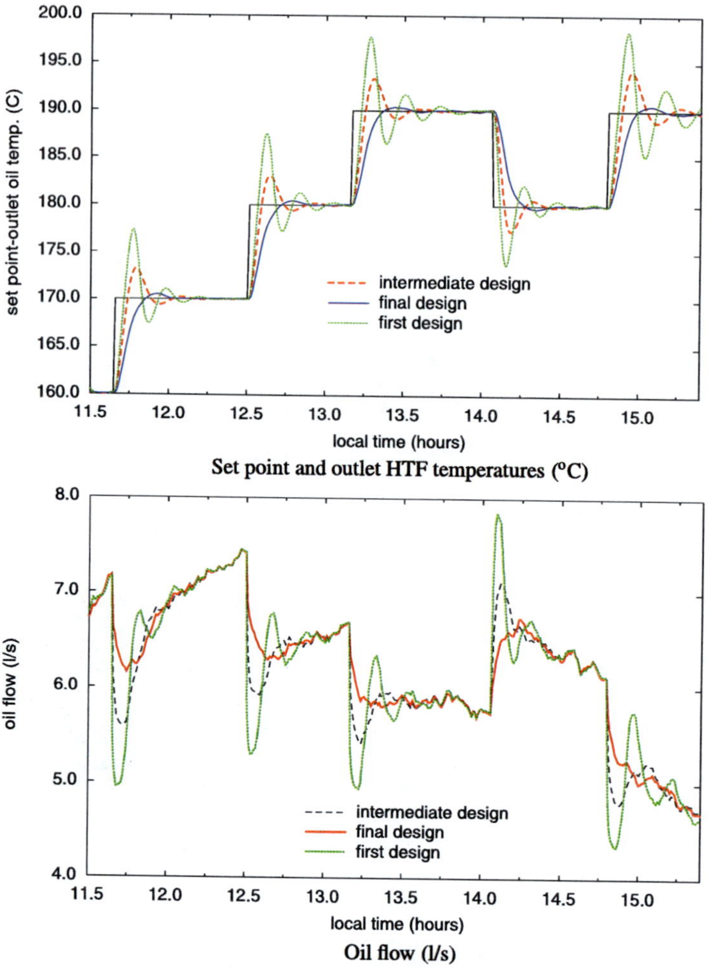

Fig. 5.64 Simulation with different designs of the FLC, [333]

As can be seen, very good results are obtained in all cases without oscillations in the system response. These curves show a small offset (less than 1.5°C) in the system response.

One possible explanation for the appearance of the offset in the output signal is the use of a wrong value of mirror reflectivity (used in calculating the effective solar irradiance) in the feedforward controller. Mirror reflectivity is usually measured once a week, thus, if dust accumulates on the mirror surfaces between measurements, the real value of mirror reflectivity can vary from the one last measured used by the feedforward controller. This fact leads to an error in the outlet signal of the feedforward controller (HTF flow demanded to the pump) that depends on the effective solar irradiance. Theoretically, as the feedforward controller is placed in the

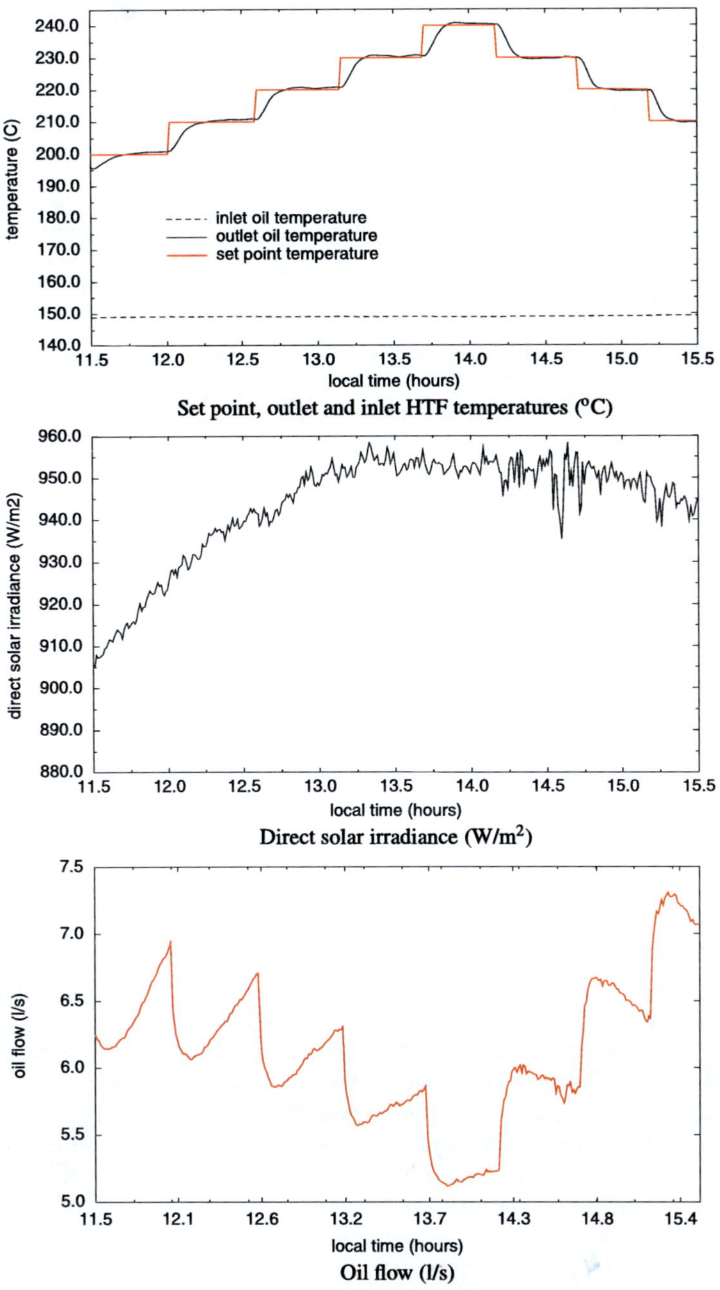

Fig. 5.65 Test carried out with the FLC controller (27/04/94), [333]

control loop, this error should be corrected for, but the integral term action is not fast enough to do so.

The dependence on effective solar irradiance can be observed in Fig. 5.65, in which the offset is positive when solar irradiance increases and negative when it decreases, achieving a value near zero at solar midday. Even though this small offset is not important in terms of heat transference, some variations in the initial implementation were performed to achieve faster responses without offset. The values used in the inference table were slightly modified and two new rules were added in the FLC implementation in order to increase the integral action. These rules take the form:

if $abs(e(k)) > 2$ **and** $e(k) > 0$ **and** $e(k-1) > 0$ **and** ... **and** $e(k-4) > 0$
then $\Delta u(k) = \Delta u(k) + \sum_{i=0}^{4} e(k-i)/100$

The same rule was also implemented changing the *greater than* sign for a *less than* sign.

Results obtained by performing these modifications are shown in Fig. 5.66, which show the outlet HTF temperature, inlet HTF temperature, set point temperature, direct solar irradiance and HTF flow. It corresponds to a test carried out the 21th December, 1994. As can be seen, very good results were obtained under all operating conditions in spite of the heavy perturbations produced by changes in the direct solar irradiance (200 W/m^2 drops) caused by clouds and in the inlet HTF temperature (45°C increment).

5.11 Neural Network Controllers (NNC)

Some of the approaches to control DSCF using NNC have been commented on in the previous sections. In [17, 18] and [42] an application of ANN identification is presented to obtain models of the free response of the solar plant to be used in the algorithm proposed by [76], see Fig. 5.46. In [179] an ANN-based indirect adaptive-control scheme is developed. The output regulation (OR) theory aims to derive a control law such that the closed-loop system is stable and, simultaneously, the tracking output error converges to zero. This technique leads to a straightforward method for solving non-linear control problems. However, the OR theory assumes perfect model knowledge. Given the ANN model plant mismatch, an on-line adaptation of ANN weights is considered in order to improve the discrepancies between the output of a previous off-line model and the actual output of the system.

By means of a Lyapunov analysis, a stability condition for weight updating is employed. In [94], the local PID controllers of a switching strategy were previously tuned off-line using an ANN approach that combines a dynamic recurrent non-linear ANN model with a pole-placement control design (Fig. 5.67). In [153] the authors used recurrent ANN aimed at obtaining a pseudo inverse of the plant to apply FLC techniques. Further improvements led to the works of [155], where a non-linear adaptive constrained MPC scheme is presented using non-linear state-space ANN and their on-line training. The identification of the ANN is performed at two levels. First, a parameterization is obtained for the selected topology by training the

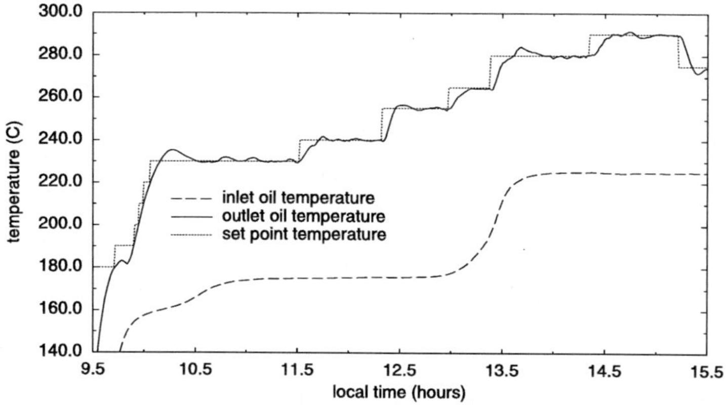

Set point outlet and inlet HTF temperatures (°C)

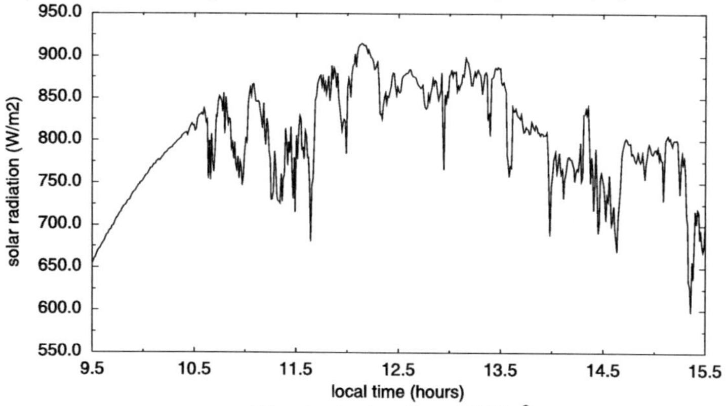

Direct solar irradiance (W/m^2)

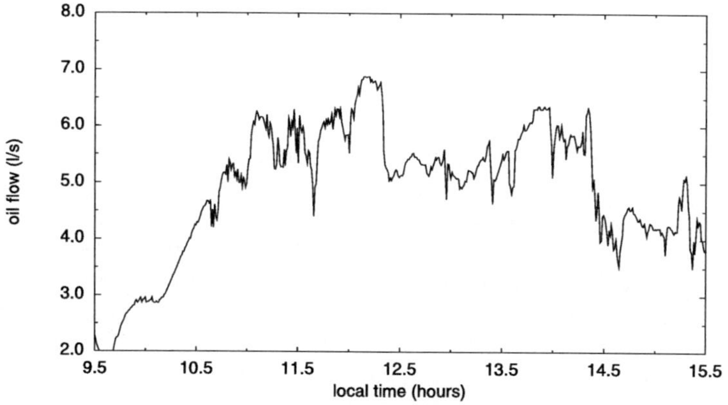

Oil flow (l/s)

Fig. 5.66 Test carried out with the FLC controller (21/12/94), [333]

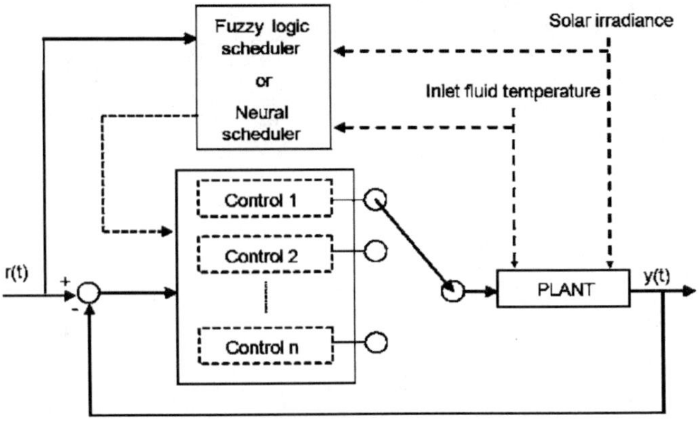

Fig. 5.67 Block diagram of neural/fuzzy switching control, [94]

ANN on a batch mode, following an on-line estimation of weights in order to eliminate any model/plant mismatch due to the quality of the off-line training set or the time variant nature of some plant parameters. In [156], another non-linear adaptive constrained MPC scheme with steady-state offset compensation is developed and implemented. The ANN training is carried out on-line by means of a distribution approximation filter approach.

In [154, 179], an NNC strategy is applied. The ANN is trained based on measured data from the plant providing a way of scheduling between a set of PID controllers, a priori tuned in different operating points by means of Takahashi rules. The scheduling variable is obtained from an ANN having as inputs the values of direct solar irradiance, inlet HTF temperature and reference (or outlet) temperatures. Thus, the scheduler implements an inverse of the plant at steady state. In [179] the modeling capabilities of a recurrent ANN to replace the unknown system and the effectiveness and stability of the OR control theory (geometric approach) are combined.

5.12 Monitoring and Hierarchical Control

Recently, hierarchical multilayer control systems are being developed [46, 106, 108, 110], aimed at automatically determining the optimal plant operating points for maximizing economic profit from the sale of the electricity produced. Early detection of faults (system malfunctions) combined with a fault tolerant control strategy can help to avoid system shut-down, breakdown and even catastrophes involving human fatalities and material damage. The first trends in the application of this kind of technique to DSCF have been performed in two directions. On one hand, in [156] the robustness of a constrained MPC was tested in the face of several faults on the actuator, sensors and system parameters. Further works [95, 96] incorporate a fault diagnosis module and a supervisory system in order to detect, identify and accommodate these types of fault. On the other hand, related works are currently being performed within the scope of data-mining and monitoring techniques [47, 215, 235, 236], aimed at providing software tools to facilitate operation and data exploitation and to predict possible failures in the system. These last works are based on the new generation of plants using water/steam as heat transfer fluid [130, 226, 383, 384, 416, 419], such as the DISS plant explained in Sect. 4.5.

In [280, 325] an optimal control formulation is suggested where the objective is to maximize net power produced when the pumping power is taken into consideration. In [105, 106] a compensator was introduced to automatically compute set points for the whole range of operating conditions of the ACUREX DSCF, looking for the maximum achievable temperature taking into account operational constraints such as the maximum constructive temperature (305°C), the saturation of the control signal (HTF flow between 2 and 12 l/s), the maximum temperature gradient between the inlet and outlet HTF temperature (80°C), and accounting for the actual values of the disturbances (mainly in direct solar irradiance, inlet HTF temperature and mirror reflectivity). An enthalpy balance is used for set point optimization purposes taking into account the aspects mentioned. The advantages of using this kind of set point optimization strategy are evident in the starting phase of the operation when the largest variation in the inlet HTF temperature occurs, due to the existence of cold HTF in the pipes and recirculation using the three-way valve until the minimum temperature to be entered at the top of the storage tank is reached.

5.12.1 Reference Governor Optimization and Control of a DSCF

This subsection describes the design and implementation of two-layer hierarchical control strategies for a distributed solar collector field, as well as representative experimental results in which the benefits of using this approach compared to current operations are highlighted [110]. The upper layer of the hierarchical strategy was implemented using two different approaches, fuzzy logic and physical model-based optimization. Both calculate the optimum plant operating point automatically, taking operating constraints into account while maximizing profit from selling the electricity generated. The lower layer uses the output generated by the upper layer as set point to automatically track the operating point despite any disturbances affecting the plant.

Hierarchical control consists of decomposing the original task into hierarchically structured subtasks and then handling each subtask with a specific control [67]. This decomposition can be spatial or functional. Due to the nature of the problem of the solar field, functional (or multilayer) hierarchical control is applied. Figure 5.68 shows the generic hierarchical control scheme planned for the ACUREX field. The upper layer is aimed at generating the trajectories that the system should follow and monitoring them (previously done manually by the operator). The lower layer involves reference tracking and disturbance rejection.

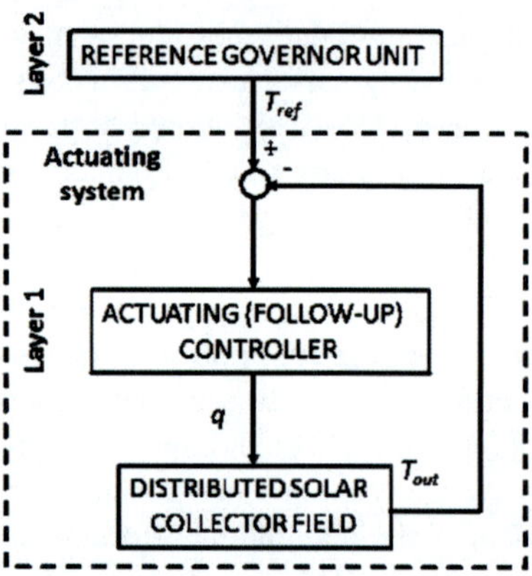

Fig. 5.68 Multilayer control structure with regulatory and reference governor layers

The lower or regulatory control layer is related to automatic control of the HTF outlet temperature (as in all previous control algorithms studied in this and the previous chapter). Using the desired output temperature (T_{ref}) calculated by a reference governor (layer 2 in Fig. 5.68), the automatic control has to regulate the HTF flow q to reach a T_{out} as near as possible to T_{ref}, in spite of disturbances affecting it (mainly solar irradiance and inlet HTF temperature). In the examples developed in this section, the PI + FF and FL algorithms studied in Sect. 5.8.1 have been used.

An upper layer in the hierarchical control system (reference governor) is useful to take into account more aspects of the operation. Some situations, such as start-up, periods when there are solar radiation disturbances or variations in the inlet temperature, require the total attention of the plant operator. Forcing the plant to unreachable temperatures can reduce the life of its materials and lead the actuators to saturation, producing undesired oscillation and, in general, deteriorated performance. On the other hand, establishing a set point that is too safe may cause production losses. This subsection shows two different approaches for automatic computation of the best set point for the lower layer T_{ref}, taking into account plant safety constraints, input conditions and, finally, attempting to minimize the final production costs.

5.12.1.1 Fuzzy Reference Governor (FRG)

The first approach to designing a reference governor consisted of 'imitating' what the plant operator does during operation. The main inputs taken into account of the plant operator (besides the actual flow rate) are solar irradiance and inlet temperature, thus, these were the signals used as inputs to the fuzzy reference governor (FRG) (see Fig. 5.69). Parameter I gives the solar energy currently available which is used to calculate the maximum solar field output temperature and the inlet temperature is related to temperature increment constraints. Another approach would be to add q as input, however, there is not enough improvement to compensate for the complexity in the design of the fuzzy logic inference mechanism. Moreover, the

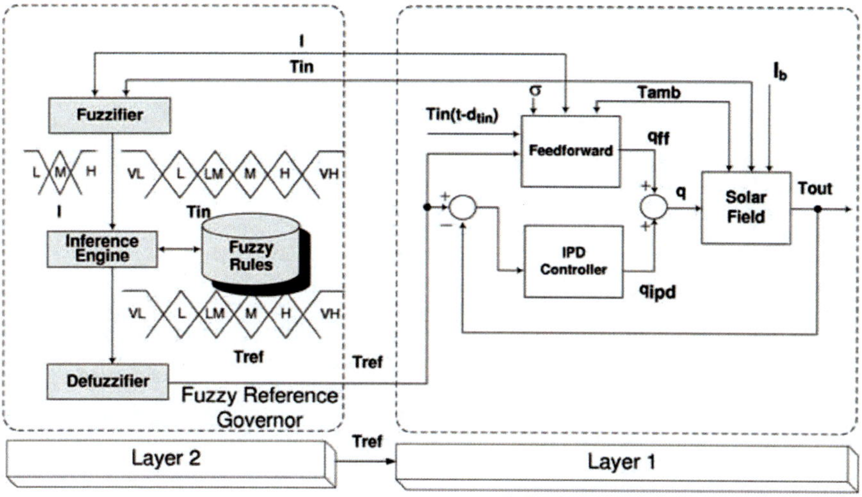

Fig. 5.69 Control structure including the FRG (courtesy of C.M. Cirre et al., [110])

design of the fuzzy sets is easier and more intuitive using T_{in} for the safety mechanism related to $\Delta T = (T_{out} - T_{in})$. The goal of the FRG is to find the maximum T_{ref} according to inputs I and T_{in} allowed by current safety constraints.

Only three membership functions were needed for direct solar irradiance since more than three made the reference governor too sensitive to changes in this variable, strongly influencing its output. The design of the inlet and reference temperature fuzzy sets was made taking into account the plant safety constraints in temperature increments. The universe of discourse of the corrected direct solar irradiance covers the most significant levels where the minimum can be set at 0 W/m^2 and the maximum at 1100 W/m^2. The membership function is triangle-shaped, the left end is L-shaped and the right Γ-shaped. Figure 5.70 shows the membership functions for inlet temperature, corrected direct solar irradiance and reference temperature, with the following linguistic labels: L (Low), M (Medium) and H (High). This number of sets is enough to establish the maximum T_{ref} that fulfills the safety constraints and can be reached with the solar radiation levels. T_{in} universe was designed using triangle-shaped fuzzy sets. The universe of T_{in} covers all the usual HTF inlet temperatures, from 15°C to 250°C (Fig. 5.70). The distribution of the membership functions takes the maximum temperature increments allowed during operation into account ($\Delta T \approx$ 70–80°C). After several tests with the DSCF simulator, this number of fuzzy sets was found to be the best because it is the smallest that covers the whole range of possible inlet HTF temperatures within the set of allowed ΔT increments. There are the same number of fuzzy sets for T_{ref} and T_{in}. T_{ref} fuzzy sets were designed for an output providing good process start-up performance within ΔT constraints. To deal with these goals, the design of the T_{in} fuzzy sets was mapped with ΔT and the fuzzy set designed for low T_{ref} temperatures usually related to start-up was enlarged (Fig. 5.70). Several shapes were tested with the plant simulator, but

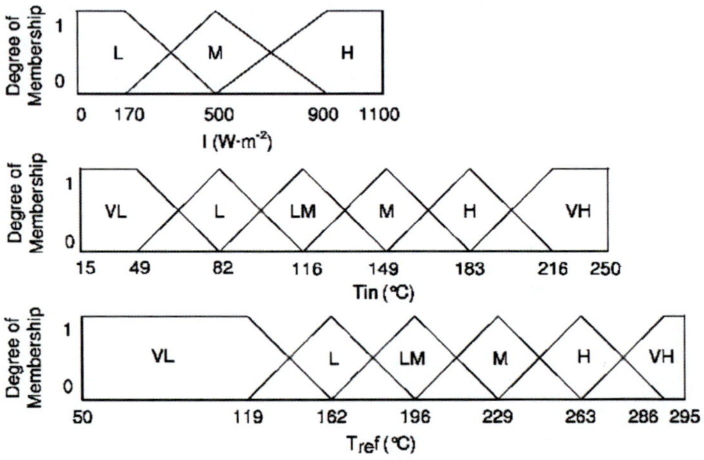

Fig. 5.70 Membership functions for inlet temperature, corrected direct solar irradiance and reference temperature (courtesy of C.M. Cirre et al., [110])

Table 5.10 FRG: rule table to describes a relationship between the inputs and the output (courtesy of C.M. Cirre et al., [110])

	VL	L	LM	M	H	VH
L	VL	VL	L	L	LM	H
M	VL	VL	L	LM	M	H
H	VL	L	LM	M	H	VH

there was no significant improvement (over the triangle-shaped membership functions). The linguistic labels mean: VL (Very Low), L (Low), LM (Low Medium), M (Medium), H (High) and VH (Very High). The FRG output is T_{ref} (Fig. 5.69). To reduce the effect of variation in solar radiation on the FRG output, I and T_{ref} are filtered with a low-pass filter. A broad ranking of different I and T_{in} was used to check output calculated by the FRG. The results are shown in Fig. 5.71.

Table 5.10 shows the rule table defined based on past experience with manual settings. In all, 18 inference rules were devised to describe the behavior of the entire system. The output of the rule-based model is computed by the max–min relational composition [207, 286].

5.12.1.2 Optimizing Reference Governor (ORG)

The optimizing reference governor (ORG) tries to find the set point for the lower layer that maximizes the energy the solar field is able to provide while minimizing costs; that is, a trade-off between the cost of electricity generation and profit from its sale. The purpose of the optimizing reference governor is to calculate the reference temperature T_{ref} (or the future T_{out} at the solar field outlet) that meets plant safety constraints while maximizing the cost function represented in Eq. (5.71), where

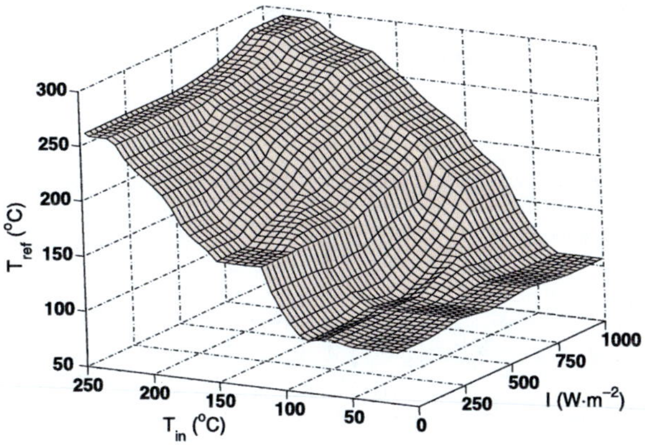

Fig. 5.71 T_{ref} surface generated by the FRG with the inputs T_{in} and I (courtesy of C.M. Cirre et al., [110])

$P_r(T_{out})$ gives the production [W$_e$] and $P_{load}(T_{out})$ represents the main energy consumption [W$_e$] (related to production costs):

$$J = P_r(T_{out}) - P_{load}(T_{out}) \tag{5.71}$$

The solution given by the minimization of the cost function $J(T_{out})$ (desired T_{ref}) must obey a set of constraints. The first is related to the physical limits of the pump: $g_1(T_{out}) + 2 \leq 0$ and $g_2(T_{out}) - 12 \leq 0$, the second to ΔT : $T_{min} \leq T_{out} \leq T_{max}$ safety conditions, where $T_{min} = T_{in} + 30$ and $T_{max} = \max\{T_{in} + 80, 300\}$ [°C] and where T_{min} and T_{max} are the minimum and maximum T_{out} achievable by the system. A constrained non-linear function solver was used to find the solution. The thermal power provided by the solar field is shown in Eq. (5.72), where $\dot{m}_{out}$ and $\dot{m}_{in}$ are the outlet and inlet mass flow [kg/s], respectively:

$$P_{th} = \dot{m}_{out} c_f(T_{out}) T_{out} - \dot{m}_{in} c_f(T_{in}) T_{in} \quad [\mathrm{W}_t] \tag{5.72}$$

In steady state, $\dot{m}_{out}$ should be equal to $\dot{m}_{in}$. As the variable manipulated in the solar field is q (volumetric HTF flow, l/s), for simplification, the volumetric flow related to $\dot{m}_{in}$, is measured and the volumetric flow at the inlet of the solar field is considered the same as the volumetric flow at the outlet ($q_{in} = q_{out} = q$).

The function $P_{th}(q, T_{ref})$ must only be calculated as a function of temperature. The simplified model of the solar field, in Eq. (4.10) in Sect. 4.3.4.2, is used to find the relationship between HTF flow q and the rest of the model parameters. In steady state, Eq. (4.10) can be substituted in Eq. (5.72), yielding the approximation in Eq. (5.73):

$$P_{th}(T_{out}) = \left[\eta_{col} G I - \frac{\widetilde{H}_l(T_{out}, T_{in}, T_a)}{L_2}\right] L n_{ope} \quad [\mathrm{W}_t] \tag{5.73}$$

As the cost function $J(T_{out})$ is designed in W$_e$, the thermal-to-electric power (W) conversion is made using historical plant efficiency reported when operating and producing electricity. The thermal storage efficiency chosen is $\eta_a = 0.98$ and the thermal-to-electric power conversion is $\eta_b = 0.22$ [341], thus obtaining the relationship between P_r and P_{th} [110].

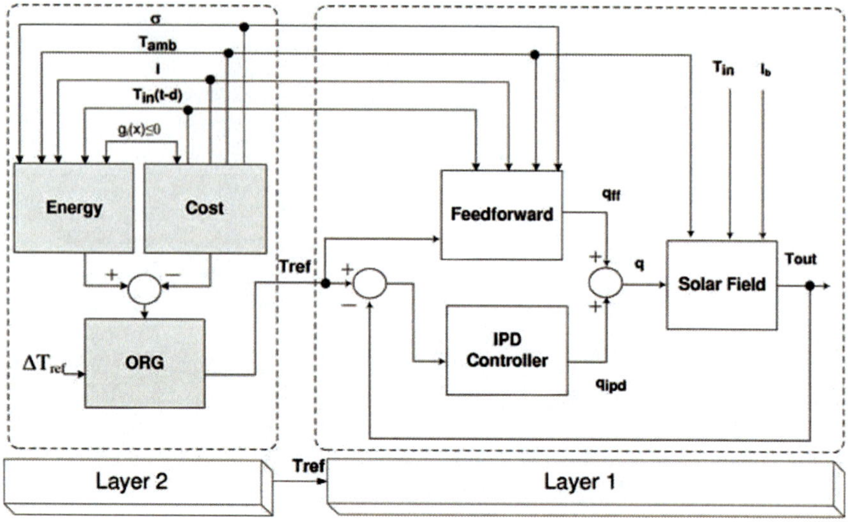

Fig. 5.72 Hierarchical control structure with ORG (courtesy of C.M. Cirre et al., [110])

The cost model includes electricity consumed by the HTF pump which is required by the control system. A function providing the relationship between the HTF flow and electricity consumed was found experimentally with the help of the lumped parameter model in Eq. (4.10). Equation (5.74) shows the cost function model [W_e]. Function F provides the HTF flow required to maintain T_{out} at a level depending on I, T_a and T_{in}. In Eq. (5.74), C_i, $i = 1, \ldots, 6$, are coefficients related to the features of the pump, HTF and physical characteristics of the solar field [110]:

$$P_{load}(T_{out}) = \left(C_1 e^{(C_2 F(T_{out}, T_{in}, T_{amb}, I) - C_3)} - C_4\right)\sqrt{3}\cos(C_5)C_6 \qquad (5.74)$$

Figure 5.72 shows the entire hierarchical optimization structure implemented. The regulation layer is the same as for the control structure shown in Fig. 5.69. An improvement in this scheme would be to limit ΔT_{ref}, thereby avoiding any significant change in the ORG output. Figure 5.73 shows an all-day test with the scheme presented in Fig. 5.69. It is worth mentioning results during start-up, where it can be seen how the T_{ref} generated by the FRG changes with T_{in} and I. Around 12:75 (in Fig. 5.73), there is a fast increase in T_{in} (60°C in 30 min). Note how T_{ref} increases very smoothly compared to T_{in}. Setting T_{ref} manually and under these conditions, the pump will probably become saturated.

Figure 5.74 shows the results with the control scheme developed for Layer 1 (Figs. 5.69 and 5.72) at the actual plant, but with T_{ref} inserted manually by the plant operator during operation (no reference governor was used). On this day, solar radiation varied greatly due to clouds and fog. The response of the automatic controller makes it possible to track the reference with errors of less than 5°C, which may be considered a good result. However, it must be mentioned that the increment in T_{in} at start-up, while maintaining the same T_{ref}, does not provide a desirable system response because the HTF pump becomes saturated. Human operators cannot change

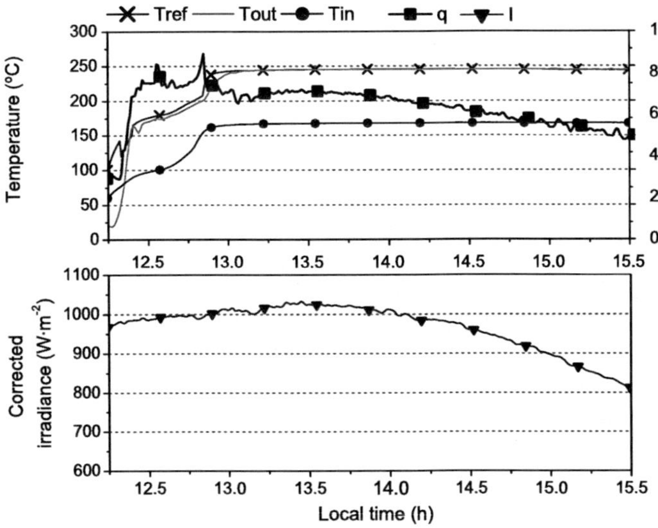

Fig. 5.73 Test developed at the real plant with the hierarchical fuzzy control scheme (courtesy of C.M. Cirre et al., [110])

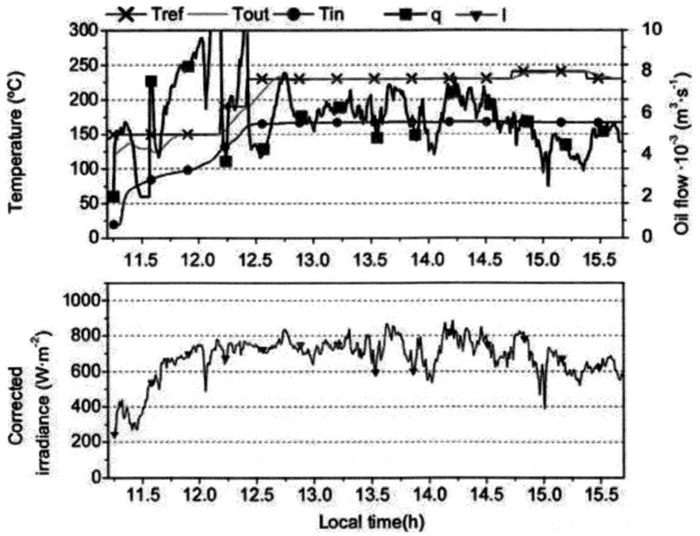

Fig. 5.74 Test developed at the real plant where the human operator impose T_{ref} and the automatic controller is formed by a feedforward in parallel with an I-PD (courtesy of C.M. Cirre et al., [110])

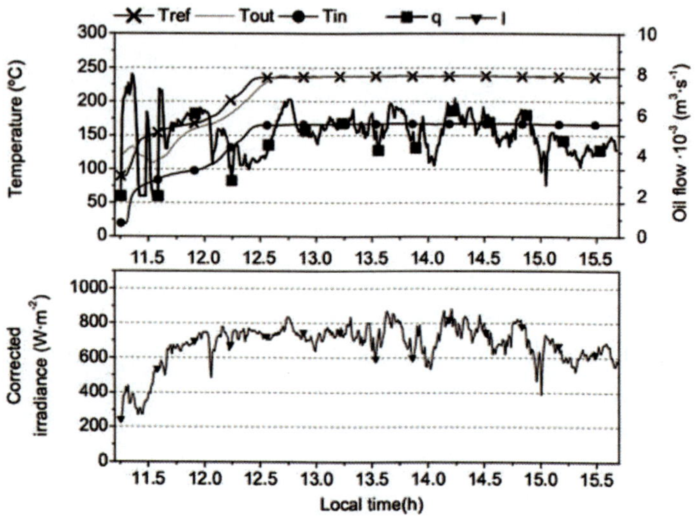

Fig. 5.75 Test developed in simulation with the input conditions presented in Fig. 5.74 and the control structure shown in Fig. 5.72 (courtesy of C.M. Cirre et al., [110])

the set point as frequently and their decisions cannot be very accurate during operation, so it is very difficult to know if the set point is the best or not as external conditions are constantly changing.

Figure 5.75 shows simulation with the ORG hierarchical control structure under the input conditions of the operation day shown in Fig. 5.74. The input conditions are the same and so is the control structure. In this case, T_{ref} was calculated using the ORG structure and, in spite of disturbances, the response of the pump is not as abrupt, the pump never becomes saturated and safety constraints are maintained at all times. A comparison of the response obtained with a manual T_{ref} established in a real test and that obtained for the same conditions with the ORG is given in Fig. 5.76. Manual maintenance of a set T_{ref} means that the plant operator has to pay close attention to operating and safety constraints at all times. As seen in this experiment, the outlet temperature could have been higher during the real test without damaging the solar collector field and this might also have been more profitable.

Figure 5.77 represents the net power calculated with the output given by the ORG hierarchical structure (NP_{ORG}) for each sample time and its respective accumulative sum $\sum NP_{ORG}$ at the bottom versus the output found with a fixed T_{ref} ($NP_{FT_{ref}}$) and the cumulative sum $\sum NP_{FT_{ref}}$ for the results in Fig. 5.76. All units are in MW_e. As may be observed in Fig. 5.77, the net power estimated with the output calculated by the ORG, is higher than the net power estimated with the fixed T_{ref} throughout the experiment. This can also be verified in Fig. 5.77, where the cumulative sum of NP_{ORG} increases faster than the cumulative sum of $NP_{FT_{ref}}$. The parts of the experiment where $NP_{FT_{ref}}$ are higher than NP_{ORG} are due to the output not remaining within the constraints.

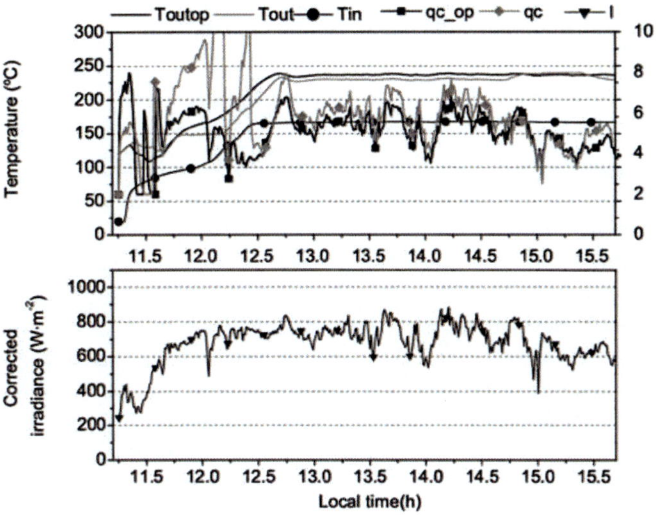

Fig. 5.76 Comparison between the experiment shown in Fig. 5.74 (with T_{ref} imposed by the human operator) and Fig. 5.75, with the ORG (courtesy of C.M. Cirre et al., [110])

Fig. 5.77 Net power calculated for the results shown in Fig. 5.76 (*top*) and their respective accumulative sum (*bottom*) (courtesy of C.M. Cirre et al., [110])

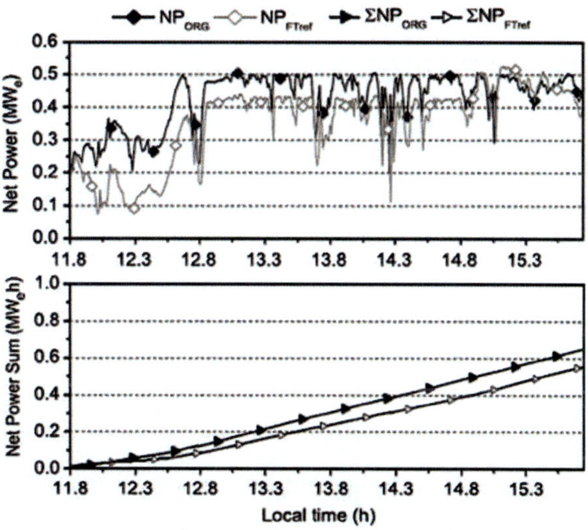

5.12.2 Hierarchical Control

The reference governor strategies explained in the previous section calculate T_{ref} at all times during operation (including plant start-up), staying within the plant safety constraints and with only two data inputs. An improvement on these schemes will be studied in Chap. 8, where other elements belonging to the DSCF system are taken into account to allow for operational planning using prediction and forecasting models (radiation, electricity demand, ...), models of the storage system and models of the power conversion system.

5.13 Summary

The main features of the different advanced control approaches used during the last 30 years to control DSCF have been outlined and are summarized in this chapter. It is difficult to demonstrate the relative merits of one controller in respect to the others since they are based on different conceptual and methodological approaches and the exact conditions in which the tests are performed are different (mainly in terms of solar radiation and inlet HTF temperature conditions) [88]. As has been mentioned, the DSCF may be described by a distributed parameter model of the temperature. It is widely recognized that the performance of PI and PID type controllers will be inferior to model-based approaches [85, 245], taking into account that the plant is highly non-linear as well as of infinite dimension [195, 196]. Even when the plant is linearized about some operating point and approximated by a finite dimensional model, the frequency response contains resonance modes near the bandwidth that must be taken into consideration in the controller in order to achieve high performance [242]. Thus, the *ideal* controller should be high-order and non-linear. The control techniques outlined in this chapter try to find a trade-off between commissioning time and performance. Different characteristics have been studied and are the basis for the selection of each technique, mainly depending on the knowledge the user has of the process and on the techniques: degree of difficulty in obtaining the model/controller tuning, degree of difficulty in the model/controller implementation, degree of acceptance by the operators, robustness, stability and performance results, use of design and/or implementation constraints, disturbance rejection capabilities, starting up of the operation, and existence of real tests [88].

Chapter 6
Control of Central Receiver Systems

6.1 Introduction

As commented on in Chap. 1, power generation by solar thermal systems consists of converting the Sun's energy into heat by means of solar concentrators that focus the solar radiation on a receiver, where it is absorbed by the working fluid. Usually this fluid is sent to a steam generator in combination with a conventional power block to generate electricity. Solar thermal power generation systems are characterized by which of three basic types of concentrator is used: (i) central receiver system, (ii) Dish/Stirling engine systems and (iii) parabolic-dish collector systems. Central receiver systems (CRS) employ an array of two-axis solar-tracking reflectors called heliostats to concentrate the solar radiation on a focal point located on top of a tower to avoid their interfering with each other. These systems are made up of the following components, which are commented on in the next section: (i) collector field, (ii) thermal storage and/or hybrid system, (iii) heat transfer system, (iv) receiver, and (v) control system. Figure 6.1 shows a flow diagram of a central receiver plant with storage.

Its heliostats and tower distinguish this type of power plant from other solar thermal power plants. The CRS is characterized by having a fixed off-axis focus (the Sun, the mirror and the focus form an angle that has a system cosine factor at all times).

If it is for electricity generation, the conventional part can be inserted at either of two points in the cycle. The solar part can send the energy already transformed into steam or gas directly to the turbine, or there can be intermediate equipment that transforms the solar output characteristics into those required by the power conversion system (steam generator).

The design and operation of a power tower plant are strongly affected by the nature of the incident solar radiation cycle. The prediction of the power to be supplied by the plant and the production costs depend in large part on the site selected and their meteorological characteristics, so there must be a long enough meteorological data series for the site selected to reduce uncertainty.

The control system is more complicated than for a conventional thermal power plant, since in addition to the power block, it must include controls for heliostat

E.F. Camacho et al., *Control of Solar Energy Systems*, Advances in Industrial Control, DOI 10.1007/978-0-85729-916-1_6,

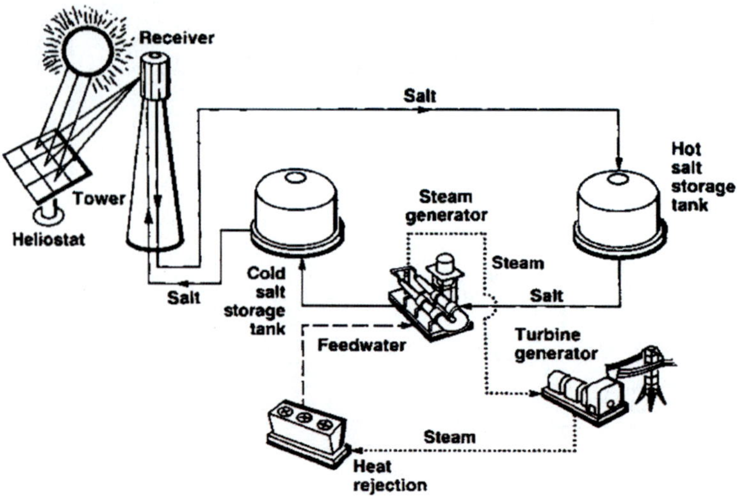

Fig. 6.1 Flow diagram of a central receiver plant with storage

field, receiver, thermal storage and steam generator. This complicates component interaction, especially during start-up and shutdown, which are the most critical periods.

Because of the high incident radiation flux (300–1000 kW/m^2), central receiver systems can work at high temperatures and be integrated into more efficient cycles by steps, they may easily be combined in hybrid systems in a wide variety of options and have the potential for generating electricity with high capacity factors through the use of thermal storage, which can today surpass 4500 equivalent hours per year.

Solar power plants can easily be represented by functional block diagrams, where the main variants are determined by the heat transfer fluids (HTF) and the solar receiver-power block interface through the corresponding heat exchange and energy storage and/or fossil hybridization systems. Four alternatives can be considered depending on the HTF: air, water/steam, molten salt and liquid sodium. And there is a fifth option: in which sodium is used as the working fluid in the receiver and molten salt is used for thermal storage (binary sodium/molten salt system).

6.2 Description of the Technology and Subsystems

6.2.1 Collector Subsystem: The Heliostat Field

The basic function of the power tower collector subsystem is interception, redirection and concentration of direct solar irradiance incident on the receiver subsystem. As has been pointed out, it consists of a field of mobile mirrors called heliostats (see Fig. 6.2) and a system that controls their movements so they track the Sun and keep

Fig. 6.2 Rear view of a heliostat focusing on the tower target (courtesy of PSA)

the solar irradiance constantly focused on the receiver while the plant is in operation. The general characteristics of the heliostat field are based on cost studies and field behavior, which attempt to minimize the annual cost of energy collected per collector. These studies must include the shape and type of receiver, the tower, the whole hot fluid piping system, as well as the collector field and its associated equipment. There are different types of field; circular, completely surrounding the tower, or being located on one side of it (in the northern hemisphere, the field is located north of the tower and in the southern hemisphere, to the south), see Fig. 6.3. In the circular field, the heliostats surround a central tower, normally displaced to the south of the field to optimize its overall efficiency.

Historically, two lines of heliostat development directed at lowering manufacturing costs without diminishing their performance are clearly differentiated. The first of these is directed at the construction of faceted heliostats and reflective surfaces of over 100 m^2, with the resulting reduction in price per m^2 of structure, mechanisms, wiring and foundations. This has some optical problems related to large surfaces and

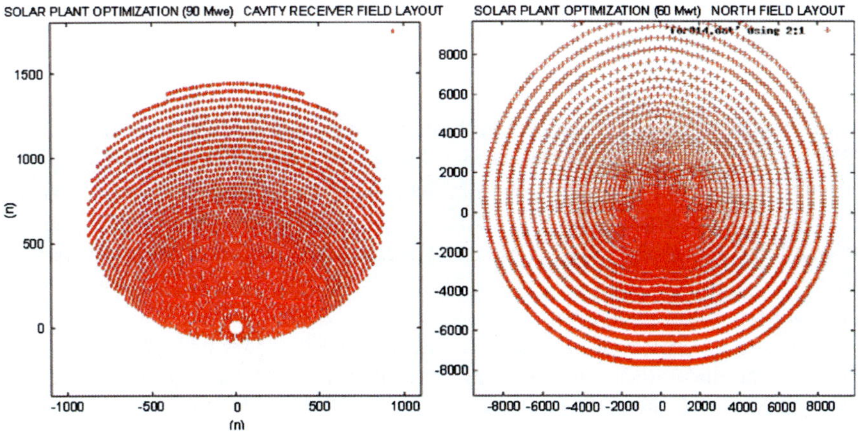

Fig. 6.3 Typical layout obtained with the NSPOC tool [306]

operational problems (for example, heliostat washing). The second line is heliostats made of new reflective materials, such as the stretched membrane (heliostats with a very thin stretched-metal surface on a ring support frame).

6.2.1.1 Characteristics

All the heliostats have a series of features in common [216, 359, 395]:

- *Facets*: Facets, which are the reflective part of the heliostat, may be different shapes and sizes. They are basically made of low-iron glass to prevent rusting, since the heliostat has to undergo outdoor conditions, on which a reflective layer of silver or aluminum is deposited, with an outer layer of protective paint. They are usually attached to the structural supports at three or four anchor points.
- *Structure*: The facets are screwed to a grid support frame attached to a horizontal cylindrical axis which is connected to the heliostat drive by a reduction mechanism. Two-axis movement, azimuth and elevation, is driven by two motors connected to the reduction mechanism. The stretched-membrane heliostat structure consists of a metal-membrane ring support frame joined to a gear wheel for positioning elevation and wheels on a rail for azimuth. Pressure inside the ring can be varied to control the focal distance of the heliostat.
- *Driver*: The typical drive mechanism usually has independent azimuth and elevation gears driven individually. These elevation and azimuth gears are the same type, with the same shape cogs and reduction rate. The first step in reduction is a planetary gear system, while the second (output) is a worm drive (worm-worm wheel). The planetary gear, which provides high reduction rates related to the heliostat aiming quality, occupies very little space. The worm gear also provides high reduction rates, as well as a high capacity for momentum, but its efficiency is reduced by the great friction it undergoes. Linear hydraulic drives are also used, but this type of positioner is less precise than worm-worm wheel drives. Stretched-membrane heliostat positioners consist of a toothed ring connected to a planetary gear which controls the elevation and wheels that move on a rail to control the azimuth.
- *Control system*: To focus properly on the desired target at all times during power tower operation, heliostat movement on both axes must be independent and fully controlled. This control can be accomplished by a centralized system based on a central computer which handles all of the tasks or by a distributed control system in which more importance is given local controls, relieving the central computer of a large part of the tasks. These tasks, regardless of which of the two versions, are:
  - Sun position calculation.
  - Heliostat position calculation (azimuth and elevation) for its location.
  - Measurement of the current heliostat position (azimuth and elevation).
  - Correction of the closed-loop position.
  - Management of heliostat communication.
  - Heliostat operating modes.

– Detection of errors and breakdowns.
– Emergency action and alarms.

The heliostat position must be corrected periodically depending on the distance of each from the aiming target in order to maintain aiming accurate to 1–2 mrad. The heliostat position is found by two coders (azimuth and elevation) which send a signal to the corresponding control, where it is compared to the position it should have according to calculations. A position adjustment command according to the result of the comparison is then sent to the drive motors.

6.2.2 Receiver Subsystem

In power tower systems, the irradiance incident on the heliostat field is reflected at all times on a receiver located at the top of a tower. The incoming energy can have up to several hundreds of MW, depending on the size of the receiver and the size of the heliostat field. The receiver absorbs the solar radiation and heats a HTF (thermal energy conversion) at temperatures that may be over 1000°C and can be used in industrial facilities as process heat, converted into electricity, or used in chemical reactions. This thermal energy is then sent to a steam generator or stored in a storage system. This takes place at high temperatures and high incident solar flux, so it must be done with the least possible loss of energy absorbed (radiation, convection), with the least consumption of electricity and avoiding loss or degradation of the transfer fluid, keeping in mind the long distances it has to cover to go up and down the 50–150 m tower.

Many solar receivers with different configurations and adapted to different HTF have been proposed and tested. There are direct exchange receivers (in which the fluid is exposed directly to the solar radiation) and indirect exchange (when a component converts the solar radiation into heat which is the convectively yielded to the HTF). Receiver construction configurations may be oriented (which only accept irradiance from a certain direction) or circular (which receive irradiance from any angle around them).

Receivers can be classified by their configuration into flat and cavity systems and by their technology as tube, volumetric, panel/film and direct absorption [359]. The heat exchange process can take place in the following basic ways:

- Through tubes that receive the irradiance on the outside, absorb the energy through their walls and transmit it to the heat fluid circulating through them. These receivers in turn may be cavity or flat and operate as an indirect recovery heat exchanger.
- Converting the heat and transferring the thermal energy by convection to the air that flows through the volume of a metal or ceramic absorber which may have different shapes. These volumetric receivers operate like a convective heat exchanger.
- Through the use of particles in fluids or jets that receive the irradiance directly in their volume or on their surface. This type of receiver operates like a direct heat exchanger.

Solar receiver designs are thus closely related to the type of plant and thermodynamic cycle. Flat receivers are absorbent surfaces directly exposed to the solar irradiance. In cavity receivers, the solar irradiance passes through an aperture to a box type structure before being absorbed by the receiver surface. Both types have different thermal loss mechanisms (spillage, convective, reflective, radiative and conductive). In tube and panel receivers, the working fluid circulates through them and is heated by conduction. Tube technologies make high temperatures or high pressures possible, but not both at the same time. In volumetric receivers, the fluid (usually air) flows through a metal or ceramic mesh and is heated convectively. Volumetric receivers can achieve the highest temperatures at pressures up to 30 bar. Finally, in direct absorption receivers, the solar irradiance is absorbed directly by the working fluid and the receiver is merely a support for it. Working fluids used up to now and at present are: air, water/steam, molten salt and liquid sodium.

The receiver is the real core of a power tower system and the most technically complex component, because it has to absorb the incident irradiance with the least loss and under very demanding concentrated solar flux conditions. A large number of configurations have been tested around the world, most of them at the PSA, with liquid sodium, molten salt, saturated steam, superheated steam, atmospheric air and pressurized air as the coolant [324].

6.2.2.1 Volumetric Receivers

Volumetric receivers consist of a set of different-shaped metal or ceramic structures, prepared to fill a volume. They may be open to the outside or have a front window. They can work at outlet temperatures of 700°C to 850°C with metal absorbers and over 1000°C with ceramic absorbers.

The volumetric receiver works by causing a fluid, usually air, to flow through the absorber volume. The radiation incident on the absorber heats it to high temperatures and the air is heated convectively as it flows through. Usually, the air that flows through the absorber is adjusted to the incident flux density distribution, so that the outlet temperature of the absorber does not form a steep gradient between two different points. When air is sucked through the volumetric matrix, the convective loss is practically nil. As the gas flows through the absorber volume, its temperature increases at the same time as the temperature of the material also increases with depth. Thus the highest temperatures are in the interior of the absorber matrix, thereby minimizing loss.

An additional advantage of the volumetric receivers is that since heat is exchanged throughout the internal volume of the matrix, working incident radiation flux can be similar to and even higher than conventional receivers. Maximum flux of 1000 kW/m^2 has been surpassed, so that receiver design sizes can be equivalent to salt or steam receivers, even though the thermal properties of air are worse.

The main advantages of a volumetric air receiver are:

- Air is an easily available fluid.
- It does not freeze.

Mounting the absorber in the 3 MW SOLAIR receiver

The 3 MW TSA receiver in operation

Fig. 6.4 A volumetric receiver (courtesy of PSA)

- There is no phase change.
- High-temperature applications are possible, with the advantage that Rankine cycles can be connected by steps and afterwards directly to a gas turbine in a closed-cycle.
- Easy operation and maintenance.
- Simplicity of the system.
- Fast response to transients from changes in incident flux.
- Does not require any special safety measures.
- Does not affect the environment.

Figure 6.4 shows a volumetric receiver in operation [387].

6.2.2.2 The Tube Receiver

The first pilot solar plants had tube receivers in which the heat transfer medium (gas or liquid) flows through the tubes to extract the solar power absorbed on the outside of the tubes. Therefore, the physical processes in this receiver are related to the use of two transfer surfaces: the solar radiation is absorbed on the outside of the tubes and the heat is extracted on the inside. The heat is transferred from the tube walls by conduction, which is intrinsically associated with a difference in temperature between two surfaces.

In these receivers, there is a wide temperature difference between the front (which is exposed to the irradiance) and the back, so the materials are subjected to strong thermal stress which often leads to their deformation or even breakage. Some solutions employed to minimize this problem consist of introducing auxiliary fossil energy on the back (gas burner, etc.) to equalize the temperatures on both sides of the tube.

The configuration may be flat or cavity, but the trend is to manufacture cavity receivers since due to the large surfaces exposed to radiation, there is higher thermal loss in the flat receivers and this loss is minimized in the cavity receivers, which may even be constructed with movable doors so they can be closed when the receiver is not in use to keep the temperature inside and minimize radiation and convection losses.

Depending on which working fluid is used, they operate in a range of up to 120 bar pressure and temperatures up to 550°C for flat receivers and 1000°C for cavity receivers, while average incident flux hardly surpasses 0.5 MW/m^2. This flux is limited by the low heat transfer coefficient between the tube and the coolant, which impedes high temperatures being reached. Its use is limited to generating steam, either directly in the receiver or introducing the working fluid in a steam generator for integration in Rankine cycles.

6.2.3 Storage Subsystem

Thermal storage, which may be considered the third specific element of a CRS, is necessary to buffer any transient from weather conditions and to accelerate start-up and makes operation possible at night when there is no energy incident on the system. Optimum dimensioning of storage capacity to meet the demand curves of a specific application is an essential part of plant design (Fig. 6.1).

In a solar power plant, the output must be kept stable at all times, regardless of the energy absorbed in the receiver, that is, regardless of fluctuations due to changes in weather. Storage allows continuous power system operation and at the same time protects the system from damage from fluctuation in the solar energy. Solar thermal plants require efficient, economical thermal energy storage to improve plant operation. These systems can increase the solar contribution to the power from the plant and at the same time increase revenues (market introduction strategies are based on hybrid solar/fossil configurations). Furthermore, it increases plant capacity by extending its daily operation without recurring to fossil fuel backup. At the same time, it allows plant production to be adapted to the power demand and simplifies plant operation. If there are discrepancies between the primary supply and secondary energy user, some system that can temporarily absorb these differences is necessary. If the discrepancies are due to change in the secondary energy demand, the problem is one of peak plant load. This can be solved by an energy storage system, because an energy storage system is cheaper than designing a plant to work at full load which is working most of the time at medium load. If the differences between energy supply and demand are due to the changes of primary energy source, the storage system must smooth out the variations in energy entering the plant and ensure specific power production. In the case of solar thermal plants, they must be equipped with a storage system and/or a fossil fuel backup supply. Due to the strong relationship between plant design and the technical solution of the storage system, the specific type of plant must be considered when selecting the storage system, including the heat transfer system, receiver operating conditions and conversion units. Only this way is it possible to define an efficient thermal storage system which best adapts to plant conditions.

6.2.4 Control Subsystem

Operation of a solar thermal power tower plant requires installation of control systems for such a diverse tasks as weather data acquisition and integration in control, automatic cloud detection system, compensation for heliostat field defects (automatic offset correction system), basic heliostat positioning for aiming at the solar receiver through interaction between the central control system with local controllers (heliostat control system), control of flux on the receiver to avoid deterioration of its components from high temperature gradients (may include flux and temperature artificial-vision measurement systems), interaction with conventional steam generation and power conversion control systems, etc. The best option, including all of the components necessary for operation and optimization of solar power tower plant functioning, must be based on hardware and software combining different control subsystems. The data from each of the subsystems are received by a central control system which coordinates and supervises functioning of the whole. Communication among the different subsystems and the central system is by a digital communications network that transmits the necessary information among them.

6.2.4.1 Central Controller, SCADA and Communications

The central control unit controls the heliostat field, supervises basic solar receiver functions and interacts with its supervisory control and data acquisition system (SCADA), which integrates the information received from local controllers in the heliostat field, operating position, other subsystems, etc. The operating position, from which the signals are sent to the various control subsystems that carry out daily routine plant operation, is integrated in it. In the first control systems developed, the central control did most of the control tasks, while local controls were only for communications for which very powerful central controls were required. With the development of microprocessors, it is possible to integrate these operations in the local controls, so the trend has reversed and central control is now in charge of communications and interaction with the operator (in addition to data acquisition), while the rest of the control operations are almost exclusively done locally in the various subsystems (distributed control).

6.2.4.2 Heliostat Field Positioning Control System

The heliostat field control system aims the heliostats at the solar receiver (normal tracking), or at the standby points (points located outside of the solar receiver where the heliostats aim before focusing on the receiver and just after defocusing in emergencies), controls tracking or sends them into stow position when they are out of service. This control is carried out by the heliostat local control boxes, which are devices located at the foot of the heliostats that control their movement and communications with the central control. Their basic purpose is to position the heliostat so that the reflected rays incident on it remain on the desired focal point.

The calculation functions for proper positioning of the heliostats and for system evaluation are [359]:

- Calculation of the Sun position and the solar vector at predetermined times, which requires permanent time synchronization with the central control and the rest of the heliostat local controls.
- Calculation of the flux that each heliostat provides based on a database built up from the original field characterization for different seasons and times of year and of the main heliostat variables (reflectivity) and irradiance.
- Provide data to the central control to facilitate the calculation of the parameters for evaluating solar system performance (effective field cosine, effective reflective surface, energy reflected by the collector field, power incident on the receiver aperture, etc.).

The local control has to solve all heliostat actions described in addition to communications with the central control using the corresponding protocols. As mentioned, the current trend is to increase the amount of local heliostat control intelligence for greater autonomy from the central control and also to employ radio-control systems that eliminate the kilometers of wiring otherwise necessary for communications.

6.2.4.3 Flux and Temperature Estimation and Measurement Subsystem

To optimize and design a CRS it is essential to know the performances of the subsystem formed by the tower and the heliostat field [147]. Modeling tools for CRS can be divided into two main categories: those dedicated to system optimization and those designed to deep analysis of the optical performances. In [147] an overview of codes for solar flux calculation dedicated to CRS applications is performed. One important topic is to select the best layout to maximize the collected solar energy or to minimize the cost of that energy. Another one is to be able to estimate the power reflected by the field and arriving on the receiver aperture (performance calculation) [147]. Thus maps on the receiver aperture have to be determined, as well as efficiency matrices on the solar field. The solar field efficiency is usually defined as the reflected power arriving on the receiver aperture divided by the product of the incident solar power by the total area of mirrors. It includes reflectivity of the mirrors, cosine factor due to the incident angle of the Sun on the heliostats, atmospheric attenuation between the heliostats and the receiver, shadowing and blocking effects and spillage of the flux around the receiver aperture. For the mean annual efficiency, heliostat availability is also taken into account. A solar field efficiency matrix is a bi-dimensional matrix giving the solar field efficiency as a function of

Sun position (azimuth and elevation). Integrated in tools for performance analysis of CSP systems, such a matrix makes it possible to assess the solar field efficiency all year long [147]. The field layout use to be optimized based on costs criteria (cost of produced kWh, cost of installed kW).

The optical components of a CRS are designed to form an image of the solar disk on a focal plane. The obtained solar spot has neither the same size nor the same luminance as the Sun, due to [147]: Sun and collector geometry, size and luminance of the Sun (varying every day by diffusion in the atmosphere), specific defects of solar facilities, optical aberrations (like heliostat astigmatism, caused by the design of the reflective surfaces), microscopic errors of reflective facets, often considered as negligible, pointing (or tracking) errors, curvature and canting errors of facets or modules.

The usual approaches to calculate concentrated solar irradiance are ray-tracing and convolution methods. The principle of ray-tracing methods (or statistical or Monte Carlo methods) is to choose randomly a bundle of rays coming from a surface 1 and then to determine which of them arrive on surface 2. The irradiance of an elementary surface is proportional to the number of impacting rays. In the case of a concentrator with one reflection, this algorithm is used twice, first between the Sun and the reflective surface with an energetic distribution corresponding to the Sun shape, then between the heliostat facet and the receiver with a statistic law for the error distribution related to the defects of the facet. Precision and calculating time increase with the number of rays and the complexity of geometry. In convolution methods (or cone optics), reflected rays from elementary mirrors are considered with error cones calculated by convolutions of normal Gaussian distributions corresponding to each error (Sun shape and heliostat errors). A systematic comparison shows that with the same hypothesis similar results can be reproduced with ray-tracing and convolution methods (as shown in [147] and references therein). Simulation errors often come from an incomplete description of reflective surfaces and Sun shape properties. Nevertheless ray-tracing methods are more flexible and are able to model non-ideal optics (non-imaging concentrators). Indeed, they have the advantage of reproducing real interactions between photons and therefore of giving accurate results for small or complex systems but they need higher calculating time and computing power. That is why it is not recommended to use ray-tracing techniques for system optimization [147].

On the other side, it is important to maintain temperature distribution on the receiver surface as uniform as possible to reduce thermal gradients in the receiver and increase its useful lifetime. At the same time incident solar flux distribution must follow a predefined pattern and not separate too much from it. Both parameters, along with the thermal performance of the absorber define the effectiveness of the receiver which is very important for commercial power generation. Artificial-vision systems (using coupled charge devices—CCD—and infrared cameras) or thermocouples installed at various points on the receiver can be used to control both distributions [150]. The information from this system is processed by the central control or by other subsystems (offset correction [45]) to calculate aiming-point modification of one or several heliostats or even to detect the presence of clouds. Knowledge of the temperatures profile makes it possible to modify receiver aiming-point coordinates and keep the required temperature distribution within a defined range.

6.2.4.4 Automatic Offset Adjustment System

The error calculated between the command sent to a heliostat and the real focusing position on each heliostat axis is called the offset or quasi-stationary error. In existing solar plants, the offset correction is a periodic, time-consuming task due to the large number of units to be corrected in a heliostat field. It is done by a specialized operator with the help of an image from a camera located in the heliostat field aiming at an auxiliary target. The operator manually changes the azimuth and elevation angles of the heliostat according to the image projected so its reflection coincides with the center of the target. This task can now be done automatically using a black and white CCD camera which determines the position of the centroid of the solar image projected by the heliostat and its deviation from the center of the target, and then depending on the heliostat coordinates, the offset is estimated and aiming is corrected [45].

6.2.4.5 Power System Control

The power system control in a power tower system is analogous to that of a conventional power plant, since they have the same systems and components (steam generator, turbine, etc.). However, while in conventional plants, the fuel can be manipulated to keep the turbine requirements constant, in the solar plant the fuel (the solar radiation) cannot be manipulated, it changes seasonally and even daily and is considered a disturbance from the point of view of control. Therefore, the PID controllers, normally used with good performance in conventional plants in a multitude of industrial processes, are sometimes insufficient for good control of the power tower systems, since they cannot provide a fast response to changes in irradiance, and results differ significantly from the set point.

This is why feedforward control (FF), model predictive control (MPC), intelligent control (IC) and other techniques are being introduced into today's control systems to increase response speed and accuracy in plant control, minimize system start-up and shutdown times, reduce operating and maintenance costs and optimize the system operating parameter set points for maximum energy efficiency.

All the information from this control system is sent to the central control for interaction with the rest of the control subsystems to keep turbine steam pressure and temperature requirements and react to any alarm tripped in the power block.

6.3 Advances in Modeling and Control of Solar CRS

The modeling and control schemes object of this and the following sections are focused on the CESA-1 facility, a central receiver solar thermal power plant belonging to CIEMAT and located at the PSA. This test-bed plant was shown in Chap. 1

(Fig. 1.13) and it is an experimental prototype for electricity generation, among other research projects.

The CESA-1 facility collects direct solar irradiance by means of a field of 300 heliostats (39.6 m^2 surface) distributed in a 330×250 m^2 north field into 16 rows. The heliostats (see Fig. 6.2) have a nominal reflectivity of 92%, the solar-tracking error on each axis is 1.2 mrad and the reflected beam image quality is 3 mrad. North of the heliostat field there are two additional test zones for new heliostat prototypes, one located 380 m from the tower and the other 500 m away. The maximum thermal power delivered by the field onto the receiver aperture is 7 MW. At a typical design irradiance of 950 W/m^2, a peak flux of 3.3 MW/m^2 is obtained. In addition, the 99% of the power is focused on a 4 m diameter circle, 90% in a 2.8 m circle.

The 80 m high concrete tower has a 100 ton load capacity. The tower is complete with a 5 ton capacity crane at the top and a freight elevator that can handle up to 1000 kg loads. For those tests that require electricity production, the facility has a 1.2 MW two-stage turbine in a Rankine cycle designed to operate at 520°C 100 bar superheated steam.

The TSA is a receiver system located on top of the CESA-1 tower at the PSA. The volumetric receiver is constituted of packages of thin wire. These packages are compacted in hexagonal cells, so that the profile offered by the receiver is that presented in Fig. 6.5(a). The main receiver characteristics are: open cycle wire pack volumetric air receiver, modular and metallic wire mesh absorber, 700°C mean hot air temperature, 2500/3000 kW nominal/maximum absorber power, 800 kW/m^2/100 kW/m^2 heat flux density maximum/rim, 3400/3000 mm aperture/absorber diameter, 30° receiver tilt angle, 0.6 air return ratio, and 86 m elevation above ground level. When the beam irradiance received from the heliostat field reaches the receiver surface, the wire packages are heated and then energy can be transferred to the air flow, as the packages are quite porous and air easily circulates through them. Thirty-six thermocouples were placed behind elements of the receiver in order to measure the temperature of the air, so that the supervision of the state of the receiver during the operation could be carried out (as the absorber temperature is close to the measured air temperature). An overview of the operation of this kind of plant may be found in [150, 170].

The power stage of the TSA system is illustrated schematically in Fig. 6.6. It is composed of an air circuit and a water–steam circuit. The air circuit supplies the power demanded by the steam generator from the receiver and the thermal storage and return to the receiver the output cold air from the steam generator.

To achieve the desirable flux distribution, a five point aiming strategy is used in the heliostat field. This heats the absorber ceramic cups, which transfer their heat to the air flow behind them by the volumetric principle and the hot air leaves the receiver at 700°C, blower G1 controls the air mass flow through the receiver. A thermal storage unit in the air circuit, which provides standby power, is charged during plant start-up and then discharged to feed the steam generator, either during cloud transients or when the power supply from the receiver is not enough to feed the steam generator at nominal conditions. Blower G2 controls the air mass flow through the steam generator and thereby, incoming power. The once-through steam

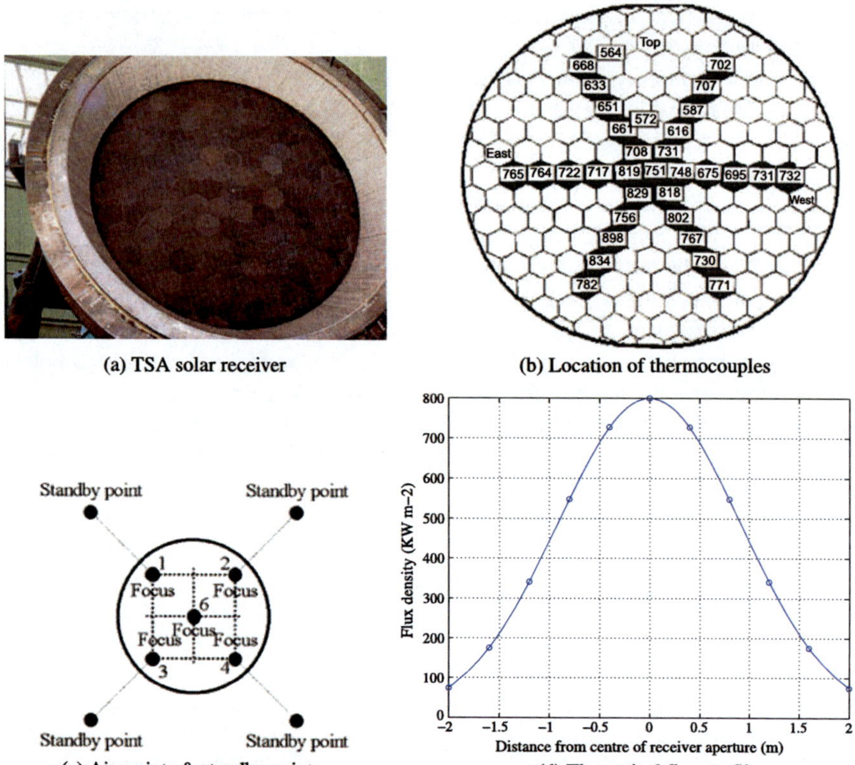

(a) TSA solar receiver (b) Location of thermocouples

(c) Aimpoints & standby points (d) Theoretical flux profile

Fig. 6.5 TSA solar receiver (courtesy of F.J. García-Martín et al., [150])

generator, which is part of the water–steam circuit, has a nominal operation of 340°C and 45 bar. The steam produced is sent to a turbine, which is part of a Rankine cycle. Steam outlet temperature is controlled by means of a feedwater pump and the steam outlet pressure by means of valve PV123 in Fig. 6.6.

6.4 The Heliostat Field Control System

As pointed out before, the CESA-1 heliostat field operation requires the utilization of approximately 180 heliostats for TSA specific field layout. These heliostats are used to concentrate the solar irradiance onto five aimpoints of the volumetric receiver. These aimpoints are described by a set of coordinates having origin at the base of the tower and will be treated in the next section. As well as the aimpoints, other points exist in the space called standby points (Fig. 6.5(c)), located outside the solar receiver onto which the heliostats are aimed at before pointing the aimpoints and just after being defocused. The use of these standby points allows the heliostats

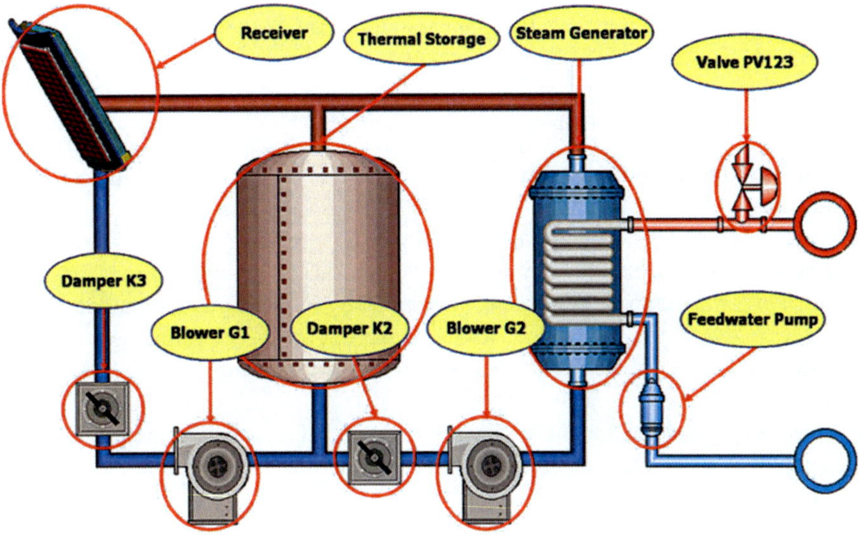

Fig. 6.6 Schematic diagram of the TSA system (courtesy of J.D. Álvarez et al., [10])

to be moved so that the concentrated irradiance is moved from the standby points into the receiver following security paths and avoiding accidents caused by the reflection of a large amount of energy over undesired points. These standby points are also used in emergency situations in which the heliostats have to be quickly defocused. The position of a heliostat can be classified into three different states: normal tracking if the heliostat is aimed at an aimpoint of the solar receiver, transferring energy to the system; phase-out tracking or defocused, if the heliostat is aiming at a standby point; and finally stow if the heliostat is out of service.

A series of factors that hinder the constant concentration inside the receiver (necessary for a good operation of the system) exist. The temporary dependence of the systems that use solar energy (the radiation intensity varies along the day) and the possible interferences (for example, for cloud presence) are among these factors. In a general way, the basic operations that a heliostat carries out are to be positioned in a fixed orientation or the tracking of a coordinate. The tracking allows the concentration of the incident solar irradiance on the heliostats in the field to be aimed in a wanted point of the space. The tracking involves the calculation of the elevation and azimuth set point values for the heliostat. The calculated values will depend on:

1. The coordinates where one wants to aim,
2. The heliostat location in the field (regarding the tower) and
3. The date and hour with the one it can obtain the position of the Sun in the sky.

Although the coordinates where it is wanted to make impact the reflected ray do not change, it is periodically necessary to recalculate the elevation and azimuth set points because of the Sun constantly changes its position in the sky. These calculations together with the necessity to be able to indicate some emergency set points quickly when a dangerous situation happens (for example the elevation of the temperature above certain limits in certain places) make that a hard real time support is required.

The following main operation modes are defined [59]:

- Waiting: The heliostat is waiting for an input message. In this mode, the heliostat may be moving to a position.
- Request: The heliostat sends an output message including position information.
- Reset: When the heliostat is in reset mode, updates its positions moving to zero position for azimuth and elevation, moving firstly in the azimuth direction. Once it get to zero-azimuth position, it begins moving in the elevation axis. In this operation mode, the heliostat does not send any output message until the reset time is reached.
- Stow: If the heliostat does not receive an input message during before the timeout, it moves to stow (zero) position. In this case, the movement is in the two axes at the same time.
- Control: The heliostat moves to the position reference specified in the input message.
- Request + Control: Request and Control modes at the same time.
- Request + Reset: Request and Reset modes at the same time.

6.4.1 Heliostat Field Simulators

In [58, 59] a real time heliostat field simulator, based on a hybrid model, using Modelica [257, 258] as the modeling language, was presented for the CESA-1 plant, mainly aimed as a tool for the enhancement of advanced control algorithms but also useful for training purposes. The developed real time heliostat field simulator is basically the union of an hybrid heliostat field model (combining the system dynamic continuous variables with operation mode discrete variables) and a wrapped model which handles the real time simulation and communication issues between the heliostat field simulator and HelFiCo (Heliostat Field Control) software [159], which is in charge of manipulating and controlling the heliostat field according to an automatic control strategy. The real time heliostat field simulator provides a virtual system, with the same response as the real plant.

6.4.2 Tracking the Sun

The problem of modeling the flux distribution in the receiver brings about the previous resolution of the following subproblems that are described below.

6.4.2.1 Computing the Solar Vector

At present, there are many algorithms to compute the solar vector (vector pointing toward the Sun). The one used in the applications shown in this book is the called PSA algorithm [56], which input is time and location (this algorithm has already been used in the PV solar tracker developed in Chap. 3). The time for the instant under consideration is given as the date (year, month and day) and the Universal Time (hours, minutes and seconds). The location is given as the longitude and latitude of the observer in degrees. Latitude is considered positive to the north and longitude to the east. The main features of this algorithm are [56]:

1. It incorporates an efficient method of computing the Julian Day (J) from the calendar date and Universal Time.
2. Memory management was improved by controlling the scope and life span of variables.
3. Speed and robustness was improved by using robust expressions for calculating the solar azimuth, which are valid for both hemispheres.
4. Accuracy was improved by modifying the simplified equations of the Nautical Almanac used by Michalsky with the introduction of new coefficients and new terms and including parallax correction.

The ecliptic coordinates of the Sun are computed from the Julian Day and then are converted to celestial coordinates using standard trigonometric expressions [56] and also to horizontal coordinates (see Chap. 1 for the definition of this coordinates). The algorithm is available in electronic form at http://www.psa.es/sdg/sunpos.htm.

Heliostats are oriented so that for any position of the Sun, the rays are reflected directed to a fixed point some distance above the level of the field. The fundamental relations governing the heliostats movement must satisfy the laws of reflection and, therefore, the angles of the incident and reflected sunlight, with respect to the normal of each mirror surface, must be equal. The defining vectors must be located in a same plane at every moment. In Chap. 2, Sect. 2.2, these defining vectors were analyzed.

6.4.2.2 Identification of the Shape of the Facets

The heliostats are composed of mirrors, or facets aligned along, required to concentrate the Sun's rays into an aimpoint defined in the receiver. The tolerances of these facets are measured using large accuracy probes or a laser before being installed at the solar field to determine their actual geometry. With this information, the surface of each facet can be built and thus the normal to each element of area can be determined.

However, it seems obvious that after the assembly process of each facet onto the heliostat, errors in the normal vectors may appear. In fact, there are multiple error sources affecting the computing of the normal vector to each element of area, from the influence of wind to noise in the test probe or laser. This is the reason of why all these errors are embedded into one, called slope error, as a Gaussian random deviation in the normal of each element of area. This value has to be adjusted from field tests and will be independent for each facet [259].

6.4.2.3 Sun Model

The Sun, as light source, must be modeled and sampled as a circular solid. In the literature there are various techniques to estimate the solar brightness distribution as a function of solar radius in steradians [72]. One of these approaches makes the assumption that the distribution of the solar intensity can be modeled as a circular Gaussian of the form:

$$z(x, y) = \frac{1}{\sqrt{2\pi\sigma^2}} \exp\left(\frac{-(x^2 + y^2)}{2\sigma^2}\right); \quad \sigma = 2.325$$

where the standard deviation is considered as one half of the mean solar radius (4.65 mrad).

Another possibility is to use brightness profiles of the circumsolar radiation (CSR) as those obtained by the Lawrence Berkeley Laboratory (LBL) [72] or the German Aerospace Center (DLR) [266].

Solar radiation, when passing through the atmosphere, is affected by scattering processes such as Mie or Rayleigh ones, which creates a halo around the solar disk. The CSR is a simple description of the distribution of solar intensity that determines the percentage of integrated intensity of the halo respect to the total [259]:

$$CSR = \frac{I_{CS}}{I_{Sun} + I_{CS}}$$

where I_{CS} represents the intensity of radiation of the halo and I_{Sun} the intensity of radiation of the solar disk. This way several profiles can be created using different values of the CSR from data captured using image cameras.

6.4.2.4 Computing the Flux Density at the Receiver

The total power reflected by a heliostat onto a receiver can be approached by

$$P_h = I A_m \cos(\theta_i) f_{at} r \quad [\text{kW}] \tag{6.1}$$

where I is the direct solar irradiance [kW/m^2], A_m the mirrors area [m^2], $\cos(\theta_i)$ the cosine of the incidence angle of the solar rays on the heliostat, f_{at} is an attenuation factor due to turbidity of the atmosphere and r is the mirror reflectivity.

At present, there exists approaches that simplify the calculation of the flux density [118, 147] due to its computational load. When a good accuracy is required, ray-tracing techniques are used.

6.5 Basic Offset Correction Techniques

6.5.1 Introduction

As will be seen in Sect. 6.7 (aiming strategies), when controlling the temperature and flux distribution in a volumetric receiver, the algorithms calculate the amount of the shift using an equation appropriate for each heliostat depending on its current temperature dependent focal length and orientation dependent aberrations in addition to beam errors and Sun size. As a first approximation, a heliostat aiming-point strategy providing a desired energy flux correlated with the air mass flow through the receiver can be selected to solve the control problem. Nevertheless, there are error sources that increase the complexity of the control system [363]: time, ephemeris equations (Sun model), site location (latitude and longitude), heliostat position in the field, time-varying astigmatism and cosine effects, processor accuracy, atmospheric refraction, control interval, and structural, mechanical and installation tolerances. Some of these error sources are systematic ones, mainly due to tolerances (joints, encoder, etc.), wrong mirrors facets alignment (optical errors), errors due to the approximations made when calculating the solar position [56], etc. Heliostat beam quality error sources are analyzed in [210]. The approximation adopted in this section to overcome some of these error sources (mainly those related to the calculation of the solar position and to tolerances) is based on the use of a Black/White (B/W) CCD camera which captures images of the Sun reflected by each of the heliostats of the field onto a target devoted to the task of offset correction [45]. The reflection of the light coming from the Sun produces a shape that continuously changes due to the Sun–Earth relative movement. The obtained images serve as feedback information that allows the automatic calculation of the distance between the center of the target and the sunbeam centroid in such a way that this error signal can be used for adjustment purposes. The use of artificial-vision algorithms permits the calculation of the center of the target and the sunbeam centroid. After the calculation of the required displacement of the motors of the heliostats (encoder steps), the system sends this information to the central control system to perform the correction. There exist several works from the late 1970s and early 1980s on the Solar One Ten Megawatt Solar Central Receiver Power Plant at Daggett, California in which the fundamental approach presented in this section was originally developed, not only aimed at compensating offsets of the heliostats but also to characterize total beam power, irradiance distribution, beam centroid, tracking error, overall mirror reflectivity and the Sun's radiance distribution, which could be used to compare the actual flux distribution with the theoretical flux distribution for each of the 1818 heliostats, at any time during the day [172, 210, 211, 361]. The approach was based on the Beam Characterization System (BCS) [51, 54, 294, 364, 370]. The BCS, originally termed the Digital Image Radiometer (DIR), was conceived and developed by McDonnell Douglas (now a wholly owned subsidiary of The Boeing Company) and installed at Solar One as part of the DOE/Sandia contract (DE-AC03-798f10499). The Solar One BCS determined the irradiance distribution of the reflected beam from a heliostat on one of four trapezoidal targets mounted below the central receiver. Each of the 1818 heliostats was sequentially moved automatically onto the target such that *image grabs* could be taken with a digitizer and the centroid of the beam determined and compared with the theoretical position. From this position difference on the target, the tracking error was determined and the heliostat tracking aim point was then corrected, based upon various algorithms. The video camera was modified to provide for background subtraction, control of black level (which can vary with temperature), elimination of automatic gain control and to allow its use for various purposes [52, 53, 362, 363]. Various algorithms were used to assess and validate data, or discard them and retake the measurements. Factors such as wind speed, mean and standard deviation of the centroid variation, solar irradiance variations, etc., were taken into account. These steps were necessary to have accurate determinations of the net irradiance distribution and thus accurate corrections for the heliostat tracking. The BCS provided a number of data displays for the plant operators and was capable of operating in both manual and automatic data acquisition modes and it was integrated with the Data Acquisition System (DAS) and the Heliostat Array Controller (HAC) such that automatic updates of heliostat aim point biases could be made and overall heliostat optical performance monitored.

Later, Sandia developed a system for use at their test facility in Albuquerque. These systems are more complete and accurate than that presented in this section, which is less ambitious and has a different objective from those previously published in the literature in the sense that its objective is simply to replace the operator in the task of performing offset correction. The system tries to imitate the way in which operators perform the offset correction, that is, by watching an image on a TV screen obtained from a typical B/W CCD camera installed in the field and trying to center the centroid of the reflected Sun image with the center of a target. That is the reason of why the system presented in this section only does offset correction based on this simple image obtained from the CCD, without including modifications to the video camera and without using more complex aiming algorithms, as the objective was only to facilitate the operation, relying on the corrections that the aiming-point strategy makes during nominal operation (that implicitly compensates for tracking errors and low accuracy). The daily operation is based on selecting several groups of heliostats that have to focus reflected solar irradiance onto fixed aimpoints defined on the receiver surface (for instance, five aimpoints are defined for CESA-1 operation with TSA configuration). The selection of these groups is based on theoretical studies to achieve desired flux and temperature distribution in the receiver (this last related to security reasons, as temperature gradients within the receiver must be bounded). During operation, an aimpoint strategy is used to try to achieve the mentioned desired flux and temperature distribution, based on the assumption that the heliostats focus the Sun rays on the pre-defined aimpoint (see Sect. 6.7). If large offsets exist, more corrections will have to be done by the aimpoint strategy, increasing the complexity of the operation with the system. Thus, the main impact of the corrections on receiver performance is that a smaller number of corrections should be carried out by the aiming-point strategy, diminishing the transients that could lead to undesired temperature gradients within the solar receiver. The section tries to explain different aspects of the developed work as: (i) the technique used to calculate the position of the centroid of the projected image of the Sun, (ii) the strategy followed when the offset is such that the centroid lies out of the view field of the camera, (iii) other problems that appear in the application (lens distortion, vibrations in windy days, Sun reflections in the tower, etc.), (iv) the software tool used by the operators, (v) etc. As has been mentioned before, the task of detecting and eliminating the heliostats offsets is performed by operators at the PSA, increasing their workload (this is a time-consuming task) and consuming hours that cannot be devoted to nominal operation, as the offsets have to be corrected under the same conditions (daylight). If high accuracy is required, some corrections should be performed during each day due to time-dependent drift [238]. After the work of [56], the aiming algorithms were changed at PSA and smaller aiming errors were obtained, thus decreasing the frequency of offset correction tasks. The development and implementation of an automatic closed-loop control system using images captured by the CCD camera to perform the automatic offsets correction became an important issue, allowing the operator to be kept out of this task and only performing the supervision of the control results.

6.5.2 The Offset Correction Problem

The problem that the systematic errors produce in the operation of the solar plant is manifested in a deviation of the projected Sun shape, which can be far from the expected point. Each heliostat has a pre-defined aimpoint to which it has to point at the beginning of the operation. As is pointed out in [361], the offset correction problem consists of comparing the actual sunbeam centroid position on a target to a command reference position to determine the error in the sunbeam centroid location. The sunbeam centroid position error is then analyzed to correlate the error to errors in the heliostats' track alignment system. In other works referenced in the introduction of this section, a DIR evaluates the heliostat by measuring the total beam power, irradiance distribution, beam centroid, tracking accuracy and overall mirror reflectivity. New coefficients are established for the heliostats' track alignment system to automatically correct for errors in the system, this eliminates the need for resurveying and field work normally associated with aligning heliostats. In the alignment program, calculations are made for the Sun's position based upon the stored time data when the measurements were made. Such factors as the Sun's azimuth and elevation are calculated. The program then calculates the orientation of the heliostat based upon the Sun's position and the command position of the sunbeam. Once the program has calculated the optimum heliostat position, this information is compared with the stored measurements made. In the error transformation routine, errors between the command position and the measured positions are used to calculate the alignment error coefficients. The aimpoint errors are corrected by changing the database stored values. A target placed below the volumetric receiver of the CESA-1 plant (Fig. 6.7) is used to test the accuracy of the heliostats positioning typically once a week to ensure a correct and safe operation. In order to do so, a B/W CCD camera (equipped with a pan-tilt mechanism and with an automatic contrast/brightness adaptation mechanism) is used by an operator who modifies the azimuth and elevation coordinates (encoder steps) of the heliostat by a trial and error procedure till the sunbeam centroid coincides with the center of the target. The commands are sent by the operator using different programs implemented in a workstation, that processes these control actions and sends the appropriate signals to the local controllers (microprocessors) of the servomotors to move the heliostats to

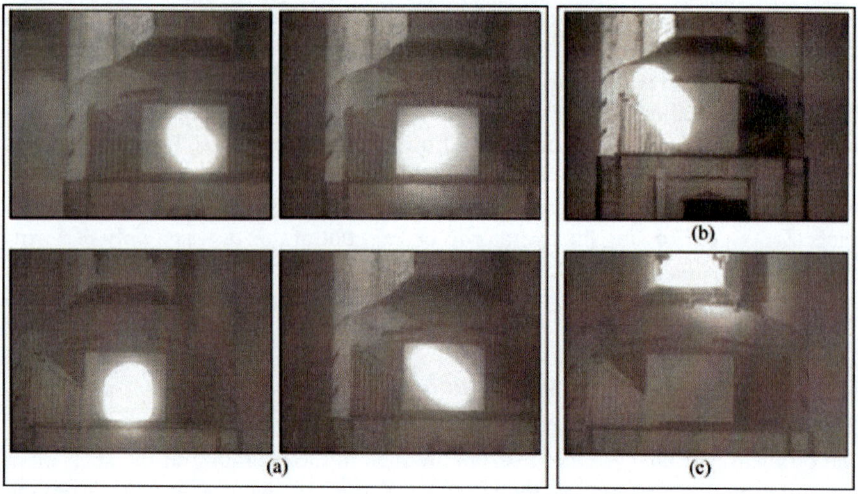

Fig. 6.7 Details of the target and different shapes of the image of the Sun projected by the heliostats: (**a**) centered ellipsoids (the shape orientation changes during the day), (**b**) ellipsoids out of the target due to aiming errors [45]

the positions indicated by the control program. The system can also be used during operation, as one heliostat can be deviated from its spot to correct its offset in real time. In the applications in which these corrections are manually performed, it is impossible or quite expensive to carry out this correction during operation, as more than one operator should be needed. In the next section, an automated system for offset compensation is explained.

Figure 6.8 shows a scheme of the architecture of the offset correction system. The corrective actions are generated in a PC devoted to the offset correction problem and are finally sent to the workstation via a serial port (see Fig. 6.8). The workstation processes these control actions and sends the appropriate signals to the local controllers of the servomotors to move the heliostats to the positions indicated by the control program. The implementation of the automatic control system allows the simplification and optimization in the operation, as the control program developed automatically decides the correct control actions, then interrogates the operator about the proposed correction (optional) and sends the corresponding commands to the workstation to achieve the desired heliostat movements.

6.5.3 Offset Adjustment Mechanism

6.5.3.1 Database

The movement of the heliostats is discrete (resolution limited by the encoders). The database of the central controller contains, for each one of the 300 heliostats of the

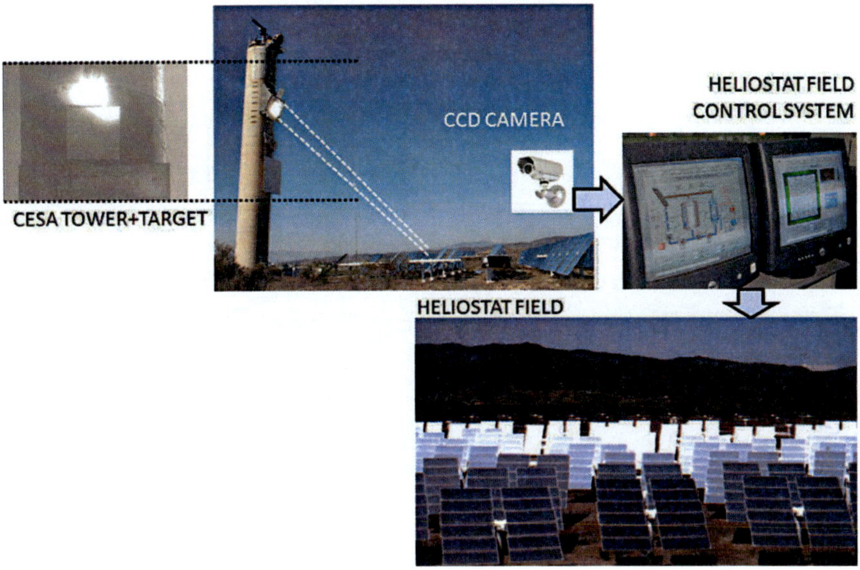

Fig. 6.8 Diagram of the offset compensation system [45]

field, the following information: row and column (position in the field), $(XYZ)_{hel}$ heliostat center of rotation related to the base of the tower, kind of heliostat (heliostats from different manufacturers are currently in use), nominal elevation and azimuth angles in encoder step units and related information not relevant for the purpose of offset compensation. The encoder resolution is limited to $\pi/4096$ rad. The nominal angles are used in the solar vector calculation in such a way that, in an ideal situation, the heliostats should aim at a target placed on the tower at the beginning of the operation. Notice that due to the discrete nature of the movement of the motors, it is impossible to exactly correct the deviations. The process followed to correct the offset is shown in Fig. 6.9.

6.5.3.2 Calculation of the Center of the Target

As the CCD camera is used for different purposes, the operator should manually adjust its position (by remotely acting on pan-tilt motors) in such a way that the captured image can show the target. The algorithm that calculates the center of the target is based on threshold detection techniques in such a way that in a first stage, the grey level of the points belonging to the target are determined. To facilitate this task, the algorithm makes use of the fact that the target should be centered (with some tolerances) in the image (as this is done by the operators acting on the cursors placed at the operation room). First, the image of the target without illumination from the heliostats is captured and the histogram (number of pixels versus grey-level intensity) of a portion of this image is calculated (using values of a rectangle

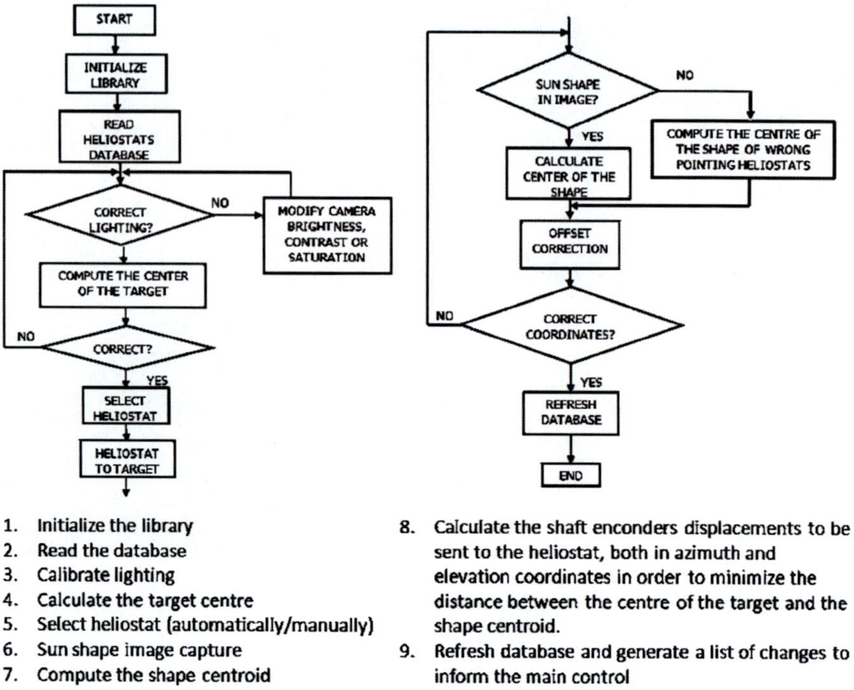

Fig. 6.9 Flow diagram of the offset correction problem, [45]

around the center of the image). Once the most representative grey-level intensity of the pixels in the selected rectangle has been obtained, two thresholds are defined to distinguish those pixels belonging to the target (intensity between both thresholds) from those of its environment. The selection of two thresholds is necessary due to the fact that the segments of the target do not produce the same intensity color (grey level), mainly due to mechanical and mounting distortions and also due to the fact that only a part of the target is being used for threshold selection purposes. These thresholds (`TH1` and `TH2`) are selected to be near the peak with the maximum value of grey intensity (`grey_max`) in the histogram: low `TH1=grey_max-SHIFT`, high `TH2=grey_max+SHIFT`. The amount of `SHIFT` (30) has been heuristically obtained by a trial and error procedure using different experiments in different operating conditions (different illumination levels) and thus is not a general recommendation, as it depends on the camera settings. The reason for using a fixed SHIFT instead of exploring the histogram is due to the fact that if the target is not adequately centered, it is possible to obtain pixels with many different intensity levels, producing a wide threshold that leads to errors in the discrimination of the target. Figure 6.10 shows one of the obtained histograms (many of them have been obtained) under hard operating conditions (early in the morning with the Sun far from the solar midday). In a second step the central row of the target is calculated, that is, the horizontal line crossing the center of the target. The same applies to the cen-

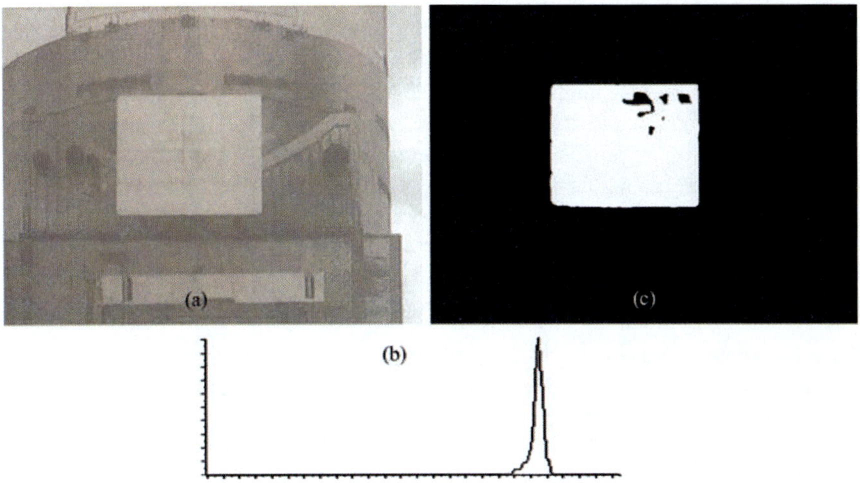

Fig. 6.10 (**a**) Example of an image of the target, (**b**) the corresponding histogram and (**c**) the result of applying the threshold detection algorithm [45]

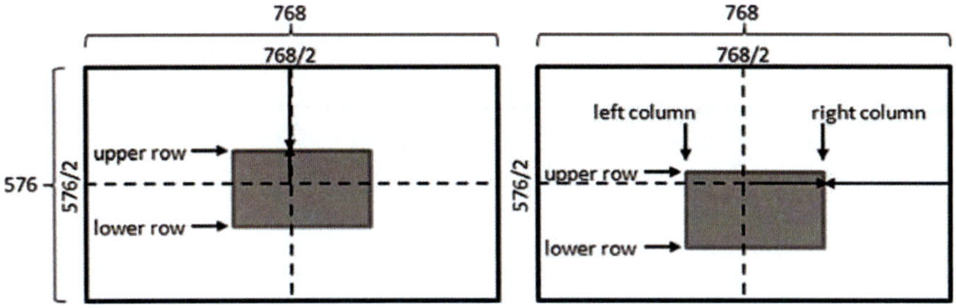

Fig. 6.11 Simple calculation of the center of the target [45]

tral column. These processes are calculated simultaneously to the application of the threshold detection algorithm (the resulting image is shown in Fig. 6.10(c)). The steps followed in the calculation of the central row and column were (Fig. 6.11):

1. Taking into account the size of the image (576×768 pixels), in a first step a vertical scan is performed taking into account only those pixels belonging to rows over the middle of the image ($< 576/2$) and those belonging to the central column ($= 768/2$). In this way, a detection of the pixels belonging to this vertical line with grey level belonging to the threshold interval is performed, starting with the central pixels (bottom–up). The discontinuity border between pixels belonging to the threshold interval and those out of the target is found in this simple way. As the dimensions of the target are known (both in XY coordinates and in pixels—the camera is not provided with any zoom mechanism and the distance between the camera and the target is fixed) and the upper horizontal line has been detected, the horizontal line containing the center of the target can be easily cal-

culated. Notice that this simple algorithm is valid for those cases in which the center of the complete image contains part of the target, which is always true because the camera is manually positioned using the pan-tilt mechanism.

2. The process followed to calculate the central column of the target is quite similar, but uses the known position of the upper row of the target in such a way that an horizontal line belonging to the image (for instance, 20 rows below the upper one) is selected and is scanned from the center of the image to the right, looking for consecutive pixels belonging to the threshold interval. The central column can thus be obtained knowing the size of the target both in meters and pixels.
3. The intersection of the central row and column provides the center of the target.

Once the center has been calculated, it is shown to the operator in order to test if the calculation is correct. If due to hard operating conditions (mainly producing changes in illumination) the center of the target is not adequately estimated, the operator has only to slightly modify the brightness and contrast of the image using a menu provided in the software tool (see next subsection).

6.5.3.3 Calculation of the Sunbeam Centroid

The next step consists of sending a heliostat from a standby point to the target (which is predefined as an aimpoint, in this case number 9). Figure 6.7 shows zones of different images obtained from the CCD camera in which: (a) the heliostat has nearly correct offset coordinates, (b) and (c) the heliostat has wrong offset coordinates. Notice that the intensity of the shape is much higher than that of the other elements in the image, in such a way that the calculation of the sunbeam centroid is simplified using, for instance, threshold detection techniques, as done in the case of the target. Depending on the selected heliostat, the time elapsed in reaching the target will vary between 18 and 60 s, depending of the kind of heliostat and its position in the field. So, an upper bound of the time the system has to wait till performing the acquisition of a new image is 60 s, this being the first approximation adopted. After several tests, another algorithm was implemented, in such a way that after the first 18 s, each 5 s the center of the centroid was calculated and compared with the previous obtained one (when it appears in the image). When the difference between two consecutive centers was small (less than a small constant selected to take into account the camera vibration due to the wind), the automatic system considered that the heliostat was in its final position. This second algorithm worked well when images were obtained around solar midday. Nevertheless, it did not give good results when direct projections of Sun rays impinged on the tower (as happens early in the morning in clear days, e.g. Fig. 6.10(a)), before the reflected shape of the Sun appeared in the image. Due to the automatic brightness/contrast adaptation mechanism of the CCD, once the reflected shape enters in the view field of the camera, the relative intensity of other Sun projections different from the main shape is largely diminished (e.g. those due to the incidence of Sun rays on the side of the tower early in the morning or late in the evening). Another problem is the case of those heliostats which projection is out of the view field of the camera and does not appear in the

Fig. 6.12 (**a**) Histogram used to calculate the sunbeam centroid, (**b**) result of the application of the segmentation algorithm, (**c**) shows the centroid position on the real image calculated using (**b**) [45]

captured image. These heliostats are labeled as *wrong pointing* ones and their offset can be corrected using a modification of the main algorithm commented later in the subsection. The calculation of the sunbeam centroid was also performed using a threshold detection technique (based on histogram information), as the intensity of the image was near 255. Different more complex image pre-processing algorithms were used, but the added complexity and processing time did not justify their use, because the obtained improvements were not relevant. In order to avoid that other Sun projections different from the main one should lead to wrong results, the pixels with intensities over the threshold are grouped according to their intensity level and the existence of neighbors with the same intensity (segmentation). The figure with largest area is selected as that corresponding to the sunbeam shape and its center is calculated. Regarding this criterion of largest area, in the application shown in this work a situation with most of the radiation falling on surfaces that are not essentially perpendicular to the line of sight of the camera does not appear due to the heliostat field–tower–target layout (this has been verified in the tests). In larger plants other criteria should be used, for instance those related with intensity weighting of the image, requiring more sophisticated hardware (cameras, frame grabbers, etc.) than those used in this subsection. Figure 6.12(a) shows the histogram of the image in Fig. 6.7(b), where the shape of the Sun projected by a heliostat can be observed. An intermediate buffer (or auxiliary file) is used to store the different elements of the image. The steps followed in the calculation of the sunbeam centroid were:

1. Determination of the threshold using the histogram.
2. Bottom–up/left–right scans are performed comparing each pixel with the threshold, in such a way that if the grey level of the pixel is below the threshold, it is saved in the auxiliary buffer as a black one (0 intensity). If the intensity of the pixel is greater than the threshold, a fixed intensity value is assigned to it (starting at 255 level) and to all the consequent pixels meeting the threshold condition, until a pixel below the threshold is found (and saved as a black one). The fixed intensity value is decremented and the algorithm continues looking for pixels fulfilling the threshold condition till finding the end of the row. This last processed row is stored into memory, in such a way that the procedure is repeated for the following row, not only comparing the intensities of the pixels with the threshold, but also with the intensities of the pixels belonging to the previous

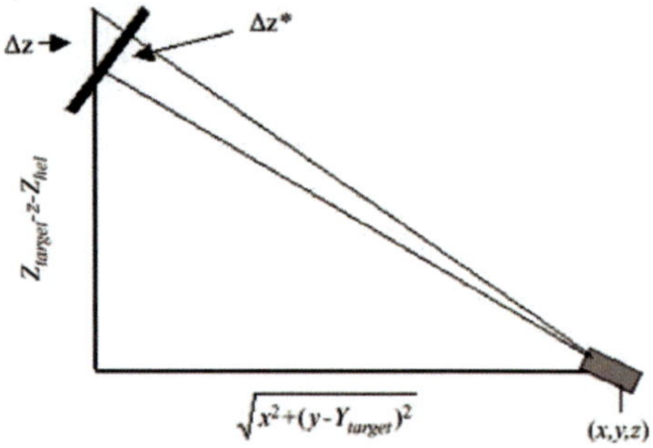

Fig. 6.13 Approximated calculation of the vertical displacement [45]

processed row, in order to group and label with the same intensity level all the adjacent neighbors (backward modification of the assigned intensity is required in the current processing line to assign the same grey level to pre-processed ones belonging to the same object in the figure). Figure 6.12(b) shows the result (negative image) of the application of the algorithm to the image shown in Figs. 6.8(b) and 6.12(c).

3. The completely processed rows are stored in an auxiliary buffer, also saving the number of pixels belonging to the used intensities (histogram), in such a way that when the process finishes, the grouped pixels with the predominant intensity should be selected as belonging to the main shape and the corresponding center will be calculated.
4. The intermediate buffer is scanned but only looking for pixels with the selected intensity (belonging to the shape after the segmentation process). The extreme pixels (up, low, left and right) are selected and an approximation to the centroid of the figure is obtained by using the center of the outer box including all the pixels of the shape (Fig. 6.12(b)), using a simple formula as the intersection of the central column $((\text{right} + \text{left})/2)$ and row of the figure $((\text{up} + \text{low})/2)$.
5. As is mentioned in the next subsection, once the centroid has been calculated, this is indicated in the image shown to the operator, in such a way that the decision of correcting the offsets is done by the operator, depending of the results given by the algorithm. In the mentioned case, in which the shape lies out of the view field of the camera, the system returns (0, 0) as the sunbeam centroid and the operator is asked about the inclusion of this heliostat in the group of *wrong pointing* ones, to be corrected with an alternative algorithm.

6.5.3.4 Offset Correction

The final correction consists of minimizing the distance between the center of the target and the sunbeam centroid. In order to do so, it is necessary to calculate the following data:

- mm/pixel relationship: as the camera does not use a zoom mechanism and the dimensions and position of the target are fixed, this is a known relationship. The inclination (Fig. 6.13) of the target (0.61 rad) and that of the camera (0.35 rad) have been taken into account in this calculation, providing a value of 3700 mm/200 pixels.

- mm/encoder steps relationship: this is calculated using the Cartesian coordinates of the heliostats related to the base of the tower.

Knowing these values, a relationship can be established between the number of horizontal/vertical pixels in the image corresponding with an azimuth/elevation heliostat movement (encoder steps). The relationship mm/encoder steps varies, not only between heliostats, but depending on the actual elevation and azimuth coordinates of each heliostat. The calculation of this relationship uses the Cartesian coordinates of each of the heliostats of the field (x, y, z stored in the database) and those of the target in relation to the base of the tower ($X_{target} = 0$, $Y_{target} = 8180$, $Z_{target} = 64200$ [mm]). The z coordinate of the heliostats stored in the database has to be augmented with their corresponding height ($Z_{hel} = 3650$ mm). In what follows, the calculations that the algorithm implements are summarized. It is important to point out that due to the actual architecture of the control system when the algorithm is implemented (in which commands can be sent to the workstation via RS-232 communication, but no information has to be received from it, as the main control program installed twenty years ago should be modified), an approximate calculation had to be carried out. The exact calculation could be performed by knowing the azimuth and elevation angles of the heliostats (related to the stow position) in real time when obtaining the solar vector each 4 s (the α angle in Fig. 6.13 should correspond in this case to the angle measured from the stow position minus $\pi/2$). As no information can be obtained from the workstation, the calculations were done based on initial angles stored in a database. The recursivity included in the algorithm helps to correct the errors induced by the approximation. The steps followed to calculate this relationship are summarized in Table 6.1. In the left column the approximated calculations used in the algorithm are included. The right column provides the equations that should be used in the case of knowing the azimuth and elevation coordinates of the heliostats in real time, obtained from the calculation of the solar vector. Once these calculations have been performed, the mm/encoder steps approximated relationships are obtained both in vertical and horizontal axes. As a simplification to find a trade-off between accuracy and performance, it has been supposed that when an elevation movement is performed, the revolution surface corresponds to a plane, and so its intersection with the plane of the target is a line. In the case of an azimuth movement (maintaining the elevation angle to a fixed value), the revolution surface can be approximated by a cone and its intersection with the plane of the target does not correspond with a horizontal line (in the plane of the target). Nevertheless, the error of approximating this movement by a horizontal one is negligible compared to other sources of errors like the discretization of the movement of the heliostats. If a high accuracy is required, the heliostat position, target center, beam on target and Sun position are required to compute tracking errors with vector math [210]. Once the relationship mm/pixel and mm/encoder steps are known (and thus pixel/encoder steps relationship), the algorithm to calculate the number of encoder steps is easy to implement and is based on successive approximations, by subsequently calculating the distance in pixels (both in vertical and horizontal axes)

Table 6.1 Offset correction calculations [45]

Vertical axis	
Approximated calculation	Exact calculation
1. Calculation of the angle between the heliostat and the center of the target. $\tan\alpha = \frac{Z_{target} - z - Z_{hel}}{\sqrt{x^2 + (y - Y_{target})^2}} \to \alpha$	1. The z coordinate of the point of the target to which a heliostat points for an elevation angle α is calculated. $\tan\alpha = \frac{Z_{target} - z - Z_{hel}}{\sqrt{x^2 + (y - Y_{target})^2}} \to \alpha$
2. Add to the calculated angle the increment corresponding to a encoder step. Obtain the height corresponding to the new angle $\alpha' = \alpha + \frac{\pi}{4096}$. $\left(\sqrt{x^2 + (y - Y_{target})^2}\right)\tan\alpha = Z_{target} - z' - Z_{hel} \to z'$	2. The point $z + \Delta z$ in the target corresponding to a heliostat elevation angle of $\alpha + \pi/4096$ is calculated (corresponding to a encoder step). $\tan\left(\alpha + \frac{\pi}{4096}\right) = \frac{z + \Delta z}{\sqrt{x^2 + y - Y_{target})^2}} \to z + \Delta z$
3. The encoder step indicates a vertical change of the shape corresponding to $\Delta z = z' - z$ in the z-axis.	3. By subtracting both quantities the mm/encoder steps relationship can be obtained in the vertical axis (Δz).
As the target is inclined 0.61 rad, the obtained increment has to be multiplied by $\cos(0.61)$ to obtain the real increment related to the plane of the target (Δz^*).	

Horizontal axis	
Approximated calculation	Exact calculation
1. Calculation of the angle between the heliostat and the center of the target. $\tan\beta = \frac{x}{\sqrt{(y - Y_{target})^2 + (Z_{target} - z - Z_{hel})^2}} \to \beta$	1. The x coordinate of the point of the target to which a heliostat points for an azimuth angle α is calculated. $\tan\beta = \frac{x}{\sqrt{(y - Y_{target})^2 + (Z_{target} - z - Z_{hel})^2}} \to x$
2. Add to the calculated angle the increment corresponding to a encoder step. Obtain the horizontal displacement corresponding to the new angle $\beta' = \beta + \frac{\pi}{4096}$. $\left(\sqrt{(y - Y_{target})^2 + (Z_{target} - z - Z_{hel})^2}\right) \times \tan\beta = x' \to x'$	2. The point $x + \Delta x$ in the target corresponding to a heliostat azimuth angle of $\beta + \pi/4096$ is calculated (corresponding to a encoder step). $\tan\left(\beta + \frac{\pi}{4096}\right) = \frac{x + \Delta x}{\sqrt{(y - Y_{target})^2 + (Z_{target} - z - Z_{hel})^2}} \to x$
3. The encoder step indicates a horizontal change of the shape corresponding to $\Delta x = x' - x$ in the x-axis.	3. By subtracting both quantities the mm/encoder steps relationship can be obtained in the horizontal axis (Δx).

between the center of the target and the sunbeam centroid and calculating the required displacement in encoder steps to minimize this distance (as the effect in mm of a encoder step is known from the previous relationships). Notice that consecutive calculations have to be performed each time a encoder step is performed, as the real displacement in mm changes with the angle of the heliostat, as has been previously mentioned. Once the number of encoder steps both in elevation and azimuth have been calculated, the system tests the effects of these actions, as accumulated errors

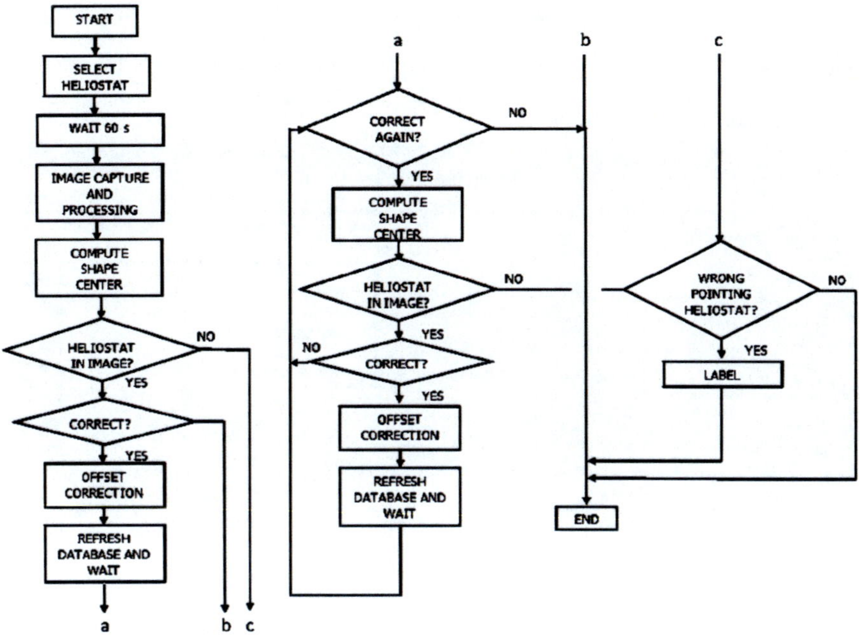

Fig. 6.14 Flow diagram of the offset correction mechanism [45]

in the calculations (approximations done in the mm/encoder steps relationships) and the discrete nature of the control signal can lead to the need of further adjustments, selected by the operator. The performed tests show that maximally three steps are needed in extreme cases. Figure 6.14 shows a flow diagram of the complete process.

Once the offset correction has been calculated, the database is modified with the new values of elevation and azimuth (requires the operator's authorization).

The correction of the offset of heliostats for which the reflected Sun shape does not appear in the view field of the image requires the modification of the previous algorithm. Three possible cases are taken into account:

1. The ellipsoid does not appear in any of the captured images each 5 s (during 60 s).
2. Only one image of the shape has been obtained in the intermediate captures.
3. Some images of the shape have been obtained in the intermediate captures.

In the first case, the search is started at an arbitrary zone of the eight neighbor zones defined with the same size of the frame captured by the camera. In the second and third cases, the zone in which the search may start can be automatically found. Heliostat displacements are thus produced and once the ellipsoid enters the image, the normal procedure for offset correction is applied.

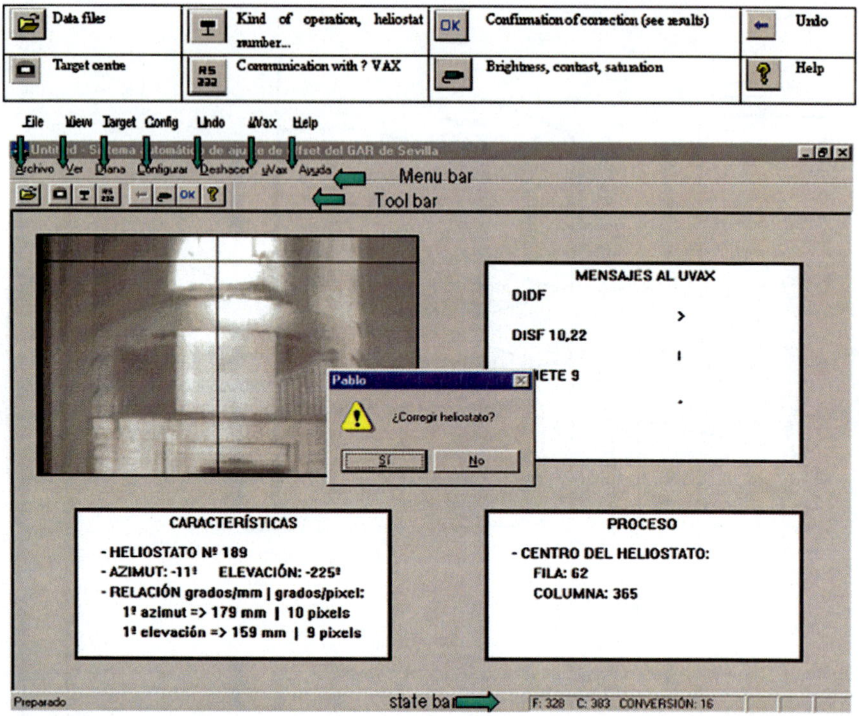

Fig. 6.15 Software tool [45]

6.5.4 Experimental Results

6.5.4.1 Software Tool and System Implementation

The system has been implemented using a windows-based software tool that allows the operators of the plant to supervise the offset correction task through a user-friendly graphic interface (in Spanish). Figure 6.15 shows the main window of the application, which is divided into four sections: (1) devoted to show real time images of the target and the reflected ellipsoid, including different processing stages of the algorithm, (2) zone of the window showing the commands sent to the workstation, (3) a window showing the main characteristics of the test: heliostat number, azimuth and elevation angles, angle/mm and angle/pixel relationships, etc., (4) results of the processing: center of the target, sunbeam centroid, etc.

The main operational problems when implementing the system were:

1. The positioning system of the camera, due to strong fluctuations produced by the wind.
2. Optical errors produced by the camera lens (superposed images due to mechanical problems).

3. Illumination variations due to the Sun displacement.
4. The implementation of the correcting system for wrong pointing heliostats.

The first two problems were mainly overcome by the plant personnel by both adjusting the mechanical fit of the camera and calibrating the lens. Nevertheless, a modification to the software was included to account for these possible error sources. In the first case, the possibility of performing a redundant calculation of the center of the target was included (using different snapshots of that) and in the second case it was noticed that the threshold algorithm discriminated between the real and the superposed image, due to the different intensity levels of both images. The third problem is due to the Sun movement and the direct exposure of the tower to Sun rays, changing from the east side near sunrise to the west side near sunset and lying out of the field of the camera during a significant portion of the operation. Due to this fact, the first version of the algorithm for calculating the sunbeam centroid sometimes provided wrong results when the Sun reflection from the tower was visible. This first version did not include the segmentation part, which was included after noticing this problem thus including two degrees of freedom for determining the real projected Sun image (intensity and size). The fourth source of problems is related to the third one previously commented on, due to the fact that the Sun reflection can confuse the algorithm when searching for the reflected Sun shape each 5 s. So, the modified algorithm was only used in the case of wrong positioning heliostats, waiting 60 s in the standard case to start the processing of the image.

6.5.4.2 Tests and Results

As a summary of the obtained results with the algorithms explained in this work, Fig. 6.16(a) shows the differences between the real target center and that calculated by the algorithm in several representative tests. The maximum obtained difference has been 7.4 cm (notice that the target dimensions are 370 cm $\times$ 366 cm). In the case of the sunbeam ellipsoid, Fig. 6.16(b) contains representative results of 30 tests performed with two heliostats (8, 11) (8, 12) located in the center of the field (east and west) from 11 to 16 hours local time. The largest errors correspond to situations in which the sunbeam centroid is far from the center of the target. As the projected shape approximates the center of the target (as a consequence of the application of the iterative algorithm in less that four steps), the results in the calculations are more precise. It is interesting to mention that in the set of performed test, a percentage of about 90% successful results (similar to those manually obtained by the operators) were obtained when correcting heliostats if the corresponding ellipsoid appears in the image. In the case of *wrong pointing* heliostats this percentage of success is about 50% (only two tests were performed).

6.6 Heliostat Beam Characterization

A heliostat is a quite complex optical system. It must provide high-quality concentrated solar beam into a receiver located on top of a tower, perhaps several hundred

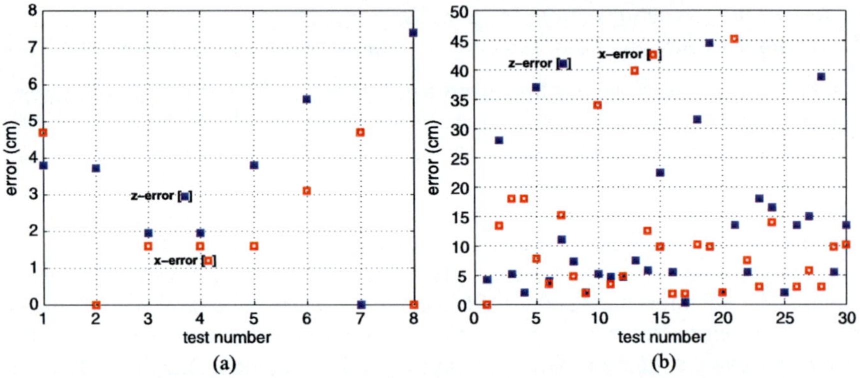

Fig. 6.16 Representative tests: (**a**) absolute value of errors in the calculation of the center of the target, (**b**) the sunbeam centroid. The tests have been performed at different local times during several days [45]

meters far and exposed to outdoor conditions, i.e. wind, dust, ... Prototype heliostats must be tested in order to evaluate performance and durability of the devices and to seek answers to a number of related questions: would the heliostat perform according to its specifications? How much power would it be capable of delivering to a receiver? What would be the overall size of the heliostat beam? Would the heliostat perform within specifications under windy conditions? Typical test campaigns study [365]: heliostat beam quality, wind effects on tracking accuracy, heliostat drives and controls, mirror module performance and durability, overall operational and maintenance characteristics and evaluation of changes in beam quality throughout the day (due to gravity effects as the heliostat elevation angle varied, changes in insolation and sunshape and changes in heliostat-to-receiver orientation).

6.6.1 The Beam Characterization System

The Beam Characterization System (BCS), see Fig. 6.17, is a primary tool used at test facilities for evaluating the performance characteristics of solar concentrators and their mirror modules or facets. It can be used for several kinds of prototype heliostat test, including beam quality, dynamic wind effects, tracking accuracy and repeatability tests. The BCS is employed to image the beam of concentrated solar energy as it is tracked by the heliostat on a flux target. The result of the BCS process is a flux map of that beam. The information obtained includes: (i) the location of both the beam centroid and peak, (ii) the flux densities over the entire beam, (iii) the peak flux, (iv) the total beam power and (v) the nominal diameter of a circle that contains all flux equal to or greater than 10% (or any specified percentage) of the peak flux.

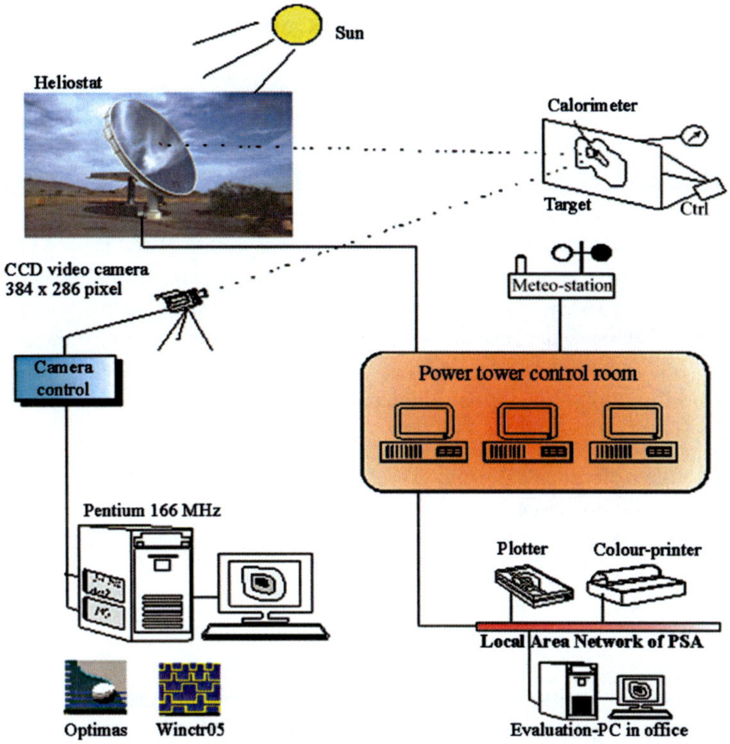

Fig. 6.17 The BCS PROHERMES at PSA [259]

The flux target used by the BCS is a white, non-specular reflective surface with unique optical properties. The relative position of the target approximates the position of a solar receiver in a central receiver plant.

Images of the heliostat beam on the target are captured using a CCD-type camera positioned at a convenient location with a normal or near-normal view of the target. The target surface is painted with a high-temperature titanium-oxide paint. Because of the diffuse (Lambertian) reflectance characteristics of the target, the intensity of the reflected light reaching the camera from each point on the surface of the target is directly (linearly) proportional to the intensity of the light reaching that same point on the target from the concentrator. This characteristic of the BCS target is essential to the measurement technique and ensures the desired result: the image of the flux beam captured by the camera is a scaled version of the actual flux incident upon the target. The target and the collector beam are imaged on the camera's sensor, digitized by a commercial frame-grabber and processed by image analysis software that is resident on the BCS' high-end computer. A flux gauge mounted on the surface of the BCS target provides a single absolute measurement of flux at one point in the BCS image. Together with the value of the picture element (pixel) intensity at that location in the image, the flux gauge reading is used to establish a conversion factor, which can be applied to the BCS image. Using this factor, the intensity value of every pixel in the image is converted into an absolute flux density value. In this manner the BCS image becomes a flux map. The analysis functions of the commercial image analysis system (IAS) produce most of the required BCS information, including the locations of the beam centroid and peak flux and the diameter of the beam. When multiple images are analyzed, such as in tracking error or dynamic wind effect tests, the IAS can compute and display the movement of the beam centroid from image to image. As stated before, the magnitude of the peak power and the total beam power is obtained by applying the conversion factor (pixel-level-to-flux-density) to the IAS's results (which are in relativistic, pixel-intensity units). The BCS also has analog data acquisition capability, which is employed to measure the flux gauge mentioned above and relevant environmental condition including wind speed, wind direction and normal incident insolation. Some other necessary tools for adequate evaluation are:

- A flux gauge (reference calorimeter)
- Heliostat simulation code
- Meteo-station (pyrheliometer, pyranometer, thermometer, wind speed and direction, humidity)
- Reflectometer
- Electrical counter.

6.6.2 A Prototype Heliostat Test Campaign

As mentioned before, the goal of a heliostat test program is to compare the classical central receiver heliostat performance aspects of a heliostat: optics, tracking, flux distribution and power consumption. In the following, all of them are briefly described.

6.6.2.1 Optics

Any irradiance distribution $I(x, y)$ projected by the heliostat onto a plane normal to the direction of propagation of the principal reflected ray can be mathematically characterized by a merit number which shows how much $I(x, y)$ distribution has been dispersed around any given reference point, usually the irradiance distribution centroid. The nature of $I(x, y)$, ideally continuous, has to be treated in discrete manner because of the digital image acquisition devices, which deliver $I(x, y)$ as a matrix of numbers representing the spatial distribution of intensity. A proper merit number that fulfills that condition is the so-called total dispersion error of intensity distribution (usually asymmetric), defined as

$$\rho_x = \sqrt{\frac{\sum I_i (x_g - x_i)^2}{\sum I_i}}, \qquad \rho_y = \sqrt{\frac{\sum I_i (y_g - y_i)^2}{\sum I_i}} \tag{6.2}$$

where $x_g = \sum I_i x_i / \sum I_i$, $y_g = \sum I_i y_i / \sum I_i$ are the mentioned coordinates of the centroid distribution. The units of ρ, initially longitudes from centroid, are converted into angular units (subtended angle by ρ from the center of the heliostat in mrad) in order to make the characterization of $I(x, y)$ independent of the plane where it was projected by the heliostat. ρ is usually called the total beam dispersion error and will be represented by the Greek letter σ_{total}.

Taking into account that the so-called image constituents, i.e. sunshape, astigmatic aberration, waviness, gravity loads influence, etc., can be assumed to be statistically independent [50] and the similar properties of σ to the standard deviation, it can be demonstrated that the total beam dispersion error of the distribution is expressible as

$$\sigma_{total}^2 = \sigma_{sun}^2 + \sigma_{aberr}^2 + \sigma_{BQ}^2 \tag{6.3}$$

where σ_{BQ} is the so-called reflected beam dispersion error. The contribution of the optic errors to the dispersion of the reflected beam can also be expressed in its own individual effects:

$$\sigma_{BQ}^2 = \sigma_{waviness}^2 + \sigma_{canting}^2 + \sigma_{grav\_loads}^2 + \cdots \tag{6.4}$$

With the help of a digital acquisition and processing system, the numbers σ_{total} and σ_{sun} can be calculated from the experimental data (images). Taking the real sunshape as an input file, a heliostat simulation program should be able to estimate σ_{aberr} with quite good approximation. So, the optical quality σ_{BQ} of the heliostat usually called the Beam Quality is

$$\sigma_{BQ} = \sqrt{\sigma_{total}^2 - \sigma_{sun}^2 - \sigma_{aberr}^2} \tag{6.5}$$

Figure 6.18 shows the isoflux lines of the reflected sunshape onto the $12 \times 12\ m^2$ white Lambertian target at PSA by a prototype heliostat. The mathematical analysis of such picture allows to unfold the σ_{BQ}.

6.6.2.2 Tracking

Heliostat tracking analysis is usually done by studying the daily and seasonal evolution of the image centroid on a reference target. As with the optical quality, a merit number can be defined to quantify the expected centroid deviations with time. An appropriate one that fulfills this condition is the so-called RMS-error of the distribution of centroid coordinates (usually asymmetric) related to the expected aiming point (x_0, y_0) defined as

$$e_{RMS_x} = \sqrt{\frac{\sum (x_0 - x_i)^2}{\sum n_i}}, \qquad e_{RMS_y} = \sqrt{\frac{\sum (y_0 - y_i)^2}{\sum n_i}} \tag{6.6}$$

where (x_i, y_i) are the successive centroid coordinates of the image and $N = \sum n_i$ the total number of computed points involved in the test. Figure 6.19 shows the angular deviation of the real impact point from the expected one (target center) for a prototype around solar noon, whereas Fig. 6.20 shows the accumulated RMS-tracking error taken in successive time intervals around noon for a prototype at PSA.

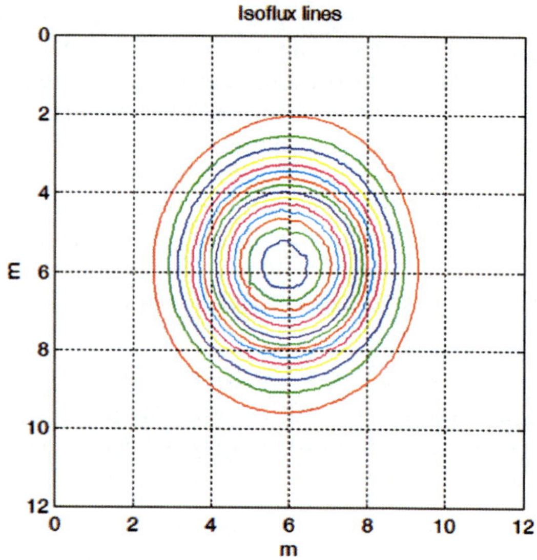

Fig. 6.18 Front view of a focal spot at noon Almería (courtesy of R. Monterreal, [259])

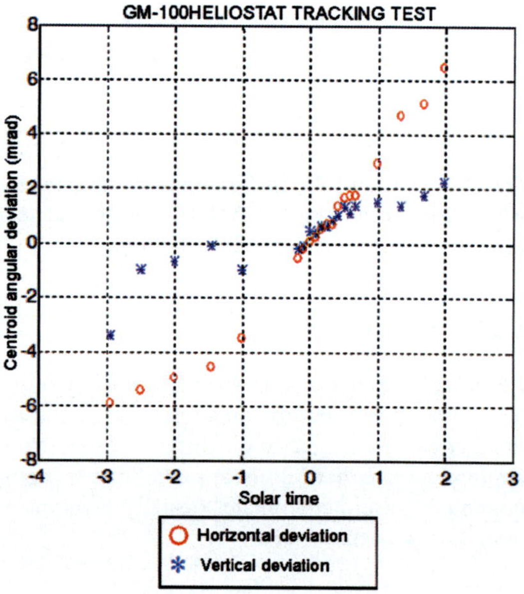

Fig. 6.19 Angular deviation from an expected point (courtesy of R. Monterreal, PSA)

6.6.2.3 Flux Measurement

Under this topic those tests involving the analysis of the concentrated heliostat beam are usually presented in physical units. For this purpose, a reference calorimeter embedded in the white Lambertian target is used. An analytical function for calibrating the relative intensity grey-level map captured by the video system has to

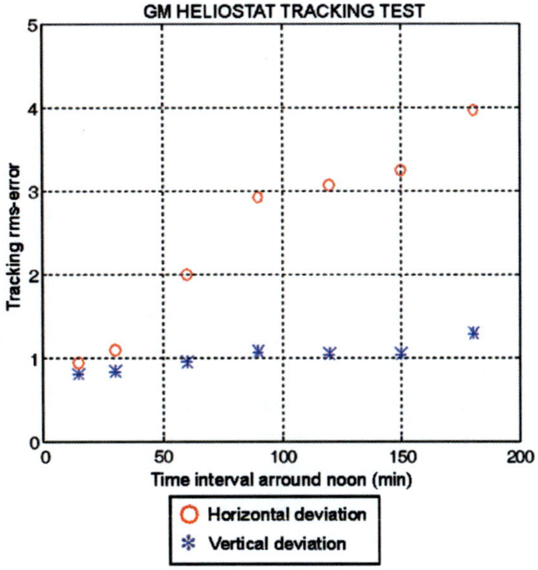

Fig. 6.20 Daily accumulated RMS-tracking error (courtesy of R. Monterreal, PSA)

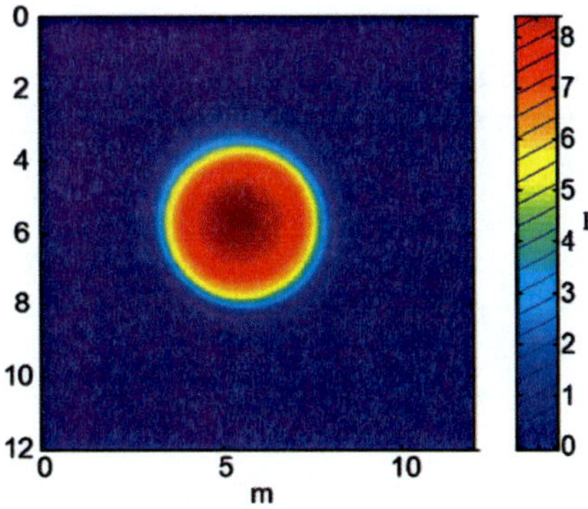

Fig. 6.21 Flux density distribution map (courtesy of R. Monterreal, PSA)

be studied. The flux density distribution map of a prototype heliostat is shown in Fig. 6.21.

6.7 Aiming Strategies

6.7.1 Introduction

One of the main problems in operating volumetric receivers is to obtain an appropriate flux distribution in order to avoid deterioration due to excessive thermal gradients. To account for this problem, this section presents the development and implementation of a heuristic knowledge-based heliostat control strategy aimed at optimizing the temperature distribution within a volumetric receiver in the PSA CESA-1 plant. The development of the PHOEBUS Technology Program Solar Air Receiver-TSA, produced a technology for optimization of the receiver and steam generator control, but this technology did not allow for achieving and maintaining an appropriate flux distribution on the TSA absorber. The development of an automatic control system to control the flux distribution within the receiver became an important issue. At the Solar Two plant in Barstow (California), two software systems were developed for the receiver protection mechanism: SAPS (Static Aim Processing System) and DAPS (Dynamic Aim Processing System). Each heliostat has been assigned to a predetermined aim level (top, center or bottom) to provide a reasonable flux density everywhere on the receiver, dropping to a low level at each end. For all heliostats assigned by the operator, SAPS calculates the amount of the shift from the equator appropriate for each heliostat depending on its current temperature dependent focal length and orientation dependent aberrations in addition to beam errors and Sun size. DAPS interrogates the receiver inlet and outlet temperatures and also computes the flux distribution on the receiver based on the current assignment of heliostats, using the appropriate reflectivity, Sun position, beam errors from all sources (entered as a sigma), temperature dependent focal lengths, insolation, and visual range. Using the known salt enthalpy, DAPS estimates the temperature at each of 21 nodes on each of the 24 panels and uses this temperature and the known (or calculated) salt flow rate to determine the allowable flux at each node of the receiver using an algorithm based on the receiver designers specifications. DAPS then compares these two values, locates nodes where the allowable is exceeded and searches for the heliostat producing the maximum flux density at that point and removes it from the active list. This process of search and removal is continued as long as excess flux exists on the receiver. In March of 1997 various issues which prevented DAPS from being used in the preliminary operation of Solar Two were resolved and DAPS has been in operation since that time [379, 390]. The control approximation adopted in this section is different from that of the Solar Two plant (as is the technology, based on an open air volumetric receiver) and is less restrictive both from the computational requirements and the instrumentation viewpoint. The task of maintaining an adequate flux distribution has been performed at the PSA by expert operators in manual mode since the system was operated. This involved important operating costs, as three skilled operators devoted to this task were necessary. As has been pointed out by [387], the adjusting of individual heliostat group aiming-point coordinates and the number of heliostats in each group to keep absorber temperatures within the desired range is the most time-consuming operation activity. On the other hand, the control of the flux distribution in the receiver constitutes a complex task, leading to frequent errors or deviations in the operation, even when an expert is operating the plant. Due to these facts, the development and implementation of an automatic closed-loop control system to perform a continuous control and supervision of the solar receiver became an important issue, allowing the operator to be kept out of this task, only supervising the control results. The objectives of the automatic control system are:

- The manufacturer of the receiver (L&C STEINMÜLLER) proposes as the control objective in steady state that the maximum and minimum temperatures in the receiver should differ less than or equal to 100°C (for preserving the integrity of the solar receiver), which in fact is an ambitious objective from the point of view of skilled operators.
- To compensate for the heliostat field deficiencies.
- To optimize the efficiency of the solar receiver.
- To minimize man-power dedication to this task and to avoid the risk of possible errors in operation.

From the control viewpoint, the system to be controlled presents high complexity, mainly due to:

- The control system involves a large number of inputs and outputs. The system inputs are the positions of each one of the 180 heliostats of the field used in this application, which have to point to a specific position on the volumetric receiver. The system outputs are the values given by forty thermocouples adhered to the surface of the volumetric receiver. Four of these thermocouples are placed on the receiver surface receiving direct solar irradiance and the other thirty-six are placed on the back side of the absorber measuring the temperature of the air flowing through the receiver.
- The difficulty of obtaining dynamic models for a system with such a high number of inputs and outputs.

Due to these problems, the design effort was concentrated on the development and implementation of a heuristic knowledge-based control strategy in order to reproduce the performance of a skilled operator during the tests, instead of using classical automatic control systems as done in other kind of solar plants [85, 150].

6.7.2 Functional Diagram of the System

Figure 6.22 shows the functional diagram of the system. The solar irradiance is reflected by the heliostat field and concentrated on the volumetric receiver surface. In consequence, the wire packages or ceramic cups which make up the receiver are heated, transferring the heat to the air that circulates through them. The nominal outlet temperature of the air leaving the volumetric receiver is 700°C and the working power depends on the number of heliostats aiming at the receiver and on the value of solar irradiance at each instant. Other working temperatures can be specified during operation in order to evaluate the performance of new elements installed in the receiver. The maintaining of an operating regime which ensures constant power energy production can be achieved by classical automatic feedback control systems of both the outlet air temperature (by using blower G1) and the mass flow which circulates through the steam generator (by using blower G2). The control of the power stage will be treated in the next section (for more details on basic control see [387]).

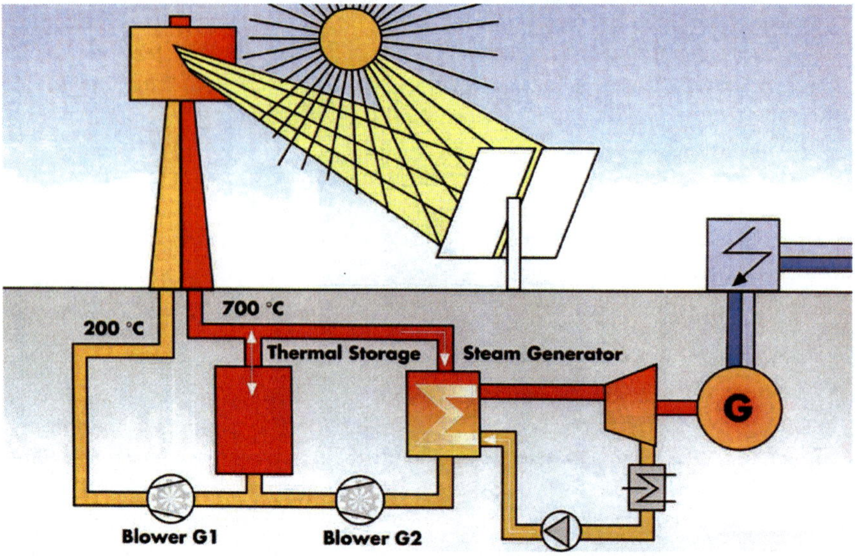

Fig. 6.22 Functional diagram of the system [150]

The main elements of the TSA receiver are shown in Fig. 6.5. As has been pointed out in the description of the receiver, the temperatures measured by the installed thermocouples can be used as feedback signals by the control strategy as they are distributed in two PCs of the control room of the CESA-1 facility in which a SCADA runs. A weighted mean (labeled `CT060`) of the value of all the thermocouples installed on the back face of the receiver is used by blower G1 (Fig. 6.22) for control purposes and is also a main reference for the control system designed and explained in this section. There is another important signal (labeled `CT015`) which constitutes the mean value of four thermocouples symmetrically placed (one at the center and the others spaced at 120° from one to the other) on the pipe through which air leaves the receiver. If the value of this signal surpasses the nominal set point (typically 800°C) by more that 10°C (the alarm threshold), a field shutdown is produced.

6.7.3 Heuristic Knowledge-Based Control System

6.7.3.1 Problem Statement

The control problem mainly consists of optimizing the efficiency and security in operation of the volumetric receiver. This problem arises due to the fact that it is not possible to concentrate the power from all the operating heliostats of the field onto one single point of the receiver, as too high temperatures could be reached at this point leading to the receiver destruction. As pointed out in [357], in concentrating solar receivers with heliostat reflectors aiming all heliostats to the same point leads to peaked non-uniform distribution. One way to overcome a non-uniform irradiance profile is the multi-aiming strategy, that is, deliberately aiming the individual heliostats to different aiming points, in such a way that the peaked irradiance profile due to central aiming is spread to a more uniform irradiance profile over a much larger aperture area. This strategy has been implemented at PSA receiver [170] using a parallel fluid flow under non-uniform irradiance with a higher mass flux in the center of the absorber in order to adapt to the higher irradiance. The design of the volumetric receiver was carried out based on an ideal flux profile, which was used by the HELIOS software package [50, 65] to obtain the number and coordinates of aimpoints and number of heliostats aiming at each one of them. The conclusion was obtained that only five aimpoints were necessary to obtain the ideal flux profile (Fig. 6.5). Four of the aimpoints are placed on the periphery of the volumetric receiver (with 22 heliostats aiming at them) and the fifth at the center (with the rest, 92 heliostats, aiming at it). The number of orifices made in the absorber elements at the center was greater than those of the periphery in order to obtain a higher mass flux according to the flux conditions. In this way, the orifices made in the absorber segments adapt the mass flow through the absorber segment to the irradiance onto the segment. At design average outlet temperature each segment should have approximately the same outlet temperature. The manufacturer proposed maximum and minimum nominal steady-state temperatures in different points of the receiver in order to guarantee the integrity of the solar receiver avoiding high temperature gradients. As mentioned before, a heliostat aiming-point strategy providing a desired energy flux correlated with the air mass flow through the receiver could be selected to resolve the control problem, but subjected to many error sources [45, 363] that produce variations in the flux distribution and non-homogeneity of the air mass flow through the receiver. As indicated by [357], an interesting point of practical value is what happens to an optimally controlled receiver under off-design conditions. It is possible that the optimal flow distribution changes; therefore, an optimum based on daily or annual performance may be different from the optimum for design conditions. In [363] an analytical method of correcting for many of the structural, mechanical and installation errors that affect an open loop heliostat tracking control system is presented. The method involves the development of an error model for the particular heliostat/concentrator design, the obtaining of beam centroid information and the use of the error model in the open loop command calculation. The closed-loop configuration of the control system presented in this subsection allows to reduce structural, mechanical and installation requirements, to account for other error sources and to obtain adequate tracking accuracy and no need for periodic updating at a low computational cost. The theoretical aimpoints configuration indicated in Fig. 6.5(c) has been taken into account to develop the control system. The aimpoints are points numbered 1, 2, 3, 4, and 6. As has been mentioned, an off-line selection of the heliostats associated to these aimpoints was performed. Deviations in the flux distribution and real air mass flow profile due to error sources mentioned above lead to a non-uniformity of the temperatures measured on the surface of the volumetric receiver. It has therefore been crucial to include an automatic control system to correct these deviations and to maintain the uniformity of the temperature profile as far as possible within the receiver during operation according to the manufacturer indications, that is, the control system must continuously adapt the incident energy flux to the real air mass flow which circulates through the receiver during operation to avoid large temperature gradients in the receiver.

6.7.3.2 Control Architecture

From the control viewpoint, the outputs of the plant are the temperatures measured by the forty thermocouples placed in the volumetric receiver and the input or control variables are the positions of the different heliostats of the field used for this application. Due to the configuration of the system for different uses when the application was developed [150] the measurements of the temperatures were introduced in two PCs running a SCADA, which were then sent via the serial port to another PC in which the heuristic knowledge-based control program used these values to obtain the control actions that were finally sent to the workstation via the serial port (see Fig. 6.23). The workstation processed these control actions and sent the adequate signals to the servomotors to move the heliostats to the positions indicated by the control program. During manual operation (without using the automatic control system), one operator must be devoted to controlling the heliostat movement by introducing commands using the workstation keyboard as a consequence of the observation of the temperature profile in the volumetric receiver on the screens of the two PCs. The implementation of the automatic control system allows the simplification and optimization in the operation, as the control program developed automatically decides the correct control actions, then sends the corresponding commands to the workstation to achieve the desired heliostat movements. The usual commands accepted by the workstation are: Aimpoint change (change the coordinates of an aimpoint), deselection or selection of heliostats individually or by groups, introduction of heliostats onto an aimpoint, change the aimpoint onto which the selected heliostats are aimed at and defocusing of selected heliostats. The few commands mentioned are the only ones necessary to execute all the possible control actions, which mainly consist of: moving an aimpoint, changing one or several heliostats from one aimpoint to another and sending heliostats to standby points due to emergency situations. At present, all these actions are performed in a unique workstation (see Fig. 6.29).

6.7.3.3 Control Algorithm

The thermocouples are grouped according to the aimpoint (focus) which they are mainly influenced by. The analysis of the state of the volumetric receiver is performed by groups of thermocouples, taking into account the deviations of temperature with respect to the nominal working point for each one of the aimpoints. Two main control actions are used: (a) Heliostat adjustment: change one heliostat from

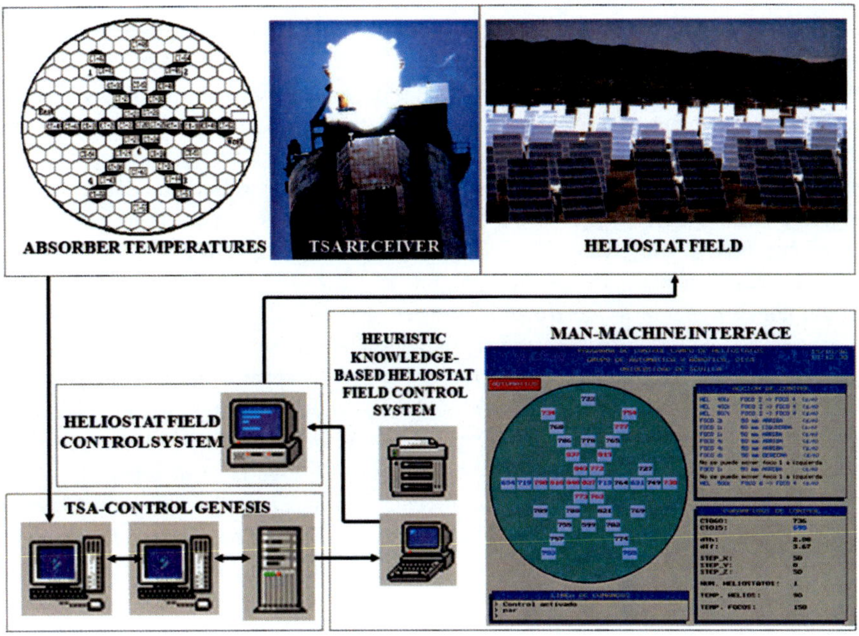

Fig. 6.23 Control architecture [150]

one aimpoint to another. (b) Aimpoint adjustment: produces an aimpoint displacement. The thermocouples located in the volumetric receiver measure the air temperature on the back of several of the hexagonal wire packages which form the receiver. Even in those cases in which the incident energy flux is equal to the theoretical ideal one, due to the fact that the air mass flow through the receiver is different from the desired one (due to manufacturing and assembly imperfections in the absorber elements and wind effects), important temperature differences appear on the receiver surface. As indicated by [357], the multi-aiming strategy may be required if the receiver's capability of absorbing and removing radiative flux is limited, e.g., due to a limitation of the mass flux occurring at high solar flux or due to instabilities in the presence of a non-uniform solar flux distribution. In [217], a study of flow instabilities is performed using a simple mathematical model. They show that the volumetric absorbers have inherent limitations and are prone to failure under certain conditions. For reasonable efficiencies control of mass flow or outlet temperature of the absorber may be required. In contrast, in [181], a modified numerical model is used and contrasted with real data. They show that up to around 800°C, the wire mesh absorber structure combines stable working conditions with high efficiency. To achieve the situation of maximum efficiency, the control algorithm should transform the energy flux profile provided by the heliostat field in order to perform an adequate correlation with the air mass flow through the receiver in order to homogenize the temperatures on the receiver. The objective of maintaining all the temperatures of the receiver elements near the values given by the manufacturer according to the ideal flux profile is a utopian situation for several reasons which are commented in the next paragraph and the objectives of the control system have to be reduced (other control policies should be adopted, as shown in [357]). As has been mentioned above, in this kind of solar receiver, an absorber element may present a temperature quite different (more than 50°C) from its neighboring elements. Due to the fact that the image projected by the heliostats over the receiver covers more than one absorber element, the heating of a *cold* element can lead the neighboring elements surpass their maximum allowable values. Moreover, the resistance to the circulation of air of an absorber element increases as its temperature augments. This is in part due to the variation of the fluid's dynamic viscosity with temperature. When the heat source increases, the temperature as well as the viscosity increase. This reduces the mass flow rate for the given pressure drop and reduces the heat removal capacity, providing a positive feedback for further temperature increase [217]. These are some of the reasons of why the homogenization of all the temperatures of the elements within the receiver becomes an impossible task and so, the control objective is to maintain the temperatures of the thermocouples in the receiver within an admissible pre-specified range (defined with the ideal objective of achieving similar temperatures in mind). A different operating range is defined for each thermocouple, depending on its specific characteristics, position in the receiver, etc. Because of the heterogeneity in the operating ranges of the thermocouples the measurements must be scaled and normalized to enable comparison and thus the identification of hot and cold zones of the receiver. This has been achieved by defining the working temperature ranges of each thermocouple. The temperature of each thermocouple is considered to be controlled if it belongs to the pre-specified range. These temperature ranges have been defined from data obtained during a great amount of previous tests. The upper limit of the temperature of each thermocouple has been selected as the maximum of the arithmetic mean values of all the measured values of the pattern tests. Notice that the pattern tests have been selected so that they correspond to the steady-state operation close to the set point (optimum operation from the manufacturer specifications), with stationary conditions of solar irradiance, without wind, etc. The same criterion has been used to select the lower limit (minimum of the arithmetic mean values). All these limits have been expressed in relation to the temperature given by the mean temperature of the thermocouples in the receiver (`CT060`), in such a way that different outlet air temperatures produce different operating ranges of the thermocouples. This is due to the fact that at the start-up, the aiming-point strategy control system is activated before the outlet temperature of the air reaches the nominal set point. In order to obtain temperature ranges valid for all operating conditions and not depending on the working power, the maximum and minimum values allowed change with the outlet air temperature for each thermocouple. Once the control program has obtained the temperatures from the SCADA, the following transformation applies to each one:

- If the thermocouple i measures a temperature that is within the pre-specified range: $t_p[i] = 0$, where $t_p[i]$ is the scaled or normalized temperature assigned to thermocouple i.

- If the thermocouple i measures a temperature that is below the pre-specified range: $t_p[i] = 100(T[i] - T_{min}[i])/T_{min}[i]$, where $T[i]$ is the temperature measured by thermocouple i and $T_{min}[i]$ is the minimum temperature allowable by the pre-defined ranges.
- If the thermocouple i measures a temperature that is over the pre-specified range: $t_p[i] = 100(T[i] - T_{max}[i])/T_{max}[i]$, where $T_{max}[i]$ is the maximum temperature allowable by the pre-defined ranges.

In this way, the percentage of deviation over the temperature ranges defined as the normal ones is obtained for each thermocouple, this being a scaled value useful for the comparison of the temperature of all the thermocouples. If the normalized temperature of thermocouple i ($t_p[i]$) is considered as *cold*, the value obtained for this index is negative. If the thermocouple is *hot*, then the value of $t_p[i]$ is positive. The control algorithm works with these normalized temperatures or indices. Notice that with this approximation a *cold* section is weighted more heavily than a *hot* one. This has been the first approximation used having in mind that actions adopted to correct deviations more heavily weighted should also help to correct the other deviations, as will be treated next (for instance, change of a heliostat from a *hot* zone to a *cold* one). Another logical possibility is to scale both temperature deviations relative to ($T_{max}[i] - T_{min}[i]$). In order to ensure the correct performance of the control program, an image of the state of the plant must be stored in memory at each sampling interval. This image contains:

- A list of the heliostats susceptible to suffering a change (those assigned to an aimpoint).
- A table with the actual position of each aimpoint on the receiver, defined by its coordinates.
- A table with the actual state of each heliostat of the field (normal tracking, defocused, etc.).
- A map of each aimpoint (focus), which consists of a 11×11 matrix which represents the receiver region which is influenced by the aimpoint. The temperatures of the thermocouples are placed on this matrix so that their position in the matrix is representative of their real position on the solar receiver with respect to the aimpoint. In this representation the aimpoint is placed in the center of each matrix (Fig. 6.24).

The first action of the control program is to build this representation of the state of the system into memory. To do this, data containing the default state of the plant (position and limits of the aimpoints, list of heliostats and the default aimpoint which they have to aim at, etc.) are used. Modifications of this first representation have to be automatically performed by the control program to account for possible changes locally performed by the operators (change of the aimpoint coordinates or the heliostats assigned to an aimpoint during the start-up to achieve a fast response during this phase, etc.) and to avoid differences between the real state of the system and its image in memory which would produce undesired actions such as the displacement of an aimpoint over its allowed limits, the erroneous change of a heliostat from an aimpoint to another, etc. Once the image of the system has been mounted, the

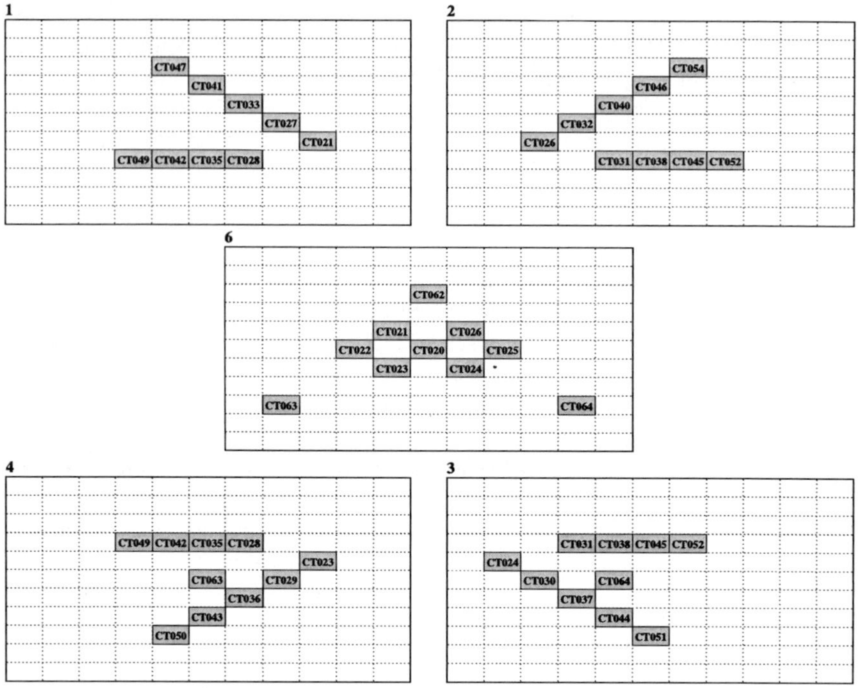

Fig. 6.24 Map of the aimpoints (focus) [150]

control program must subsequently analyze the situation of the solar receiver, infer from it the adequate control action, verify if the control action can be performed and send the corresponding commands to the workstation to achieve the adequate heliostat movements. As has been mentioned, two different control actions have been implemented: heliostat adjustment and aimpoint adjustment. The first action can be performed to compensate for a situation in which the average temperature of one aimpoint (focus) is much higher than the mean temperature of another one. In this case, one heliostat is changed from the *hot* aimpoint to the *cold* one. The second action leads to movements of the aimpoint of a fixed magnitude in the three directions of the space (so that it belongs to the target plane, with existing limits in the allowable displacement to avoid the aimpoint moving out of the receiver) and it is carried out in order to compensate for a situation in which an important internal disequilibrium within an aimpoint influence region exists. In this case, the aimpoint is displaced from the *hot* zone to the *cold* region. The analysis of the state of the receiver consists of determining which are the *hot* and *cold* zones. Once the disequilibria have been detected, the associated control actions are ordered taking into account the magnitude of the deviations, that is, the first action executed is that which corrects the major temperature difference existing within the receiver and so on. If a control action cannot be applied, the control program automatically selects the following one to be implemented. To carry out all this analysis, the control algorithm uses the indices or homogenized temperatures calculated for each thermocouple, as has been explained above. This index is indicative of the temperature of each thermocouple and can be used for comparison purposes. The control algorithm uses two pre-specified parameters denominated dT_h and dT_f, which are thresholds that indicate if a situation is susceptible of generating a control action. The first one serves as a threshold to indicate that a heliostat adjustment is necessary and the second one to produce an aimpoint adjustment. The reason for including thresholds in the control actions is to create control deadbands that prevent excessive cycling. The values of parameters dT_h and dT_f were experimentally determined with the help of a skilled operator. The procedure is simple: starting with a fixed value of these parameters, their values are reduced if a necessary control action (from the viewpoint of the operator) is not selected by the control program. The lower limit of these parameters is given by the appearance of limit cycles, indicating the necessity of using higher values of the thresholds. A trade-off solution can easily be found. In order to analyze the state of the solar receiver the first action is to determine the average temperature of each one of the aimpoints (foci), adding the indices associated to the thermocouples belonging to the matrix representative of each aimpoint and dividing the sum by the number of thermocouples associated to that aimpoint. This average value can produce a positive number, indicating that the aimpoint is *hot*, a negative number, indicating that the aimpoint is *cold*, or zero, indicating that all the thermocouples belonging to the aimpoint are within the allowed range (note that heterogeneity within a determinate aimpoint influence region will be treated later). Then, the aimpoints are ordered in descending order in relation to the value of the calculated average. To compensate for a situation in which an aimpoint has an average temperature much higher than that of another aimpoint of the receiver, the most convenient action is to change one heliostat from the *hot* aimpoint to the *cold* one. To determine if the disequilibrium existing between two aimpoints should be corrected by a heliostat adjustment, the mean index calculated for the *cold* aimpoint is subtracted from the mean index calculated for the *hot* one. If this index exceeds the threshold, the control action is accepted. Notice that the only case in which this control action cannot be implemented is that in which the *hot* aimpoint does not have heliostats aiming at it. This would seem to be a ridiculous situation, but it could be produced under certain circumstances during operation with the system on days with high level irradiance in the case of aimpoint number 6 (the central aimpoint), as the influence zone of the other aimpoints can be such that the temperature at the center of the receiver would achieve high values. A comparison is performed between the ordered aimpoints and the normalized temperature of the coldest aimpoint; if the threshold dT_h is exceeded, the control action is accepted. Each one of the accepted control actions has an associated characteristic value calculated taking into account which percentage the threshold dT_h is surpassed at. Once the previous process has been carried out, the control program analyzes the inner part of the aimpoints using the matrix descriptive of each one. The matrix is first divided into upper and lower parts and a calculation of the average value of the indices associated to the thermocouples contained in each part is performed. The mean value corresponding to the lower part is subtracted from the one associated to the upper part. If the absolute value is higher than the threshold dT_f then, in a first approximation, a vertical aimpoint movement is necessary from the *hot* part to the *cold* part. To characterize this possible control action, the percentage in which the threshold is surpassed is calculated. The same process is performed by dividing the matrix into the right and the left part, calculating the average of the indices for each one of these parts, subtracting them and calculating if the threshold dT_f is exceeded and at what percentage. Finally, the maximum of these percentages is taken as a characteristic of the control action. As a result of this process, a list of possible actions for aimpoint adjustment is generated, each one characterized by an index and ordered in descending value of this index. All the possible control actions oriented both to a heliostat adjustment and to an aimpoint adjustment are reordered in a unique list using the associated index, in such a way that the first one to be carried out is that which corrects the highest disequilibrium within the receiver temperatures. If the first action cannot be accomplished, the algorithm tries to execute the following one and so on. If the selected action is to move a heliostat from one aimpoint to another, it is convenient to determine which of the heliostats aiming at the *hot* aimpoint is selected to be changed to another aimpoint. In order to avoid certain disorder in the heliostat field, a selection method is applied searching for: (1) one heliostat which, aiming at the *hot* aimpoint, belongs by default to the *cold* one. (2) If search (1) does not match, a heliostat belonging to a aimpoint other than the *hot* one is searched for. (3) If searches (1) and (2) do not match, any heliostat is changed from the *hot* aimpoint to the *cold* one. In the case of an aimpoint adjustment, there are situations in which this action cannot be accomplished, because the aimpoint can be situated at coordinates lying within the limits allowed in order to avoid interferences of the different aimpoints and to avoid the aimpoints moving out of the target plane. Also, emergency situations can be produced during daily operation, when any of the temperatures of the solar receiver exceeds a prespecified value. The solar receiver can be damaged and to avoid this situation the control algorithm sends all the heliostats to the defocused position (standby points), finalizing the operation. The values of the temperatures in the receiver are measured and analyzed every second. The control actions are discretely implemented, that is, each control action is implemented and the following one is implemented after a predetermined sampling time. The value of the sampling time for heliostat adjustment actions is 80 s and for aimpoint adjustment is 150 s. These values have been chosen close to the corresponding temperature settling times produced by each kind of control action to avoid over-correction. Once the most adequate control action has been selected, it can be effectively implemented. The control program allows the plant operation manager to select two operation modes: supervisory automatic control or fully automatic control. In the first case, the control program asks the operator for confirmation of the proposed control action. In the second case, the selected control action is directly implemented. Finally, once the control action has been implemented, the image of the system in memory is refreshed by modifying the representative tables of the system to account for the new state of the system. The flow diagram of the control algorithm is shown in Fig. 6.25.

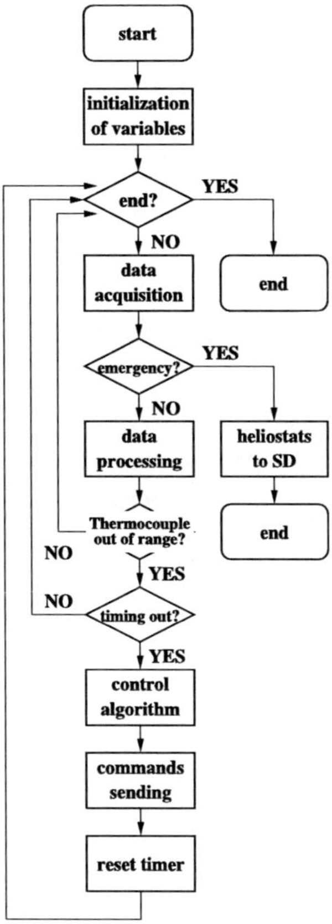

Fig. 6.25 Flow diagram of the control algorithm [150]

6.7.4 Experimental Results

The man–machine interface of the control system includes an image of the solar receiver with the temperatures of the different thermocouples (which are measured and changed on-line), a window in which the last ten proposed control actions appear and another window in which the value of the different control parameters and a command line can be found. Different commands are admitted in order to start automatic or manual control operation, fully or supervised automatic control operation, on-line modification of values of control parameters, etc. Before commenting on different results, it is important to point out again that the final objective of the developed control algorithm is to maintain the difference between the maximum and minimum temperatures in the receiver below 100°C, keeping the temperatures of the different thermocouples within pre-specified ranges. As illustrative results, the evolution of the temperature measured by different representative thermocouples of the solar receiver is shown for one selected test performed on March 14th, 1997 with changes in direct solar irradiance of less than 100 W/m^2 during the test. The criterion used for comparison purposes is the following: for each thermocouple

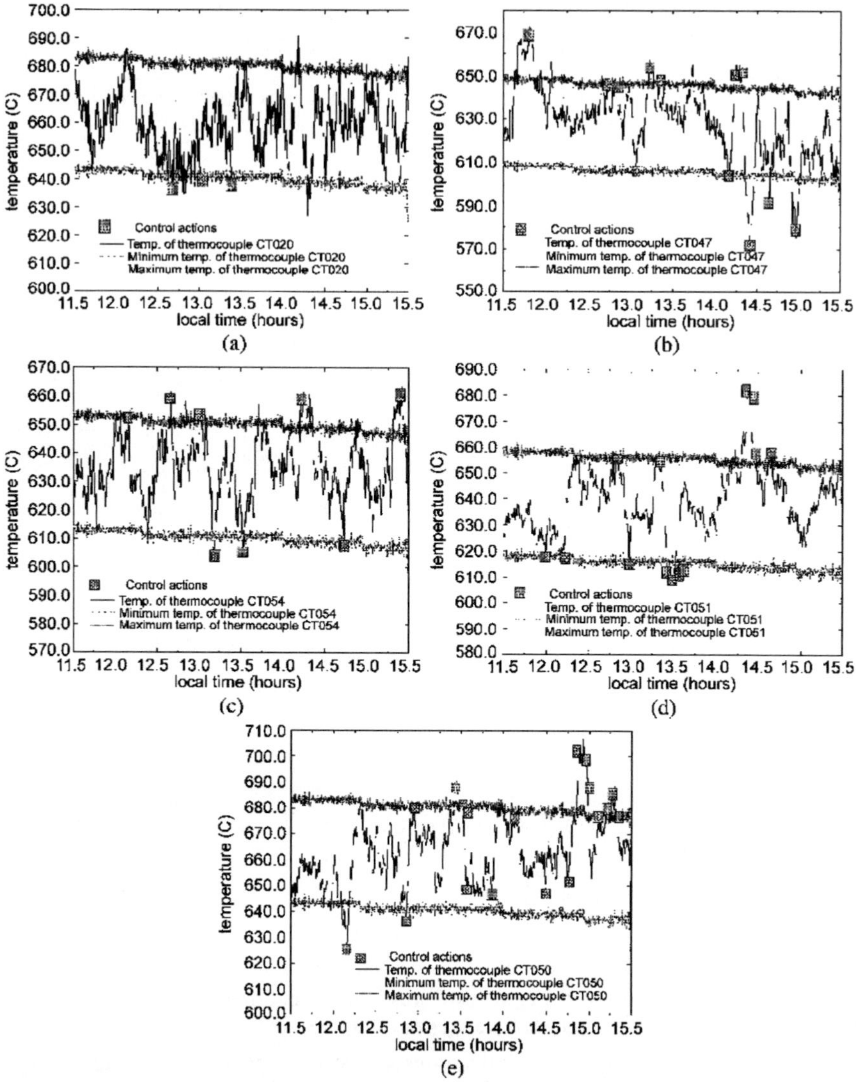

Fig. 6.26 Temperature of representative thermocouples under automatic operation (14/03/97) [150]

both the maximum and minimum temperatures achieved during the operation are selected. The first criterion is given by the average value of this range for all the thermocouples. This index indicates the variability of the mean temperature range of the thermocouples in the receiver. On the other hand, the difference between the maximum and minimum temperatures within the receiver every 5 s throughout the test is calculated. The average of these values is also calculated and used as an index indicating the homogeneity of the different temperatures in the receiver during the test. Figure 6.26(a) corresponds to the evolution of `CT020` thermocouple during a daily operation. This thermocouple is placed at the center of the volumetric receiver, belonging by default to aimpoint number 6. As can be seen, the figure represents the value of the temperature measured and the maximum and minimum temperatures within which the temperature of this thermocouple can be considered as *normal* (remember that these limits change with the outlet air temperature indicated by `CT060`). The moments in which a control action was implemented to correct deviations of the temperature of this thermocouple are indicated by a solid square box. Notice that at each sampling time, the most important corrective control action is implemented and thus those moments at which the temperature of the thermocouple is out of range and no control action has been implemented correspond to the application of a control action which corrects other more important deviations in the receiver. The previous comments are applicable to all the figures presented. Another example of the results achieved with the proposed control scheme is presented in Fig. 6.26(b) in which the evolution of the temperature of thermocouple `CT047` placed near the border of the receiver belonging to aimpoint number 1 is represented. Figures 6.26(c), 6.26(d) and 6.26(e) represent the temperature of thermocouples `CT054`, `CT051` and `CT050` placed near the border of the receiver and belonging to aimpoints numbers 2, 3, and 4, respectively. In every case, it can be seen that the control algorithm implements control actions in order to maintain the temperature of the different thermocouples within their pre-specified ranges in such a way that the distribution of temperatures within the solar receiver is also controlled. In the following, the evolution of the difference between the maximum and minimum temperatures in the receiver during operation are presented for different tests (in which similar experimental conditions have been used). As has been previously indicated, this index gives a measure of the homogeneity of the temperatures in the receiver during a test, allowing for a comparison of the results obtained by the implementation of the heuristic knowledge-based control scheme with manual control operation. The first result corresponds to the operation on March 14th, 1997 (Fig. 6.27(a)). The arithmetic mean value of this index during the test was of 92°C and the maximum difference in temperature during operation was below 100°C about 74.61% of the test time. On the other hand, the value of the difference between the maximum and minimum temperatures measured by a thermocouple during the whole test was obtained for all thermocouples, providing an average value of about 79°C (notice that this index depends upon the value of solar irradiance). Figure 6.27(b) shows the results obtained on March 6th, 1997. The maximum difference in temperature in average was 84°C, this index being below 100°C for 90.23% of the operation time. In this case, the value of the difference between the maximum and minimum temperatures measured by a thermocouple during the whole test was also obtained for all thermocouples, providing an average value of 62°C. After a period of tests carried out in order to adjust the different control parameters, the heuristic knowledge-based automatic control system performance was evaluated with data from operation during 1997, from February 21st to March 14th. During this evaluation period, eight tests covering about twenty operation hours

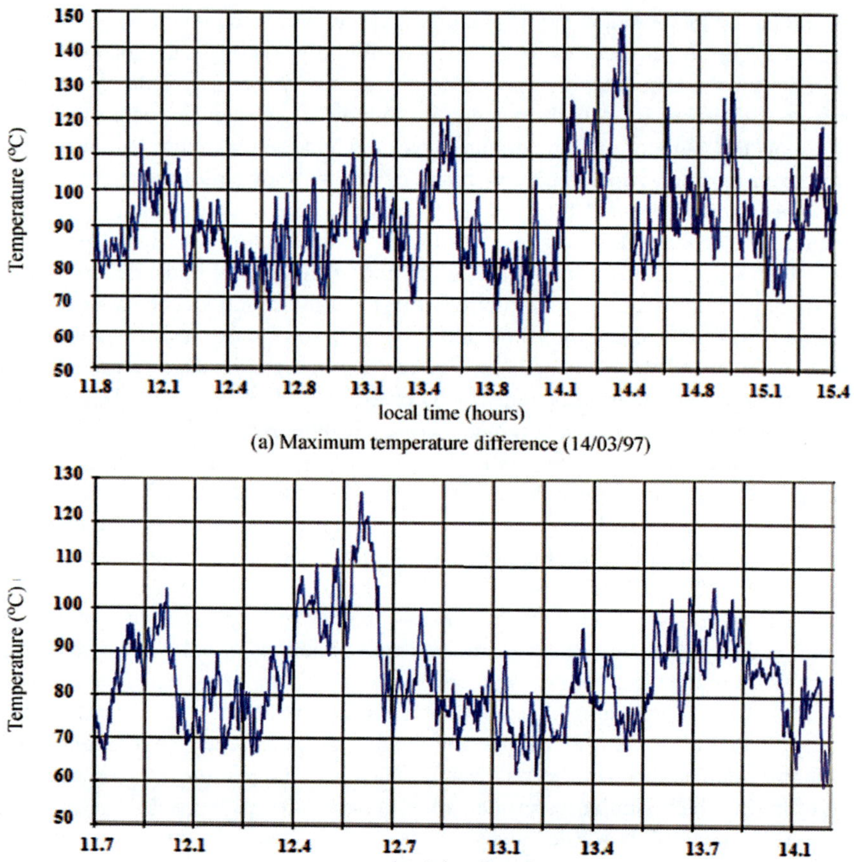

(a) Maximum temperature difference (14/03/97)

(b) Maximum temperature difference (06/03/97)

Fig. 6.27 Tests showing the maximum temperature difference within the receiver under automatic operation [150]

were performed. During these 20 h, the maximum temperature difference existing within the receiver, in average, was of about 89°C. The percentage of time during which this maximum temperature difference in average remained below 100°C was of about 78%. The temperatures of the thermocouples varied within a range of about 74°C in average. In order to perform a comparison of the results obtained with the heuristic knowledge-based automatic control scheme with the operation under manual control, a representative test manually performed on September 7th, 1995 is included (Fig. 6.28). This corresponds to a typical test, with similar experimental conditions than those under which tests shown in Fig. 6.27 were performed (number of heliostats used, set point temperature, irradiance conditions, etc.). The maximum temperature difference within the solar receiver during the test is shown in Fig. 6.28, which in average was about 140°C, evolving below 100°C for only

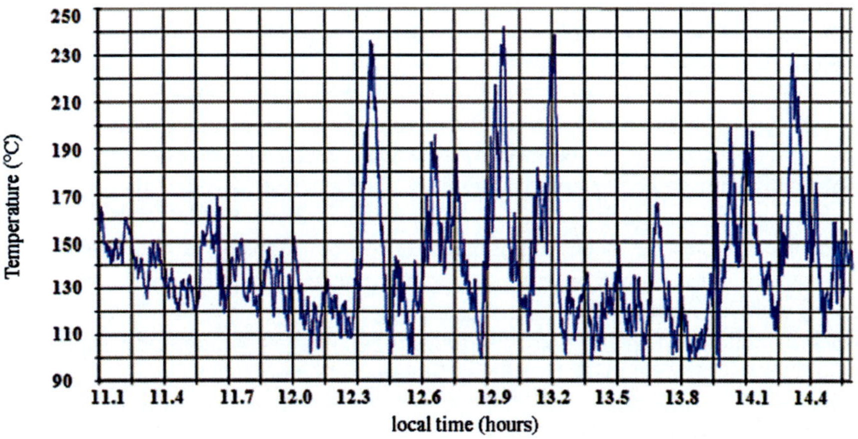

Fig. 6.28 Maximum temperature difference within the receiver under manual operation (07/09/95) [150]

0.25% of the total duration of the test. As can be understood, the implementation of the automatic control scheme not only allows the operator to avoid this task, but also widely optimizes the operation of the system, reducing the arithmetic mean value of the maximum temperature difference about 50°C and considerably augmenting the time during which this difference remains below 100°C.

6.8 Power Stage Control

6.8.1 Introduction

In many electricity generation facilities, whether based on fossil fuels or renewable energies, a buffer or energy storage system is used as a backup source of power for a heat exchanger or steam generator when the primary power supply is not ready to use or is subjected to strong changes or oscillations. This is the case of renewable power plants supplied by wind or solar energy as the primary energy source. Under primary energy shortfalls, the system must be kept at standby if the primary power support is going to be available soon, or stopped and restarted as soon as possible if the shortfall lasts too long. This stop–restart sequence is a very expensive process in terms of cost and time and it is therefore highly desirable that this situation arise as little as possible. To reduce the frequency of stops and restarts, a standby power buffer, is added to the system. This buffer, usually thermal storage, is loaded during plant start-up with excess power and it is used when the primary power support is not able to feed the system with the minimum energy needed to keep it at operating point or other desirable state. This standby power buffer is able to keep the plant in normal operation for a limited period of time, when appropriate switching controllers are used to diminish the oscillation associated with the change in supply.

In spite of the great advantage of adding a power buffer to the system, from both modeling and control viewpoints, the system dynamics are modified due to the presence of discrete events associated with the changes between the different energy sources. Thus, the nature of the process combines continuous and discrete dynamics so that it can be modeled by a hybrid model useful for control purposes. Compared with purely continuous systems, the switched nature of these systems makes them more difficult to describe, analyze and control [227]. The modeling of hybrid systems has been traditionally carried out by using several continuous models, one for each operation mode and the transitions among models are based on the discrete events of the system (producing discrete transitions between modes where continuity and stability among transitions cannot be guaranteed). So, hybrid models, which involve both continuous and discrete dynamics in a unique model, come up as a natural way to model this kind of system. It ensures, at the same time, continuity and stability among the different plant operation modes.

A wide range of hybrid renewable energy systems were reviewed in [125] and several matters related to hybrid renewable energy systems, such as simple static models (based on mathematical equations), costs and feasibility are also described. However, in the works cited in [125] the term hybrid system refers to any energy system with more than one supply source; normally one of them is a conventional diesel generator and the other a renewable energy supply source, or even, the two are renewable energy supply sources, usually solar and wind energy source.

In order to control the power stage represented in Fig. 6.29, a mixed logical dynamical (MLD) model, which is a modeling tool for the representation of hybrid systems, it is used to represent the combined continuous and discrete dynamics of the plant. The model results are compared with real data acquired from the plant in Sect. 6.8.3.

6.8.2 *Object-Oriented Modeling of the TSA System*

In [405] a model of the thermohydraulic part of the system is developed assuming a known input radiation power in the receiver as a consequence of the irradiance reflected in the heliostat mirrors and the aiming-point strategy previously presented. The Modelica language has been used to develop these models including the ThermoFluid library [128, 257]. The work analyzes each of the components of the thermohydraulic circuits of air and water–steam and explains the modeling assumptions, trying to justify each one as they are oriented to get, by means of the symbolic manipulations that Dymola tool performs, a not too high index DAE system for the complete model, in which the number of non-linear algebraic loops is minimized. For this purpose, all the components are classified, following the modeling methodology derived from the Finite Volume Method (FVM) [287], in control volumes (CV in ThermoFluid nomenclature) and flow models (FM in ThermoFluid nomenclature). In some cases information about the future control system architecture to be implemented is introduced in the modeling phase. An example of components

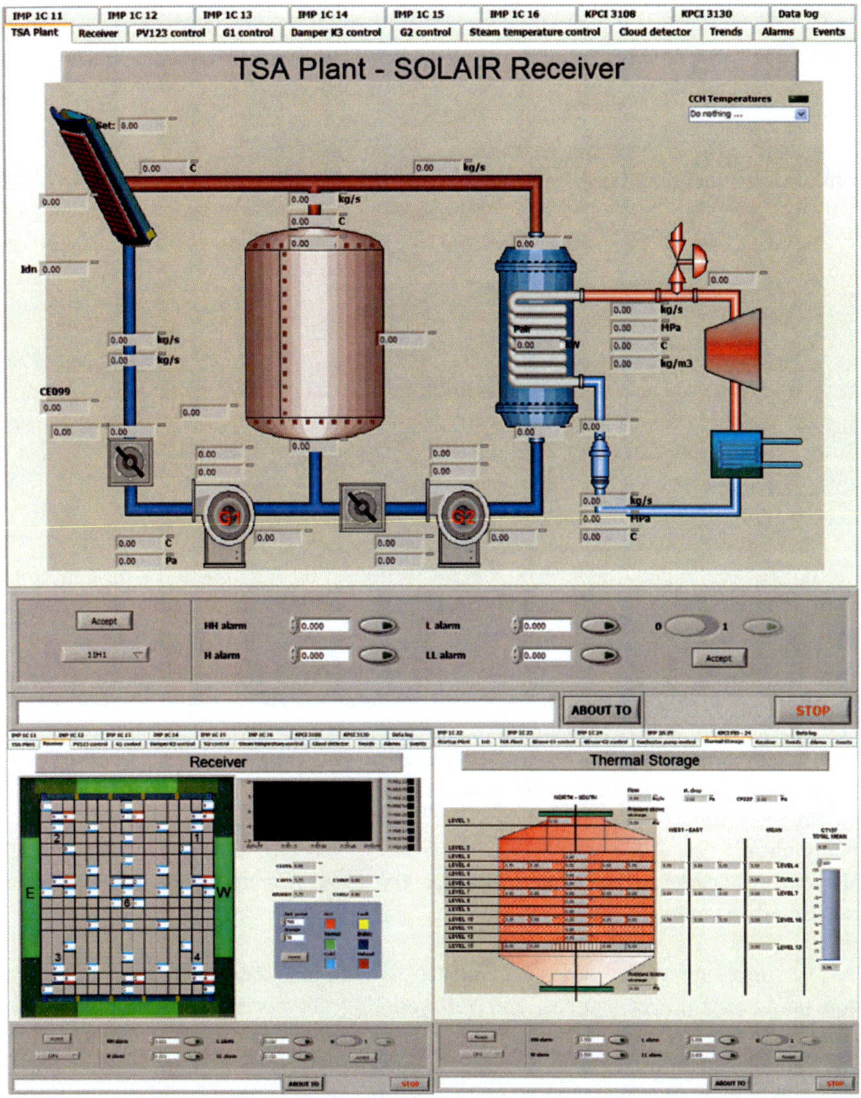

Fig. 6.29 SCADA system of the TSA layout (courtesy of PSA)

that are modeled using this kind of information are the blowers in the air circuit, in which a cascade control will help avoid the multivariable non-linear dependence of the constitutive equations and let consider them like quasi ideal flow rate generators. Due to the existence of components whose internal implementation may vary depending on the modeling hypotheses, the polymorphism and the Modelica language constructs replaceable/redeclare have been specially used in some of them, for example in the evaporator.

The following components are modeled: blowers, storage tank, solar receiver, evaporator, control valve, sensors, pipes and connections. All of them are directly instantiated and parameterized, or inherited from ThermoFluid classes. The air circuit is composed of a solar receiver, tubes, tank and evaporator; and the water–steam circuit with a water pump, a control valve, tubes and reservoirs with boundary conditions. For more details of the development of the model see [405].

6.8.3 Hybrid Modeling of the TSA System

The TSA system has been described in Fig. 6.6 and its components enumerated in the previous subsection. It is composed of an air circuit and a water–steam circuit. The air circuit supplies the power demanded by the steam generator from the receiver and the thermal storage and return to the receiver the output cold air from the steam generator. At the same time, in the water–steam circuit, the water which is used to cool the hot air into the steam generator flows by the feedwater pump, whereas the produced steam leaves the system through the pipes.

To operate the plant, solar irradiance is reflected by the heliostat field and concentrated onto the receiver's ceramic absorber cups following the aimpoint strategy explained in Sect. 6.7. The hot air leaves the receiver at 700°C and blower G1 controls the air mass flow through the receiver. The thermal storage unit in the air circuit, which provides standby power, is charged during plant start-up and then discharged to feed the steam generator, either during cloud transients or when the power supply from the receiver is not enough to feed the steam generator at nominal conditions.

Blower G2 controls the air mass flow through the steam generator and thereby, incoming power. The once-through steam generator, which is part of the water–steam circuit, has a nominal operation of 340°C and 45 bar. The steam produced is sent to a turbine, which is part of a Rankine cycle. Steam outlet temperature is controlled by means of a feedwater pump and the steam outlet pressure by means of valve PV123.

The simplified modeling work mainly deals with the pipes connecting the steam generator to its power sources, and not modeling the system power supplies (the volumetric receiver and the thermal storage), but the control-related problems which arise, due to the hybrid nature of the TSA system. Clearly the steam generator can be fed by the energy from the receiver, from the thermal storage, or even by both simultaneously. The power supply depends on the relationship between the speeds of blowers G1 and G2. Signal $\dot{m}_{12}$, which is the difference between blower G1 air mass flow ($\dot{m}_1$) and blower G2 air mass flow ($\dot{m}_2$), is used to determine the direction of air through the thermal storage system (see Fig. 6.30). When mass flow from blower G1 is equal to the mass flow from blower G2, all the power supplied to the steam generator comes from the receiver. Otherwise, if blower G1 mass flow is greater than that of blower G2, the power from the receiver feeds both the steam generator and thermal storage. Finally, if blower G2 mass flow is greater than that of blower G1, the power to the steam generator comes from both the receiver and thermal storage.

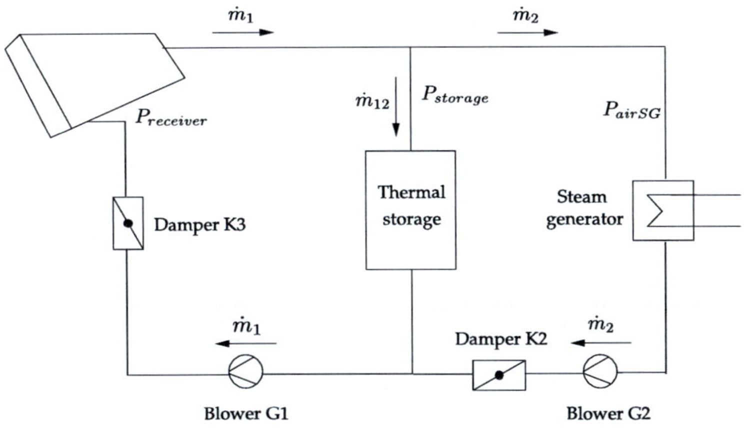

Fig. 6.30 Air flow direction based on the signal $\dot{m}_{12}$ (courtesy of J.D. Álvarez et al., [10])

So, the source from which the power is supplied to the steam generator can change dynamically according to the relationship between the speeds of G1 and G2. This system feature cannot be modeled with traditional dynamic models, so it would be necessary to introduce the system's discrete nature by means of either '*switch*' blocks or '*if-then-else*' rules. However, from a control point of view this particular solution does not ensure stability in the transitions between the different dynamics. With hybrid models, such as that described in this work, there is no problem in handling both discrete and continuous dynamics in the same model; moreover, stability is ensured with the use of this kind of model.

6.8.3.1 TSA Air Mass Flow Distribution Dynamic Model

The TSA air mass flow distribution dynamic model is presented in two parts [10]. In the first one, the general equation which relates the output variable (steam generator input power) to the input and disturbance variables is calculated. In the second, the simplest case, in which the steam generator input power comes from only one source, is validated with real data by means of transfer functions.

Model validation when the steam generator input power comes from different sources and can change dynamically is explained in the following section with using a hybrid model.

Air Circuit Model

In order to model the air circuit, a model of the pipes which form a part of it must be constructed. Figure 6.31 shows the scheme of a single pipe in which a fluid is cooled while the outside temperature of the pipe is kept constant. Inside the pipe, the hot fluid which flows through it feeds a heat exchanger or a steam generator. The result

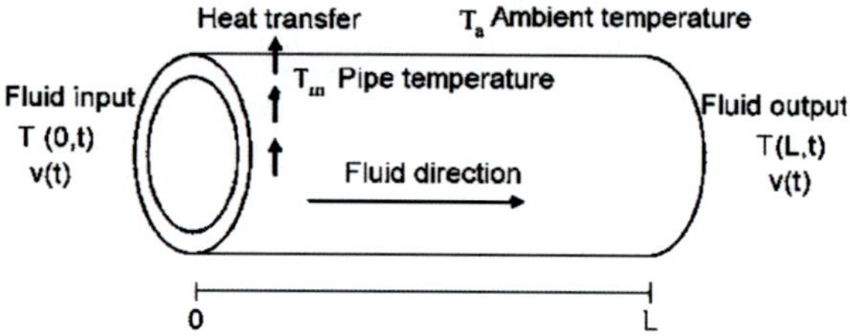

Fig. 6.31 Pipe schematic diagram

of applying the law of conservation of energy to the pipe shown in Fig. 6.31 is very similar to a tubular heat exchanger where the internal fluid is heated by saturated steam at constant temperature, with the difference that the heat is transferred in the opposite direction. The differential energy balance of the fluid flow through the pipe over the volume element of length Δx is given by

$$vA_f\rho_f c_f T_f - vA_f\rho_f c_f\left(T_f + \frac{\partial T_f}{\partial \ell}\Delta\ell\right) - \pi D_f H_t \Delta\ell (T_f - T_m)$$
$$= \frac{\partial}{\partial t}(A_i \rho_f \Delta\ell c_f T_f) \tag{6.7}$$

where the terms involved are summarized in the table of variables and parameters defined at the beginning of the book. Similarly, an energy balance for the pipe wall over the volume element $\Delta\ell$ can be written as follows:

$$\pi D_f H_t \Delta\ell (T_f - T_m) - \pi D_m H_l \Delta\ell (T_m - T_a) = A_m \Delta\ell \rho_m c_m \frac{\partial T_m}{\partial t} \tag{6.8}$$

where the terms are also those summarized in the table of variables and parameters defined at the beginning of the book.

From the analysis above, after linearizing Eqs. (6.7) and (6.8) in terms of the deviation variables around a steady-state value $\{v_s, T_{m_s}, T_{a_s}\}$, it is possible to find transfer functions relating the output variable (the outlet fluid temperature, $T_{out} = T(L,t)$) to the input or disturbance variables (the incoming fluid temperature $T_{in} = T(0,t)$, the ambient temperature (temperature outside pipes), $T_a(t)$, and the fluid velocity, $v(t)$). As in Chap. 4, the easiest way to find these transfer functions is to analyze single input–single output relationships while keeping the rest of input/disturbances constant. In the first case, if the fluid velocity and the ambient temperature are considered constant, the relationship between the input and output fluid temperatures is given by

$$\frac{\tilde{T}_{out}(s)}{\tilde{T}_{in}(s)} = e^{-\frac{a(s)}{v_s}L} \tag{6.9}$$

where $\tilde{T}_{out}(s)$ and $\tilde{T}_{in}(s)$ represent the Laplace transform of the deviation variables $\tilde{T}_{out} = T_{out} - T_{out_s}$ and $\tilde{T}_{in} = T_{in} - T_{in_s}$, respectively; subscript 's' denotes the steady-state value; $a(s) = s + \frac{1}{\tau_1} - \frac{\tau_2}{\tau_1(\tau_{12}\tau_2 s + \tau_{12} + \tau_2)}$ and $\tau_1 = \frac{A_f\rho_f c_f}{\pi D_f H_t}$, $\tau_{12} = \frac{A_m\rho_m c_m}{\pi D_f H_t}$, $\tau_2 = \frac{A_m\rho_m c_m}{\pi D_m H_l}$. On the other hand, if the temperature and the velocity of the fluid entering the pipe do not vary, the transfer function relating the outgoing fluid temperature to the ambient temperature becomes

$$\frac{\tilde{T}_{out}(s)}{\tilde{T}_a(s)} = \frac{b(s)}{a(s)}\left(1 - e^{-\frac{a(s)}{v_s}L}\right) \tag{6.10}$$

where $\tilde{T}_a(s)$ represents the Laplace transform of the deviation variable $\tilde{T}_a = T_a - T_{a_s}$ and $b(s) = \frac{\tau_{12}}{\tau_1(\tau_{12}\tau_2 s+\tau_{12}+\tau_2)}$. Finally, if both input fluid temperature and ambient temperature do not vary, the transfer function relating the fluid velocity to its output temperature is given by

$$\frac{\tilde{T}_{out}(s)}{\tilde{v}(s)} = G(s) = -\frac{(T_{a_s} - T_{s0})}{c}e^{\frac{L}{c}}\frac{(s+f_1)}{s(s+(f_1+g))} \times \left(1 - e^{-\frac{L}{v_s}(s+g(\frac{s}{s+f_1}))}\right) \tag{6.11}$$

where parameters c, f_1 and g are given by $c = v_s\tau_1(1+\frac{\tau_2}{\tau_{12}})$, $f_1 = \frac{\tau_{12}+\tau_2}{\tau_{12}\tau_2}$ and $g = \frac{\tau_2}{\tau_1(\tau_{12}+\tau_2)}$. A more detailed procedure of this model can be found in [8], as it has been the same used for modeling parabolic trough collectors in Chap. 4. It is possible to verify that Eq. (6.9) represents almost pure delay. On the other hand, although Eqs. (6.10) and (6.11) have a step input response similar to a first order system, a frequency study shows that these transfer functions have resonance dynamics modeled by terms between bracket $(1 - e^{-\frac{L}{v_s}sQ(s)})$, $Q(s)$ being a transfer function with different values for Eqs. (6.10) and (6.11). The frequencies, in which this resonance dynamics appears, are inversely proportional with the value of L/v_s, these resonance modes being found on lower frequencies for higher values of L/v_s and vice versa. However, for this case L/v_s has a low value, which allows us to consider the response of Eqs. (6.10) and (6.11) as a first order system.

The equation that relates the temperature to steam generator air input power is (see the table of variables and parameters defined at the beginning of the book for parameter definitions)

$$\frac{dP}{dt} = \frac{d(\dot{m}C_F T_f)}{dt} = \frac{d(v\rho_f A_f c_f T)}{dt} \tag{6.12}$$

If ρ_f, A_f and c_f are considered constant at an operating point, the following result is obtained:

$$\frac{dP}{dt} = \rho_f A_f c_f \frac{d(T_f v)}{dt} = \rho_f A_f c_f \left(T_f \frac{dv}{dt} + v\frac{dT_f}{dt}\right) \tag{6.13}$$

The terms $T dv/dt$ and $v dT/dt$ in Eq. (6.13) are non-linear. The Taylor series expansions at point $(v = v_s, T_f = T_{f_s})$ are used to obtain the linear approximation:

$$\frac{dP}{dt} = \rho_f A_f c_f \left(T_{f_s}\frac{dv}{dt} + v_s\frac{dT_f}{dt}\right) \tag{6.14}$$

Representing the function in terms of deviation variables around steady-state conditions and applying the Laplace transform:

$$\frac{d(P-P_s)}{dt}=\rho_f A_f c_f\left(T_{f_s}\frac{d(v-v_s)}{dt}+v_s\frac{d(T_f-T_{f_s})}{dt}\right)$$
$$\rightarrow\quad \frac{d\tilde{P}}{dt}=\rho_f A_f c_f\left(T_{f_s}\frac{d\tilde{v}}{dt}+v_s\frac{d\tilde{T}_f}{dt}\right)$$
$$\rightarrow\quad s\tilde{P}(s)=\rho_f A_f c_f\left(T_{f_s}s\tilde{v}(s)+v_s s\tilde{T}_f(s)\right)$$
$$\Rightarrow\quad \tilde{P}(s)=\rho_f A_f c_f\left(T_{f_s}\tilde{v}(s)+v_s\tilde{T}_f(s)\right) \tag{6.15}$$

where $\tilde{T}_f=T_f-T_{f_s}$, $\tilde{v}=v-v_s$, $\tilde{P}=P-P_s$, and the dependent variables $\tilde{T}_f(s)$, $\tilde{v}(s)$, and $\tilde{P}(s)$ in the previous equations represent the Laplace transform of the deviation variables. Equation (6.15) yields two equations, one of them relating the fluid power to the fluid velocity when the fluid temperature is constant and the other relating the fluid power to its temperature when the fluid velocity does not vary:

$$\frac{\tilde{P}(s)}{\tilde{v}(s)}=\rho_f A_f c_f T_{f_s} \tag{6.16a}$$
$$\frac{\tilde{P}(s)}{\tilde{T}_f(s)}=\rho_f A_f c_f v_s \tag{6.16b}$$

Considering that the temperature is related to the velocity by Eq. (6.11), it is possible to find an overall equation which involves the fluid power and its velocity in the following way:

$$\frac{\tilde{P}(s)}{\tilde{v}(s)}=\rho_f A_f c_f\left(T_{f_s}+v_s G(s)\right) \tag{6.17}$$

However, when the steam generator input power comes from both receiver and thermal storage, two pipe models are necessary in Eq. (6.12), one for each branch and the resulting power is as shown in the following equation:

$$\frac{dP}{dt}=\frac{d(\dot{m}_1 c_{f_1}T_{f_1}+\dot{m}_2 c_{f_2}T_{f_2})}{dt}=\rho_{f_1}A_{f_1}c_{f_1}\frac{d(v_1 T_{f_1})}{dt}+\rho_{f_2}A_{f_2}c_{f_2}\frac{d(v_2 T_{f_2})}{dt} \tag{6.18}$$

The subscripts '1' and '2' denote the receiver branch and the thermal storage branch, respectively. For a better understanding of the model, the model inputs, disturbances and outputs delimiting the model boundary conditions are listed below:

- $\dot{m}_2$: air mass flow from blower G2 [kg/s], which is the only operator-controllable variable.
- $\dot{m}_1$: air mass flow from blower G1 [kg/s], which is responsible for the air mass flow through the receiver, so it is considered a boundary condition disturbance signal.
- T_{in1}: mean receiver air outlet temperature [°C]; this signal is supposed to be controlled by blower G1 air mass flow, but in this case it is considered a disturbance that merges the influence of the aiming-point strategy [149], solar irradiance, and ambient temperature changes that are difficult to model and changes air mass flow

in blower G1, so this measurable temperature is used as a boundary condition both for modeling and control.

- T_{in2}: thermal storage outlet temperature [°C], which depends on air mass flow from blowers G1 and G2, the mean receiver air outlet temperature (and thus on the aiming-point strategy and disturbances) and the heat exchanger state, in such a way that this measurable signal is considered a boundary condition disturbance signal in the hybrid model and related control algorithms.
- T_a: ambient temperature [°C], which is the constant temperature outside the pipes, which cannot be manipulated and is therefore considered another disturbance.
- P: steam generator air input power [kW], which is the model output signal.

Equation (6.18) must be modified such that its inputs change with signal $\dot{m}_{12}$, as described previously. That is, if $\dot{m}_{12}$ is equal to or greater than zero, the receiver branch mass flow is equal to $\dot{m}_2$, whereas the thermal storage branch mass flow is zero. On the contrary, if $\dot{m}_{12}$ is less than zero, the receiver branch mass flow is equal to signal $\dot{m}_1$, whereas the thermal storage branch mass flow is the difference between signals $\dot{m}_2$ and $\dot{m}_1$.

Modeling the Simplest Case

The simplest case is when the steam generator input power comes only from the receiver, $\dot{m}_{12} \geq 0$ (the primary supply) and no power is transferred to or from thermal storage (standby power). In this case, signals $\dot{m}_1$ and T_{in2} are not model boundary conditions; so the model has one input (signal $\dot{m}_2(s)$), two disturbances (signals $T_{in1}(s)$ and $T_g(s)$) and one output (signal $Power(s)$), as shown in Fig. 6.32. The model has been validated with real TSA system data saved by the SCADA during typical operating sequences. Figure 6.33 shows a typical operation day, where solid lines represent model data and dotted lines represent real data. As well as showing the model output variable, steam generator air input power, Fig. 6.33 shows both input and disturbance signals, like mean receiver air outlet temperature and air mass flow. Ambient temperature is not shown in this figure because it does not change significantly during plant operation. In the operation with date and label 2004-05-07 (left charts in Fig. 6.33), the input variable which has a wider range is the input temperature; its changes, during operation, are reproduced by steam generator inlet power, changes that, at the same time, are captured by the model. On the opposite, in the other operation day, labeled 2004-06-18 (right charts in Fig. 6.33), the input variable which changes most frequently, and inside a wider range, is the mass flow; its changes are reproduced by incoming power variable and, the same as in the previous case, these changes are captured by the model. So, there is excellent agreement between the real data and the modeled steam generator input power dynamics.

On the other hand, in order to handle all the possible cases (power from different sources), the model must be modified according to the logical $\dot{m}_{12}$ conditions. '*Switch*' blocks must therefore be added to allow inputs to be switched to the appropriate pipe model. The new model scheme is shown in Fig. 6.34, where the power from the two different supplies is added to calculate the overall steam generator

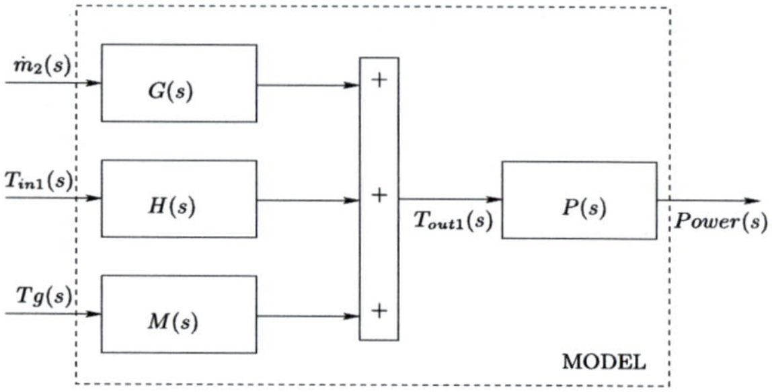

Fig. 6.32 Model scheme when the steam generator inlet power only comes from the receiver (courtesy of J.D. Álvarez et al., [10])

input power and it is possibly observed as each branch has its own transfer functions relating mass flow with output temperature ($G(s)$ and $N(s)$ with the structure of Eq. (6.11) divided by $A_i \rho$ to pass from mass flow to velocity), input temperature with output temperature ($H(s)$ and $J(s)$ with the structure of Eq. (6.9)) and ambient temperature with output temperature ($M(s)$ and $R(s)$ with the structure of Eq. (6.10)) and another one to relate output temperature with output power too ($P(s)$ and $T(s)$ with the structure of Eq. (6.16b)).

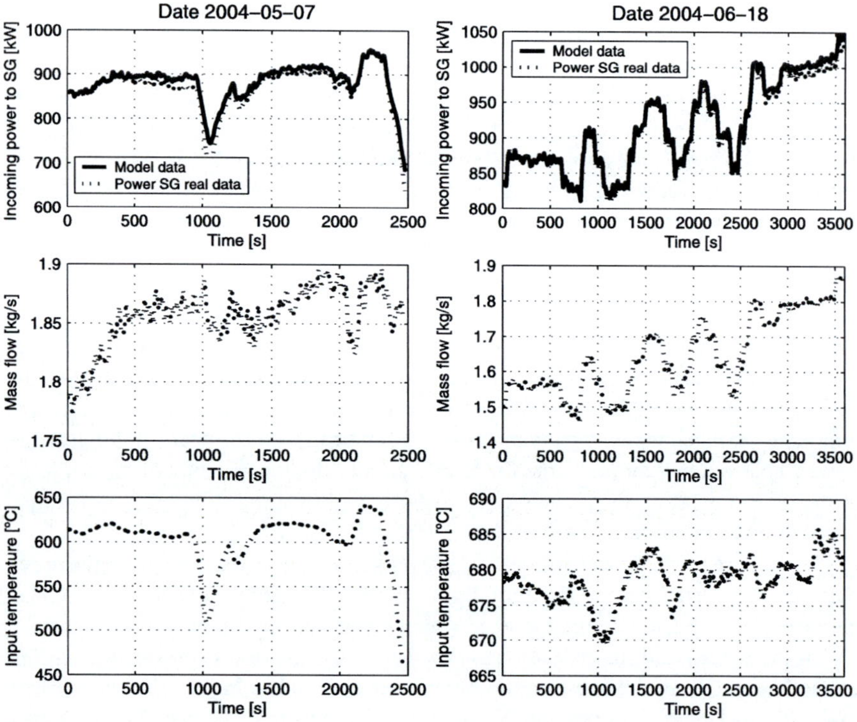

Fig. 6.33 Comparison of model results versus real data when air mass flow only comes from the receiver (courtesy of J.D. Álvarez et al., [10])

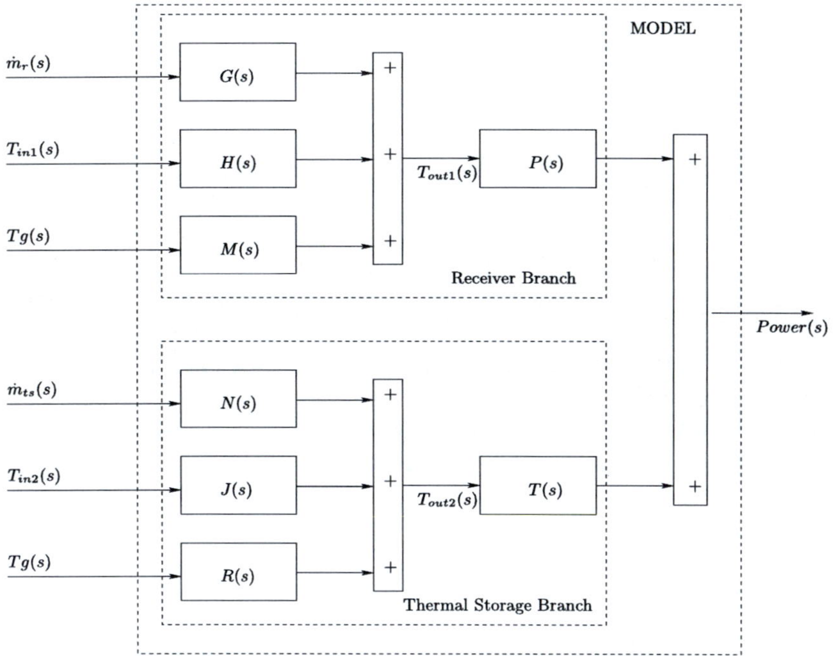

Fig. 6.34 Model scheme when the steam generator inlet power comes from both the receiver and the thermal storage (courtesy of J.D. Álvarez et al., [10])

As modeling of whether the addends in Eq. (6.18) are active or not is based on the $\dot{m}_{12}$ signal, several discrete events which capture the dynamic changes must be virtually implemented. Hybrid models allow these logical conditions to be systematically combined with the different dynamics using a single overall model. This kind of model is suitable for designing future hybrid control algorithms, for example, using the well-known MATLAB hybrid toolbox [32]. The following subsection describes how the different dynamics of the TSA system can be represented with a hybrid model.

6.8.3.2 Hybrid Model

Hybrid systems are loosely defined as dynamic systems whose state has two components, one that evolves in a continuous set such as $\mathbb{R}$ (typically, according to a differential or difference equation) and another one that evolves in a discrete set such as $\mathbb{N}$ (typically, according to some transition logic-based rule). Also, complex, hierarchically organized systems, where discrete planning algorithms interact on a higher level with continuous control algorithms and are processed at a lower level, are another example of hybrid systems [33]. Perhaps the simplest model of a hybrid system is [227]

$$\begin{aligned} \dot{x}(t) &= f_{\sigma(t)}\big(x(t)\big), & x &\in \mathbb{R}^n, \\ \sigma(t) &= \lim_{\tau \to t^-} \phi\big(x(\tau), \sigma(\tau)\big), & \sigma &\in \mathbb{N}, \end{aligned} \tag{6.19}$$

where x and σ denote the continuous and discrete state components, respectively. Usually, $\{f_p \colon p \in \wp\}$ is a family (normally finite) of sufficiently regular functions from $\mathbb{R}^n$ to $\mathbb{R}^n$ that is parameterized by some index set $\wp$ and $\sigma : [0, \infty) \to \wp$ is a piecewise constant function of time, called a switching signal. In specific situations, the value of σ at a given time t might just depend on t or $x(t)$, or both, or may be generated using more sophisticated techniques such as a hybrid feedback with memory in the loop. Standard assumptions in this context are that the solution $x(t)$ is continuous everywhere and that there are finite switches in finite time [227]. The particular value for σ may be chosen by some higher process, such as a controller, computer or human operator, in which case the system is said to be controlled. It may also be a function of time or state or both, in which case the system is said to be autonomous. A more general formalism is that hybrid systems are hierarchical systems consisting of dynamic components at the lower level, governed by upper level logical/discrete components [64, 164].

The most common class of hybrid systems is known as Discrete Hybrid Automata (DHA), which is described by the interconnection of linear dynamic systems and automata providing the discrete part of the system and a set of logical rules [373].

DHA is able to generalize many oriented computational models for hybrid systems and is therefore the starting point for solving problems of complex hybrid system analysis and synthesis. From this representation, an abstract representation in a set of constrained linear difference equations involving integer and continuous variables may be found that yield the equivalent Mixed Logical Dynamical (MLD) model. This kind of model is very useful for optimization problems and for optimal control purposes [33].

The MLD modeling tool is a relevant framework for the representation of hybrid systems. From this framework, it is possible to model the evolution of continuous variables through linear dynamic discrete-time equations, the discrete variables through propositional logic statements and automata and the interaction of both. The main idea of this approach is to embed the logical part in the state equations by transforming Boolean variables into 0–1 integers and by expressing the relationships as mixed-integer linear inequalities. Therefore MLD systems are able to model a broad class of systems arising in many applications: linear hybrid dynamic systems, hybrid automata, non-linear dynamic systems where the non-linearity can be approximated by a piecewise linear function, and some classes of discrete event systems. The general MLD representation is given by

$$\mathbf{x}(k+1) = \Phi\mathbf{x}(k) + \mathbf{G}_1\mathbf{u}(k) + \mathbf{G}_2\varsigma(k) + \mathbf{G}_3\mathbf{z}(k)$$

$$\mathbf{y}(k) = \mathbf{H}\mathbf{x}(k) + \mathbf{D}_1\mathbf{u}(k) + \mathbf{D}_2\varsigma(k) + \mathbf{D}_3\mathbf{z}(k) \tag{6.20}$$

$$\mathbf{E}_2\varsigma(k) + \mathbf{E}_3\mathbf{z}(k) \leq \mathbf{E}_1\mathbf{u}(k) + \mathbf{E}_4\mathbf{x}(k) + \mathbf{E}_5$$

where $\mathbf{x}(k) \in \mathbb{R}^{n_c} \times \{0,1\}^{n_l}$ is a vector of continuous and binaries states, $\mathbf{u}(k) \in \mathbb{R}^{m_c} \times \{0,1\}^{m_l}$ are the inputs, $\mathbf{y}(k) \in \mathbb{R}^{p_c} \times \{0,1\}^{p_l}$ the outputs, $\varsigma(k) \in \{0,1\}^{r_l}$, $\mathbf{z}(k) \in \mathbb{R}^{r_c}$, represent auxiliary binary and continuous variables, respectively, which are entered when transforming logical relationships into mixed-integer linear inequalities and Φ, $\mathbf{G}_1$, $\mathbf{G}_2$, $\mathbf{G}_3$, $\mathbf{H}$, $\mathbf{D}_1$, $\mathbf{D}_2$, $\mathbf{D}_3$, $\mathbf{E}_1$, $\mathbf{E}_2$, $\mathbf{E}_3$, $\mathbf{E}_4$, $\mathbf{E}_5$ are matrices of suitable dimensions.

There are other techniques to model hybrid systems, like Linear Complementarity (LC) systems, PieceWise Affine (PWA) systems, Extended Linear Complementarity (ELC) systems, and Max-Min-Plus-Scaling (MMPS). However, it has been proved that the resulting different hybrid models, using the different techniques enumerated previously, could be considered mathematically equivalent under well-determined suppositions [173, 174]. So, the MLD framework has been chosen, due to its wide use in modeling hybrid systems and to be able to use the MLD model, in the future, to develop a controller for the plant [61, 126, 422].

The following describes the different steps to obtain a MLD model of the TSA system.

Operating Modes

As mentioned above, hybrid dynamics deals precisely with systems that result from the interconnection of differential equations with logic-based decision rules. These logic-based decision rules cause the hybrid systems to have different operating modes as commented on in previous sections. The different TSA system operating modes are based on the difference between blower G1 air mass flow ($\dot{m}_1$) and the blower G2 air mass flow ($\dot{m}_2$). These operation modes are summarized as follows:

1 Receiver 100% / Thermal storage 0% / Steam generator 100%: In this operating mode, all the power supplied by the receiver is sent to the steam generator. The air mass flow from the two blowers is the same, so it does not go through thermal storage in either direction.

2 Receiver 0% / Thermal storage 100% / Steam generator 100%: No air mass flow is supplied by blower G1, therefore, all the power supplied to the steam generator comes from thermal storage and this power depends exclusively on the blower G2 air mass flow.

3 Receiver 100% / Thermal storage 100% / Steam generator 0%: In this configuration, all the receiver outlet power is used to charge thermal storage. Blower G1 is running and blower G2 is stopped, so there is no air mass flow through the steam generator branch.

Table 6.2 Air mass flow through the system branches in the different operation modes (courtesy of J.D. Álvarez et al., [10])

	Receiver branch	Thermal storage branch
Mode 1	$\dot{m}_1 = \dot{m}_2$	$0 = \dot{m}_2 - \dot{m}_1$
Mode 2	$\dot{m}_1 = 0$	$\dot{m}_2 = \dot{m}_2 - \dot{m}_1$
Mode 3	$\dot{m}_2 = 0$	$0 = \dot{m}_2$
Mode 4	$\dot{m}_1$	$\dot{m}_2 - \dot{m}_1$
Mode 5	$\dot{m}_2$	0

4 Receiver $n\%$ / Thermal storage $(100 - n)\%$ / Steam generator 100%: Thermal storage is discharging and the power supplied to the steam generator comes from both the receiver and the thermal storage. The blowers are running at different speeds, but blower G2 speed must be greater than blower G1 speed.
5 Receiver 100% / Thermal storage $n\%$ / Steam generator $(100 - n)\%$: This operating mode, which charges thermal storage, is used when the receiver outlet power is higher than power demanded, so the part not needed is sent to thermal storage. So, all the power to steam generator comes from the receiver branch. For this, the two blowers are switched on and the air mass flow supplied by blower G1 must be greater than the blower G2 air mass flow.

Logical Rules and Automata

From the section above, it is easy to create some Boolean rules to detect the system modes based on the difference between the mass flows in the two blowers. These Boolean rules can be described in the following way:

1 $\dot{m}_1 = \dot{m}_2 \neq 0 \rightarrow (equalG1G2 \wedge \neg zeroG1 \wedge \neg zeroG2)$
2 $\dot{m}_2 \neq 0$ and $\dot{m}_1 = 0 \rightarrow (zeroG1 \wedge \neg zeroG2)$
3 $\dot{m}_1 \neq 0$ and $\dot{m}_2 = 0 \rightarrow (zeroG2 \wedge \neg zeroG1)$
4 $\dot{m}_2 > \dot{m}_1$ and $\dot{m}_1 \neq 0 \rightarrow (G2gtG1 \wedge \neg zeroG1)$
5 $\dot{m}_1 > \dot{m}_2$ and $\dot{m}_2 \neq 0 \rightarrow (G1gtG2 \wedge \neg zeroG2)$

where *equalG1G2* means that the air mass flow is the same in blowers G1 and G2, *zeroGi* indicates that the air mass flow in blower Gi is 0 and *GjgtGi* means that the air mass flow in blower Gj is greater than that in blower Gi.

With these Boolean rules and the different operating points, system automata, such as those shown in Fig. 6.35(a), can be found. In each different operating point, the air mass flow through the branches changes as shown in Table 6.2, where the mass flow in the branches which goes to the heat exchanger is equal to blower G2 air mass flow, $\dot{m}_2$, for all the operating modes.

However, careful examination of the DHA in Fig. 6.35(a) reveals that modes 2 and 3 are particular cases of modes 4 and 5, respectively. Hence, the previous automata can be reduced to a new one with only three modes to simplify the representation. The reduced automata are shown in Fig. 6.35(b) and the air mass flow through the branches is shown in Table 6.3. The dynamics of different states of these automata is described in the next section.

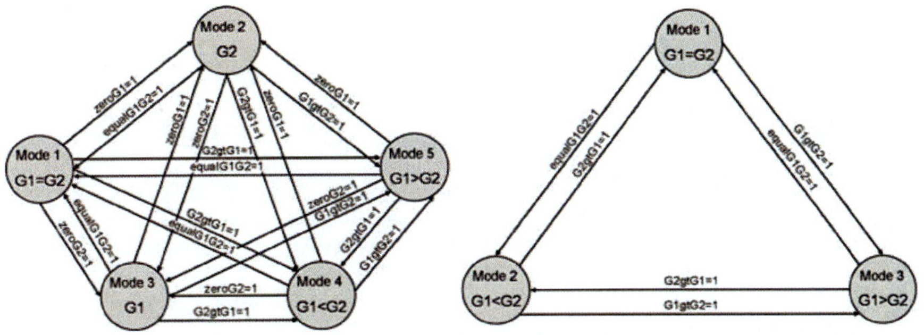

Fig. 6.35 Discrete hybrid automata (DHA) and reduced DHA (courtesy of J.D. Álvarez et al., [10])

Table 6.3 Air mass flow through the system branches for the reduced DHA (courtesy of J.D. Álvarez et al., [10])

	Receiver branch	Thermal storage branch
Mode 1	$\dot{m}_1 = \dot{m}_2$	$0 = \dot{m}_2 - \dot{m}_1$
Mode 2	$\dot{m}_1$	$\dot{m}_2 - \dot{m}_1$
Mode 3	$\dot{m}_2$	0

TSA Steam Generator Inlet Air Power Dynamics

As described above, the TSA steam generator inlet air power dynamics can be represented by the pipe model in Eq. (6.17). However, and as also described above, two pipe models are necessary, one for the receiver branch and another one for the thermal storage branch, where the inputs of these models change with the operating mode as described in Eq. (6.18). So, from the discrete-time version on z-domain of the transfer function in Eq. (6.17), these two pipe models can be represented by their equivalent space-state representation as follows:

$$\begin{aligned} \dot{\mathbf{x}}_r &= \mathbf{A}_r\mathbf{x}_r + \mathbf{B}_r\mathbf{u}_r, & \dot{\mathbf{x}}_{st} &= \mathbf{A}_{st}\mathbf{x}_{st} + \mathbf{B}_{st}\mathbf{u}_{st} \\ \mathbf{y}_r &= \mathbf{C}_r\mathbf{x}_r + \mathbf{D}_r\mathbf{u}_r, & \mathbf{y}_{st} &= \mathbf{C}_{st}\mathbf{x}_{st} + \mathbf{D}_{st}\mathbf{u}_{st} \end{aligned} \tag{6.21}$$

where the subscripts r and st denote the receiver branch and the thermal storage branch, respectively; $\mathbf{x}$ and $\mathbf{y}$ represent the fluid power states and outputs due to the velocity and $\mathbf{A}$, $\mathbf{B}$, $\mathbf{C}$ and $\mathbf{D}$ are matrices of corresponding size.

Consider the following extended state vector including $\mathbf{x}_r$, $\mathbf{x}_{st}$:

$$\mathbf{x}_g = [\mathbf{x}_r\ \mathbf{x}_{st}]^T \tag{6.22}$$

It is possible to include the dynamics of the two branches in a single overall model represented by Eq. (6.23):

$$\begin{aligned}\dot{\mathbf{x}}_g &= \mathbf{A}_g\mathbf{x}_g + \mathbf{B}_g\mathbf{u}_g \\ \mathbf{y}_g &= \mathbf{C}_g\mathbf{x}_g + \mathbf{D}_g\mathbf{u}_g\end{aligned} \tag{6.23}$$

where $\mathbf{A}_g$, $\mathbf{B}_g$, $\mathbf{C}_g$, $\mathbf{D}_g$ are diagonal matrices easily found from Eq. (6.21) as

$$\begin{aligned}\mathbf{A}_g &= \begin{bmatrix}\mathbf{A}_r & 0 \\ 0 & \mathbf{A}_{st}\end{bmatrix}, & \mathbf{B}_g &= \begin{bmatrix}\mathbf{B}_r & 0 \\ 0 & \mathbf{B}_{st}\end{bmatrix}, \\ \mathbf{C}_g &= [\mathbf{C}_r \quad \mathbf{C}_{st}], & \mathbf{D}_g &= [\mathbf{D}_r \quad \mathbf{D}_{st}]\end{aligned} \tag{6.24}$$

On the other hand, the input model $\mathbf{u}_g$ is a variable, including system dynamics, which changes in the following way:

$$\mathbf{u}_g = \big[\underbrace{\big(\dot{m}_1\alpha_d + \dot{m}_2(1-\alpha_d)\big)}_{\mathbf{u}_r} \ \underbrace{\big((\dot{m}_2 - \dot{m}_1)\alpha_d\big)}_{\mathbf{u}_{st}}\big]^T \tag{6.25}$$

where $\alpha_d \in [0, 1]$ is the discrete variable allowing the commutation between operating modes. The value of α_d at each sampling time is found directly from the Boolean rules and is described as

$$\alpha_d = \begin{cases}1 & \text{if } (equalG1G2 \vee G2gtG1) \\ 0 & \text{if } G1gtG2\end{cases} \tag{6.26}$$

Therefore, in view of the above, the TSA steam generator inlet air power dynamics can be modeled as a hybrid system, where the input is switched based on the relationship between blower G1 air mass flow, $\dot{m}_1$ and blower G2 air mass flow, $\dot{m}_2$. Hence, the whole hybrid description of the system can be expressed as Eq. (6.20):

$$\begin{aligned}\dot{\mathbf{x}}_g &= \begin{bmatrix}\mathbf{A}_r & 0 \\ 0 & \mathbf{A}_{st}\end{bmatrix}\begin{bmatrix}\mathbf{x}_r \\ \mathbf{x}_{st}\end{bmatrix} + \begin{bmatrix}\mathbf{B}_r & 0 \\ 0 & \mathbf{B}_{st}\end{bmatrix}\begin{bmatrix}\dot{m}_1\alpha_d + \dot{m}_2(1-\alpha_d) \\ (\dot{m}_2 - \dot{m}_1)\alpha_d\end{bmatrix} \\ \mathbf{y}_g &= [\mathbf{C}_r \quad \mathbf{C}_{st}]\begin{bmatrix}\mathbf{x}_r \\ \mathbf{x}_{st}\end{bmatrix} + [\mathbf{D}_r \quad \mathbf{D}_{st}]\begin{bmatrix}\dot{m}_1\alpha_d + \dot{m}_2(1-\alpha_d) \\ (\dot{m}_2 - \dot{m}_1)\alpha_d\end{bmatrix} \\ \text{s.t.} \quad & \alpha_d = (\dot{m}_2 \geq \dot{m}_1)\end{aligned} \tag{6.27}$$

In the next paragraph an abstract representation of the model in Eq. (6.27) is found as the equivalent MLD model useful for control purposes.

Mixed Logical Dynamical (MLD) Model

As mentioned above, MLD representations are the standard way to model hybrid systems for control purposes. The rules and properties proposed in [174, 253] are used to translate the TSA hybrid system into the MLD representation using the MATLAB hybrid toolbox [32]. The resulting MLD model for the discrete version of the system is shown in Eq. (6.28):

$$\begin{aligned}\mathbf{x}(k+1) &= \Phi\mathbf{x}(k) + \mathbf{G}_3\mathbf{z}(k) \\ \mathbf{y}(k) &= \mathbf{H}\mathbf{x}(k) \\ \mathbf{E}_2\varsigma(k) + \mathbf{E}_3\mathbf{z}(k) &\leq \mathbf{E}_1\mathbf{u}(k) + \mathbf{E}_5\end{aligned} \tag{6.28}$$

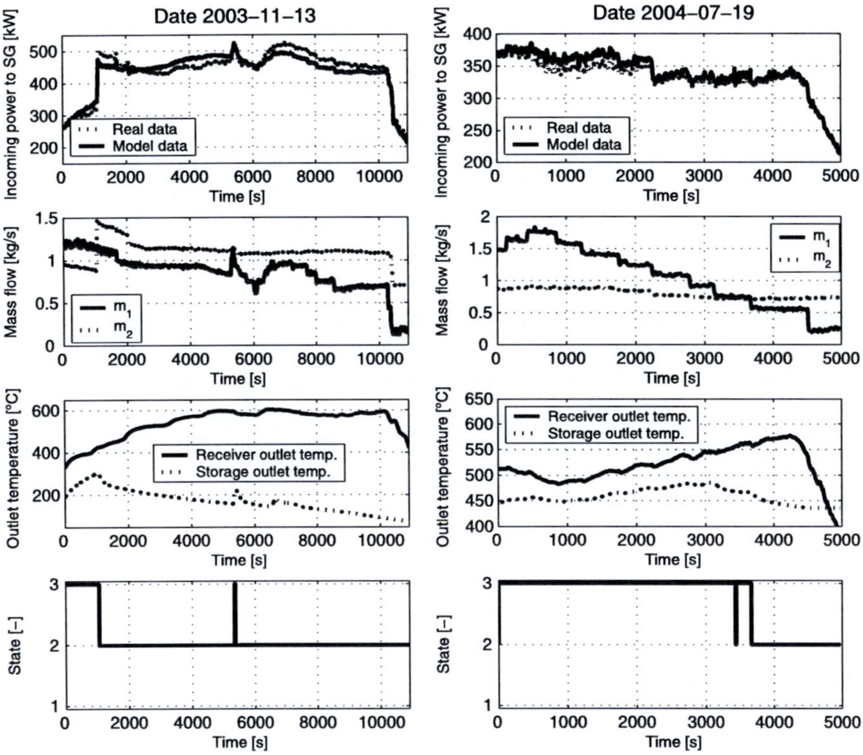

Fig. 6.36 Comparison of MLD results versus real data (courtesy of J.D. Álvarez et al., [10])

where values of matrices $\boldsymbol{\Phi}$, $\mathbf{H}$, $\mathbf{G}_3$, $\mathbf{E}_1$, $\mathbf{E}_2$, $\mathbf{E}_3$, and $\mathbf{E}_5$ are presented in the Appendix and in [10].

The MLD model was validated by comparing simulations with real data. The results of the comparison are shown in Fig. 6.36 with model inputs, output and disturbance signals as in Fig. 6.33, as well as model state signal. It may be observed that the MLD model faithfully captures the system dynamics in the different operating modes. The MLD model also ensures continuous transitions between the operating modes, which is a very important feature for control purposes.

Figure 6.36 shows simulations corresponding to two different operation days. At the beginning of the first operation day (date and label 2003-11-13) the air mass flow from blower G1, $\dot{m}_1$, is greater than the air mass flow supplied by blower G2, $\dot{m}_2$, so the MLD model is in mode 3 according to the DHA shown in Fig. 6.35(b). After that, $\dot{m}_2$ increases and exceeds the value of $\dot{m}_1$, passing the MLD model to mode 2. Finally in the middle of the operation day $\dot{m}_1$ is again greater than $\dot{m}_2$, for a short time, and this causes the MLD model to go into mode 3, $\dot{m}_1$ quickly drops to less than $\dot{m}_2$ and the MLD model goes back into mode 2 again. It is important to notice that, in this operation, where most of the time the plant is in mode 2 ($\dot{m}_1 < \dot{m}_2$), the changes in $\dot{m}_1$ are reproduced by the steam generator inlet power, because through the thermal storage branch there is an air mass flow with the value of $(\dot{m}_2 - \dot{m}_1)$, according to Table 6.3.

On the second operation day (date and label 2004-07-19), as the same that the previously one, $\dot{m}_1$ is greater than $\dot{m}_2$ at the beginning, so the MLD model is in mode 3. Afterwards $\dot{m}_1$ value progressively decreases and for the time period over 3200–3700 s both $\dot{m}_1$ and $\dot{m}_2$ have almost the same value, and this fact causes the MLD model to go from mode 3 to mode 2 and vice versa for a couple of times. Finally, the last 1500 s, $\dot{m}_2$ is greater than $\dot{m}_1$ and MLD model remains in mode 2. In this operation day, with different features from the previous one, most of the time the plant is in mode 3, where $\dot{m}_1 > \dot{m}_2$, so, the changes in $\dot{m}_1$ do not affect to the steam generator inlet power since through the thermal storage branch there is not any air mass flow, according to Table 6.3. This system feature can only be reproduced by a hybrid model, like the MLD one developed. Therefore, this MLD model can be used for control or design purposes. A great advantage to design a controller based on this kind of hybrid model is that the controller is able to take into account the different operating modes and discrete events which are used to switch between modes.

6.8.4 Hybrid Control of the TSA System

The MLD model allows developing a MPC controller for the system. Due to restrictions in the use of the CESA-1 installation, only simulation tests are shown in this section [9]. Using the Hybrid Toolbox [32] it is possible to develop a receding horizon MPC controller solving a Mixed-Integer Programming (MIP) optimization problem [33]. For those cases in which on-line optimization is not possible, multiparametric programming may be useful for optimal PWA controller synthesis. The optimal control problem can be formulated as follows:

$$\min_{\{\mathbf{u},\varsigma,\mathbf{z}\}_0^{N-1}} J\left(\{\mathbf{u},\varsigma,\mathbf{z}\}_0^{N-1},\mathbf{x}(k)\right)$$

$$\triangleq \left\|\mathbf{Q}_{xT}\left(\mathbf{x}(k+N|k)-\mathbf{x}_r\right)\right\|_p + \sum_{j=1}^{N-1}\left\|\mathbf{Q}_x\left(\mathbf{x}(k+j)-\mathbf{x}_r\right)\right\|_p$$

$$+ \sum_{j=0}^{N-1}\left\|\mathbf{Q}_u\left(\mathbf{u}(k)-\mathbf{u}_r\right)\right\|_p + \left\|\mathbf{Q}_z\left(\mathbf{z}(k+j)-\mathbf{z}_r\right)\right\|_p + \left\|Q_y\left(y(k+j)-r\right)\right\|_p \tag{6.29}$$

such that

$$\begin{cases} \mathbf{x}(0 \mid k) = \mathbf{x}(k) \\ \mathbf{x}(k+j+1 \mid k) = \mathbf{A}\mathbf{x}(k+j \mid k) + \mathbf{B}_1\mathbf{u}(k) + \mathbf{B}_2\varsigma(k+j \mid k) + \mathbf{B}_3\mathbf{z}(k+j \mid k) \\ \mathbf{y}(k+j \mid k) = \mathbf{C}\mathbf{x}(k+j \mid k) + \mathbf{D}_1 u(k) + \mathbf{D}_2\varsigma(k+j \mid k) + \mathbf{D}_3\mathbf{z}(k+j \mid k) \\ \mathbf{E}_2\varsigma(k+j \mid k) + \mathbf{E}_3\mathbf{z}(k+j \mid k) \leq \mathbf{E}_1\mathbf{u}(k) + \mathbf{E}_4\mathbf{x}(k+j \mid k) + \mathbf{E}_5 \\ \mathbf{u}_{min} \leq \mathbf{u}(k+j) \leq \mathbf{u}_{max}, \ j = 0, 1, \ldots, N-1 \\ \mathbf{x}_{min} \leq \mathbf{x}(k+j \mid k) \leq \mathbf{x}_{max}, \ j = 1, \ldots, N \\ \mathbf{y}_{min} \leq \mathbf{y}(k+j) \leq \mathbf{y}_{max}, \ j = 0, 1, \ldots, N-1 \\ \mathbf{S}_{xx}(k+N \mid k) \leq \mathbf{T}_x \end{cases} \tag{6.30}$$

where $\mathbf{y}(k)$ is the system output in discrete time k, $\mathbf{r}$ is the reference, $\mathbf{x}_r$, $\mathbf{u}_r$ and $\mathbf{z}_r$ are the references to states, inputs and auxiliary variables, N is the prediction horizon, $\mathbf{x}(k+j \mid k)$ is the state predicted for time $k+j$ at sampling time k resulting from input $\mathbf{u}(k+j)$, $\mathbf{u}_{min}$, $\mathbf{u}_{max}$, $\mathbf{y}_{min}$, $\mathbf{y}_{max}$ and $\mathbf{x}_{min}$, $\mathbf{x}_{max}$ are the limits for inputs, outputs and states, respectively; $\{\mathbf{x}\colon \mathbf{S}_x\mathbf{x} \leq \mathbf{T}_x\}$ is a polyhedrical subset of the final state space $\mathbb{R}^n$. The weighting factors in Eqs. (6.29) and (6.30) have the same interpretation as those of a typical GPC controller: $\mathbf{Q}_y$ and $\mathbf{Q}_x$ have the same effect: δ and they should equal 1, $\mathbf{Q}_u$ is the control signal weighting factor, equivalent to λ in GPC. $\mathbf{Q}_{xT}$ is a weighting factor penalizing the state at the end of the horizon ($j = N$) if this is far from the reference.

6.8.4.1 Illustrative Results

The variable to be controlled is the air power at the input of the steam generator, while the manipulated variable is the mass flow provided by blower G2 $\dot{m}_2$. In this case, the mass flow given by blower G1 $\dot{m}_1$, that is, the manipulated variable of the controller described in the previous subsection, is considered as a disturbance, because its use is only devoted to maintain a desired air temperature at the receiver outlet.

Figure 6.37 shows two illustrative simulations. It can be seen how the closed-loop time constant (relating air power to blower G2 mass flow) is quite small (when $\dot{m}_2 > \dot{m}_1$). This feature allows rejecting disturbances (mainly those coming from changes in $\dot{m}_1$ trying to control the air temperature at the receiver outlet). The transfer functions relating blower G1 mass flow and outlet receiver temperature are encapsulated into the hybrid model of the air circuit, so that the controller is able to compensate for their changes. When $\dot{m}_2 < \dot{m}_1$ all the power comes from the receiver (Table 6.2) using a flow at the steam generator input equal to $\dot{m}_2$. This is the main reason of why a multivariable controller has not been used, because it could not account for the different operating modes.

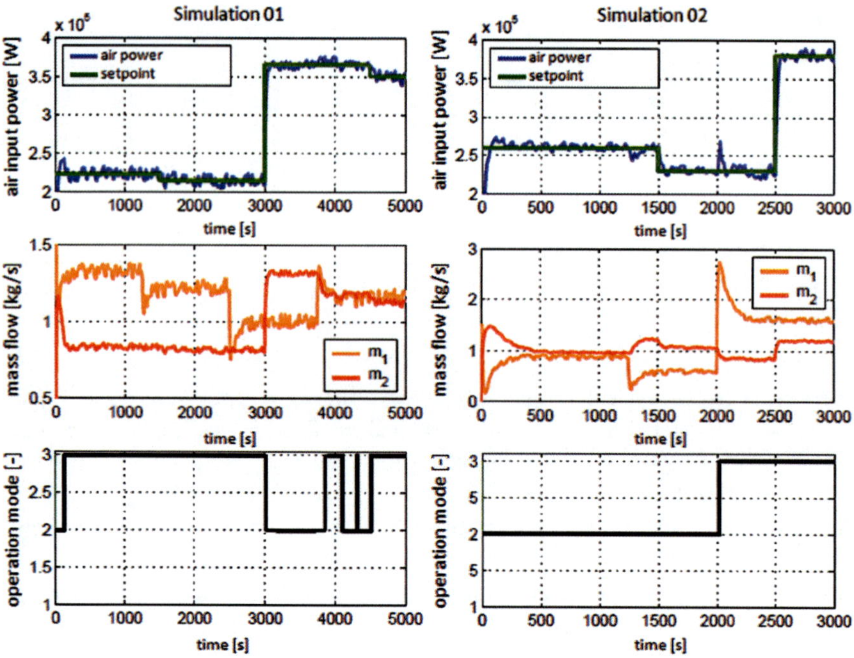

Fig. 6.37 Results with the hybrid controller (courtesy of J.D. Álvarez, [9])

6.8.5 Steam Generator Control

The TSA steam generator has a once-through configuration to transfer energy from hot air to water. This heat exchanger has been modeled following the same approaches explained in Chap. 4, obtaining both non-linear and black-box linear models obtained from Taylor series expansions and parametric identification validating use of multisine signals [66, 255, 316, 317]. Figure 6.38 shows a comparison between the frequency response of the non-linear model and that of a linear model of order 20 (also reduced order models have been obtained for controller synthesis purposes). Notice that resonances are not so important in this case and robust PID controllers are able to control this kind of system [277].

6.9 Summary

This chapter has dealt with modeling and control problems associated to thermosolar plants with central receiver system. All the components of a typical installation have been explained, as well as the associated control schemes: control of the heliostat field (and offset correction problems), control of the receiver and control of the power stage. Illustrative examples of control solutions for the CESA-1 plant of

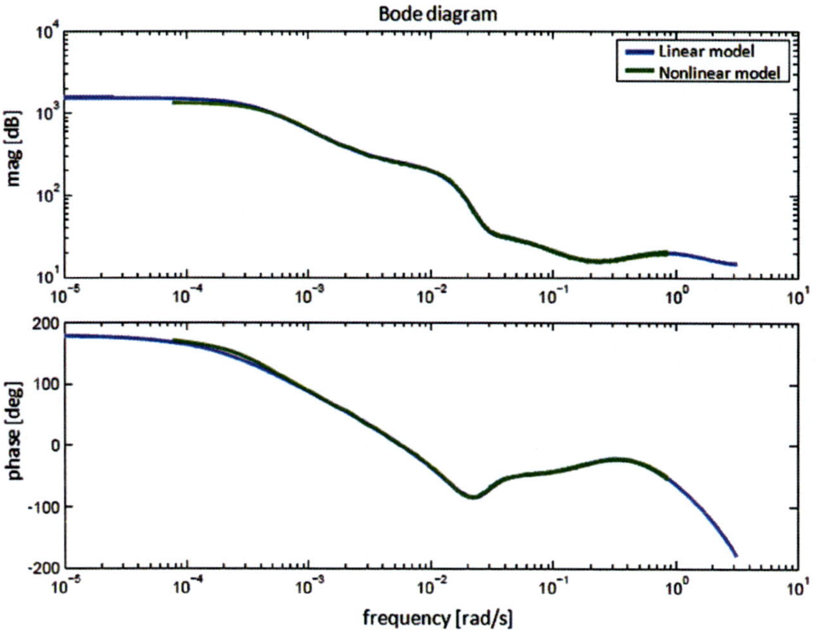

Fig. 6.38 Comparison of the frequency response of non-linear and linear models of the steam generator (courtesy of J.D. Álvarez, [9])

the PSA have been included. These algorithms have served as a reference for the industrial facilities that have been put into operation recently.

Chapter 7
Other Solar Applications

7.1 Introduction

For the last twenty years, solar energy has been used for a wide range of applications. In [205, 371], a review of solar thermal technologies is carried out. Two of the most promising fields for solar energy are that of solar furnaces for material testing and treatment and solar-based refrigeration systems.

A solar furnace is a high concentrating facility made up of a collector system with tracking (usually with a varying number of flat-faceted heliostats) and a static parabolic concentrating system at the focal spot of which a high percentage of the solar energy collected by the collector system is concentrated within a small area. One attenuator (shutter) can be used between the collector system and the concentrator to control the amount of energy used for heating samples placed at the focal spot. A test table, movable in three dimensions, is placed in the area of the focal spot within the test zone. Solar furnaces have received great attention since the 1950s [141, 158, 265, 302, 375]. An excellent description and overview of different solar furnaces can be found in [240], where open loop computer-based control systems are used [157].

Tests in a solar furnace usually aim at improving the mechanical properties such as hardness and wear resistance by melting and/or casting different samples (steel, cast-iron, ceramic composites such as alumina, etc.) [320, 368, 403], sintering [5] or processing of silicon cells [140, 376], by means of heating the samples following many different temperature patterns. The study of physical-thermal properties of materials at high temperatures can be made with innovative treatments impossible to carry out using conventional heating processes, improving the results of industrial material treatments such as laser surface tempering in the case of materials which have to work under very severe conditions.

Due to the complexity and diversity of sample materials and temperature trends such research plants are usually manually controlled by expert operators. Obviously, the efficiency of the operations depends on the operators' skill and, therefore, the presence of a properly designed automatic control system would have the advantage of providing adequate results for different operating conditions. In Sect. 7.2 different modeling and control approaches applied to solar furnaces are developed.

E.F. Camacho et al., *Control of Solar Energy Systems*, Advances in Industrial Control,

DOI 10.1007/978-0-85729-916-1_7,

7.2 Solar Furnaces

7.2.1 Introduction

This section presents automatic control system strategies for controlling the temperature of a solar furnace. As has been pointed out, solar furnaces are manually controlled by skilled operators due to the variety of sample materials and temperature profiles. In recent years, different control strategies have been developed to allow automatic control of solar furnaces, ranging from adaptive control [13, 43, 120, 121], fuzzy logic control [220, 221] and predictive control [48]. Many of these techniques are based on the physical model developed in [43].

This section shows the results obtained in the application of proportional-integral (PI) and fuzzy logic controllers (FLC) to a solar furnace. In the case of PI controllers, both fixed and adaptive versions of the controllers have been developed, incorporating feedforward (FF) action, anti-windup and slew-rate constraint handling mechanisms.

From the control viewpoint, a solar furnace is a system which presents several interesting characteristics making the control problem a difficult task:

- The characteristics of the samples are quite different depending on their nature (steel, alumina, etc.). Obtaining a fixed parameter controller which allows different samples to be controlled becomes a difficult task.
- The dynamic characteristics of each sample greatly depend on the temperature and introduce a high non-linearity which makes the behavior of the system controlled change with the operating conditions.
- The control specifications are quite severe (rate of temperature increase, rate of temperature decrease, variable step changes, etc.) and have to be achieved with small errors.
- The system suffers from strong disturbances caused by solar irradiance variations (slow variations due to the daily cycle or fast and strong variations due to passing clouds), which make the exact reproduction of the conditions of a determined test impossible.
- Limitations exist in the maximum temperature achievable by the materials and different constraints (non-linearities) in the actuator (amplitude, slew rate, etc.).

Recently, there have been extensions to solar-driven thermochemical processes [293], in which high-temperature process heat is supplied by concentrated solar energy, providing an efficient route for fuel and material production. In this case, a linear feedback controller was implemented using an optimal control design method (LQG/LTR).

7.2.2 The Solar Furnace at the PSA

The PSA Solar Furnace is mainly devoted to material treatment. Samples of different kinds of material have been quenched or sintered at its high flux focal spot, improv-

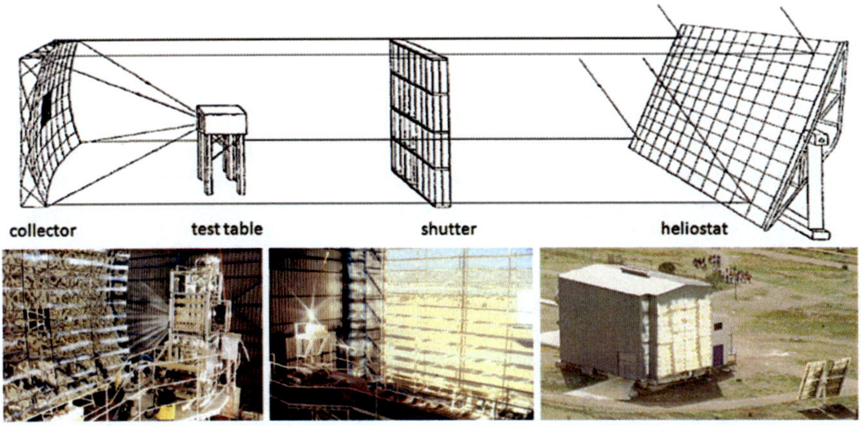

Fig. 7.1 Solar furnace (courtesy of PSA)

ing their hardening properties and wear resistance. Many research institutions have carried out several test campaigns on metallic and ceramic samples [321].

The PSA solar furnace facility structure and layout is summarized below. Essentially, it is composed of a huge heliostat to reflect sunshine onto the furnace chamber. Its main components include a 150 m^2 heliostat which reflects sunlight onto the concentrator disk, with 90% reflectivity and controlled azimuth and pitch positioning mechanism, plus an implemented Sun tracking algorithm. There is also a louvered shutter (control actuator) made out of 30 steel parallel panels which allow the amount of incoming light to be adjusted onto the concentrator, with a dimension of 11.5 m × 11.2 m, and 15896 positions between 0° (open) and 55° (closed). At the back of the chamber, a concentrator convex mirror gathers most of the incoming sunlight from the outdoor heliostat onto a 22 cm diameter focal spot. The physical properties to be highlighted are that it has a 98.5 m^2 reflecting area with 94% reflectivity and a concentration peak of 3000 MW/m^2. Samples are placed on top of a precision 3-D adjusting table, capable of accurate positioning on three axes (space dimensions) with a work area comprising (0.86, 0.6, 0.5) m in dimensions X, Y, Z, respectively.

Figure 7.1 shows a model representation of the plant layout depicting the solar radiance concentration process and the main components involved. It also shows an outer view of the PSA solar furnace. It can be seen how the heliostat concentrates solar irradiance onto the shutter which controls the fraction of beam irradiance which goes into the furnace for material testing purposes. The fraction of the solar irradiance collected by the heliostats is incident on a static parabolic receiver to concentrate the irradiance onto a small surface. The sample is situated at the focal point and the mobile table serves to position it to collect maximum energy before starting the tests. Once the sample has been placed within the focus on which solar irradiance is concentrated, the table stands at a fixed position. The main system difficulties lie in the disturbances present and persistent in every solar power plant, i.e. the energy source itself is an endless disturbance (due to its daily and yearly cycles, as well as passing clouds). In addition, each distinct sample brings its own physical properties such as heating-up and down ratios or light absorption capacity. If all this were not enough, the aforementioned actuator adds a variable slew-rate nonlinearity to its intrinsic physical aperture limitation. From the control viewpoint it is a single-input single-output (SISO) system where the output variable is the sample temperature measured by welded thermocouples, while the input variable is the degree of aperture at the shutter that regulates the amount of sunlight incident on the concentrator.

When the task of controlling the temperature of the sample is manually performed by a skilled operator, efficiency in the operation and the results obtained depend on human capabilities. The development of an automatic control system for these kinds of plant presents many advantages, such as simplifying operation of the whole system. This is important, because the operator has to perform many tasks before and during operation, and test different kinds of sample and for different operating specifications.

7.2.3 Dynamical Models of the System

As a first step in the development of an automatic control scheme for the solar furnace, a simplified model of the system was obtained to be used for control design purposes. In this case, the model was obtained from first principles and compared to input/output data measured in the system. A simplified energy balance is introduced in the following. For the sake of clarity, the dependence on time of different variables has not been explicitly written. Figure 7.2(a) shows a schematic diagram of the energy balance (conservation principle), which is given by $P_i - P_r - P_c = dE/dt$, where E is the thermal energy of the sample, P_i is the input power that the sample receives (this term takes into account the incident flux attenuation due to mirror reflectivity, shutter aperture, etc.), P_r are radiation losses and P_c convection losses. The following paragraphs show the mathematical expressions of the different terms of the energy balance.

7.2.3.1 Input Thermal Power

The heliostats reflect a fraction $I_s = r_h I$ of the direct solar irradiance I coming from the Sun (r_h: mirror reflectivity). The shutter aperture limits the reflected solar irradiance which goes through it. The attenuation which the incident flux suffers is equal to the percentage of the total area of the panels (blind) which is open. The shutter opening is performed using an AC motor which rotates the axle to which the panels are linked between 0° (fully open) and 55° (fully closed). The variable given by the control program is the aperture percentage. As can be seen in Fig. 7.2(b) the relationship between the aperture angle (α) and the aperture (A) is $A = L_s(\sin\alpha_0 - \sin\alpha)$, where L_s is the length of the panel, α_0 is the angle at which

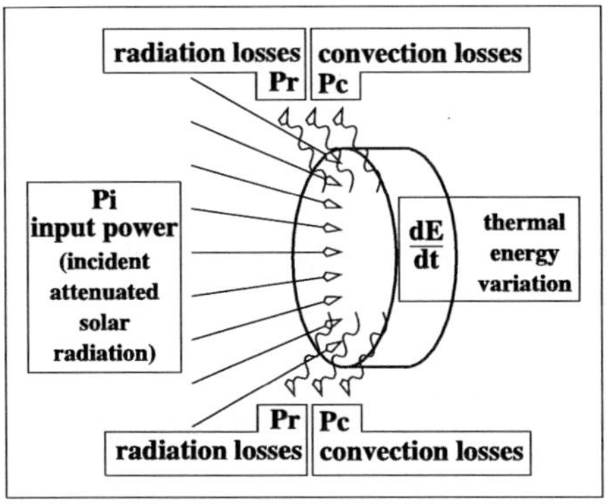

(a) Schematic of the energy balance

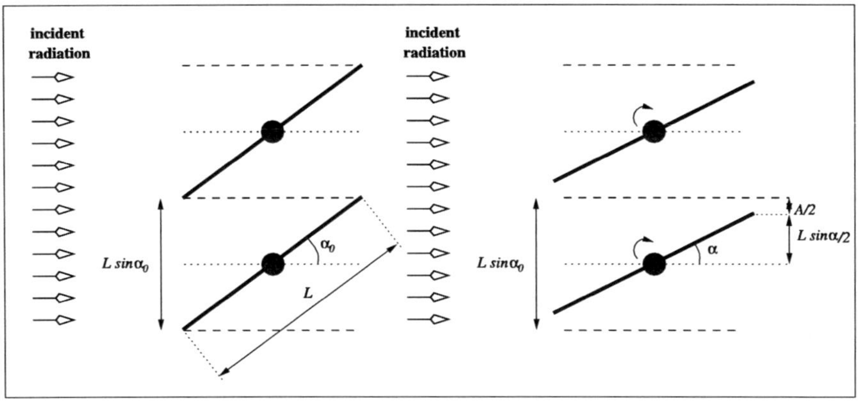

(b) Shutter aperture

Fig. 7.2 Energy balance and shutter aperture [43]

the shutter is completely closed (zero aperture) and α is the angle which indicates the aperture percentage. The encoders are such that the control signal to the motor which moves the shutter is the percentage of rotated angle (U) in respect to α_0, that is, $\alpha = (1-U/100)\alpha_0$, the aperture A related to the control signal U is given by $A = 1-(\sin[(1-U/100)\alpha_0]/\sin\alpha_0)$, where U is the input to the system. This conversion introduces a non-linearity, as the attenuation of the input flux does not linearly vary with the input but follows a sinusoidal relationship. The power density behind the shutter is $I_c = I_s A = I r_h[1-(\sin[(1-U/100)\alpha_0]/\sin\alpha_0)]$. This power density is collected by the concentrator in proportion to its surface S_c and is projected toward the focus with different losses due to its reflectivity r_c. The power obtained at the concentrator output is $P = I_c S_c r_c = I r_h r_c S_c[1-(\sin[(1-U/100)\alpha_0]/\sin\alpha_0)]$. The energy received by the sample depends upon its surface (S_s). Due to the flux distribution at the focus of the system, 90% of the power is concentrated within a circumference with a diameter of about 20 cm. Supposing that the energy flux is uniformly distributed within the focus (this is obviously an approximation, as in reality the distribution is of Gaussian type), the input power is

$$P_i = \frac{r_h r_c S_c}{S_{f(90\%)} \sin\alpha_0} I S_s \alpha_a \left[\sin\alpha_0 - \sin\left[(1 - U/100)\alpha_0\right]\right]$$

being $S_{f(90\%)}$ and α_a the focus area and absorption capacity of the sample, respectively.

7.2.3.2 Radiation and Convection Thermal Losses

The radiation losses depend on the emissivity of sample α_e, which is a parameter that determines the capacity of the sample to irradiate energy. The irradiated power is $P_r = \alpha_e \sigma S_s (T^4 - T_a^4)$ where σ is the Stephan–Boltzmann constant,[1] T_a is the environmental temperature and T is the temperature (supposedly uniform) of the sample given in Kelvin degrees. As can be seen, the previous relationship is of a non-linear nature.

Convection losses depend linearly on the sample temperature and on a constant α_c which indicates the capacity of the sample to interchange heat with the air. This constant depends on the position of the sample and on the properties of the air: temperature, viscosity, etc. The power lost by convection mechanisms can be approximated by $P_c = \alpha_c S_s (T - T_a)$.

7.2.3.3 Energy Balance

As has been previously pointed out, the energy balance is given by $P_i - P_r - P_c = dE/dt$ (where $E = m_s c_s T$, m being the mass of the sample, c_s the specific heat and T the temperature in Kelvin degrees). By substituting each one of the terms, we have

$$\begin{aligned} \frac{d(m_s c_s T)}{dt} &= \frac{r_h r_c S_c}{S_{f(90\%)} \sin\alpha_0} I S_s \alpha_a \left[\sin\alpha_0 - \sin\left[(1 - U/100)\alpha_0\right]\right] \\ &\quad - \alpha_e \sigma S_s \left(T^4 - T_a^4\right) - \alpha_c S_s (T - T_a) \end{aligned} \tag{7.1}$$

The energy variation in time is null at the equilibrium point (U_0, T_0). If a small perturbation around the equilibrium point is produced ($U = U_0 + u$ and $T = T_0 + \xi$), a linearized model which reproduces the behavior of the system around a determined operating point can be obtained and some consequences can be interpreted. Notice that all parameters which determine the thermal characteristics of a body depend on the temperature. Nevertheless, if small deviations from the equilibrium point are supposed, constant values of the parameters can be used. In reality, this is another focus of non-linearities in the system as, depending on the operating point, different values of these parameters are obtained. By merging Eq. (7.1) and the one

---

[1] $\sigma = 5.67 \cdot 10^{-8}$ W/(m^2 K^4).

that which can easily be obtained for the equilibrium point and by linearizing the sinusoidal term and all those terms with high powers, the following relationship can be obtained:

$$m_s c_s \frac{d\xi}{dt} = \frac{r_h r_c S_c \alpha_0}{100 S_{f(90\%)} \sin\alpha_0} I S_s \alpha_a \cos\big[(1 - U_0/100)\alpha_0\big] u \\ - 4\alpha_e \sigma S_s T_0^3 \xi - \alpha_c S_s \xi \tag{7.2}$$

The fraction in the first term in Eq. (7.2) is constituted of characteristic parameters of the system, independently of the type of material. By a few calculations, this term can be embedded in a constant $K_1 = (r_h r_c S_c \alpha_0)/(100 S_{f(90\%)} \sin\alpha_0) = 24.95$ for this system so that, if null initial conditions are supposed and the Laplace Transform is applied to this linearized model ($u(t) \rightarrow U(s)$; $\xi(t) \rightarrow T(s)$), the following transfer function can be obtained:

$$\frac{T(s)}{U(s)} = \frac{K_1 I S_s \alpha_a \cos[(1 - U_0/100)\alpha_0]}{m_s c_s s + 4\alpha_e \sigma S_s T_0^3 + \alpha_c S_s} \tag{7.3}$$

which represents a first-order model $T(s)/U(s) = K/(1 + \tau s)$, where

$$K = \frac{K_1 \alpha_a I \cos[(1 - U_0/100)\alpha_0]}{4\sigma\alpha_e T_0^3 + \alpha_c} \quad \text{and} \quad \tau = \frac{m_s c_s}{S_s(4\sigma\alpha_e T_0^3 + \alpha_c)} \tag{7.4}$$

Some simple but important conclusions can be drawn from this simplified model:

- The gain of the system depends proportionally on the solar irradiance and the linearized system time constant is independent of it.
- The variation in the gain of the system with the cosine of a term involving the initial aperture angle and the equilibrium operating point aperture models the shutter non-linearity, so that, the higher the aperture of the shutter the higher the system gain.
- Both the system gain and the characteristic time constant depend inversely on the cube of the temperature. Furthermore, samples with high emissivity and convection loss rate values (great capacity of the material to lose energy) have lower values of the system gain and time constant than those samples with low emissivity and convection loss rate values.
- The value of the area of the sample exposed to the concentrated solar irradiance does not influence the value of the gain of the simplified model. Nevertheless, the value of the time constant is influenced by the area of the sample, so that for samples of the same material, the larger the area of the sample the smaller the corresponding time constant.
- Those samples with high specific heat lead to an increase in the time constant of the system.
- The ambient temperature does not influence the linearized transfer function of the system.

Notice that all these conclusions are valid under the assumptions made to obtain the linearized model. As indicative values, Table 7.1 shows typical gain and time constant ranges estimated using data from different tests performed on different

Table 7.1 Example of off-line estimated parameters, K in °C/%, τ in s

	Zone A (200–400°C)	Zone B (400–600°C)	Zone C (600–800°C)	Zone D (800–1000°C)
Small steel sample	$K = [70, 50]$	$K = [55, 30]$	$K = [35, 20]$	$K = [25, 15]$
	$\tau = [100, 85]$	$\tau = [90, 45]$	$\tau = [45, 30]$	$\tau = [40, 25]$
A 316-L steel	$K = [140, 100]$	$K = [105, 65]$	$K = [70, 50]$	$K = [55, 35]$
	$\tau = [120, 100]$	$\tau = [105, 80]$	$\tau = [85, 70]$	$\tau = [75, 60]$
White zirconia	$K = [65, 50]$	$K = [55, 45]$	$K = [50, 30]$	$K = [40, 20]$
	$\tau = [130, 125]$	$\tau = [125, 120]$	$\tau = [120, 115]$	$\tau = [115, 100]$
Silicon carbide	$K = [80, 60]$	$K = [60, 40]$	$K = [45, 20]$	$K = [20, 10]$
	$\tau = [150, 130]$	$\tau = [135, 125]$	$\tau = [130, 120]$	$\tau = [120, 100]$

kinds of sample under various temperature conditions (as commented at the end of the section).

Another aspect relevant to modeling the system refers to the shutter activating mechanism (AC motor) which introduces two non-linearities: one of saturation type with minimum (0%) and maximum (100%) aperture and the other of slew-rate type, where the aperture rate response of the shutter is fixed at 5% per second, taking 20 s to open from 0% to 100%. These non-linearities influence the behavior of the different control schemes. As is well known, the existence of saturation non-linearities within control loops incorporating integral action can cause the integral term to achieve undesired values producing long-lasting oscillations when set point changes are performed (if adequate anti-windup mechanisms are not included). The effect of the slew-rate constraints is also harmful from the control viewpoint, as the limitation of the speed of response can also lead to undesired performance of the controlled system.

It is difficult to validate these types of model, because the thermodynamic characteristics of the materials tested vary with temperature and quite often information including tables with these characteristics (specific heat, emissivity, absorption coefficient, etc.) is not provided for such a wide range of temperatures as that covered in these types of application. Different simplified approaches using input/output data from manual tests have been used to validate these models. These approaches have been based on open loop step and pseudo random binary sequence (PRBS) tests used to obtain input/output data for model validation purposes. Step tests are useful to validate first order models (usually carried out by the reaction-curve method) as it is simple to estimate the gain and characteristic time constant of the system by inspection of the response of the system. They are also useful for studying the variation of the system characteristic parameters under operating conditions, such as changes in working (temperature) conditions, solar radiation, etc. For example, Fig. 7.3 shows open loop step tests performed with a small steel sample. A zoom of several zones of the plot has been included showing both the real response and that obtained by using first-order models with the estimated parameters shown in Table 7.2. These characteristic parameters have been obtained for different operating

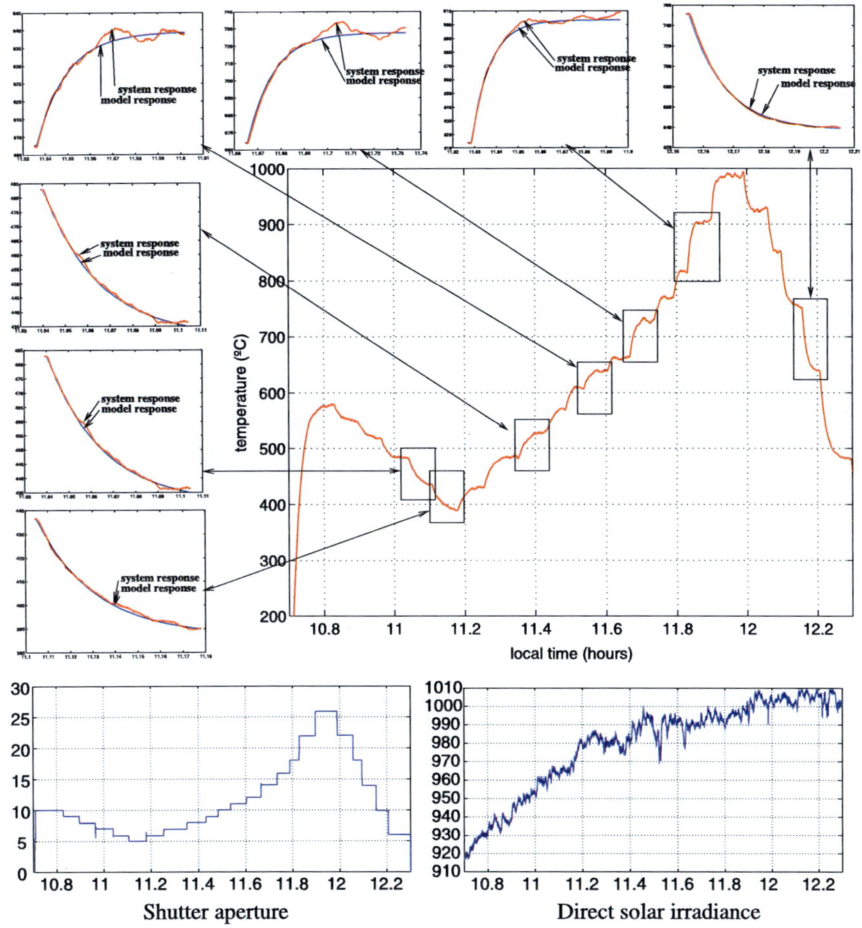

Fig. 7.3 Step tests for model validation [43]

conditions (given by mean temperature T_0 in degrees Celsius, mean shutter aperture U_0 and mean direct solar irradiance I_0) by identifying each of the step responses shown in Fig. 7.3. This table provides information for model validation purposes (notice that these values are subject to errors as they have been estimated from input/output data). As can be seen, slightly different dynamic behavior is obtained when heating the samples than in the case of cooling them. The values of both the estimated steady state gain and characteristic time constant decrease as temperature increases and for a similar value of the temperature conditions, the value of the static gain increases as the mean solar irradiance augments. The static gain should also increase as the shutter aperture augments, but the corresponding temperature increase counteracts this effect. These conclusions are in accordance with those obtained from the energy balance.

Table 7.2 Estimated model parameters for a steel sample

Heating-up steps					Cooling-down steps				
T_0	U_0	I_0	K	τ	T_0	U_0	I_0	K	τ
410	5.9	981	55	80	413	5.0	966	55	92
458	6.9	980	53	74	459	5.9	962	47	89
507	8.0	981	50	73	500	7.0	956	36	88
550	8.9	992	46	70	535	9.0	946	31	77
591	10.0	984	37	45	562	6.0	1003	40	60
625	11.0	991	32	46	565	10.0	937	30	62
651	12.0	988	30	42	696	9.9	1005	28	42
698	14.0	991	28	40	801	13.9	1005	25	40
751	15.9	993	21	38	887	18.0	1002	20	31
794	18.0	992	23	32	957	22.0	1003	18	29
864	21.9	996	23	31					
950	25.9	1002	20	25					

More comments can be made on model validation. The inverse of the time constant can be written as (see Eq. (7.4))

$$\frac{1}{\tau} = a_1 T_0^3 + a_2 \tag{7.5}$$

with $a_1 = \frac{4\sigma\alpha_e S_c}{m_s c_s}$ and $a_2 = \frac{\alpha_c S_c}{m_s c_s}$. If the coefficients α_e, α_c and c_s are considered to be constant (this is a reasonable assumption for small temperature ranges), the coefficients a_1 and a_2 can be computed by a linear regression algorithm using the data given in Table 7.2. Figure 7.4 shows the plot of the values of the identified time constants (inverse of Eq. (7.5)). The deviations observed are due to noise and the dependence of coefficients α_e, α_c and mainly c_s on temperature (which was ignored) and are small enough to justify the assumptions made to obtain the simplified model.

As has been previously mentioned, other types of test performed for model validation purposes are PRBS ones in order to obtain input/output data with enough dynamic information for model parameter identification purposes. For example, Fig. 7.5 shows one of these tests around 750°C (also including the response obtained with the identified model). A standard least-squares identification method was used in order to obtain the values of the parameters of the model. In the case shown in this example, supposing a first-order model structure, the values of the identified parameters were: $K = 25$°C/% and $\tau = 38$ s.

7.2.3.4 Improvements to the Model and Sintering Tests

The basic model presented in this section can be modified to take into account special requirements of a test, as is the case of copper sintering, where a sample is

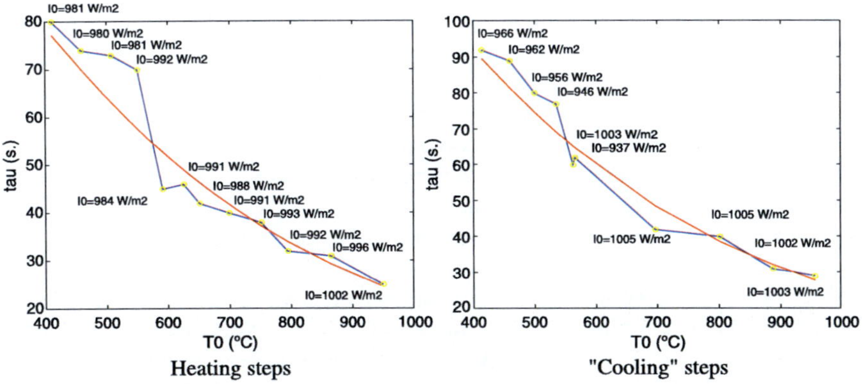

Fig. 7.4 Real and approximated relationships between τ and T_0 [43]

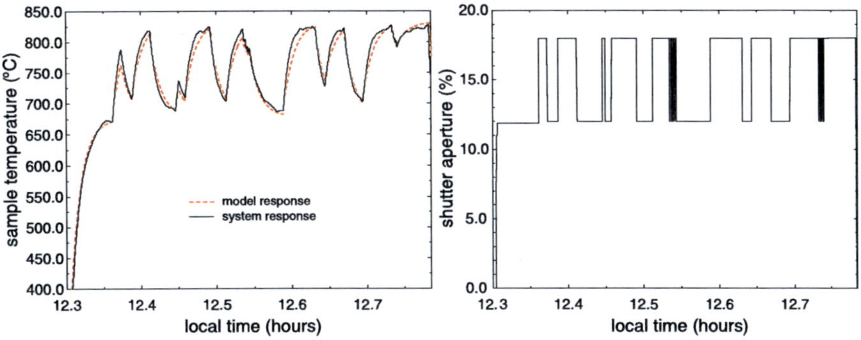

Fig. 7.5 Model validation using PRBS signals [43]

exposed to long high-temperature set points so that small particles bond together and the aggregate shrinks resulting in a decrease in surface area and energy; allows for this an ulterior study on a number of topics, such as volume diffusion of atoms. A lengthier discussion on this topic that has been around in the field of material processing since the mid-1950s can be found in [2]. Sintering is a well-known complex material processing technique. It involves multiple simultaneous physical processes, including various diffusion paths (along surface and through bulk lattice), vapor transport (evaporation and condensation), particle rigid body motions (translation and rotation) and grain growth through boundary migration [151, 212]. The sintering kinetics depend on interactions among these processes as well as the specifics of sample microstructure. Computer modeling of the sintering process to account for the competing phenomena of either grain growth or neck formation and densification is an active subject of research [394] and, currently, it is limited to partial solutions imposing ad hoc assumptions dependent on specifics, i.e. shape, size, relative location and crystallographic orientation of individual particles.

In the case treated in this section, the thermal profile is modeled using the energy balance given by Eq. (7.1) including the effect of conduction along the dimension of incident irradiance. Test sample structure consists of a multi-layered copper spool of straightened 2-mm-diameter, 50-mm-long wires placed in a specifically designed container. The sample and its container are placed inside a vacuum chamber which is provided with a gas preparation system for working in a controlled atmosphere. In sintering tests a slightly reductive atmosphere of HYD45 gas (5% H_2–95% N_2) is employed to prevent the sample from corruption, i.e. rusting or outer layer film decomposition.

Manual tests are performed under heating temperatures ranging from 850 to 1050°C and for varied exposure times. In general, with temperatures around 1000°C and for periods of 2 to 180 min, necks are formed at a limited number of contacts, the length of which depends on the treatment time.

The previous formulas included the sample effect by means of three unknown parameters, viz., absorption, emissivity and convection indices. It is known that these empirical factors are somewhat variable depending on temperature. Therefore, after a careful investigation of dimensionless Nusselt numbers and the ventilation system, it can be safely concluded that this convection process follows a forced, turbulent type. From these converging facts, the convection index has been approximated as $\alpha_c = c_c(T - T_a)^{\frac{1}{3}}$, where c_c is to be estimated as a constant factor.

Once T has been obtained, bearing in mind that the sample is mainly an interconnected mesh of copper wires, a simple estimation of the temperature gradient conduction [323] along the longitudinal dimension may take the form of a differential equation where the temperature of an intermediate region T_2 is the target, T_3 being the temperature at the base (non-irradiated) of the sample. The actual position of the target layer is an input option, by means of variable distances (d_1, d_2) (Fig. 7.6). The sample conduction index also needs to be taken into account and since the temperature gradient can reach several hundred degrees, a distinct parameter for each layer has been chosen (c_1, c_2). In short, by substituting terms and denoting $T = T_1$ (Fig. 7.6), the model formulas could be written as

$$m_s c_s \frac{dT_1}{dt} = \frac{r_h r_c S_c}{S_f} I S_s \alpha_a \left[1 - \frac{\sin(1 - U/100)\alpha_0}{\sin \alpha_0}\right] - c_c S_c (T_1 - T_a)^{\frac{4}{3}} - \alpha_e \sigma S_s \left(T_1^4 - T_a^4\right) \quad (7.6)$$

$$m_2 c_s \frac{dT_2}{dt} = \frac{c_1}{d_1}(T_1 - T_2) - \frac{c_2}{d_2}(T_2 - T_3) \quad (7.7)$$

There are so far six unknown parameters, viz., α_a, α_e, c_c, c_s, c_1, and c_2, that can be found by identification using, for instance, least squares (LS) methods or genetic algorithms (GA). There are some physical data about them that may narrow down the search. First, α_a and α_e are both thermodynamic properties of the sample that must be of the same order of magnitude with $\alpha_a > \alpha_e$. Regarding the specific heat of the sample (c_s), some useful data was gathered. The sample is mainly composed of copper wires as well as some amount of refrigerant gas. This gas, used for preventing sample corruption and rusting, is slightly reductive and is composed of 95% nitrogen and 5% hydrogen, which gives a compound specific heat of 2423.3 [J/(kgK)].

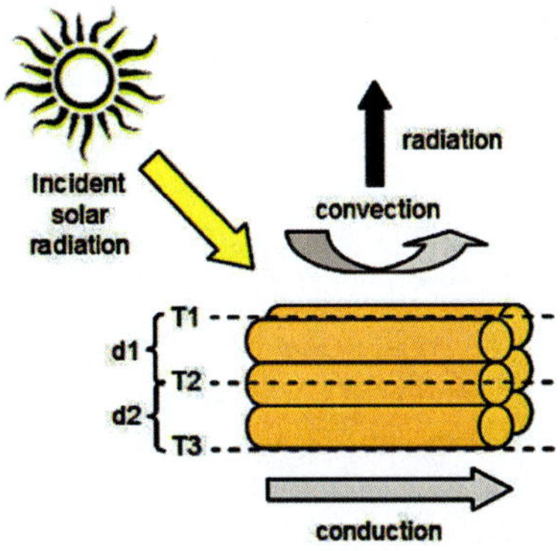

Fig. 7.6 Model improvement estimating conduction parameters (courtesy of D. Lacasa et al., [220])

By contrast, that of copper is considerably lower, being 383 [J/(kgK)] at ambient pressure. As a result, the sample's specific heat is somehow bounded and known to be closer to that of copper. Regarding the conduction parameters (c_1, c_2), these could be safely assumed to be very similar because none of the sample elements vary, which signifies that any existing change can be considered to be due to a temperature difference.

Since the aim was to arrive at an approximate thermal model for control purposes, a practical cross-validation with a separate data set from that of the real test was also carried out. These tests represent both manned and automatic controlled experiments performed on vessels interweaving actual samples and thermocouple sensors. Steep variable temperature gradients between the front and back sections are shown along the entire length down the midplane of the sample. The energy disposition at the vacuum chamber is graphically examined through infrared scanning. The model has yielded satisfactory results at simulation, given that it exhibits a good tracking of the real system trends. In addition, extremely rare peak errors are kept under 35°C, which account for less than 3% of the output variable range. Figure 7.7 shows a couple of comparative charts with simulated vs. real data.

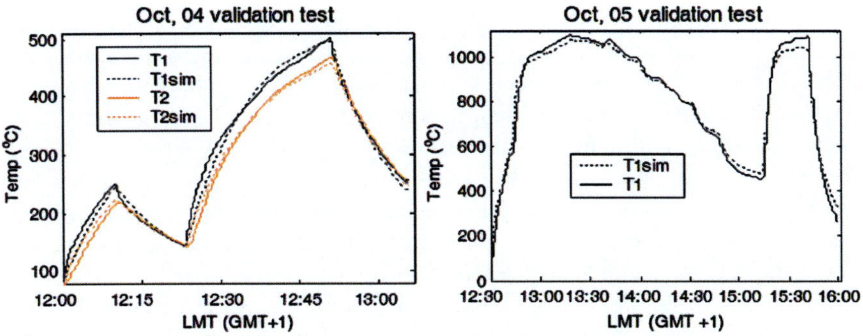

Fig. 7.7 Augmented model validation tests (courtesy of D. Lacasa et al., [220])

7.2.4 *Simple Feedforward and Feedback Control Schemes*

Many different temperature profiles are required for a wide class of tests [321] covering melting experiences, production of a super alloy wear resistant coating for high-temperature applications, etc. The *ideal* specifications for operation in these kinds of plant can be translated into classical control specifications as follows:

- The steady state error for a ramp input should be as small as possible as temperature profiles of this type (or more complex ones) are frequently requested.
- A fast settling time after a step in the reference temperature is desirable. The required settling time (depending on the kind of sample) is approximately that obtained by a very skilled operator and of the same order as the characteristic time constant of selected samples.
- A minimal overshoot for the closed-loop system is desirable. This requirement comes from the fact that there are no active cooling mechanisms and it is sometimes necessary to work at temperatures close to melting point and so overshoots or other types of oscillatory response would destroy the sample.

A sampling time of 3 seconds has been used for control purposes and relevant signals are stored every second by the data acquisition system running in the central control computer.

7.2.4.1 Feedforward Control (FF)

As has been demonstrated in other types of solar plant [82, 83, 85], the use of feedforward controllers is fundamental for systems subjected to measurable disturbances (as is the case of direct solar irradiance). In the case treated in Chap. 4, Sect. 4.4.1, both parallel and feedforward configurations have been used following the basic control schemes shown in Fig. 4.15.

The most important disturbance acting on the furnace is the solar irradiance. The continuous system transfer function can be written as: $G(s) = KI/(1 + \tau s)$, where the solar irradiance proportionally affects the linearized system gain. Thus, if a block with a transfer function $G_{ff} = I_{ref}/I$ is placed in the forward path (series feedforward controller in Fig. 4.15(b)), the resulting global transfer function is theoretically *independent* of the solar irradiance values, having a gain constant (if the supposition of the operation around an operating point is fulfilled) corresponding to a solar irradiance level (input energy) equal to I_{ref}. In this way, the series feedforward controller is simply a variable gain only dependent on the solar irradiance. The value of I_{ref} has been taken close to the mean value of the solar irradiance during different tests and is equal to 900 W/m^2. This is a valid approximation both for steady state and transient conditions, as the transfer function which relates changes in the sample temperature to changes in solar irradiance has, theoretically, the same characteristic polynomial as the system transfer function (when considering linear operation and without non-linearities introduced by the actuator). This is due to the fact that a decrease in the solar irradiance level has the same effect on the output temperature as a diminishment in the shutter aperture. Notice that all these considerations are based on several simplifications made when obtaining an approximated transfer function of the system and that G_{ff} could also contain a low-pass filter.

7.2.4.2 Fixed PI Controllers with Anti-windup Action

As is usual in the development of automatic control systems [21, 24], the first step in the design is to try to implement classical PID control schemes. As the plant can be approximated as a first-order system and taking into account the system specifications, the decision was taken to tune a PI controller given by $C_{PI}(s) = K_P[1+1/(T_I s)]$ implemented using the anti-windup mechanism shown in Fig. 4.19 [22].

In many control problems, the existence of an actuator constraint of the slew-rate type is not accounted for during the design step and the obtained controller is detuned when implemented to obtain a generally robust and well-damped performance. As the system is modeled as a first-order one, a common solution in the control of these kinds of system to obtain fast responses from simple root locus analysis is to use low integral times and high gain controllers. This is not always an adequate solution when implementation issues are considered, as the output of the controller must be realistic and take the constraints into account. In [43] a simulation is performed showing how the behavior of a controlled system deteriorates with the existence of slew-rate constraints.

From the root locus analysis viewpoint and taking the characteristic time constant of the controlled system to be at a determined location, two cases can be considered, depending on the position of the PI controller zero (depending on the value of T_I).

In CASE A ($T_I \leq \tau$), for low gains, the expected type of response is an overdamped one (two real poles), but the closed-loop system dynamics is quite slow. For intermediate gains, two complex poles are obtained and so oscillations are produced in the closed-loop response which, in general, are not desirable. For high gains, two real poles are again obtained but the zero of the controller is closer to the origin so that the output of the closed-loop response could surpass the reference after a step set point change. As can be seen, it seems to be impossible to completely fulfill all the specifications. As a trade-off design, the controller can be designed for this last situation (the controller zero is closer to the origin than real closed-loop poles). From a simple analysis of the closed-loop transfer function, the condition to obtain real closed-loop poles is that

$$T_I \geq \frac{4\tau K K_P}{(K K_P + 1)^2} \quad \text{with } T_I \leq \tau \tag{7.8}$$

An approach is adopted to express the limitations in the design of the controller imposed by the actuator non-linearities. As this device moves at a rate of 5 units per second, the idea is to limit the dynamics of the input signal to the actuator at the design stage of the algorithm in such a way that the rate of variation of this signal does not exceed that of the shutter. In this sense, the PI controller can be designed so as to incorporate low-pass action according to the actuator limitations in such a way that all frequencies higher than a determined cut-off frequency ω_0 are attenuated at least in 20 dB (that is, the bandwidth of the PI controller is limited at the design stage). The worst-case input signal for the actuator is one sinusoidal signal with amplitude equal to 100 and with a maximum slope equal to 5. Thus, if a sinusoidal signal $100\sin(\omega_0 t)$ is introduced in the actuator input, the derivative of this signal in $t = 0$ gives the maximum rate variation which has to be limited to 5, leading to a value $\omega_0 = 1/20$ rad/s. This approximation introduces another criterion for the PI design. The analytical expression of the module of the frequency response of a PI controller is given by $|C_{PI}(j\omega)| = K_P/(T_I\omega)\sqrt{T_I^2\omega^2 + 1}$. For $\omega = \omega_0$, the gain of the PI controller must be less or equal to -20 dB. By performing a few operations, another relationship between T_I and K_P can be obtained:

$$T_I \geq \frac{20K_P}{\sqrt{0.01 - K_P^2}} \tag{7.9}$$

In this way, two inequalities (7.8) and (7.9) relating T_I and K_P are obtained. The set of possible PI parameters which fulfill both inequalities is represented in Fig. 7.8. The upper curve represents Eq. (7.8), that is, the curve delimits the region over which the closed-loop poles are real poles. For a proportional gain less than $1/K$, the poles are on the right side of the zero. For upper gains, the poles are placed on the left side of the controller zero. One point belonging to this curve corresponds to a PI controller which produces a critically damped closed-loop system (with respect to the situation of the closed-loop poles). The lower curve is given by Eq. (7.9) and represents the limitation in the controller bandwidth to adjust the controller dynamics to the shutter constraints. The region within which the controller parameters must be situated (shadowed in Fig. 7.8) is below the line $T_I = \tau$ and above the lines representing Eqs. (7.8) and (7.9).

One possible approach for obtaining two relationships between the system model parameters and the controller parameters is to select point C given by the intersection of these last curves (notice that this is the less conservative approach, not accounting for unmodeled dynamics/modeling errors), leading to the minimum integral time within the possible range. The parameters corresponding to this point can be found by solving the following equality:

$$\frac{4\tau K K_P}{(K K_P + 1)^2} = \frac{20K_P}{\sqrt{0.01 - K_P^2}} \tag{7.10}$$

Both iterative methods and approximations can be used. If an analytical solution is convenient (for instance to be used in adaptive control schemes) an approximated upper bound on the curve representing Eq. (7.9) can be found in such a way that the solution is given by a quadratic function. It can be demonstrated [43] that an upper bound of the function on the right side of Eq. (7.10) is given by $2K_P/(0.01 - K_P^2)$,

Point	K_P	T_I
A	$1/K$	τ
B	$\sqrt{\frac{0.01\tau^2}{400+\tau^2}}$	τ
C	$\frac{\sqrt{K\tau(0.02K^2+0.04K\tau-\tau)}-K}{K(K+2\tau)}$	$\frac{4\tau K K_P}{(K K_P+1)^2}$

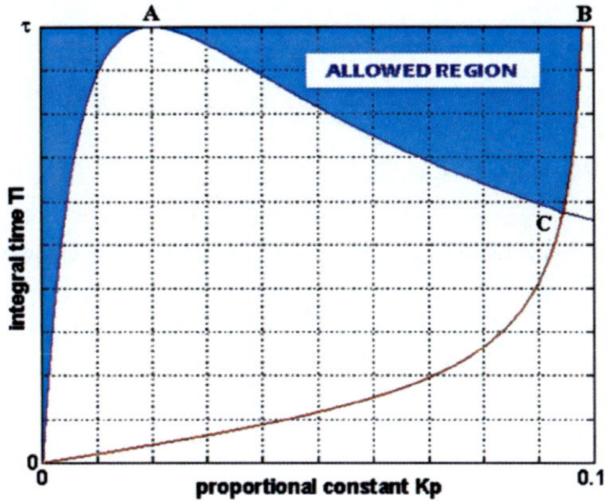

Fig. 7.8 Relationships between T_I and K_P [43]

that is, $T_I \geq 2K_P/(0.01 - K_P^2)$ fulfills Eq. (7.9). In this way, the controller parameters can be found by solving the equality

$$\frac{4\tau K K_P}{(K K_P + 1)^2} = \frac{2K_P}{0.01 - K_P^2}$$

which leads to the solution

$$K_P = \frac{\sqrt{K\tau(0.02K^2 + 0.04K\tau - 2)} - K}{K(K + 2\tau)} \quad \text{and} \quad T_I = \frac{4\tau K K_P}{(K K_P + 1)^2}$$

Simulations with this design method for the solar furnace can be found in [43].

Another possibility in this case is to select an interior point of the shadowed region represented in Fig. 7.8. For conservative design purposes (to implicitly account for modeling errors-process/model mismatch), an interior point of the shadowed region in Fig. 7.8 can be chosen as follows: K_P located half way between the values of the proportional gain constant of points A and B in Fig. 7.8; T_I given by the value of the integral time constant of point A minus one third of the difference between the values of the integral time constant of point A and that of point C in Fig. 7.8. So, the following formulas can be used:

$$K_P = \frac{1}{2}\left[\frac{1}{K} + \sqrt{\frac{0.01\tau^2}{400 + \tau^2}}\right] \quad \text{and} \quad T_I = \frac{1}{3}\left[2\tau + \frac{4\tau K K_P}{(K K_P + 1)^2}\right] \tag{7.11}$$

The second design possibility given by root locus analysis (CASE B) introduces the controller zero between the origin and the system pole ($T_I \geq \tau$). This seems to be a poor solution as the closed-loop system pole is slower than the open loop one, although the existence of a closed-loop zero near the origin makes the closed-loop dynamics fast enough to fulfill the settling time requirements. From the actuator constraint viewpoint, the same considerations made in previous paragraphs can be made here and controller parameters can easily be found. Nevertheless, this second approach is more sensitive to modeling errors, as changes in system gain lead to pronounced changes in settling time [43].

7.2.4.3 Plant Results with Fixed PI Controllers

The control algorithm previously developed (PI control including series feedforward control, anti-windup mechanism and designed to cope with slew-rate constraints) was implemented at the PSA solar furnace and different results are shown in this section. The tests were performed under very different solar irradiance conditions and with samples of different materials. Only tests corresponding to the less conservative approach of CASE A are presented in this subsection.

As a first example, Fig. 7.9 shows the results obtained in a test with a sample made of silicon carbide. The representative test was performed on July 30th, 1996 and consisted of a set of steps with an amplitude of 50°C. As can be seen, the behavior obtained in this case is highly dependent on the operating point, having different responses as the set point temperature increases (diminishing the overshoot and increasing the rise time), leading to unacceptable performance (this being one of the reasons why an adaptive control approach was implemented, as shown in the next section). As the temperature increases, the gain and time constant of the system diminishes producing a slower controlled system. Notice that the properties of the samples (specific heat, emissivity, etc.) are also influenced by operation at high temperatures.

Other representative tests were performed using a white zirconia sample, which has different dynamic characteristics to metal or silicon carbide samples. The results of the test performed on November 7th, 1996 are shown in Fig. 7.10 in which the white zirconia sample was used. The set point temperature profile consisted of two ramps with different slopes. As can be seen, the tracking characteristics of the closed-loop system were adequate below 1200°C. However, at higher temperatures the tracking error increased, mainly due to the plant non-linearities because a higher shutter aperture is needed as the set point temperature increases.

In conclusion, the use of a well-tuned PI controller (including feedforward action, anti-windup and slew-rate constraint handling mechanisms) provides adequate results from the plant staff viewpoint for a wide range of tests. Nevertheless, two main drawbacks can be indicated:

- The controller must be tuned for each type of sample whose dynamic parameters (characteristic gain and time constant) have to be a priori estimated.

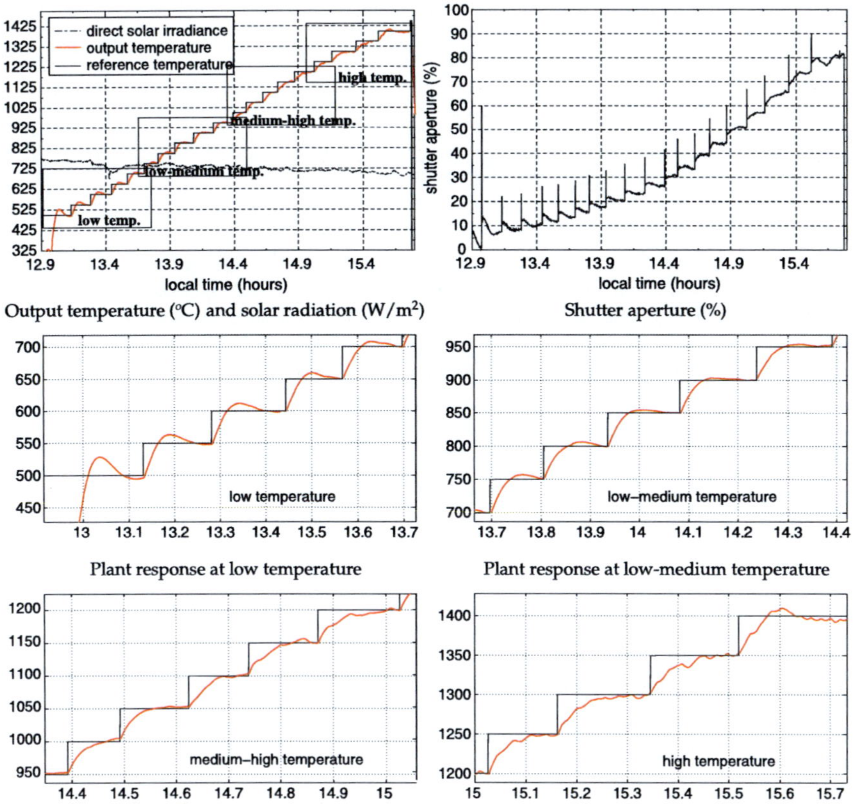

Output temperature (°C) and solar radiation (W/m²) — Shutter aperture (%)

Plant response at low temperature — Plant response at low-medium temperature

Plant response at medium-high temperature — Plant response at high temperature

Fig. 7.9 Test with the PI controller (30/07/96) [43]

- When using a fixed parameter controller, the controlled system set point tracking capabilities are only acceptable within a small operating range of the whole test.

These drawbacks justify the inclusion of other kinds of control approach such as adaptive control, dealt with in the next subsection.

7.2.5 Adaptive Control (AC)

An adaptive control scheme seems to be a possible solution to solve the problem of plants with different types of dynamics. Adaptive control has previously been used at other kinds of solar plant [83, 85], as shown in Chap. 5. This control technique provides a framework in which the controller parameters are adapted to the diverse operating conditions and different plant dynamics. The control scheme implemented consists of a self-tuning controller [25] in which the parameters of a

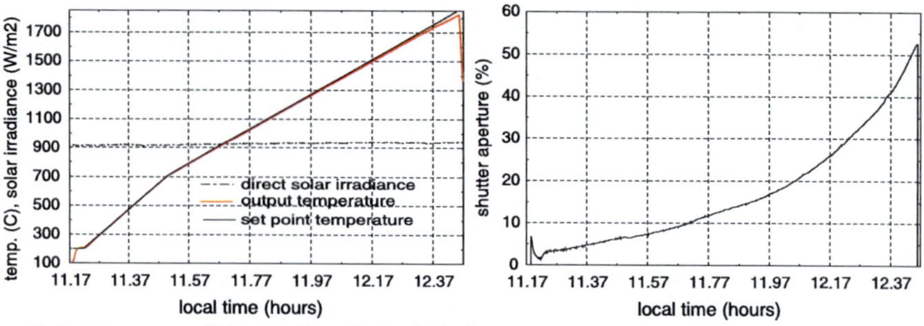

Output temperature (°C) and solar radiation (W/m²) Shutter aperture (%)

Fig. 7.10 Test with the PI controller (07/11/96) [43]

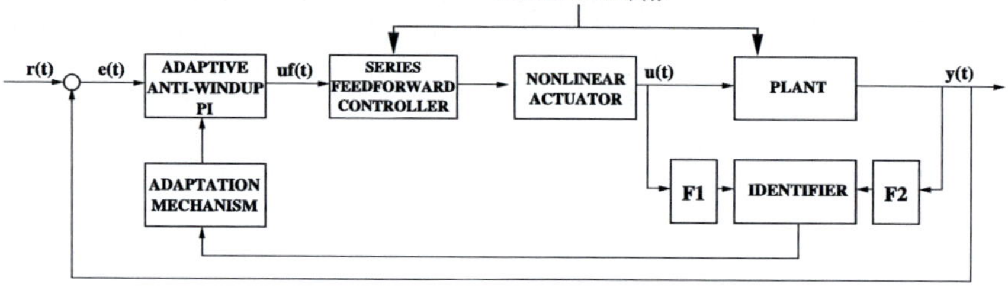

Fig. 7.11 Adaptive control scheme [43]

simplified model of the plant are identified on-line. From these parameters, the controller parameters are changed according to an adaptation mechanism that relates controller parameter changes to system parameter changes. The adaptive controller also incorporates feedforward action, anti-windup mechanism, slew-rate actuator constraint handling and identifier data prefiltering. The control scheme is shown in Fig. 7.11.

Different problems arise when using adaptive control at solar plants [85], the main ones being:

- The system excitation is usually poor as identification is performed in a closed-loop configuration and the input signals to the plant have few frequency components.
- The existence of two time scales (one for closed-loop system dynamics and the other for identified parameter dynamics) is not always ensured, mainly when coping with slow processes.

To deal with these problems, different variations to classical adaptive control schemes have been introduced. As previously mentioned, the main advantages of using a series feedforward action [82, 85] are both the compensation for solar irradiance variations and the preservation of the validity of simple SISO system models of the plant used for estimation purposes (considering the feedforward controller as a part of the estimated plant). This is valid if the actuator dynamics are linear (or with static non-linearities of saturation type). Nevertheless, in this case due to the slew-rate constraints, if the feedforward controller is included as part of the estimated plant, problems can arise in the estimation algorithm as the input signal to the identifier may not correspond to the real input to the system. To cope with this problem, the input signal to the identifier loop is the real plant input $u(t)$ and different filters can be introduced in the identification loop (scale filters F_1 and F_2 in Fig. 7.11). Notice that one possibility is to make $F_1 = 1$ and to multiply the output signal of the plant by the same filter used for feedforward control before entering the identifier ($F_2 = I_{ref}/I$). In this sense, if solar irradiance augments, the output of the plant which is introduced into the identifier is scaled (diminished) by a factor which, theoretically, makes the values entering the identifier independent of the solar irradiance variations. Another possibility is to make $F_2 = 1$ and to multiply $u(t)$ by the inverse of the feedforward filter ($F_1 = I/I_{ref}$). Notice that the feedforward action can also include the inverse of the non-linearity introduced by the conversion from aperture angle to aperture percentage represented in Fig. 7.2. Scale filters F_1 and F_2 can additionally include a first-order low-pass filter to introduce smoother signals within the identifier not affected by the noisy natural changes in solar irradiance.

The models used for estimation purposes are of first-order type and the controller is the PI developed in the previous subsection. At each sampling time, the adaptive control strategy consists of estimating the linear model parameters using input–output data from the process, adjusting the PI controller parameters, calculating the control signal and supervising the correct behavior of the controlled system. The parametric identification algorithm used is the LS one with a fixed forgetting factor, which has to identify only two parameters as the dead-time of this plant is negligible (the model of the plant is a first-order one in discrete time: $G(z^{-1}) = (bz^{-1})/(1 - az^{-1})$, $a = e^{(-T_s/\tau)}$ and $b = K(1 - a)$. The control laws obtained in the previous section have been used as an adaptation mechanism, as they relate plant parameters to PI controller parameters.

Finally, three basic supervisory mechanisms have been implemented to overcome the mentioned drawbacks of adaptive control schemes. The first one is based on limiting the maximum and minimum values of the estimated parameters, thus avoiding a wrong identification leading to dangerous behavior of the controlled system. The values of the possible range of variation of the identified parameters have been obtained from a dynamic study of the plant and from experience in operating with different types of material (see Table 7.1). The second mechanism consists of filtering the estimated system parameters obtained using a first-order low-pass filter with a cut-off frequency such that the variation of the estimated parameters is guaranteed to be slower than the plant dynamics (to ensure the existence of two time scales). In this sense, the pole of the filter can be placed in the upper limit of the allowed range introduced by the first supervisory mechanism. Finally, the third supervisory mechanism consists of stopping the identifier in cases in which the dynamic information entering the identifier is poor for identification purposes, as sometimes happens when using slow ramps as inputs to the system.

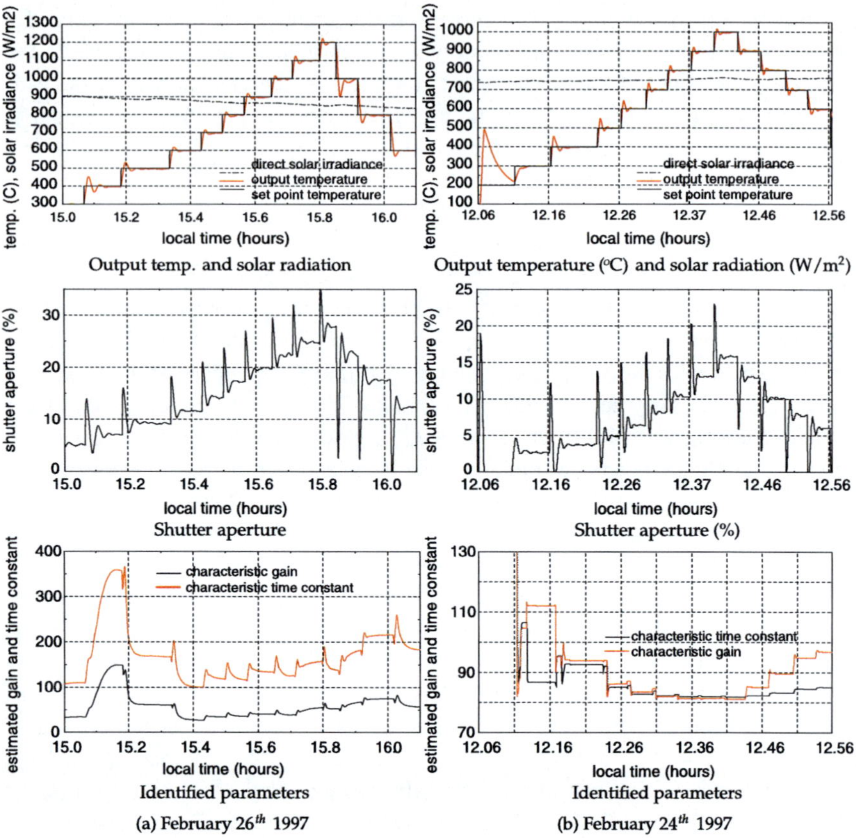

Fig. 7.12 Tests with the adaptive PI controller [43]

7.2.5.1 Experimental Results

Several tests have been performed using the adaptive control algorithm for three kinds of material: A316-L steel (high gains and small time constants), white zirconia (low gains and high time constants) and a small steel sample (low gains and low time constants). In all cases, the characteristics of the materials change with temperature and solar irradiance levels.

The first two tests shown in this subsection were performed using the CASE A, less conservative design approach (minimum integral time), which does not account for modeling errors. Figure 7.12(a) represents a test with the white zirconia sample in which both the evolution of the main variables of the test (temperature, direct solar irradiance and shutter aperture) and the evolution of the estimated parameters of a first-order model of the plant are shown.

Filtering the estimated parameters avoids the induction of sudden changes in the controller parameters after a step change in the reference temperature. The filtered

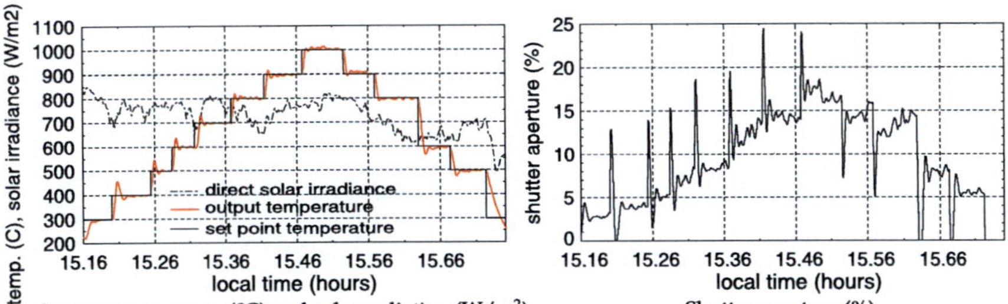

Output temperature (°C) and solar radiation (W/m²) Shutter aperture (%)

Fig. 7.13 Test with the adaptive PI controller (25/02/97), [43]

estimated values converge to adequate values after a transient which depends on the filter used, as can be seen in Fig. 7.12(a) in which the initial values of the parameters differ considerably from the real ones. In the same figure, it can be seen that the trend of the estimated parameter approaches that expected of the dynamical study of the plant as both the gain and fundamental time constant of the system decrease as the temperature of the sample augments and augment as the temperature decreases. Undesirable initial transients due to a wrong initial estimation of parameters can be avoided using a conservative fixed PI controller during the transient in which the estimated parameters approximate the appropriate ranges by initially introducing a sequence of set point changes to excite the system in order to provide the identifier with dynamic information for identification purposes, or by using some kind of open loop auto-tuning capability. As the estimated parameters achieve the adequate values, the overshoot obtained after a step response decreases (34% at 500°C and between 13% and 24% in the rest of the test). The rise times are between 31 and 40 s, which constitutes quite good behavior.

Other similar tests have been performed with an A316-L steel sample, obtaining adequate results which lead to the same considerations made above. For example, Fig. 7.12(b) shows the results obtained on February 24th, 1997, again using the less conservative design approach. After an initial transient during which identifier excursions occur due to the wrong selection of the initial values of the selected model, the controlled system performance improves as new dynamical information enters the identifier.

A test performed on February 25th, 1997 is shown in Fig. 7.13 with the same type of sample. The relevance of this test comes from the fact that the solar radiation conditions during the test were such that manual operation under said circumstances was impossible, even for a skilled operator. The good characteristics of set point tracking and disturbance rejection are in part due to the series feedforward action, which compensates for changes in solar irradiance within a band of 200 W/m^2 allowing the output temperature to be maintained within a band of 10°C around the reference.

As can be seen from the previous tests, several of the experimental results exhibit a high degree of overshoot and/or oscillation (notice that one of the control objectives was minimal overshoot). There are several causes which contribute to this

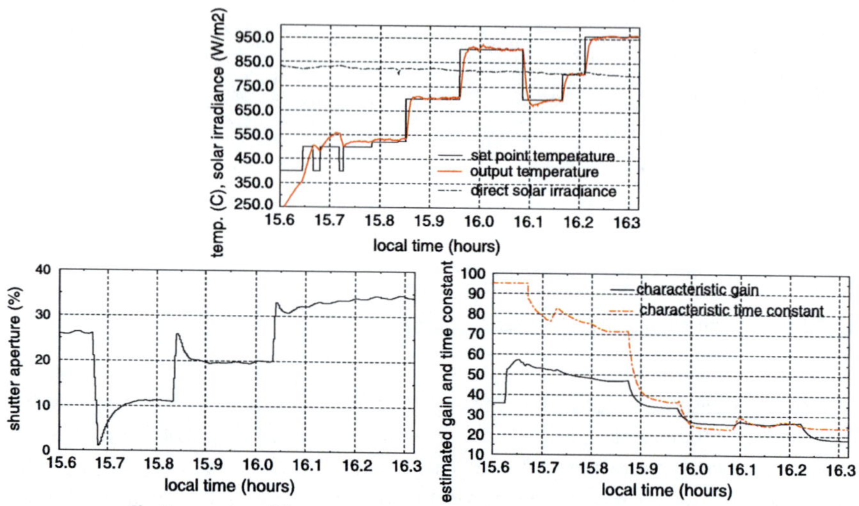

Shutter aperture (%) Identified parameters

Fig. 7.14 Test with the adaptive PI controller (22/07/98), [43]

undesirable kind of behavior. First of all, the design approach chosen in this case (CASE A, less conservative approach) leads to a closed-loop system that is quite sensitive to modeling errors, which can produce undesirable behavior especially when using a small value of the integral time constant in the PI controller (as in this case). Moreover, the effect of the controller zero can also lead to an undesired overshoot, or other aspects such as the use of type B thermocouples which does not provide for any measurement below 100°C (this fact can produce errors in the integral part of the controller at the starting phase of the operation). From the identification viewpoint, it has been pointed out that the wrong selection of the initial values of the estimated parameters can produce undesirable divergences in the estimated parameters, leading to a deterioration in performance of the controlled system. In the first tests shown with the adaptive controller, results show spikes in parameter estimates due to a wrong selection of the filters in the identifier (Fig. 7.12(a)). These problems can be avoided by a careful selection of the filters and the supervisory mechanisms (as in the second test in Fig. 7.12(b)).

In order to improve the results with the adaptive controller, some tests were performed using the more conservative design approach (also explained in the section devoted to the CASE A design approach), in which the adaptation law is given by Eqs. (7.11) and by a careful selection of the supervisory mechanisms. Figure 7.14 shows the results obtained with this approach with a small steel sample. Several set point changes were produced at the beginning of the operation to provide the identifier with dynamic information so that it was able to *detect* the kind of sample and the parameters of the supervisory mechanisms could be adequately fixed. Notice that another way of characterizing the samples at the beginning of the tests is to perform a classical open loop auto-tuning process. As can be seen in Fig. 7.14, the behavior of the controlled system was quite acceptable, showing similar response characteristics under different operating conditions, with rise times of about 40 seconds and small overshoot (of less than 5°C, except in one of the steps). The evolution of the estimated parameters is also shown in the figure. It can be seen that the evolution is as expected from studies performed in previous sections and spikes have been avoided in the identified parameters.

7.2.6 Fuzzy Logic Control (FLC)

As has been commented in the previous section, adaptive control is prone to overshoots when applied to on different sample materials and it is difficult to reproduce smooth responses such as those achieved by skilled operators. For this reason, FLC seems to be a good approach for the task by reproducing the expert knowledge needed, following the same approaches described in Chap. 4. Copper sintering is the main process studied in this section, although the control algorithm is valid for other kinds of sample.

From the control viewpoint, the worst-case scenario appears when the temperature is approaching the current set point too quickly. This fact is proven and corroborated by experts who themselves find it extremely difficult and cumbersome to manually correct this inconvenient situation. In order to accomplish a good design, three different approaches were initially taken, tested and cross-validated with one another and with previous manual tests. The final controller was implemented from out of these initial approaches, making use of their best qualities. A Fuzzy Associative Memory (FAM) was found to be most suitable.

7.2.6.1 Input and Output Variables

The FLC architecture of this work takes as inputs the error and the error derivative of the controlled variable, the sample temperature. Hence, they can be defined as

$$\begin{aligned} e(k) &= T_{ref}(k) - T(k) \quad [^\circ\mathrm{C}] \\ \frac{de(k)}{dt} &\approx \frac{[e(k) - e(k-1)]}{T_s} \quad [^\circ\mathrm{C/s}] \end{aligned} \tag{7.12}$$

In temperature control, the error derivative parameter is as important as the temperature itself and even more important when working with small samples (fast dynamics). Temperature variation forces changes in the system response because there is a high correlation between the capacity of the sample to absorb heat and its temperature. These changes imply modifications in the system time constant, pole allocation and hence, strong variations in system speed. Keeping apace with error significance, its derivative gives a relative measure of what the error would be at the next sample time. Thus, it offers crucial information to the fuzzy controller about trend and how intense the controller's reaction must be in order to adjust the system to reach the next equilibrium state. The output variable is the shutter aperture. The fuzzy logic control scheme used is the same as that in Fig. 5.54(a), where the increment of the shutter aperture is computed at each sampling time based on the error and the error derivative values. The non-linearities present in the system have been added in rule base design [221].

7.2.6.2 Simulation-Based Test Procedure

Since expert operators usually work full time on a particular kind of sample for a number of subsequent test campaigns, they are able to develop and adjust their ability to operate a plant more or less smoothly under the new conditions a new sample entails after several probing tests. However, their knowledge is often tacit and not easily translated to control engineering requirements. Consequently, interviews turned out to be more of a task of polling and enquiring. These polls consisted of a battery of graphically aided questions about supposed cases that presented distinct plant states. This polling technique was only moderately helpful and, as a result, a more powerful approach was devised involving a great deal of use of the hand-controlled tests already performed on the sample, plus experience on the author's side in applying classical control strategies [44, 85] to solar plants. As commented on in previous sections, observation of step responses applied to the plant indicates that, under nominal steady state conditions, unclear dissimilar zones of temperature could be found and approximated by first-order transfer functions, enabling use of linear system control theory, as previously made in [44].

7.2.6.3 Universe of Discourse of Error Variable

Both input variables are bounded by the temperature range variation in a test, which is a little over 1050°C (as the melting point of copper is 1084.62°C). However, by empirical evidence, it was agreed that a 100°C to 200°C range above and below a given temperature should suffice for control objectives as, outside this range, the human operator will select a certain constant control signal. Figure 7.15 displays the universe of discourse of the error variable according to three different partitions. The first row presents candidate approaches, while the second one records the final approach. Candidates have been grouped to save space and for ease of comparison. X-axis values correspond to the midpoint of triangular membership functions (degrees Celsius). In this case, membership function edges and midpoints vary among controllers, though their meanings are equal, so plain names have been assigned, namely Zero (0°C), VSmall (5°C), Small (10°C), Medium (20°C), Large (30°C) and VLarge (100°C), standing for increasing relative magnitude of error (values pertaining to final approach midpoints). For the negative part of the universe, they are preceded by the minus sign. The triangular functions not only vary in their midpoint placement and base width but they also vary in number from one controller to another.

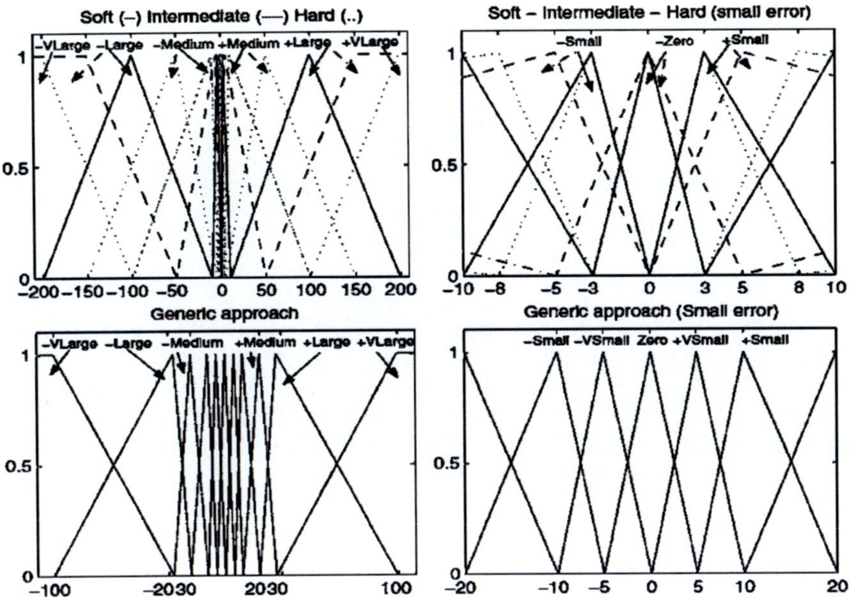

Fig. 7.15 Universe of discourse of the error variable (courtesy of D. Lacasa et al., [221])

7.2.6.4 Universe of Discourse of Error Derivative Variable

Although the experts' capability to operate the plant was more than proven, their knowledge was not very useful because plant operators tend to work by experience, hence, they simply do not know about such explicit numerical data. As a result, a script process that allowed for user defined error derivative interval declaration was applied to a number of test logs yielding basic statistics of this variable e.g., maximum and minimum magnitude and relative frequency of predefined intervals. Figure 7.16 shows two graphs; the first includes three different candidate partitions of the universe, tuned by simulation and tested on the plant; the second has the generic controller's membership functions corresponding to fuzzy sets that make up the universe of discourse of the error derivative variable. Furthermore, this variable is most prone to rapid changes, thus the sampling rate is a concern. If compared to most submodels' response times, it is sufficient for acceptable results, however, it is clearly not the best choice. A reason to believe this assumption is that a human operator makes continuous small changes in the control signal (shutter aperture) when fast approaching set point. It is this very capability that avoids unacceptable overshoots in manual control mode.

7.2.6.5 Universe of Discourse of Output Variable

The output variable has been selected as an increment of the shutter aperture. The discretization of the output universe into fuzzy sets plays a vital role in the final

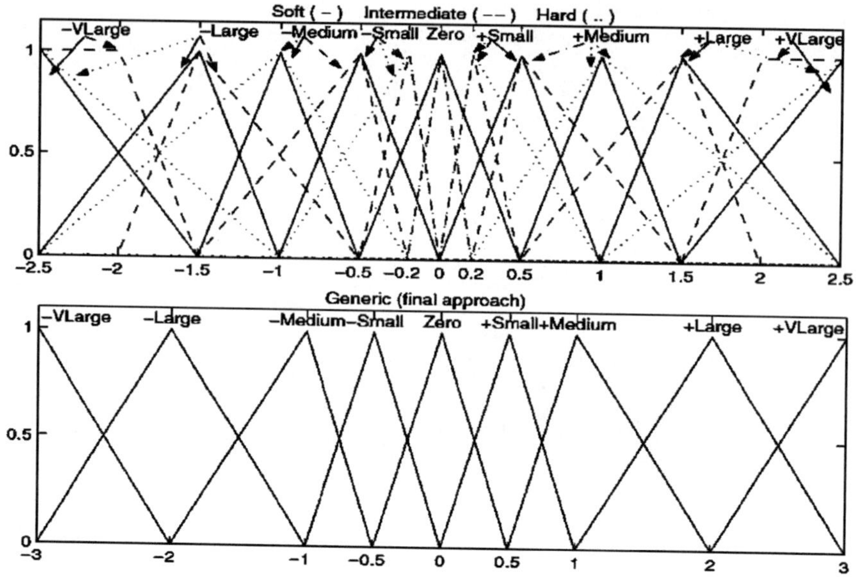

Fig. 7.16 Universe of discourse of the error derivative variable (courtesy of D. Lacasa et al., [221])

behavior of the controlled system. Moreover, the plant presents a non-linearity of slew-rate type in the actuator, which basically means that its speed is constrained and varies with direction (opening or closing) and beginning and end positions. Therefore, much attention and time was devoted to its study, i.e. the actuator underwent rigorous tests of opening/closing rates at a series of positions, throughout the whole range available. These test records were normalized to a second, which produced a time look-up table where the worst case recorded was a 7% aperture per second. With regards to the least significant change to have a direct effect on the sample temperature, 0.1% was common among experts, though only valid when the temperature is reaching set point. Accordingly, the output universe was divided into fuzzy sets in a triangular partition fashion, i.e. Z (0%), VS (0.25%), S (0.5%), M (1%), L (3%), VL (5%), XL (7%) and their negative counterparts with the same value and opposite sign.

7.2.6.6 Rule Base and Method of Inference

Two features left to give life to this FLC are knowledge and reasoning capabilities. The former will come in the form of a rule base while the latter will fire rules and work with them to return, in the end, an incremental crisp value to add to the current state of the output variable. In this case (two input–single output system), the rules take the form: IF *error* is positive very large (+VLarge) AND *change in error* is negative medium (-Medium) THEN control signal increment must be medium (+M).

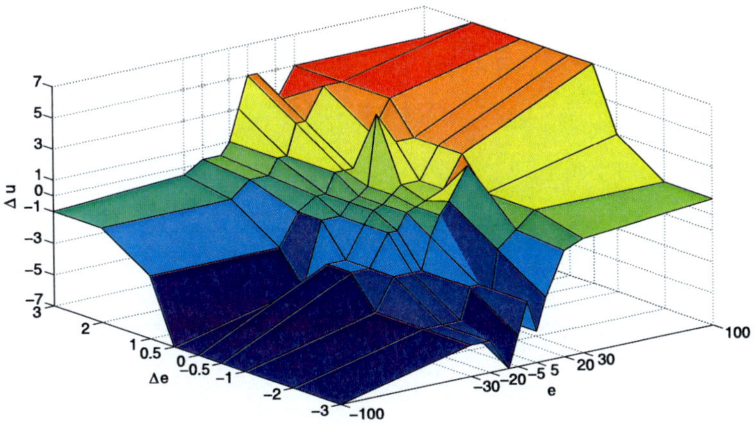

Fig. 7.17 Fuzzy control surface (courtesy of D. Lacasa et al., [221])

Because of the partial matching attribute of fuzzy control rules and the fact that the antecedents of rules overlap, two in each input dimension, up to four rules can fire at the same time. Since a triangular partition rule base has been used, both input and output universes have been subdivided into overlapping fuzzy triangular membership functions [367]. Assuming that the *algebraic product* with AND connective is employed, each triggered rule i proposes a control action increment (Δu_i) with a degree of membership $\mu_{\Delta u_i} = \mu_{e_i}\mu_{\Delta e_i}$. As a mechanism of fuzzy inference, the compositional rule *sup-min* is used. The combination of these rules in order to produce a non-fuzzy control action follows Tsukamoto's defuzzification method for computing the aggregated output of inference rules due to its computation simplicity. A two-input fuzzy knowledge base, so-called 2-dimensional FAM, may be graphically represented in a surface contour plot as in Fig. 7.17. It is not symmetrical because the dynamics of the system is much faster when the temperature is rising than when falling.

7.2.6.7 Plant Results

The fuzzy controller was tested at the plant in the aforementioned versions, i.e. soft, intermediate and hard approaches, each of them varying most of the parameters of the fuzzy paradigm, such as partition of the input universes and knowledge rule base. Quality, not quantity, should be stressed, thus two experiments were devoted to tuning the initial simulation-based candidate controllers which yielded the final generic controller. Another fact to be noted is the ease of implementation and deployment of these kinds of system. Additionally, more than decent results were obtained from the very early tests performed at the plant. Figure 7.18 corresponds to a step-response test where a wide range of sample temperature is covered (from ambient to 600°C). As can be seen, good set point tracking is obtained in any of the approaches. For instance, because of the integral effect it copes effectively with large banks of passing clouds, as displayed in the third row (hard approach). Moreover, every approach

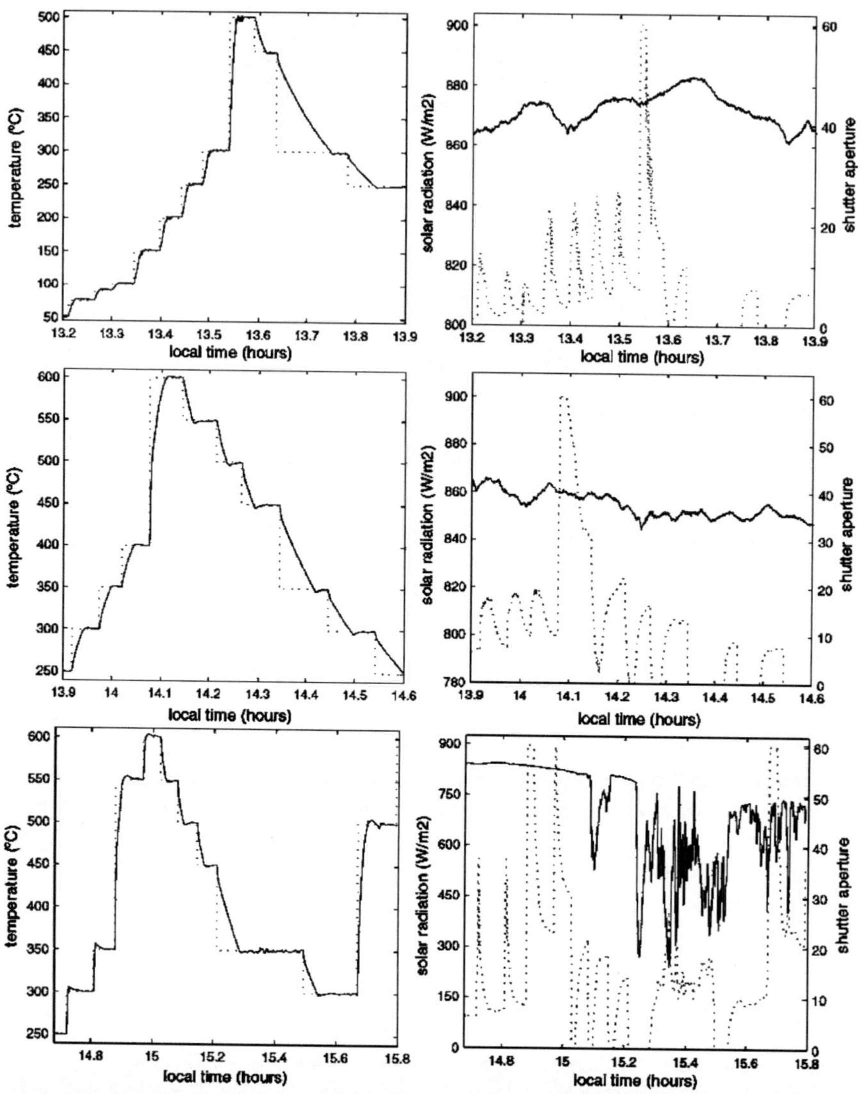

Fig. 7.18 Soft, Intermediate and Hard approaches tested on December 21st, 2004 (courtesy of D. Lacasa et al., [221])

meets the 10°C maximum overshoot requirement established by the experts. Offsets are kept to within 1°C and can hardly be improved due to sensor noise (thermocouples) and the large sampling rate (1 sample per second). It would be fair to say about the usefulness of fuzzy control that it is a fallacy that under persistent banks of passing clouds, manual operation becomes unmanageable and stressful because many consecutive changes in aperture are needed to maintain reference, whereas

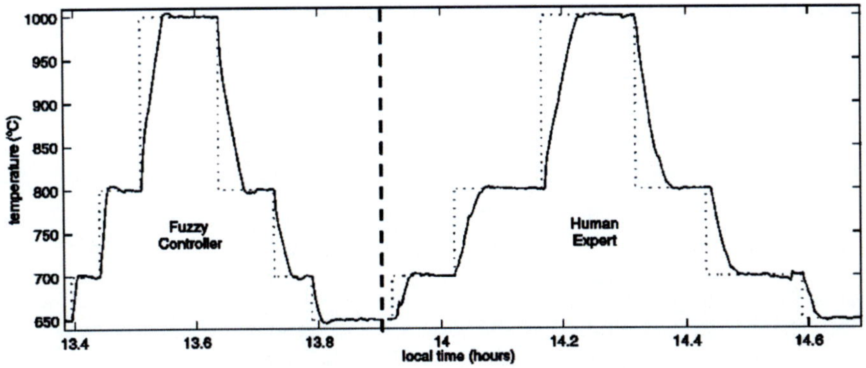

Fig. 7.19 Final controller vs. Expert Operator. Test performed on May 5th, 2005 (courtesy of D. Lacasa et al., [221])

fuzzy logic strategy can handle it more extensively and with better performance than a human operator. Another proof of the viability and convenience of applying this technique would be a direct comparison of the final controller with a human expert. With this aim, a test was devised to put the controller and a human operator face to face under very similar conditions and with identical set point planning.

Figure 7.19 graphically portrays the result. The outcome of the test can be summarized as great performance from both automatic (FLC) and human operation at the plant. Overshoot was kept under 6% of step amplitude, although the expert provided a smaller overshoot (around 2%). Conversely, FLC outperformed the expert in rise time. Table 7.3 reflects and elaborates on these records. In summary, faster responses with smaller rise times were obtained at the cost of a slightly larger overshoot that did not break the rules and needs set by experts in the process. More importantly, very good results were attained in spite of heavy and continuous solar perturbations, as the controller proved useful not only against scattered clouds but also with large banks of passing clouds. From the operator's point of view, the more stringent testing requirements are, the more stressful and demanding a task manual controlled operation becomes. Therefore, the automatic control architecture here designed based on fuzzy logic makes a worthy and helpful companion to all operators, releasing them from the monotonous part of set point tracking so that they can focus on monitoring activities, thus providing more safety and accuracy to tests.

7.2.7 Summary

An application of automatic control strategies to a solar plant for material treatment has been shown. This kind of plant is manually operated by skilled operators and so the development of an automatic control strategy aimed at achieving adequate results throughout the wide range of operating conditions under which these plants operate, supposes an important improvement toward facilitating operation and obtaining the desired performance. The first control scheme presented is based on a

Table 7.3 Fuzzy vs. human statistics (courtesy of D. Lacasa et al., [221])

	Step	Rise time [s]	Rise rate [°C/s]	Overshoot [%]
Fuzzy control	650–700	42	1.1906	5.99
	700–800	47	2.1276	4.59
	800–1000	136	1.4706	2.45
	1000–800	149	−1.3423	−1.92
	800–700	107	−0.9345	−3.95
	700–650	53	−0.9434	−5.50
Operator	650–700	113	0.4545	5.94
	700–800	175	0.5714	3.33
	800–1000	234	0.8547	0.80
	1000–800	220	−0.9000	−1.14
	800–700	190	−0.5025	−1.12
	700–650	102	−0.4901	−4.04

PI controller which incorporates feedforward compensation, an anti-windup mechanism and actuator slew-rate constraint handling, both in fixed and self-tuning configurations. It has been developed and applied to the control of a solar furnace for samples of different materials under extreme temperature profiles. Different plant results have been shown and both advantages and drawbacks of the scheme have been commented on.

A fuzzy controller has been applied to the process of copper sintering in a solar furnace. The process can be briefly described as prone to high perturbation and strong changes in dynamics caused, respectively, by meteorological phenomena and shifting pattern requirements in the interest of research. Additional disturbing elements are the variable reflectivity index at the main collector heliostat, a hard non-linearity presented by the actuator, namely of slew-rate type and a shift in the physical properties of the sample while sintering. Notwithstanding, the fuzzy controller yielded good results throughout the whole range of operating conditions. Programming and deployment time were negligible in comparison with simulation and testing. Contrasting with expert human manual operation, the fuzzy approach outperformed it in terms of rise time, while a human operator could achieve smaller overshoots.

7.3 Solar Refrigeration

7.3.1 Introduction

Air conditioning consumes a lot of electric energy. An important factor of conditioning systems is the relationship between solar irradiance and the ambient temperature and the refrigeration demand (related to temperature). Most people use their air

Fig. 7.20 Solar plant

conditioning units when it is hot and this high ambient temperature usually occurs together with high solar radiation during daytime. In a passive solar energy system such as solar collectors, solar radiation is the main energy source and it is therefore, appropriate as the energy source for a cooling system [284].

There are many advantages coming from the use of solar energy for air conditioning systems. The use of renewable energies instead of oil or coal, which cause much more pollution, is always desirable and innovative. Furthermore, solar radiation for cooling is particularly attractive, since the more sunshine the better the conditions for the plant to work properly for cooling purposes.

Among the multiple methods existing for refrigeration using solar radiation [338], one of the most successful is based on an absorption machine which produces cold water at the output from hot water previously heated by the Sun at the input. These types of system are difficult to control because the solar radiation cannot be manipulated and there are great disturbances due to changes in environmental conditions and the presence of dead-times due to fluid transportation, variable cooling demand, etc. [85].

A picture from the plant located at the System Engineering and Automatic Control Department of the University of Seville can be seen in Fig. 7.20. It is similar to that of the CIESOL building at the University of Almería shown in Chap. 1 in Fig. 1.16. Both plants have been used to perform experiments that will be shown in the following sections.

The solar air conditioning system of the University of Seville consists of a collector field that produces hot water which feeds an absorption machine as described previously. The main components of the system which are depicted in Fig. 7.21 can be described as follows:

- The main source of energy is solar irradiance (I) which is used by the solar collectors to heat the circulating water. The field is composed of 151 m^2 of flat collectors and supplies a nominal power of 50 kW.
- The accumulation system is made up of two tanks (2500 l) working in parallel and it stores warm water. This water, which should be at a lower temperature

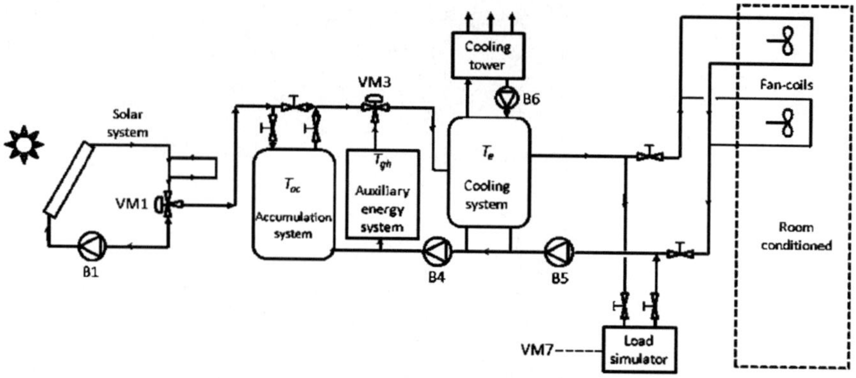

Fig. 7.21 Schematic diagram of the air conditioned plant

($T_{st} = T_{ac}$), mixes with that returning from the collector field. The mixture is controlled by the three-way valve, VM1, according to the following rule:
 - VM1 at 0%: The water from the collector field is totally recirculated to the collector field.
 - VM1 at 100%: The water is totally transported to the storage water tanks.
- The cooling system is an absorption machine that works with water as a cooling fluid and it requires its inlet temperature (T_{sc}) to be within the range of 75–100°C to work adequately. Different processes (absorption, evaporation, etc.) take place to produce cold water.
- An auxiliary energy system consisting of a gas-fired heater can be used to complement the energy supplied by the collector field when there is not enough solar radiation.
- A load simulator (a heat pump) that allows tests for different load profiles to be performed.

The plant operates with two different energy sources (solar and gas), which can be combined or used independently. Furthermore, thermal energy coming from a storage tank can be added to the system. The plant can be re-configured on-line, manipulating open/close valves and pumps (on/off) to allow the selection of the components for energy supply.

The plant evolves among several operating modes throughout daily operation. There are many operating possibilities but only 13 operating modes make sense: (1) recirculation, (2) loading the tanks with hot water, (3) using the solar collectors only, (4) using the solar collectors and a gas heater, (5) using a gas heater only, (6) using the tanks and a gas heater, (7) using the tanks only, (8) loading the tanks and using a gas heater, (9) recirculation and using a gas heater, (10) using the solar collectors and loading the tanks, (11) using the solar collectors and a gas heater and loading the tanks, (12) using the tanks and a gas heater to feed the absorption machine and recirculation in the solar collectors, and (13) using the tanks to feed the absorption machine and recirculation in the solar collectors.

The control objective is to supply chilled water to the air distribution system at the demanded temperature, minimizing auxiliary energy (gas) consumption and fulfilling operational constraints in the absorption machine. Furthermore, the stored energy in the tanks at the end of the day, is taken into account since it can be used on the following day when the solar irradiance is low. The primary energy source (solar radiation) cannot be manipulated and has to be treated as a measurable disturbance. This means that the control system must keep the cooling machine working at the desired operating point and this is achieved by keeping the machine inlet water temperature at the given set point. The inlet water is the mix of water coming from the solar system, the storage tanks and the gas-fired heater, when additional energy is needed. Additionally, the temperature of the water in the solar system can be controlled by adjusting the water flow inside the solar field.

7.3.2 Controllers for the Solar Air Conditioning Plant

The main problem encountered when controlling a solar energy process is that the primary source of energy cannot be manipulated and, from a control point of view, the solar irradiance acts as a perturbation. The control research community has contributed considerably in this field by designing advanced controllers for solar processes [85]. Some controllers applied to the solar air conditioning plant are described in the following. They are classified according to the operating modes considered by the controller, i.e. single and multiple operating modes.

7.3.2.1 Single Operating Mode

Many controllers have been tested at the actual plant. In [119] a robust controller based on the H_∞ mixed sensitivity problem was applied to regulate the generator inlet temperature. The controller includes a feedforward action to deal with the measurable disturbances. In [275] the Smith predictor generalized predictive control (SP-GPC) is used to control the generator inlet temperature. The controller includes the robustness filters and a feedforward action and also considers system constraints. A robust sliding mode predictive control is developed in [148]. It is based on the idea of a combination of model predictive control and sliding mode control. The main idea is to introduce the prediction of the sliding surface into the control objective function of a model predictive control. The resulting control law has few parameters, all with clear meaning for tuning. The proposed controller was applied to control the output temperature of the collector subsystem at a solar air conditioning plant. It has variable time delay with non-linear behavior that produces minimum and non-minimum phase response depending on the operating point. Validation at a real plant showed an enhanced ability to handle set point changes and disturbance rejection. In [285] a switching control procedure was developed to control the solar field of the air cooling plant taking into account changes in plant dynamics. Several control systems composed of IMC-based PID controllers and feedforward compensators were designed for each operation region and a continuous switching mechanism for the overall control system able to achieve a constant temperature at the absorption machine inlet in spite of changes in solar irradiance was defined.

7.3.2.2 H_∞ S/T Mixed Sensitivity Problem-Based Controller Design

The H_∞ control theory has received a lot of attention in the last decade within the research community due to the robustness characteristics supplied by its controllers. These features make it of a priori interest to be used in controlling such solar refrigeration systems. The basic idea is to minimize the ratio between the energy of the error vector and the energy of the exogenous signals [353]. The sub-optimum solution of the problem based on the S/T or S/KS/T mixed sensitivity problem for building up the generalized plant, allows the controller to be obtained by just designing a nominal model and some suitable weighting matrices.

In this application [119], the H_∞ control is used to regulate the absorption machine inlet temperature which establishes the evaporator and absorber pressures. It is also worth mentioning that the controller contains a feedforward action to treat system disturbances. The results of the application are tested both in simulation and in real plant experiments.

The feedback controller design problem for this system can be formulated as an H_∞ optimization problem with suitable features of disturbance rejection and robustness. The optimal H_∞ problem has not been solved yet but there is a solution for the suboptimal problem where the value of the energy ratio is decreased as much as possible by means of an iteration process. This synthesis process is used in this work and implemented in various well-known software packages [26].

A configuration for building up the generalized plant is the S/T mixed sensitivity problem [353], which is described in Fig. 7.22, where $P(s)$ is the generalized plant and $K(s)$ is the controller. The terms $W_S(s)$ and $W_T(s)$ constitute weighting functions which allow to specify the range of frequencies of most importance for the corresponding closed-loop transfer function.

Once the nominal model $G_N(z)$ has been chosen, the magnitude of the multiplicative output uncertainty can be estimated as follows:

$$\left|E_{o,i}\left(e^{j\omega T}\right)\right| = \left|G_i^*\left(e^{j\omega T}\right) - G_N\left(e^{j\omega T}\right)\right| \cdot \left|G_N\left(e^{j\omega T}\right)^{-1}\right| \tag{7.13}$$

where $G_i(z)$ stands for the different non-nominal systems at each operation point where the controller is required to work effectively.

As shown in [281], the weighting function $W_T(s)$ must be designed with the following conditions: stable, minimum phase and with module greater than the maximum singular value of the uncertainty previously calculated for each non-nominal model and frequency, that is,

$$\left|W_T(j\omega)\right| \geq \bar{\sigma}\left(E_{o,i}\left(e^{j\omega T}\right)\right) \quad \forall\omega,\ \forall i \tag{7.14}$$

In the case of $W_S(s)$, the following form is proposed:

$$W_S(s) = \frac{\alpha_W s + \omega_S}{s + \beta_w \omega_S} \tag{7.15}$$

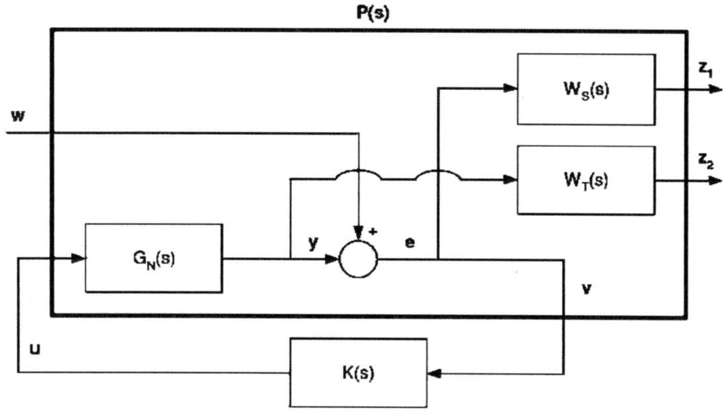

Fig. 7.22 S/T mixed sensitivity problem

where each of the parameters is designed in the following way:

- α_W is the function gain at high frequency. A suitable value is approximately about 0.5.
- β_W is the function gain at low frequency. A suitable low value for these parameter is 10^{-4}.
- ω_S is the crossover frequency of the function; as an initial value, a decade below the crossover frequency of the function $W_T(s)$ previously designed. It is proposed to change ω_S according to the expression $\omega_S = 10^{(\kappa_W - 1)}\omega_T$ in order to shape the desired speed of the output response [335]. The parameter κ_W is employed to vary the value of ω_S once the value of $\omega_T(s)$ has been obtained. The initial value of ω_S is obtained for κ_W equal to zero, while a value of this parameter equal to one shows that ω_S is equal to ω_T. Therefore, the final selection of this frequency is determined by an adimensional parameter where the value must be higher as the desired response speed increases [281].

7.3.2.3 Application in the Solar Plant

Although the real process has non-linear behavior, as is the case of many industrial processes, if working variables are kept close to a particular operating point, a linear model computed considering small changes around this point will be sufficient. It will be necessary to model VM1–T_{sc} and I–T_{sc} transfer functions. Both VM1–T_{sc} and I–T_{sc} dynamics are similar to a classic first-order system with a certain delay, so an *ARX* 1 2 0 structure (with a sampling time of 40 s) is selected to model them. Enough data were obtained from several experiments to identify VM1–T_{sc} and I–T_{sc} characteristics. These data are subject to a parametric model identification process implemented in a well-known software packet [229]. The obtained

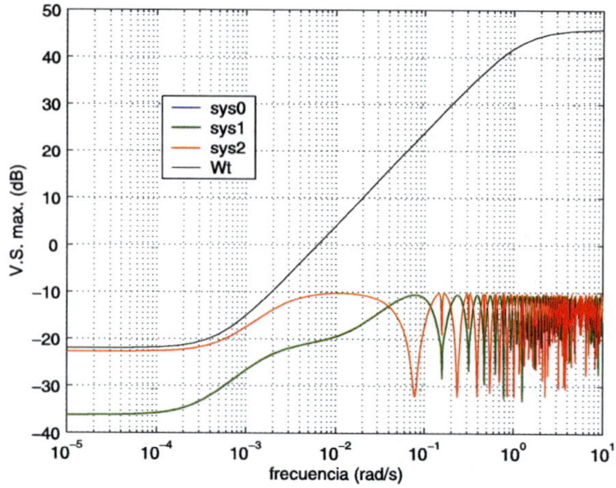

Fig. 7.23 Multiplicative output uncertainties and weighting function $W_T(s)$ [335]

VM1–T_{sc} transfer function is shown in Eq. (7.16), which corresponds to a nominal 50% valve opening:

$$G_N(z) = \frac{0.001119z - 0.08281}{z - 0.9264} z^{-3} \tag{7.16}$$

In the I–T_{sc} model identification experiment, a climatologic phenomenon such as clouds must be expected to produce enough variability in irradiance to give valid information. The estimated transfer function is

$$G_d(z) = \frac{0.002049z - 0.000785}{z - 0.9836} z^{-2} \tag{7.17}$$

These dynamic models, despite their simplicity, are considered good enough for control purposes.

In order to design the controller the nominal working point has been taken as opening the VM1 valve to 50%. The corresponding uncertainty diagram and the associated $W_T(s)$ function (Eq. (7.18)) are shown in Fig. 7.23:

$$W_T(s) = \frac{10^{-0.1}(2000s + 1)}{0.8s + 1} \tag{7.18}$$

Taking into account that the crossover frequency of function $W_T(s)$ is about $6.3 \cdot 10^{-3}$ rad/s, the function $W_S(s)$ is taken as shown in Eq. 7.19:

$$W_S(s) = \frac{0.5s + 10^{-0.25} \cdot 6.3 \cdot 10^{-3}}{s + 10^{-4} \cdot 10^{-0.25} \cdot 6.3 \cdot 10^{-3}} \tag{7.19}$$

Finally, a bilinear transformation of the nominal model has been applied to build up the generalized plant as a previous step to synthesize the controller. Thus, with these weighting functions, the controller is designed by means of suitable functions for the solution of the H_∞ suboptimal problem.

Several simulation experiments were carried out on a suitable model of the system. Simulations were performed in Simulink environment including some features

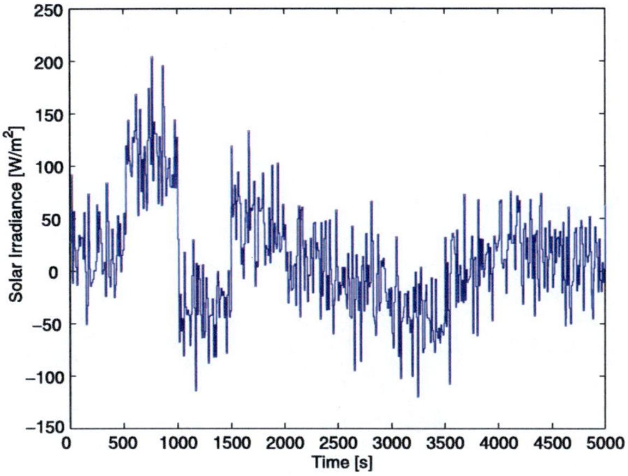

Fig. 7.24 Irradiance pattern used in simulations [335]

of the real process such as the solar irradiance effect and saturation in the VM1 valve action. Figure 7.24 shows the irradiance profile used and Fig. 7.25 shows the obtained response, in terms of output temperature (T_{sc}) and corresponding reference.

Some experiments to show the performance of the controller have been performed on the real system. A client–server data base access system (OPC) is used between the environment where the plant management system (CUBE) is implemented and the program which implements the control system (Matlab/Simulink). In this way, some numerical problems, such as controller fragility, are avoided.

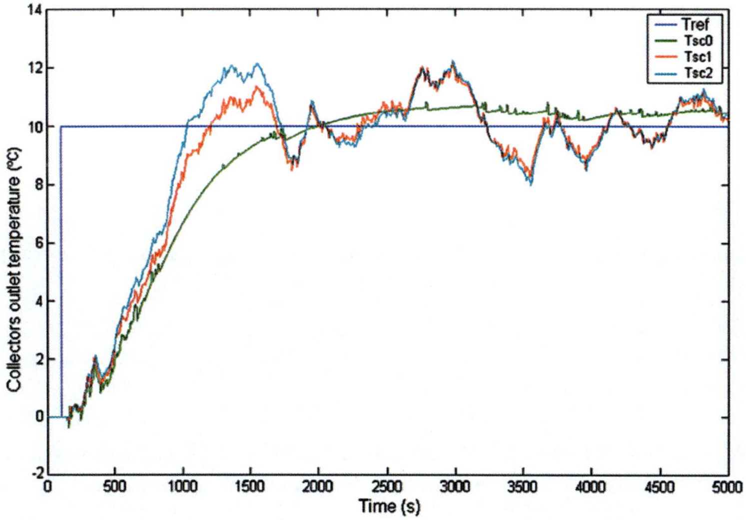

Fig. 7.25 Simulated tracking response in irradiance conditions in Fig. 7.24 [335]

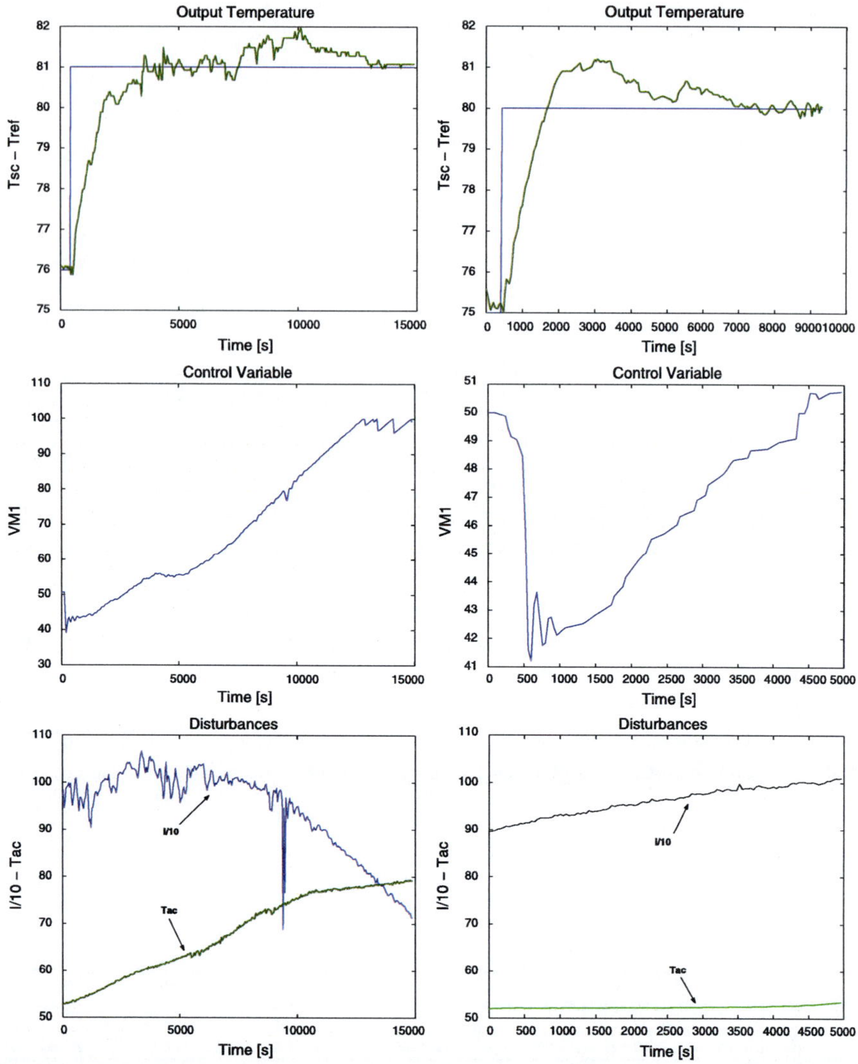

Fig. 7.26 Experimental results in VM1–T_{sc} control with: **(a)** *(left)* $\kappa_W = 0.2$ ($t_{rt} \simeq 3000s$ and $OS \simeq 0\%$); **(b)** *(right)* $\kappa_W = 0.8$ ($t_{rt} \simeq 1200s$ and $OS \simeq 20\%$) [335]

Figures 7.26(a) and 7.26(b) show the behavior of the output temperature (T_{sc}) together with its reference, the control action (VM1) and the disturbance variables, both solar irradiance (I) and temperature of water from accumulators ($T_{st} = T_{ac}$). Parameter κ has been modified in the different experiments and the approximated values of rising time (t_{rt}) and overshoot (OS) are also shown.

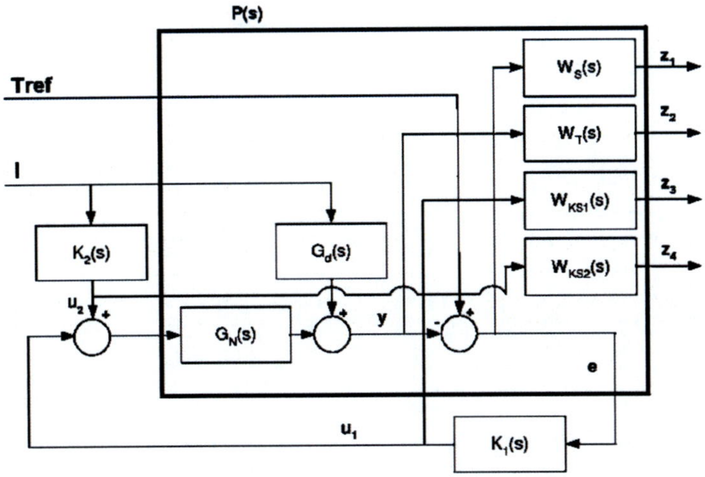

Fig. 7.27 Implemented feedforward control loop scheme, [335]

The effect of the κ_W value on the response speed and overshoot is clear in the figures. As κ_W increases, a faster response is obtained, although the overshoot of the response increases too.

Given the simple structure of the controller and despite the effect of the disturbances, the overall performance, especially regarding disturbance rejection (Fig. 7.26(a)), is fairly good, although this issue was not considered in the original control design.

An improvement in this control scheme will be made by a compensation loop of the disturbance introduced in the system by solar irradiance (I). The compensation mechanism will consist of a feedforward loop of the disturbances which will be included in the H_∞ calculus by means of a suitable algorithm. On introducing the feedforward loop in the structure that was previously shown for the VM1–T_{sc} control, the resulting scheme is like the one shown in Fig. 7.27.

The weighting functions $W_S(s)$ and $W_T(s)$ design methodology is exactly the same as that presented in previous sections. The main difference observed in this structure is that now, besides error signal and output, the control signals (u_1 and u_2) are also weighted (because of numerical problems in the synthesis algorithm) by means of $W_{KS_1}(s)$ and $W_{KS_2}(s)$, respectively, which have been taken equal to unity.

In Fig. 7.28, the response of the simulated system at each operating point can be observed. The simulations have been made for controllers calculated for different values of parameter κ_W in $W_S(s)$ weighting function.

In Fig. 7.29, the results obtained in an experiment at the real plant is shown. In these figures, an important improvement in overall control performance is observed thanks to including the feedforward loop. This fact is especially evident in Fig. 7.29, between time instants 7000 and 10500 approximately, where a proper response of the controller is observed when an important disturbance (caused by clouds) in the

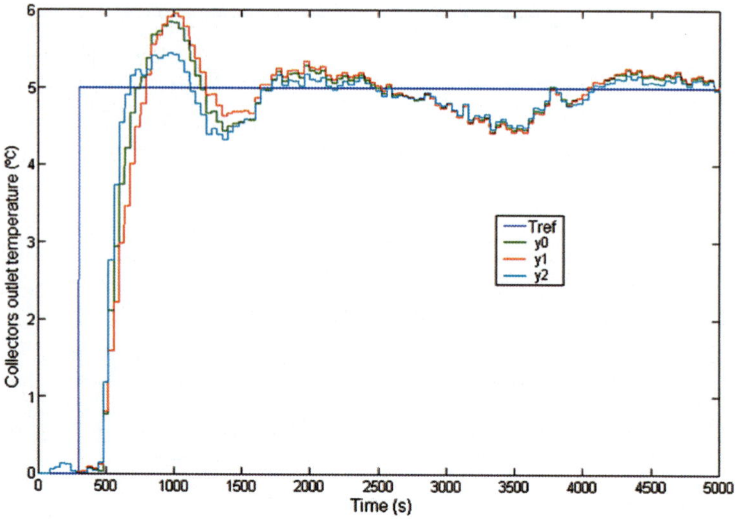

Fig. 7.28 Simulated tracking response in irradiance conditions in Fig. 7.24 [335]

irradiance pattern occurs. Other characteristics like rising time or overshoot are conserved and even improved from the previous controller.

7.3.3 Multiple Operating Modes

The plant was used as a benchmark within the network of excellence HYCON funded by the European Commission under FP6. The benchmark exercise [412] consisted of comparing the results obtained by each controller under simulation and the results of the controller working for one day at the real plant. The simulation results had to be obtained for two days with the given environmental conditions corresponding to a clear day followed by a day with scattered clouds. The following quantities were to be measured: mean square error of evaporator temperature tracking, energy consumed by the gas heater, value of the tank temperature at the end of each day. In the literature, there have been many contributions devoted to the development of techniques for modeling, analyzing and controlling hybrid systems but very few papers describing applications of hybrid control techniques. Most of the application papers use simulation models or very simple laboratory processes. The plant dealt with here, however, is a real solar refrigeration plant of certain complexity.

Different MPC approaches to the global operating control of this plant were presented for the benchmark exercise and four were selected to be published in special issue of the European Journal of control devoted to the benchmark exercise. The first approach can be seen in its extended version in [411], where a hierarchical scheme

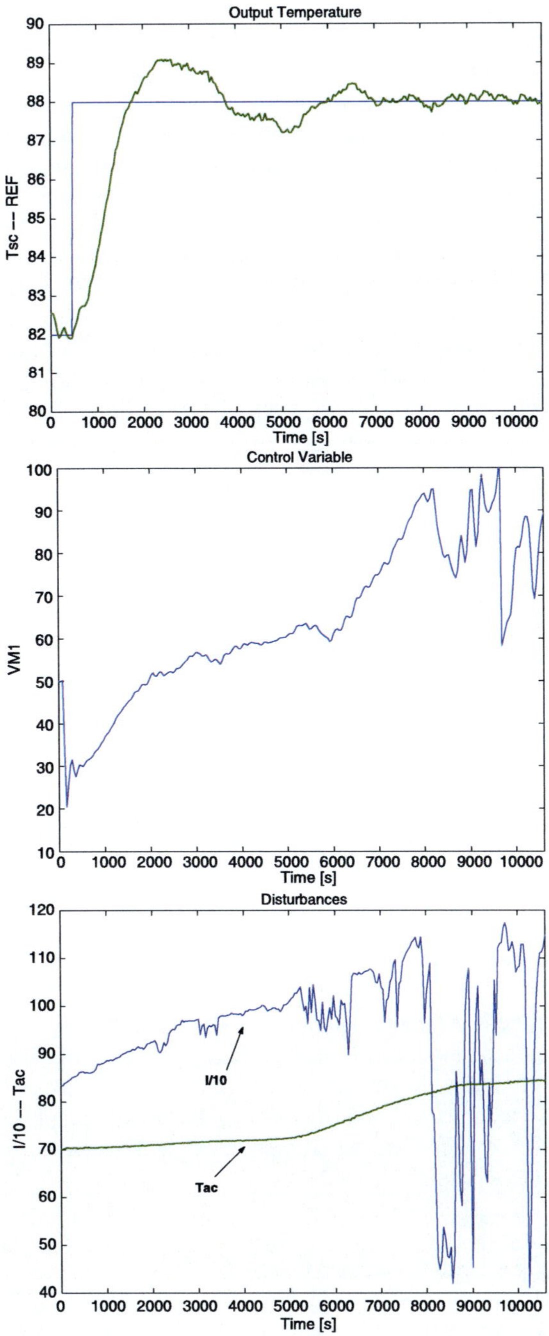

Fig. 7.29 Experimental results in VM1–I–T_{sc} control with $\kappa_W = 0.5$ ($t_{rt} \simeq 1100s$ and $OS \simeq 17\%$), [335]

is presented. The higher level selects the operating mode based on an integer optimization problem with variable weights and the lower level regulates the continuous variables using a set of MPC. The extended version of the second approach was presented in [355]. The supervisory control scheme was designed from process insight gained from a thorough analysis of the energetic and dynamic aspects of the system. The discrete process inputs are adjusted by switching between a set of operating modes and the settings of the continuous inputs are chosen by a look-up table in each operating mode. The third approach was presented in [322] and it is based on a model predictive control strategy developed to deal efficiently with the mixed discrete-continuous nature of the process. A novel approach incorporating an internal model with embedded logic control is used to transform the hybrid problem into a continuous non-linear one where NMPC can be applied. The fourth approach [251] proposes a multi-layer hybrid controller consisting of a high-level supervisor that decides the optimal operating mode on-line using a hybrid model predictive control strategy, a static lower-level controller defining proper set points for the chosen mode and existing standard low-level controllers that ensure robust tracking of such set points. Recently, in [414], a transition graph-based predictive algorithm was applied to the solar air conditioning plant. This algorithm switches between the operating modes of the plant on the basis of minimizing the cost functions, the performance cost function associated to each operating mode and the switching cost function associated to the transition between operating modes, to find the optimal solution to the configuration problem and the optimal control signal of the plant throughout the day. The transition graph was used to reduce the computational cost of the optimization problem.

7.3.3.1 Hybrid Model of the Solar System

The same approach as that has been commented on in Chap. 6, Sect. 6.8.4 in the case of the CRS has been applied to control the overall installation using a hybrid MPC control scheme. As shown in [284], a hybrid model of the system composed of a solar collector field, a gas heater and two tanks connected in series to store hot water can be developed (see Fig. 7.21). These systems are used as energy sources for an absorption machine whose objective is to obtain chilled water for the fan-coil system. All these components are included or excluded from the cooling system by means of switching valves and activating or deactivating signals. Switching among different configuration modes in real time increases the challenge of the system to be modeled or controlled, due to the mix of both discrete signals and continuous dynamics. The idea is to develop models for each energy source and to integrate them into a state-space model used to characterize the hybrid model, as in Sect. 6.8.4.

Solar Field Collector Model

The solar collector model used is the same as that explained in Chap. 4, Sect. 4.3.4.2, calibrated for the specific solar field layout at this plant [284].

Tank Model

The accumulation system, consisting of two separate tanks connected in series, was simplified by a lumped-parameter model. The tanks can be considered as a buffer and thus, they can be modeled as a transport delay between the inlet and the outlet temperature which depends on the flow. Equation (7.20) models the relationship between the tank outlet temperature and the tank input temperature, ambient temperature and fluid flow:

$$\begin{aligned}\rho_f c_f V_{st}\frac{dT_{stout}(t)}{dt} &= A_{st}\rho_f v(t)c_f\big(T_{stin}(t)-T_{stout}(t)\big)\\ &\quad + H_{st}S_{st}\big(T_a(t)-T_{stout}(t)\big)\end{aligned} \tag{7.20}$$

where ρ_f is the water density, c_f is the water specific heat, V_{st} is the tank volume [m^3], T_{stout} is the accumulation tank water outlet temperature [°C], A_{st} is the accumulation tank area [m^2], v is the fluid velocity in the system [m/s], T_{stin} is the accumulation tank water inlet temperature [°C], T_a the ambient temperature [°C], H_{st} is the tank's loss coefficient per surface unit [W/(m^2 °C)] and S_{st} is the tank's surface [m^2].

Simple Gas Heater Model

The gas heater behavior presents a double stage response when it is turned on. The first stage increases the temperature to 40% of the gas heater performance and after exactly 5 min, the gas flow opens totally, increasing the temperature at a second stage with a higher gas heater performance. This behavior was dealt with using a combination of first-order systems by means of four transfer functions divided into two sets, identifying the relationship between the gas heater inlet and outlet temperature and between the water flow and the outlet temperature when the gas heater is turned on and after 300 s.

The global transfer function (7.21) relates the output variable, $T_{gh,out}$ [°C], to the input temperature $T_{gh,in}$ [°C], fluid flow rate q [m^3/h], gas heater on signal steps G_{Hon} [–] and gas heater off signal steps G_{Hoff} [–].

$$\begin{aligned}T_{gh,out}(s) &= \frac{K_{T_{gh,in}}}{\tau_{T_{gh,in}}s+1}e^{-st_{d_{T_{gh,in}}}}T_{gh,in}(s)+\frac{K_F}{\tau_F s+1}q(s)\\ &\quad+\frac{K_{G_{Hon}}}{\tau_{G_{Hon}}s+1}\big(1+e^{-st_{d_{G_{Hon}}}}\big)e^{-st_{d_{G_{Hon2}}}}G_{Hon}(s)\\ &\quad+\frac{K_{G_{Hoff}}}{\tau_{G_{Hoff}}s+1}e^{-st_{d_{G_{Hoff}}}}G_{Hoff}(s)\end{aligned} \tag{7.21}$$

where K_i, τ_i, t_{d_i} are the static gains, time constants and time delays, respectively, for the different first-order transfer functions and $T_{gh,in}$, q, G_{Hon} and G_{Hoff} subscripts refer to the values for input temperature, mass flow, gas heater on signal steps and gas heater off signal steps, respectively.

Table 7.4 Operating point conditions (courtesy of M. Pasamontes et al., [284])

Input	Description	Nominal value	Units
$T_{in,0}$	Collectors inlet temperature	65	°C
$T_{out,0}$	Collectors outlet temperature	75	°C
I_0	Irradiance	600	W/m^2
$T_{a,0}$	Ambient temperature	20	°C
v_0	Fluid velocity	0.8333	m/s
$T_{stin,0}$	Tank's inlet temperature	75	°C
$T_{stout,0}$	Tank's outlet temperature	75	°C
q_0	Flow rate	6	m^3/h
$T_{gh,in,0}$	Gas heater inlet temperature	75	°C
$T_{gh,out,0}$	Gas heater outlet temperature	79	°C

Space-State Linear Models

Once the models for the different subsystems are developed, they are linearized around a specific operating point (Table 7.4), converted into state-space form and discretized.

The resulting multiple inputs, single output (MISO) system and the input vector $\mathbf{u}(k)$ are shown in (7.22) and (7.23), respectively.

$$\begin{aligned} \mathbf{x}(k+1) &= \begin{bmatrix} A_c & 0 & 0 \\ 0 & A_{st} & 0 \\ 0 & 0 & A_{gh} \end{bmatrix} \mathbf{x}(k) + \begin{bmatrix} B_c & 0 & 0 \\ 0 & B_{st} & 0 \\ 0 & 0 & B_{gh} \end{bmatrix} \mathbf{u}(k) \\ y(k) &= [\, C_c \quad C_{st} \quad C_{gh} \,] \mathbf{x}(k) \end{aligned} \tag{7.22}$$

$$\begin{aligned} \mathbf{u}(k) = \big[& I(k) \quad T_{in}(k) \quad T_a(k) \quad v(k) \quad T_{stin}(k) \quad T_a(k) \quad v(k) \quad T_{gh,in}(k) \quad q(k) \\ & G_{Hon}(k) \quad G_{Hoff}(k) \big]^T \end{aligned} \tag{7.23}$$

where $y(k)$ is the system output temperature and the subscripts c, st and gh refer to solar collectors, storage tanks and the gas heater, respectively. Every matrix and vector has been adjusted for suitable dimensions and $\mathbf{x}(k)$ is initialized to zero.

Plant Hybrid Configuration Modes

The space-state model obtained in the previous section only allows the system response to be simulated when all the cooling system components work together but this model is not valid for all the other plant configuration modes. The on–off valve position and system dynamics change at each different plant configuration mode (Fig. 7.30). The presence of both continuous and discrete dynamics defines the cooling process as a hybrid system. Changes in the valve position will cause changes in the system configuration modes and, therefore, in the system dynamics. These transitions among system configuration modes must be ensured to be continuous. More information about the nature of hybrid systems can be found in [33].

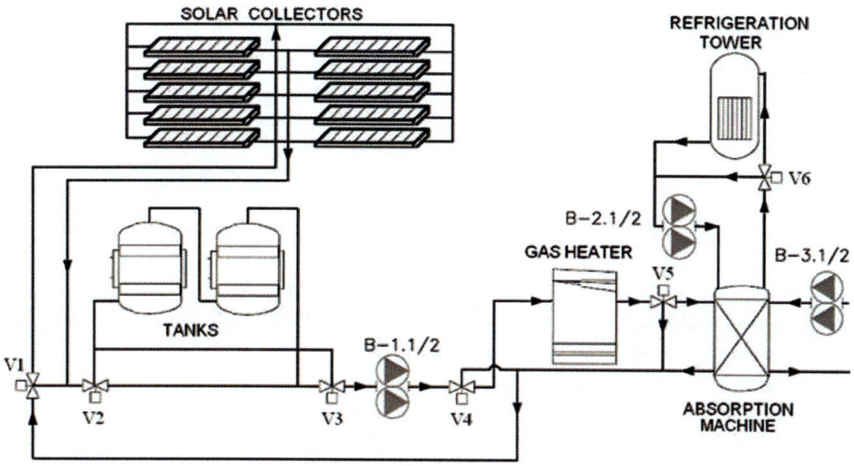

Fig. 7.30 Solar cooling plant scheme (courtesy of M. Pasamontes et al., [284])

The valves and signals, all discrete, which influence the system configuration mode during operation, are:

1. Gas heater activation/deactivation signal (referenced as G_H). Values $= \{1, 0\}$, respectively.
2. Valve V3 state signal. This defines the tank's operating mode between no tank use (V3 $= 1$) and load/download (V3 $= 0$) depending on the plant operation mode.
3. Valve V1 state signal. This defines the collector's operating mode between not using the collectors (V1 $= 0$) and using them (V1 $= 1$). Note that V1 is continuous (0–100%), but it will be used as discrete, allowing only 0% and 100% values.

The combination between these logical variables (V1, V3, and G_H) and the state-space model defined in (7.22) results in the next state-space hybrid model:

$$\begin{aligned}\mathbf{x}(k+1) &= \begin{bmatrix} A_c * \mathrm{V1} & 0 & 0 \\ 0 & A_{st} * \neg\mathrm{V3} & 0 \\ 0 & 0 & A_{gh} * G_H \end{bmatrix} \mathbf{x}(k) \\ &\quad + \begin{bmatrix} B_c * \mathrm{V1} & 0 & 0 \\ 0 & B_{st} * \neg\mathrm{V3} & 0 \\ 0 & 0 & B_{gh} * G_H \end{bmatrix} \mathbf{u}(k) \\ y(k) &= [\, C_c * \mathrm{V1} \quad C_{st} * \neg\mathrm{V3} \quad C_{gh} * G_H \,]\, \mathbf{x}(k)\end{aligned} \tag{7.24}$$

Simplifying the operation models described at the beginning of the section, it is possible to distinguish the following seven different operating modes in this hybrid system (see Fig. 7.31):

1. Operation mode s1: Solar collector output to absorption machine. The solar collector output water is sent to the absorption machine.

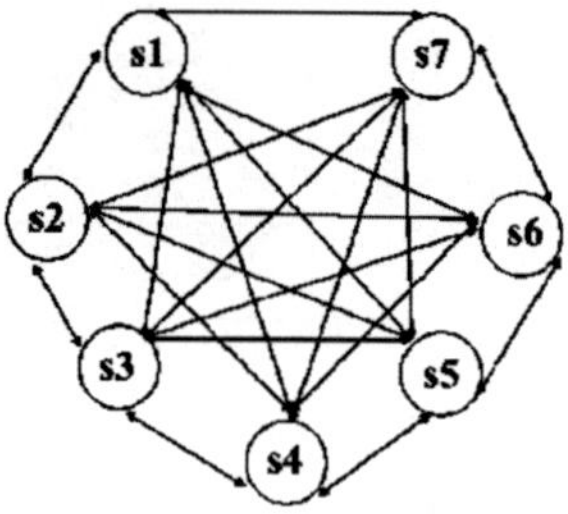

Fig. 7.31 Cooling System FSM (courtesy of M. Pasamontes et al., [284])

2. Operation mode s2: Tank output to absorption machine. Hot water from the tanks is sent to the absorption machine.
3. Operation mode s3: Gas heater output to absorption machine. The gas heater output water is sent to the absorption machine.
4. Operation mode s4: Solar collector and tank output to absorption machine. Hot water from the solar collector output pulls the tank water which finally arrives at the absorption machine. This fact allows the tanks to be used as a buffer.
5. Operation mode s5: Solar collector and gas heater output to absorption machine. The solar collector output water is sent to the gas heater (turned on) and from there it arrives at the absorption machine.
6. Operation mode s6: Tank and gas heater output to absorption machine. The hot water from the tanks is sent to the gas heater (turned on) and from there, it is sent to the absorption machine. It helps to reduce the gas heater output temperature oscillation.
7. Operation mode s7: Solar collectors, tanks and gas heater output to absorption machine. The hot water from the solar collector goes into the tanks and from there, it enters the gas heater (turned on), finally arriving at the absorption machine.

7.3.4 System Hybrid Model

The different system configuration modes can be defined according to the signal values for every state, as shown in (7.25) and the transitions among states is defined in an FSM (Finite State Machine) such as shown in Fig. 7.31.

$$\begin{aligned}
\mathrm{s1} &= \mathrm{V1} \wedge \mathrm{V3} \wedge \neg G_H \\
\mathrm{s2} &= \neg\mathrm{V1} \wedge \neg\mathrm{V3} \wedge \neg G_H \\
\mathrm{s3} &= \neg\mathrm{V1} \wedge \mathrm{V3} \wedge G_H \\
\mathrm{s4} &= \mathrm{V1} \wedge \neg\mathrm{V3} \wedge \neg G_H \\
\mathrm{s5} &= \mathrm{V1} \wedge \mathrm{V3} \wedge G_H \\
\mathrm{s6} &= \neg\mathrm{V1} \wedge \neg\mathrm{V3} \wedge G_H \\
\mathrm{s7} &= \mathrm{V1} \wedge \neg\mathrm{V3} \wedge G_H
\end{aligned} \tag{7.25}$$

Table 7.5 Transitions among states (courtesy of M. Pasamontes et al., [284])

From	To	Events	From	To	Events
s1	s2	$\neg$V1 $\wedge$ $\neg$V3	s2	s1	V1 $\wedge$ V3
s1	s3	$\neg$V1 $\wedge$ G_H	s3	s1	V1 $\wedge$ $\neg G_H$
s1	s4	$\neg$V3	s4	s1	V3
s1	s5	G_H	s5	s1	$\neg G_H$
s1	s6	$\neg$V1 $\wedge$ $\neg$V3 $\wedge$ G_H	s6	s1	V1 $\wedge$ V3 $\wedge$ $\neg G_H$
s1	s7	$\neg$V3 $\wedge$ G_H	s7	s1	V3 $\wedge$ $\neg G_H$
s2	s3	V3 $\wedge$ G_H	s3	s2	$\neg$V3 $\wedge$ $\neg G_H$
s2	s4	V1	s4	s2	$\neg$V1
s2	s5	V1 $\wedge$ V3 $\wedge$ G_H	s5	s2	$\neg$V1 $\wedge$ $\neg$V3 $\wedge$ $\neg G_H$
s2	s6	G_H	s6	s2	$\neg G_H$
s2	s7	V1 $\wedge$ G_H	s7	s2	$\neg$V1 $\wedge$ $\neg G_H$
s3	s4	V1 $\wedge$ $\neg$V3 $\wedge$ $\neg G_H$	s4	s3	$\neg$V1 $\wedge$ V3 $\wedge$ G_H
s3	s5	V1	s5	s3	$\neg$V1
s3	s6	$\neg$V3	s6	s3	V3
s3	s7	V1 $\wedge$ $\neg$V3	s7	s3	$\neg$V1 $\wedge$ V3
s4	s5	V3 $\wedge$ G_H	s5	s4	$\neg$V3 $\wedge$ $\neg G_H$
s4	s6	$\neg$V1 $\wedge$ G_H	s6	s4	V1 $\wedge$ $\neg G_H$
s4	s7	G_H	s7	s4	$\neg G_H$
s5	s6	$\neg$V1 $\wedge$ $\neg$V3	s6	s5	V1 $\wedge$ V3
s5	s7	$\neg$V3	s7	s5	V3
s6	s7	V1	s7	s6	$\neg$V1

The FSM defines the possible transitions among system states and the logic variable values triggering the different transitions. The outputs define the state variables (s1 to s7) values at every sampling time. Transitions are listed in Table 7.5.

System MLD Model

A Discrete Hybrid Automata (DHA) is a connection of a finite state machine (FSM) and a switched affine system (SAS) through a mode selector (MS) and an event generator (EG) [373].

A set of linear dynamic systems defining the cooling system has been integrated into a state-space model. Their interconnection has been defined as a set of states, defined as an automaton (see Fig. 7.31) and the transition among these states has been defined according to a set of logical rules according to Table 7.5. This allows a DHA for the system to be defined. From this representation, an abstract representation in a set of constrained linear difference equations involving mixed-integer and continuous variables may be found that yield the equivalent Mixed Logical Dynamical (MLD) model. Using the composed state-space model as defined in (7.22), the system was defined in HYSDEL, a tool for modeling used with the Matlab hybrid toolbox for obtaining the MLD model for simulation and control. More information about HYSDEL can be consulted in [373].

The MLD modeling framework has been chosen from among others hybrid system techniques due to its extensive use in modeling hybrid systems and because the resulting model can easily be used in the future to develop controllers. From this framework, it is possible to model the evolution of continuous variables through linear dynamic discrete-time equations, the discrete variables through propositional logic statements and automata and the interaction of both. The main idea of this approach is to embed the logical part in the state equations by transforming boolean variables into 0–1 integers and by expressing the relationships as mixed-integer linear inequalities. The MLD model, which is introduced in [373], related to the previous state-space model, is

$$\begin{aligned}\mathbf{x}(k+1) &= \mathbf{A}\mathbf{x}(k) + \mathbf{B}_1\mathbf{u}(k) + \mathbf{B}_3\mathbf{z}(k)\\ \mathbf{y}(k) &= \mathbf{C}\mathbf{x}(k)\\ \mathbf{E}_2\varsigma(k) + \mathbf{E}_3\mathbf{z}(k) &\leq \mathbf{E}_1\mathbf{u}(k) + \mathbf{E}_5\end{aligned} \tag{7.26}$$

where $\mathbf{x}(k) \in \mathbb{R}^{n_c} \times \{0,1\}^{n_l}$ is a vector of continuous and binaries states, $\varsigma(k) \in \{0,1\}^{r_l}$, $\mathbf{z}(k) \in \mathbb{R}^{r_c}$ represent auxiliary binary and continuous variables, respectively, $\mathbf{y}(k) \in \mathbb{R}^{p_c} \times \{0,1\}^{p_l}$ is the output (in this case, the absorption machine inlet temperature) and $\mathbf{u}(k) \in \mathbb{R}^{m_c} \times \{0,1\}^{m_l}$ are the inputs, including both discrete (V1, V3 and V4 valves and gas heater state signals) and continuous ones (irradiance, ambient temperature, flow and the system inlet temperature). Finally, $\mathbf{A}$, $\mathbf{B}_1$, $\mathbf{B}_3$, $\mathbf{C}$, $\mathbf{E}_1$, $\mathbf{E}_2$, $\mathbf{E}_3$, $\mathbf{E}_5$ are matrices of suitable dimensions.

Model Validation

Every model developed in this work has been individually tested and validated with data from the real system, but only the full hybrid model will be presented here to save space. Two experiments of 400 and 150 min are presented. The latter experiment is so short because it has been prepared to force changes in the operation modes of the plant.

The first validation test uses real data collected on 28/08/08. Figures 7.32(a) and 7.32(b) show the environment conditions for the validation test and the system volumetric flows, respectively. Figures 7.32(c) and 7.32(d) show the real system and simulated output temperature and the operation mode, respectively.

The experiment started when the system output temperature was close to the operating point. For the first 190 min, the irradiance increases from 480 W/m^2 to 600 W/m^2 and the ambient temperature stays high (solid and dashed lines in Fig. 7.32(a), respectively). Valve V1 stays open at 50%, setting a stable volumetric flow. During this first half, the system configuration mode is set to s1 (see Fig. 7.32(d)), where the solar collectors are fixed in the system configuration.

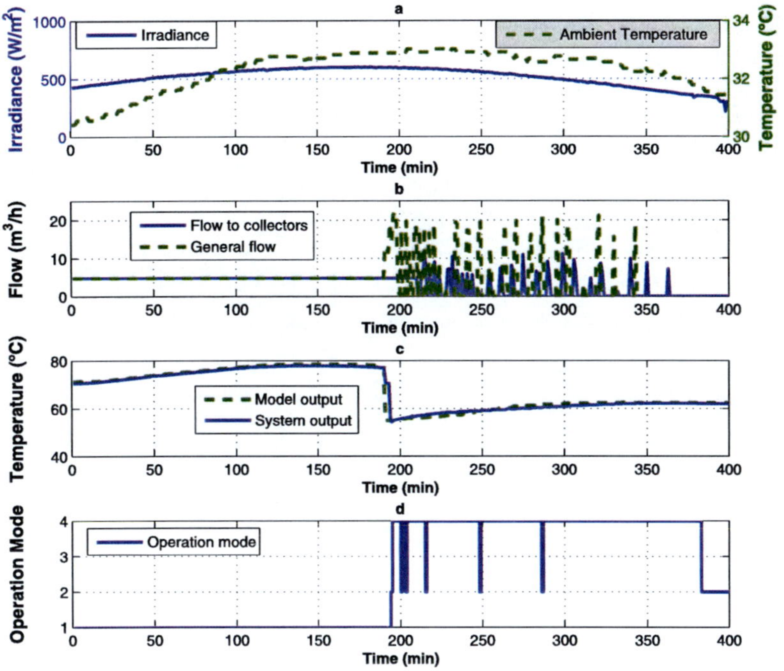

Fig. 7.32 Hybrid model results. Date 28/08/08 (courtesy of M. Pasamontes et al., [284])

After the first 190 min approximately, the radiation starts to decrease and the tanks with water stored with a temperature of around 52°C are connected into the system configuration. This water is pulled out to the pipes, cooling down the water temperature in the primary circuit, as can be observed in Fig. 7.32(c), where the solid line is the real temperature output and the dashed line the simulated one. In this second half, the system configuration mode changes between modes s2 and s4 (see Fig. 7.32(d)), where tanks are fixed and the solar collectors are included and excluded from the system configuration.

As can be seen in Fig. 7.32(c), the real and simulated temperatures are similar and the most discording point is found in the mode change, where the model response is faster than the real system one, causing a faster temperature decrease in the model temperature output.

The second test uses real data collected on 07/09/08 during the afternoon. Figure 7.33 shows how the system is forced to change among modes more frequently than usual. Figures 7.33(a) and 7.33(b) show the environment conditions for the validation test and the system volumetric flows, respectively. Figures 7.33(c) and 7.33(d) show the real system and simulated output temperature and the operation mode, respectively.

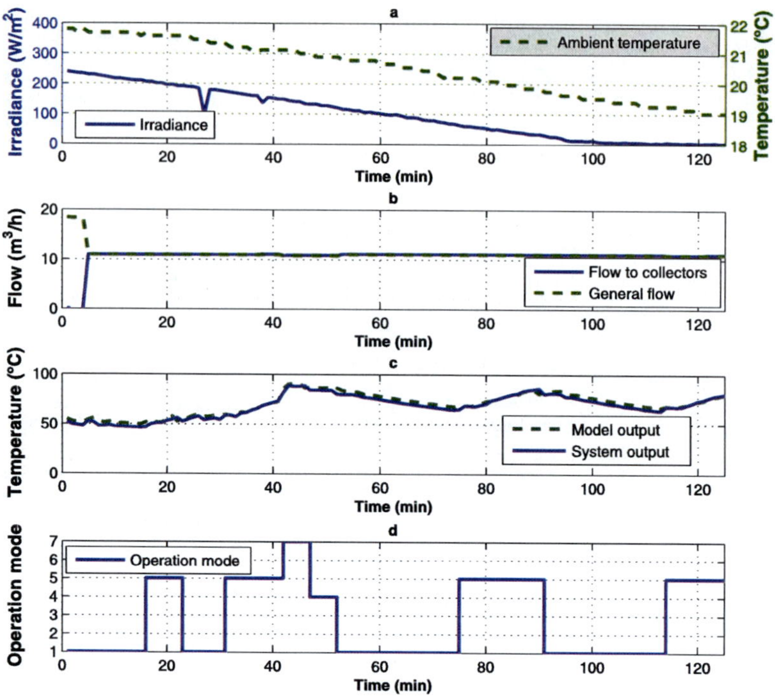

Fig. 7.33 Hybrid model results. Date 07/09/08 (courtesy of M. Pasamontes et al., [284])

Modes s1, s4, s5 and s7 define the system behavior. The gas heater was frequently turned on during the experiment, causing increases in the system output temperature. The changes in the operation mode are shown in Fig. 7.33(d).

At the start of the experiment, the system output temperature is around 50°C and only the solar collectors are connected (mode s1). From minute 15 to minute 42, the system configuration mode swaps from s1 to s5 and back again, producing a low, but continuous, increase in the system output temperature. After 42 min, the system configuration includes solar collectors, tanks and gas heater for 6 min. The temperature of the tank is 88°C, so a fast increase in the system output temperature is produced when it is included in the system.

Next, the system operation mode is changed to s4, turning off the gas heater for 7 min and then the operation mode changes from s4 to s1, excluding the tanks from the system and causing a temperature decrease for the next 23 min.

Finally, the system configuration mode swaps between s1 and s5, where solar collectors are fixed and the gas heater is included and excluded from the system configuration. The effect on the gas heater on the output temperature can be observed in Fig. 7.33(c).

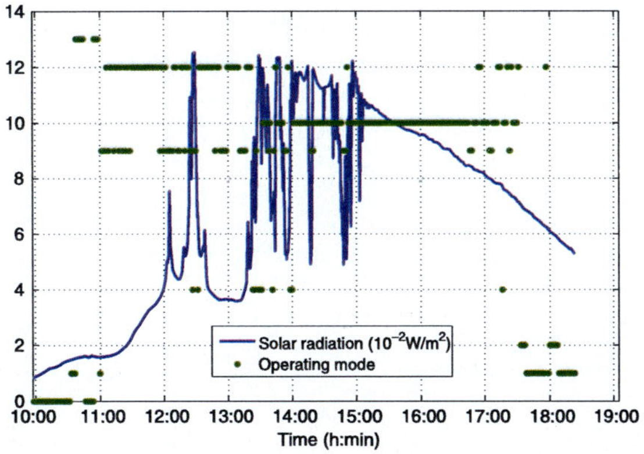

Fig. 7.34 Operating mode and solar irradiance (courtesy of D. Zambrano et al., [411])

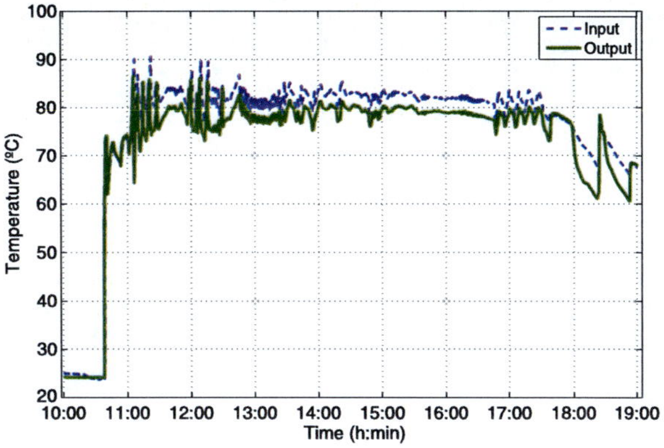

Fig. 7.35 Inlet and output temperatures of the generator (courtesy of D. Zambrano et al., [411])

7.3.5 Hybrid Control Results

Using the hybrid model descriptions of the plant, different MPC control approaches can be applied, as those commented on in Chap. 6, Sect. 6.8.4. As an example, some of the results obtained using a hierarchical scheme [411] can be seen in Figs. 7.34 and 7.35. Figure 7.34 shows the operating modes throughout the day and the solar irradiance. It can be seen that the Hybrid MPC system decides the appropriate operating mode. Figure 7.35 shows the inlet and outlet temperatures at the generator. Notice that the oscillations observed are due to the on–off gas heater, the behavior is smooth when the solar field produces enough energy.

7.3.6 Summary

Different control strategies have been applied to the air conditioning plant and as an example, a robust controller based on the H_∞ mixed sensitivity problem for the control of a solar refrigeration plant has been designed. The resulting method is simple and it is limited to shaping only one parameter for each of the outputs considered. A basic feedback structure and a more complex one including a disturbance feed-forward compensation loop have been designed. These controllers have been tested using the real plant, concluding that the performance achieved justifies the use of such a kind of regulator.

Chapter 8
Integrated Control of Solar Systems

8.1 Introduction

This chapter deals with the upper control level of solar power plants. Models for predicting solar irradiance and electrical loads, as well as models of the energy storage systems and power conversion systems, are needed to generate optimal operating modes and the corresponding set points, which are then sent to the lower level controllers.

8.2 Operational Planning of Solar Plants with Parabolic Trough Collectors

The price of electrical energy is needed to determine the best operating modes and hourly electrical production. This changes with demand. Prediction models for electrical demand are useful for decision making and for the design of hierarchical control schemes capable of optimizing electrical production, being necessary to model all plant subsystems; that is, the solar collector field, the thermal storage system and the power conversion system (PCS). Electrical demand may change substantially depending on the date and environmental conditions.

Since optimal scheduling of solar power plants depends greatly on predictions about solar radiation and other weather conditions, optimal scheduling windows for this type of plant are limited by the realistic predicting windows of weather conditions. There are three well differentiated levels: weekly planning, daily planning and tracking [144].

- *Weekly planning*. The planned production for the week is made at this stage. The weekly plan is made taking into account the weather forecast for the following seven days, the predicted network loads and prices, as well as the scheduled maintenance operations.
- *Daily planning*. The objective of this stage is to determine the planned plant production schedule for the day. The decision mechanism has to take into account

E.F. Camacho et al., *Control of Solar Energy Systems*, Advances in Industrial Control, DOI 10.1007/978-0-85729-916-1_8,

the state of the storage system, the weather prediction for the day as well as the price of electricity for the day.

- *Tracking*. The objective at this level is to control the planned production by generating the appropriate set points to the different subsystems. At a lower level, different controllers such as those described in Chaps. 4 and 5 control the fundamental process variables such as the outlet HTF temperature of the solar field subsystem. In this case, the mission of the set point optimizer is to determine the optimal outlet oil temperature for the given solar radiation level, inlet oil temperature and ambient temperature which optimizes the energy collected and minimizes ancillary energy consumption. The outlet HTF temperature set point is sent to the lower level controller, which regulates the HTF temperature as indicated in Chaps. 4 and 5.

8.2.1 Subsystems Modeling

Planning and control of solar plants calls for dynamical models of their main parts. This requires modeling the following elements: the solar field (Chap. 4), the thermal storage system and the power conversion system (PCS). A prediction model of the solar radiation (Chap. 2), electrical tariffs and electrical demands are also needed for optimal operation planning. Important variables to be considered are the temperature of the inlet HTF ($T_{in_{Tur}}$) and of the outlet HTF ($T_{out_{Tur}}$) of the power block, as well as the HTF flow (q_T) going from the storage system to the PCS. Other variables with an economic profile in the objective function appear, such as operational cost C_{op}, or electrical selling price tariff e_1, or net benefits, B.

Solar radiation, ambient temperature and electricity selling prices are considered as perturbations. Predictions of these variables are needed for control. The field flow, (q) and the flow toward PCS, q_T are the manipulated variables. The outlet HTF temperature and generated power are the main variables to be controlled. The temperature of the hot HTF feeding the PCS can be predicted using a model of the storage tank. The overall hierarchical control scheme computes the temperature and power references T_{ref} and P_{ref} to maximize benefits while fulfilling the plant operating constraints [19]. The main prediction models are treated in what follows.

8.2.1.1 Solar Irradiance

Direct solar irradiance is the most important variable in thermal solar power plants, as the electrical energy generated depends basically on it. Different models for solar irradiance prediction and forecasting have been explained and developed in Chap. 2, for both short and long term forecasting.

8.2.1.2 Ambient Temperature

Ambient temperature affects the thermal losses of the solar field, heat storage tanks and turbines. It also affects the electrical demand and it may affect the prices of electrical energy. Low ambient temperature increases the need for heating and, therefore, demand. Similarly, high ambient temperature increases the need for air cooling and, therefore, electrical energy demand and energy prices. Ambient temperature predictions can be obtained from weather forecasting institutions and corrected for by using current measured variables for short term predictions, as done in the case of solar irradiance in Chap. 2.

8.2.1.3 Market

The prices of electrical energy on the intraday market depend on supply and demand. The offer curve used to be well known because it depended on only a few generating agents who, in the past, know their production capacities well in advance (except for unforeseen failures). The renewable energy market, especially the solar energy market, is affected by policy mechanisms designed by governments to encourage the use of renewable energy sources, such as the feed in tariffs (FIT). In some countries, Spain for example, the owners or promoters of these installations can use two different energy selling modes: (i) To sell all the energy to an electrical utility at fixed prices. (ii) To sell the electricity on the intraday market.

The renewable energy installations can choose, for periods of no less than one year, their preferred option. The energy selling mode needs to be modeled to compute the income in the objective function.

8.2.1.4 Prediction Model for Electricity Prices

Electricity prices are set by bilateral contracts between electrical utilities and by the pool where producers and consumers submit bids consisting of a set of quantities of energy and prices which they want to sell or buy and the corresponding minimum or maximum prices. The contracted quantities and corresponding prices are determined by a central operator who takes into account the offering and selling bids. Usually the demand and offer curve crossover points determine the price of electricity. On some occasions, prices are fixed by the regulating agent using matching of buyers and sellers by a mechanism; the highest buying price (e_b) is matched to the lowest selling price (e_s) at the mid point $((e_b + e_s)/2)$. After the matched packet has been assigned, the same procedure is repeated with the remaining selling and buying offers. Clearing prices are announced by the central operator. For example, the pool prices in California are announced at www.calpx.com. In Spain the prices of the intraday market are announced at www.omel.es.

In some cases, electrical utilities want to diminish their price uncertainties and establish longer term contracts. To establish contracts, producers need to know how much energy is demanded and how much it is going to cost, similarly, consumers need to know how much energy they are going to consume in the future. Accurate electrical energy price forecasts are fundamental for producers and consumers to negotiate good contracts and to establish appropriate bidding strategies on the short term market.

Autoregressive models tuned by historical data have frequently been used in the past to predict commodity prices and have also been used to predict electrical energy weekly prices [144]. Other methods have also been used, but it should be said that the price of electricity on a deregulated market is so volatile that it is very difficult to predict an accurate market price from models derived from historical data.

8.2.1.5 Costs

Similarly to price models, energy cost predictions are fundamental for producers and consumers in order to establish their bidding and contract strategies as well as operating their installations. The cost model basically consists of transforming the model of electrical consumption to a cost model. Parasitic loads and personnel costs have to be added. For instance, operating cost for the SEGS trough plants have been estimated to be around 0.03 euros/kWh [416].

8.2.1.6 Solar Field

The solar field is the solar energy collector mechanism and, therefore, one of the most important parts of a solar energy plant. Different models of solar collector fields have been developed in Chap. 4 that are useful for optimization purposes (mainly those based on first principles).

8.2.1.7 Storage System

The main types of storage system have been analyzed in Chap. 1. In electrical grid energy sources, stored energy has to be used to balance the mismatch between the energy production from intermittent energy sources and the demand. Simplified models of oil and molten salt accumulators will be briefly studied in this chapter.

8.2.1.8 Power Conversion System

There are different methods for modeling the PCS, many of them based on manufacturer specifications or historical data. A PCS is typically composed of a steam generator, preheater, steam turbine and refrigeration tower. The steam generator is fed by the hot HTF.

8.2.2 Non-committed Production

If prices are constant, the best option would be to produce as much electrical energy as possible. Storage can be used in those situations where more solar energy can be collected than converted and delivered. This occurs because turbines and generators are designed for nominal conditions which are usually below maximum solar conditions. In some cases, due to faults or maintenance operations some parts of the energy conversion system of the plant are not fully operative and it is not possible to convert all the energy collected from the solar system and it has to be either lost or stored.

If prices vary with the time of day or if production is carried out with penalization errors, energy storage can be used to maximize profits. The energy power balance is given by the following equation in discrete time:

$$P_r(k) = \gamma_{te}\big(P_{solar}(k) - P_{s+}(k) + P_{s-}(k)\big) - P_{load}(k) \tag{8.1}$$

where $P_r(k)$ is the energy delivered to the grid during sampling period k. Similarly, $P_{solar}(k)$ is the solar energy generated, $P_{load}(k)$ is the energy consumed by local electric loads, $P_{s+}(k) \geq 0$ is the energy stored at time interval $[k-1,k]$ and $P_{s-}(k) \geq 0$ is the energy extracted from the energy storage and delivered to the grid and γ_{te} is the thermal to electrical power conversion factor.

The total energy stored ($P_{stored}(k)$) can be modeled by

$$P_{stored}(k+1) = P_{stored}(k) + P_{s+}(k) - \beta_{st}P_{s-}(k) - \alpha_{st}P_{stored}(k) \tag{8.2}$$

where β_{st} is the energy storage-conversion coefficient (i.e. the efficiency of the storage system) and α_{st} is the proportion of the stored energy lost in each time period due to thermal losses. The total energy stored $P_{stored}(n)$ can be written as

$$P_{stored}(n) = P_{stored}(0) + \sum_{i=1}^{n}\big(P_{s+}(i) - \beta_{st}P_{s-}(i)\big)\alpha_{st}^{n-i} \tag{8.3}$$

$P_{solar}(k)$ basically depends on solar radiation during the sampling period and on the operating conditions of the plant. For example, if the plant is starting-up, some of the solar energy is needed to warm the plant systems up to operative stage, while if the plant is already generating electricity, no energy will be needed for warming up.

This model considers the plant to be connected to an infinite grid where power can be injected or extracted as necessary. Usually, the power sold to network $P_r(k)$ is bounded as $P_{r,min} \leq P_r(k) \leq P_{r,max}$. The purchase of energy from a grid is only allowed for dealing with an inability to feed the local load with power available in the system.

The benefit scheduled for period k is given by $B_{spot}(k) = e_1(k)P_{contract}(k)$, where $e_1(k)$ is the electricity price at that period and $P_{contract}(k)$ is the power scheduled (which was offered on the daily market).

The benefit for the time interval k can be computed as

$$B(k) = P_r(k)e_1(k) + P_{stored}(k)e_{st}(k) \tag{8.4}$$

where $e_1(k)$ is the price of energy for period k, $e_{st}(k)$ is the price of energy stored.

The benefits obtained in the course of one day will be

$$B(24) = \sum_{i=1}^{24} P_r(i)e_1(i) + P_{stored}(24)e_{st}(24) \tag{8.5}$$

The problem to be solved at time period k is to determine the sequence $\mathbf{P_r} = [P_r(k), P_r(k+1), \ldots, P_r(k+N-1)]$ which maximizes

$$\begin{aligned} J(N) = {} & \sum_{j=1}^{N} P_r(k+j)e_1(k+j) + P_{stored}(k) \\ & + \sum_{i=1}^{N} \big(P_{s+}(k+i) - \beta_{st} P_{s-}(k+i)\big)\alpha_{st}^{N-i} \end{aligned} \tag{8.6}$$

subject to

$$0 \le P_{stored}(k+j) \le P_{max}$$

$$0 \le P_{s+}(k+j) \le P_{smax}$$

$$0 \le P_{s-}(k+j) \le P_{smax}$$

$$P_{rmin} \le P_r(k+j) \le P_{rmax}$$

$$P_r(k) = \gamma_{te}\big(P_{solar}(k+j) - P_{s+}(k) + P_{s-}(k)\big) - P_{load}(k)$$

for $j = 1, \ldots, N$. $P_{solar}(k+j)$ is not yet known and has to be estimated using models such as those developed in Chap. 2. The decision variables are $P_r(k+j)$, $P_{s+}(k+j)$ and $P_{s-}(k+j)$ for $j = 0, \ldots, N-1$. The resulting problem can be solved with a linear programming (LP) algorithm. Once the solution is found only the first element of the sequence is used at time period k.

8.2.3 Committed Production

In this case the plant managers have to take two types of decision. First, they have to decide the production for the day and then, they have to decide the energy to be delivered and stored at each time period.

The benefit for the time interval k can be computed as

$$\begin{aligned} B(k) = {} & P_r(k)e_1(k) + \big(P_{s+}(k) - \beta_{st} P_{s-}(k)\big)e_{st}(k) \\ & - \big|P_r(k) - P_{contract}(k)\big|e_2(k) \end{aligned} \tag{8.7}$$

where $e_1(k)$ is the price agreed for period k, $e_{st}(k)$ is the price of the energy stored and $e_2(k)$ is the penalty paid for contract deviation.

The benefits obtained in the course of one day will be

$$B(24) = \sum_{i=1}^{24} P_r(i)e_1(i) + P_{stored}(24)e_{st}(24) - \left|P_r(i) - P_{contract}(i)\right|e_2(i) \quad (8.8)$$

There are two types of decision that have to be taken. First, at the beginning of the day, to decide the offer of $P_{contract}(k)$ for the next 24 hours and, once $P_{contract}(k)$ has been agreed, the next problem is to determine $P_r(k)$. Determination of $P_{stored}(k)$ can be done by maximizing Eq. (8.8) subject to

$$0 \leq P_{stored}(k+j) \leq P_{max}$$

$$0 \leq P_{s+}(k+j) \leq P_{smax}$$

$$0 \leq P_{s-}(k+j) \leq P_{smax}$$

$$P_{rmin} \leq P_r(k+j) \leq P_{rmax}$$

$$P_r(k) = \gamma_{te}\big(P_{solar}(k+j) - P_{s+}(k) + P_{s-}(k)\big) - P_{load}(k)$$

The optimization problem can be formulated as

$$\begin{aligned} \max_{P_r, P_{s+}, P_{s-}, P_{contract}} & \sum_{j=k}^{k+23} P_r(j)e_1(j) + \Bigg[P_{stored}(k)\alpha_{st}^{24} \\ & + \sum_{i=k}^{k+23} \big(P_{s+}(i) - \beta_{st}P_{s-}(i)\big)\alpha_{st}^{N+k-i} \Bigg] e_{st}(k+N) \\ & + \sum_{j=k}^{N+k-1} \left|P_r(j) - P_{contract}(j)\right|e_2(j) \end{aligned} \quad (8.9)$$

subject to

$$0 \leq P_{stored}(k) \leq P_{max}$$

$$P_{smin+} \leq P_{s+}(k) \leq P_{smax+}$$

$$P_{smin-} \leq P_{s-}(k) \leq P_{smax-}$$

$$P_{rmin} \leq P_r(k) \leq P_{rmax}$$

$$\left|P_r(k) - P_{contract}(k)\right| \leq P_d$$

$$P_r(k) = \gamma_{te}\big(P_{solar}(k+j) - P_{s+}(k) + P_{s-}(k)\big) - P_{load}(k)$$

The main decision variable at this stage is the energy to be committed for the day. However, predicted delivered energy and predicted stored energy can be considered as decision variables. The solar energy available at any given time is not known at this stage, however, since it cannot be manipulated it has to be estimated, together with the price of energy throughout the day. The resulting optimization algorithm can be solved by an LP algorithm.

This optimization is based on finding the best sequences of the control variables $P_{contract}(k)$, $P_{s+}(k)$, $P_{s-}(k)$ and $P_r(k)$. This problem is usually solved for a 24 hour period. Once the problem has been solved and the $P_{contract}(k)$ has been determined for the next 24 hours, the following problem can be solved:

$$\begin{aligned}\max_{P_r,P_{s+},P_{s-}} &\sum_{j=k}^{k+N-1} P_r(j)e_1(j) + \Bigg[P_{stored}(k)\alpha_{st}^N \\ &+ \sum_{i=k}^{k+N-1} \big(P_{s+}(i) - \beta_{st}P_{s-}(i)\big)\alpha_{st}^{N+k-i}\Bigg]e_{st}(k+N) \\ &+ \sum_{j=k}^{k+N-1} \big|P_r(j) - P_{contract}(j)\big|e_2(j) \end{aligned} \tag{8.10}$$

subject to

$$0 \le P_{stored}(k) \le P_{max}$$

$$P_{smin+} \le P_{s+}(k) \le P_{smax+}$$

$$P_{smin-} \le P_{s-}(k) \le P_{smax-}$$

$$P_{rmin} \le P_r(k) \le P_{rmax}$$

$$\big|P_r(k) - P_{contract}(k)\big| \le P_d$$

$$P_r(k) = \gamma_{te}\big(P_{solar}(k+j) - P_{s+}(k) + P_{s-}(k)\big) - P_{load}(k)$$

Notice that the optimization problem described only considers the random nature of solar power and energy prices by using their expected values. Furthermore, the solar power produced depends on the solar radiation and on the plant operating conditions. The solar power produced cannot be generated instantaneously, the systems need some time to start up and to change from one operating point to another. Dynamic models of the different subsystems are needed, the following section presents some of these models.

8.2.4 Prediction Models

This section analyzes some prediction models needed for scheduling production and optimal control of solar power plants.

The most important variable in a solar plant is solar radiation. Different models for solar radiation prediction and forecasting have been explained and developed in Chap. 2, both for short and long term forecasting. Different models of solar collector fields have been developed in Chap. 4 that are useful for optimization purposes (mainly those based on first principles).

Ambient temperature affects the thermal losses of the solar field, heat storage tanks and turbines. Ambient temperature also affects electrical demand. Low ambient temperature increases the need for heating and, therefore, electrical demand. Similarly, high ambient temperature increases the need for air cooling and electrical energy demand. Hourly temperature prediction can be obtained from weather forecasting services. The weather predictions can be corrected for by using current measured variables.

8.2.4.1 Storage System

Molten Salt Accumulators

Different thermal energy storage systems are used in solar plants. In some cases molten salt tanks are used. The basic idea is to use the heat from salt fusion; that is, the amount of heat required to melt a solid salt at its melting point into a liquid without increasing its temperature. The heat capacity of a body is defined as the heat needed to increase the temperature of the body by one degree and since the temperature is not increased during melting, the heat capacity is very high at this stage. The liquid salt releases the same amount of heat when it solidifies.

The energy stored at a molten salt energy storage system can be modeled by the following equations:

$$c_s\big(E_{st}(t)\big)\frac{dT_{st}}{dt} = P_{s+} - P_{s-} - L_{st}(T_{st} - T_a)$$

$$0 \le P_{s+} \le \bar{P}_{s+}$$

$$0 \le P_{s-} \le \bar{P}_{s-}$$

$$\bar{P}_{s+} = H_{st+}(q_{st}, T_{stin} - T_{st})$$

$$\bar{P}_{s-} = H_{st-}(q_{st}, T_{st} - T_{stin})$$

where $c_s(E_{st}(t))$ is the thermal capacity of the molten salt accumulator, T_{st} is the temperature of the salts, T_{stin} is the temperature of the fluid entering the accumulator. P_{s+} and P_{s-} are the power stored and extracted from the accumulator, respectively. $L_{st}(T_{st} - T_a)$ are the rate of energy losses in the accumulator which can be expressed as a function of $T_{st} - T_a$. The function $H_{st+}(q_{st}, T_{stin} - T_{st})$ corresponds to the maximum power that can be stored, which depends on the temperature differences between the inlet oil flow and the temperature of the accumulator. The maximum power that can be extracted from the accumulator can be computed as a function, $H_{st-}(q_{st}, T_{st} - T_{stin})$, of the oil accumulator flow (q_{st}) and the difference of temperature between the molten salt and the inlet oil flow.

The thermal capacity of the molten salt accumulator, $c_s(E_{st}(t))$, can be expressed as a function of the stored energy $E_{st}(t)$. When the stored energy is below a certain level, the thermal capacity corresponds to that of the salt in a solid state. When the salt is melting, because enough energy has already been stored, the incoming energy is used to melt the salt without increasing its temperature. This can be modeled as a very high accumulator thermal capacity. Once all the salt is melted, the thermal capacity of the accumulator corresponds to that of the salt in a liquid state.

The outlet HTF temperature, T_{stout}, can be computed from the following power balance equation:

$$\rho_f c_f q_{st}(T_{stin} - T_{stout})) = P_{s+} - P_{s-} - L_{st}(T_{st} - T_a) \tag{8.11}$$

Hot Oil Accumulators

In other cases, hot oil is directly stored in a tank (thermocline). Thermal stratification of the oil in the tank allows energy to be stored at different temperatures. As an example, in the ACUREX field, the storage tank is connected to the solar field and to the PCS by means of two pipe circuits placed at the top and bottom of the tank (see Fig. 8.1). The heated oil stored in the tank is used to generate the steam needed to drive the steam turbine of the PCS.

The storage system can be used in different modes; the first mode of operation is only used when there is not enough solar radiation, then hot oil is taken from the tank to the steam generator. The second mode of operation is useful when there are large variations in solar radiation due to clouds and the storage tank is used to smooth down the oil temperature disturbances. The power conversion system is run using the thermal energy stored inside the tank together with the solar field, but the oil from the field is sent to the bottom of the tank. This is done in order to avoid temperature fluctuations at the top of the tank. The third mode of operation is used when the level of solar radiation is high enough; then the oil from the field is sent to the top of the tank to be used by the PCS. The lower part of Fig. 8.1 shows a detailed description of the geometric characteristics of the tank. Note how the oil entrance and exit contain several diffusers used to avoid disturbances in the oil stratification. Furthermore, the position of the thermocouples in the oil and in the wall of the tank is shown in the figure. These thermocouples provide temperature measurements that can be used in the identification part of the modeling procedure.

A grey-box model for the storage tank of the ACUREX field was developed in [19]. The grey-box approach for building models [230] stems from the fact that it is best to take advantage of the a priori knowledge of a system. This knowledge is usually expressed in terms of a set of ordinary or partial differential equations obtained from first principles. For some systems such equations are not completely known and data have to be used to fill in the gap via an identification procedure.

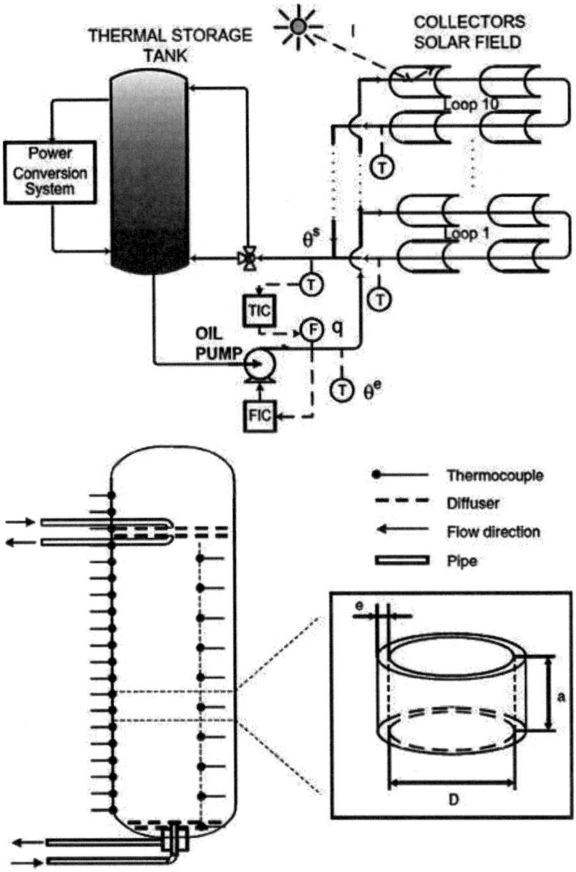

Fig. 8.1 *Top*: schematic layout of the SPSS plant at the PSA. *Bottom*: diagram of the tank showing the location of the thermocouples and pipes. The *inset* shows a generic discrete control volume used for modeling (courtesy of M.R. Arahal et al., [19])

A simultaneous perturbation stochastic approximation (SPSA) optimization procedure was used in [19] to adjust the parameters of the model to the observed data. The SPSA algorithm [356] provides an estimation of the gradient of an objective function to be optimized, making it appropriate for high-dimensional optimization problems. An interesting feature is that SPSA can be used in situations where the objective function is contaminated by noise. Furthermore, the gradient approximation is deliberately different from the alleged true gradient and this provides a means of escape from local minima while retaining the desired local convergence property. In the present case, the objective function is a measure of the simulation error given by the model.

Model Structure The model structure for the thermal storage tank corresponds to a discrete-time set of first order equations [19]. This structure has been chosen to achieve the main goals set for the model:

- Long term prediction capabilities that allow it to be used in the upper layer of a hierarchical control strategy.

- Adequate representation of the distributed energy content of the tank. The PCS does not operate in the same conditions for the whole range of temperatures resulting from stratification. For this reason, the model must reflect the temperature gradient and its changes during charge or usage periods.
- Low computational load during its use in simulation. In this way optimization in the upper layer of the hierarchical controller can be run frequently and in enough depth to provide a quasi-optimum solution in different scenarios.
- Low dependence on the sample time in order to be able to use historical data coming from different experiments.
- Good convergence capabilities in order to diminish the influence of a partially known or noise corrupted initial state.

The basic principles acting on a storage tank are heat and mass transfer laws; Thus, a first principle model seems a good choice. However, it is a well known fact that some parameters such as heat transfer coefficients among interfaces are difficult to measure. This is specially true for systems whose distributed nature cannot be overlooked which is the case of the storage tank, since stratification of temperatures along the vertical direction affects the spatial distribution of the oil parameters. Even in this case, a set of partial differential equations adequately adjusted to the particularities of the tank will yield an excellent model. Unfortunately, the computational load of running such a model in the many simulations needed for the hierarchical control scheme completely rules out this choice. Traditional grey-box approaches assume that the structure of the model is given directly as a parameterized mathematical function partly based on physical principles. In the present case a computer program or algorithm serves as well for the same purpose and has the advantage of being directly the same object later used by the hierarchical controller.

In the remainder of the section the a priori knowledge is presented together with a spatial discretization to produce a simulation algorithm that is in fact the model of the tank.

Spatial discretization: For purposes of modeling the oil and wall of the tank will be divided into sections to form control volumes and control annular sections, respectively. This spatial discretization will follow the particular arrangement of thermocouples along a vertical rod placed inside the tank, yielding ten oil volumes. The thermocouples of the wall are located at different heights and their number is different (20 instead of 10). To match the temperature measurements on the wall with the temperatures of the discrete annular sections a simple interpolation procedure was used [19]. The inset in Fig. 8.1 shows a diagram of the oil volumes and wall annular sections considered. The geometric parameters are the interior diameter D_{st}, the wall thickness t_{st} and the height h_{st} which is the same for all volumes except the lower and upper ones. Other geometrical features such as surfaces can be obtained from the above parameters.

Heat transfer models: For each volume a number of models are considered to describe the heat transfer. A simplifying assumption is that conditions (i.e. temperature θ_f, density ρ_f and specific heat c_f of oil on a given volume are homogeneous). This introduces a source of error in the model since the temperature profile can be very steep, causing conditions within one volume to vary appreciably from bottom to top. This is unavoidable since there are no other measurements than those provided by the thermocouples. In the following the different models are introduced [19]:

- *Transport*. During operation oil moves along control volumes causing changes in their energy content. The different modes of operation: charge of the tank with hot oil from the field, simultaneous charge and discharge and discharge with or without recirculation of oil cause different values of the net flow through the tank. A flow q is considered to be positive when it goes from top to bottom which is the normal situation during charge of the tank.
- *Conduction*. The energy flow between adjacent oil volumes and between adjacent wall segments due to conduction is modeled in the usual way as a linear function of the temperature increment. The distances among volumes centers are known and depend on the particular disposition of thermocouples. The thermal conductivity k_f of oil is computed from tables using the average temperature of the volume θ_f^o. For the wall section the thermal conductivity is considered constant.
- *Convection*. The convection mechanism is the trickiest in this model. It is difficult to model since it involves effects such as turbulence. The detailed modeling of such phenomenon is absolutely out of the question due to the limitations imposed on the computing load for the final model. The effects produced by convection are, however, easy to describe: when hot oil enters the bottom of the tank there is a quick mix with the cooler layers above that homogenizes the temperature profile very efficiently. From this observation the energy variation in the volumes due to convection is modeled using: (i) a coefficient that determines the amount of energy that the recirculating flow of hot oil from the collectors yields to the tank and (ii) a set of equations that ensure that this energy is efficiently distributed over the layers above. In this way the simplicity of the model is kept while producing a mechanism that performs well.

Simulation algorithm: The simulation algorithm is based in computing the changes in temperature over time for each oil volume (θ_f^i) and wall segment (θ_m^i). The transition from a generic discrete time k to the next $(k+1)$ is governed by the above described heat transfer mechanisms. For each volume $i = 1, \ldots, 10$ the energy change due to transport (ΔE^t), conduction among oil volumes (ΔE^{co}), conduction among wall segments (ΔE^{cm}), convection (ΔE^v), losses from oil to wall (ΔE^w) and from wall to the ambient (ΔE^a) is computed yielding a pair of discrete-time equations:

$$\rho_f^i c_f^i \theta_f^i|_{k+1} = \rho_f^i c_f^i \theta_f^i|_k + \frac{T_s}{V_f^i}(\Delta E^t + \Delta E^{co} + \Delta E^v + \Delta E^w)|_k \quad (8.12)$$

$$\rho_m^i c_m^i \theta_m^i|_{k+1} = \rho_m^i c_m^i \theta_m^i|_k + \frac{T_s}{V_m^i}(\Delta E^w + \Delta E^{cm} + \Delta E^a)|_k \quad (8.13)$$

The sampling time $T_s = 120$ s has been selected to provide a good balance between representation of the observed behavior of the temperatures of the volumes and the computational load that it will imply for the simulation. The volume of the discrete elements of oil (V_f^i) and metal (V_m^i) are computed from geometrical parameters. In [19] the complete model adjustment is explained using the SPSA algorithm, showing experimental results.

8.2.4.2 Power Conversion System

There are different methods for modeling the power conversion systems. The model presented in this section has been obtained from data taken from the power conversion system of the DCS project [190]. The PCS is composed of the steam generator, preheater, steam turbine, and refrigeration tower. The steam generator is fed by thermal hot oil coming from the hot oil storage tank. The steam powers the turbine and electrical generator (Stal-Laval) with a nominal power of 577 kW.

Power Loss Model

Losses are important when designing a plant-wide control scheme. Overall losses can be computed as the difference between the input and the output power. The input power to the PCS is the thermal power of the hot oil (P_{HTF}), which depends on the inlet oil temperature ($T_{in_{Tur}}$), outlet oil temperature ($T_{out_{Tur}}$), and flow (q_T) as indicated by Eq. (8.14):

$$P_{HTF} = q_T\left[1.882 \cdot 10^{-3}\left(T_{in_{Tur}}^2 - T_{out_{Tur}}^2\right) - 0.795(T_{in_{Tur}} - T_{out_{Tur}})\right] \tag{8.14}$$

The gross output power of the PCS (P_{gross}) can be obtained as the difference between the thermal power delivered to the steam generator (P_{HTF}), the thermal power delivered to the refrigeration systems (P_{rs}) and the thermal losses inside the PCS (L_{PCS}).

The steady state energy balance equation is given by Eq. (8.15):

$$P_{gross} = P_{HTF} - L_{rs} - L_{PCS} \tag{8.15}$$

The following equation was obtained for PSA from experimental data:

$$L_{PCS} = 1.56 \cdot 10^{-8} P_{HTF}^3 - 9.31 \cdot 10^{-5} P_{HTF}^2 + 0.27 P_{HTF} - 136 \tag{8.16}$$

The thermal power delivered to the refrigeration systems were modeled using linear approximations for two operating points at the PCS 290°C (in Eq. (8.17), with a correlation coefficient of 0.99) and 280°C (in Eq. (8.18), with a correlation coefficient of 0.98):

$$L_{rs} = 0.678 P_{HTF} + 136 \tag{8.17}$$

$$L_{rs} = 0.769 P_{HTF} - 43 \tag{8.18}$$

Table 8.1 Average solar plant efficiency for three operating years

Element	Performance (min, %)	Performance (max, %)
$\bar{\eta}_{solar}$	22.0	51.5
$\bar{\eta}_{st}$	60.6	98.2
$\bar{\eta}_{PCS}$	11.3	22.9

Table 8.2 Power distribution at the PCS

Net electrical power	500 kW
Parasites	77 kW
Gross electrical power	577 kW
Gross efficiency	$\eta_{gross} = 19.13\%$
Oil thermal power	$P_{HTF} = 3016$ kW
Thermal losses	$L_{PCS} = 259$ kW
Refrigeration losses	$L_{rs} = 2180$ kW

Performance-Based PCS Model

Table 8.1 shows efficiency data of the elements: solar collector field, storage tanks and power conversion part. The yearly efficiency coefficients have been obtained from the monthly efficiency data.

This type of information is useful to evaluate the best and worst cases for making decisions taking risks into account. The steam turbine can be operated at 290°C or at 280°C. The return temperature of the generator is about 70°C lower than the inlet temperature PCS operating temperatures between 287°C and 292°C, which are considered to be in the high range while operating temperatures between 277°C and 282°C are considered low range. The gross process efficiency (in Eq. (8.19)), denoted by η_{gross}, was computed (Eq. (8.14)) by a least squares fitting. The same happens with the net efficiency $\eta_{net} = P_r / P_{HTF}$, where P_r is the generated power:

$$\eta_{gross} = -2.44 \cdot 10^{-5} P_{gross}^2 + 3.07 \cdot 10^{-2} P_{gross} + 9.54 \quad (8.19)$$

$$\eta_{net} = -3.52 \cdot 10^{-5} P_{net}^2 + 4.07 \cdot 10^{-2} P_r + 5.43 \quad (8.20)$$

Table 8.2 shows power distribution at the PCS.

PCS Start-up

The energy needed to start up the plant depends on the time elapsed since it was in operation and the required energy is between 200 and 1200 kWh_t. The start-up phase includes the following steps:

1. Increase the pressure to 25 bar.
2. Preheat the pipes and turbine.
3. Start the turbine and reach the nominal rotating speed (100% rpm).

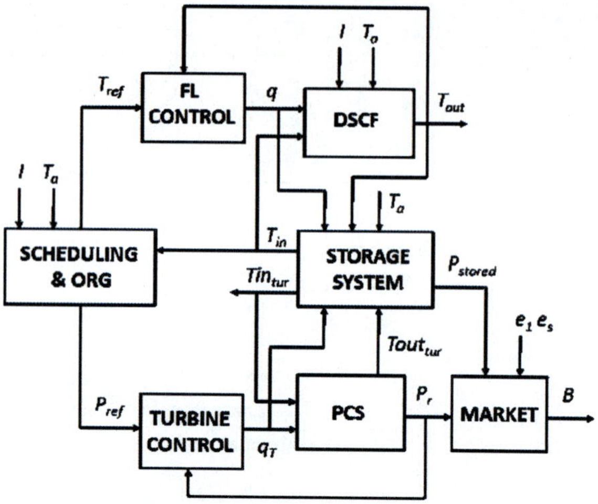

Fig. 8.2 Scheme of elements and their interconnection to perform diary planning of electricity production (courtesy of C.M. Cirre, [103])

The energy needed to start up the plant is given by

$$P_{start} = 1200 - 1000e^{(0.304t - 0.307 \cdot 10^{-4}t^2)} \quad [\mathrm{kW}_t] \tag{8.21}$$

8.3 Simulation Experiments

All the models mentioned in the previous sections can be used for daily planning of electricity production. As a simple example, a simulation experiment is shown in this section on applying the hierarchical daily production planning structure shown in Fig. 8.2. This application can use real data from the installation or a model of direct solar irradiance shown in Chap. 2, the hot oil storage system model briefly explained in Sect. 8.2.4.1, the distributed parameter model of the solar field explained in Chap. 4, Sect. 4.3.2, the optimizing reference governor explained in Chap. 5, Sect. 5.12.1.2, and the feedback linearization controller developed in Chap. 5, Sect. 5.8.1, to regulate the outlet temperature. In this way, different operation alternatives can be evaluated as a function of the state of the storage tank and the weather and thus it is possible to estimate the net electric power supplied to the network.

Figure 8.3 shows a simulation using real data of the field. Subindex ‘sim’ corresponds with the simulated trends. Depending on q_T, the evolution of the thermoclines of storage tank can be observed, as well as the net power P_r.

8.4 Summary

Basic concepts involved in the operational planning of solar plant with PTC have been treated in this chapter. Different levels of the hierarchical control problem involved have been defined within time scales of weekly planning, daily planning

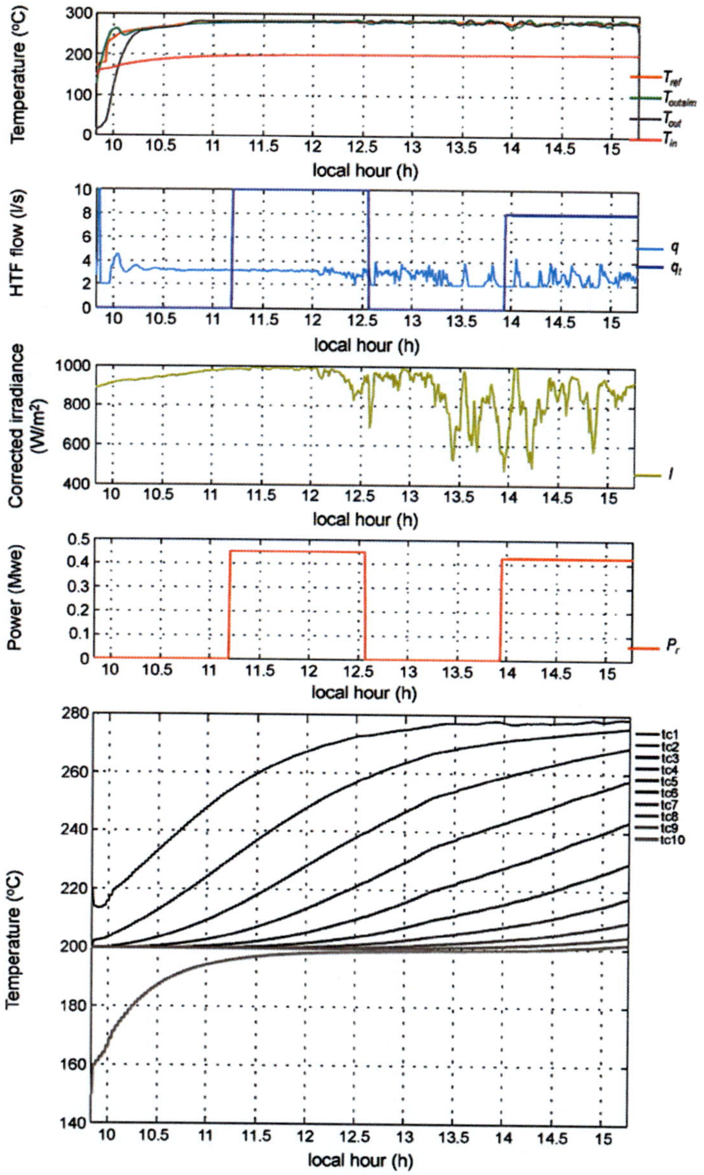

Fig. 8.3 Simulation test of a hierarchical control scheme (courtesy of C.M. Cirre, [103])

and tracking. Many static and dynamic models are required to fulfill the operational planning objectives: solar irradiance, ambient temperature, market, electrical prices, costs, solar field, storage systems and power conversion systems. Both Committed and Non-committed production cases have been studied from an optimization viewpoint and several illustrative simulation results have been provided.

Appendix
MLD Model Matrix Values

The values of matrices $\boldsymbol{\Phi}$, $\mathbf{H}$, $\mathbf{G}_3$, $\mathbf{E}_1$, $\mathbf{E}_2$, $\mathbf{E}_3$ and $\mathbf{E}_5$ from Eq. (6.28) are the following:

$$
\boldsymbol{\Phi} = \begin{bmatrix}
0 & 0 & 0 & 0 & 0 & 0 & 0 & 0 & 0 & 0 & 0 & 0 & 0 & 0 & 0 & 0 & 0 & 0 & 0 \\
0 & 0 & 0 & 0 & 0 & 0 & 0 & 0 & 0 & 0 & 0 & 0 & 0 & 0 & 0 & 0 & 0 & 0 & 0 \\
1 & 0 & 2.90 & -2.80 & 0.90 & 0 & 0 & 0 & 0 & 0 & 0 & 0 & 0 & 0 & 0 & 0 & 0 & 0 & 0 \\
0 & 0 & 1 & 0 & 0 & 0 & 0 & 0 & 0 & 0 & 0 & 0 & 0 & 0 & 0 & 0 & 0 & 0 & 0 \\
0 & 0 & 0 & 1 & 0 & 0 & 0 & 0 & 0 & 0 & 0 & 0 & 0 & 0 & 0 & 0 & 0 & 0 & 0 \\
0 & 0 & 0 & 0 & 1 & 0 & 0 & 0 & 0 & 0 & 0 & 0 & 0 & 0 & 0 & 0 & 0 & 0 & 0 \\
0 & 1 & 0 & 0 & 0 & 0 & 2.99 & -2.98 & 0.99 & 0 & 0 & 0 & 0 & 0 & 0 & 0 & 0 & 0 & 0 \\
0 & 0 & 0 & 0 & 0 & 0 & 1 & 0 & 0 & 0 & 0 & 0 & 0 & 0 & 0 & 0 & 0 & 0 & 0 \\
0 & 0 & 0 & 0 & 0 & 0 & 0 & 1 & 0 & 0 & 0 & 0 & 0 & 0 & 0 & 0 & 0 & 0 & 0 \\
0 & 0 & 0 & 0 & 0 & 0 & 0 & 0 & 1 & 0 & 0 & 0 & 0 & 0 & 0 & 0 & 0 & 0 & 0 \\
0 & 0 & 0 & 0 & 0 & 0 & 0 & 0 & 0 & 1 & 0 & 0 & 0 & 0 & 0 & 0 & 0 & 0 & 0 \\
0 & 0 & 0 & 0 & 0 & 0 & 0 & 0 & 0 & 0 & 1 & 0 & 0 & 0 & 0 & 0 & 0 & 0 & 0 \\
0 & 0 & 0 & 0 & 0 & 0 & 0 & 0 & 0 & 0 & 0 & 1 & 0 & 0 & 0 & 0 & 0 & 0 & 0 \\
0 & 0 & 0 & 0 & 0 & 0 & 0 & 0 & 0 & 0 & 0 & 0 & 1 & 0 & 0 & 0 & 0 & 0 & 0 \\
0 & 0 & 0 & 0 & 0 & 0 & 0 & 0 & 0 & 0 & 0 & 0 & 0 & 1 & 0 & 0 & 0 & 0 & 0 \\
0 & 0 & 0 & 0 & 0 & 0 & 0 & 0 & 0 & 0 & 0 & 0 & 0 & 0 & 1 & 0 & 0 & 0 & 0 \\
0 & 0 & 0 & 0 & 0 & 0 & 0 & 0 & 0 & 0 & 0 & 0 & 0 & 0 & 0 & 1 & 0 & 0 & 0 \\
0 & 0 & 0 & 0 & 0 & 0 & 0 & 0 & 0 & 0 & 0 & 0 & 0 & 0 & 0 & 0 & 1 & 0 & 0 \\
0 & 0 & 0 & 0 & 0 & 0 & 0 & 0 & 0 & 0 & 0 & 0 & 0 & 0 & 0 & 0 & 0 & 1 & 0
\end{bmatrix},
$$

$$
\mathbf{G}_3 = \begin{bmatrix}
1 & 1 & 1 & 0 & 0 & 0 \\
0 & 0 & 0 & 1 & 1 & 1 \\
0 & 0 & 0 & 0 & 0 & 0 \\
0 & 0 & 0 & 0 & 0 & 0 \\
0 & 0 & 0 & 0 & 0 & 0 \\
0 & 0 & 0 & 0 & 0 & 0 \\
0 & 0 & 0 & 0 & 0 & 0 \\
0 & 0 & 0 & 0 & 0 & 0 \\
0 & 0 & 0 & 0 & 0 & 0 \\
0 & 0 & 0 & 0 & 0 & 0 \\
0 & 0 & 0 & 0 & 0 & 0 \\
0 & 0 & 0 & 0 & 0 & 0 \\
0 & 0 & 0 & 0 & 0 & 0 \\
0 & 0 & 0 & 0 & 0 & 0 \\
0 & 0 & 0 & 0 & 0 & 0 \\
0 & 0 & 0 & 0 & 0 & 0 \\
0 & 0 & 0 & 0 & 0 & 0 \\
0 & 0 & 0 & 0 & 0 & 0 \\
0 & 0 & 0 & 0 & 0 & 0
\end{bmatrix},
$$

$$
\begin{aligned}
\mathbf{H} = [&478811 \quad 215613 \quad 38398 \quad -112802 \quad 110413 \quad -36008 \quad 1924 \\
&-3847 \quad 1923 \quad 0 \quad \cdots \quad 0 \quad -1740 \quad 3479 \quad -1739]^T,
\end{aligned}
$$

E.F. Camacho et al., *Control of Solar Energy Systems*, Advances in Industrial Control, 387
DOI 10.1007/978-0-85729-916-1, © Springer-Verlag London Limited 2012

$$
\mathbf{E}_2 = \begin{bmatrix}
-20 & 0 & 0 & 0 & 0 \\
20 & 0 & 0 & 0 & 0 \\
0 & -20 & 0 & 0 & 0 \\
0 & 20 & 0 & 0 & 0 \\
1 & 1 & -1 & 0 & 0 \\
-1 & 0 & 1 & 0 & 0 \\
0 & -1 & 1 & 0 & 0 \\
-1 & 1 & 0 & -1 & 0 \\
1 & 0 & 0 & 1 & 0 \\
0 & -1 & 0 & 1 & 0 \\
1 & -1 & 0 & 0 & -1 \\
-1 & 0 & 0 & 0 & 1 \\
0 & 1 & 0 & 0 & 1 \\
0 & 0 & 10 & 0 & 0 \\
0 & 0 & 10 & 0 & 0 \\
0 & 0 & -10 & 0 & 0 \\
0 & 0 & -10 & 0 & 0 \\
0 & 0 & 0 & 0 & 10 \\
0 & 0 & 0 & 0 & 10 \\
0 & 0 & 0 & 0 & -10 \\
0 & 0 & 0 & 0 & -10 \\
0 & 0 & 0 & 10 & 0 \\
0 & 0 & 0 & 10 & 0 \\
0 & 0 & 0 & -10 & 0 \\
0 & 0 & 0 & -10 & 0 \\
0 & 0 & 20 & 0 & 0 \\
0 & 0 & 20 & 0 & 0 \\
0 & 0 & -20 & 0 & 0 \\
0 & 0 & -20 & 0 & 0 \\
0 & 0 & 0 & 0 & 20 \\
0 & 0 & 0 & 0 & 20 \\
0 & 0 & 0 & 0 & -20 \\
0 & 0 & 0 & 0 & -20 \\
0 & 0 & 0 & 0 & 0
\end{bmatrix}, \quad
\mathbf{E}_3 = \begin{bmatrix}
0 & 0 & 0 & 0 & 0 & 0 \\
0 & 0 & 0 & 0 & 0 & 0 \\
0 & 0 & 0 & 0 & 0 & 0 \\
0 & 0 & 0 & 0 & 0 & 0 \\
0 & 0 & 0 & 0 & 0 & 0 \\
0 & 0 & 0 & 0 & 0 & 0 \\
0 & 0 & 0 & 0 & 0 & 0 \\
0 & 0 & 0 & 0 & 0 & 0 \\
0 & 0 & 0 & 0 & 0 & 0 \\
0 & 0 & 0 & 0 & 0 & 0 \\
0 & 0 & 0 & 0 & 0 & 0 \\
0 & 0 & 0 & 0 & 0 & 0 \\
0 & 0 & 0 & 0 & 0 & 0 \\
-1 & 0 & 0 & 0 & 0 & 0 \\
1 & 0 & 0 & 0 & 0 & 0 \\
-1 & 0 & 0 & 0 & 0 & 0 \\
1 & 0 & 0 & 0 & 0 & 0 \\
0 & -1 & 0 & 0 & 0 & 0 \\
0 & 1 & 0 & 0 & 0 & 0 \\
0 & -1 & 0 & 0 & 0 & 0 \\
0 & 1 & 0 & 0 & 0 & 0 \\
0 & 0 & -1 & 0 & 0 & 0 \\
0 & 0 & 1 & 0 & 0 & 0 \\
0 & 0 & -1 & 0 & 0 & 0 \\
0 & 0 & 1 & 0 & 0 & 0 \\
0 & 0 & 0 & -1 & 0 & 0 \\
0 & 0 & 0 & 1 & 0 & 0 \\
0 & 0 & 0 & -1 & 0 & 0 \\
0 & 0 & 0 & 1 & 0 & 0 \\
0 & 0 & 0 & 0 & -1 & 0 \\
0 & 0 & 0 & 0 & 1 & 0 \\
0 & 0 & 0 & 0 & -1 & 0 \\
0 & 0 & 0 & 0 & 1 & 0 \\
0 & 0 & 0 & 0 & 0 & -1 \\
0 & 0 & 0 & 0 & 0 & 1
\end{bmatrix}, \quad
\mathbf{E}_1 = \begin{bmatrix}
1 & -1 \\
-1 & 1 \\
-1 & 1 \\
1 & -1 \\
0 & 0 \\
0 & 0 \\
0 & 0 \\
0 & 0 \\
0 & 0 \\
0 & 0 \\
0 & 0 \\
0 & 0 \\
0 & 0 \\
-1 & 0 \\
1 & 0 \\
0 & 0 \\
0 & 0 \\
-1 & 0 \\
1 & 0 \\
0 & 0 \\
0 & 0 \\
0 & -1 \\
0 & 1 \\
0 & 0 \\
0 & 0 \\
1 & -1 \\
-1 & 1 \\
0 & 0 \\
0 & 0 \\
1 & -1 \\
-1 & 1 \\
0 & 0 \\
0 & 0 \\
0 & 0 \\
0 & 0
\end{bmatrix}, \quad
\mathbf{E}_5 = \begin{bmatrix}
0 \\ 20 \\ 0 \\ 20 \\ 1 \\ 0 \\ 0 \\ 0 \\ 1 \\ 0 \\ 0 \\ 0 \\ 1 \\ 10 \\ 10 \\ 0 \\ 0 \\ 10 \\ 10 \\ 0 \\ 0 \\ 10 \\ 10 \\ 0 \\ 0 \\ 20 \\ 20 \\ 0 \\ 0 \\ 20 \\ 20 \\ 0 \\ 0 \\ 0 \\ 0
\end{bmatrix}
$$

References

1. Abdallah, S., Salem, N.: Two axes Sun tracking system with PLC control. Energy Convers. Manag. **45**(11–12), 1931–1939 (2004)
2. Alexander, B., Balluffi, R.: The mechanism of sintering of copper. Acta Metall. **5**, 667–677 (1957)
3. Allidina, A.Y., Hughes, F.M.: Generalized self-tuning controller with pole assignment. IEE Proc. Part D **127**, 13–18 (1980)
4. Allwright, J.C.: On min–max model-based predictive control. In: Clarke, D.W. (ed.) Advances in Model-Based Predictive Control. Oxford University Press, London (1994)
5. Almeida, F., Shohoji, N., Cruz, J., Guerra, L.: Solar sintering of cordierite-based ceramics at low temperatures. Sol. Energy **78**, 351–361 (2005)
6. Almorox, J., Hontoria, C.: Global solar radiation estimation using sunshine duration in Spain. Energy Convers. Manag. **45**, 1529–1535 (2004)
7. Álvarez, J.D., Gernjak, W., Malato, S., Berenguel, M., Fuerhacker, M., Yebra, L.: Dynamic models for hydrogen peroxide control in solar photo-fenton systems. J. Sol. Energy Eng. **129**, 37–44 (2007)
8. Álvarez, J.D., Yebra, L.J., Berenguel, M.: Repetitive control of tubular heat exchangers. J. Process Control **17**(9), 689–701 (2007)
9. Álvarez, J.D., Yebra, L.J., Berenguel, M.: Control Strategies for Thermosolar Heat Exchangers. CIEMAT, Madrid (2008) (in Spanish)
10. Álvarez, J.D., Guzmán, J.L., Yebra, L.J., Berenguel, M.: Hybrid modeling of central receiver solar power plants. Simul. Model. Pract. Theory **17**(4), 664–679 (2009)
11. Álvarez, J.D., Yebra, L.J., Berenguel, M.: Adaptive repetitive control for resonance cancellation of a distributed solar collector field. Int. J. Adapt. Control Signal Process. **23**, 331–352 (2009)
12. Álvarez, J.D., Costa-Castelló, R., Berenguel, M., Yebra, L.J.: A repetitive control scheme for distributed solar collector field. Int. J. Control **83**(5), 970–982 (2010)
13. Andrade da Costa, B., Lemos, J.M.: An adaptive temperature control law for a solar furnace. Control Eng. Pract. **17**, 1157–1173 (2009)
14. Ångström, A.: Solar and terrestrial radiation. Q. J. R. Meteorol. Soc. **50**, 121–126 (1924)
15. Arahal, M.R.: Identification and control using neural networks. PhD Thesis, University of Seville, Seville, Spain (1996) (in Spanish)
16. Arahal, M.R., Berenguel, M., Camacho, E.F.: Nonlinear neural model-based predictive control of a solar plant. In: Proc. of the European Control Conf., ECC'97, Brussels, Belgium, vol. TH-E I2, 1997
17. Arahal, M.R., Berenguel, M., Camacho, E.F.: Comparison of RBF algorithms for output temperature prediction of a solar plant. In: Proc. of CONTROLO'98, Coimbra, Portugal, 1998

E.F. Camacho et al., *Control of Solar Energy Systems*, Advances in Industrial Control,
DOI 10.1007/978-0-85729-916-1,

18. Arahal, M.R., Berenguel, M., Camacho, E.F.: Neural identification applied to predictive control of a solar plant. Control Eng. Pract., 333–344 (1998)
19. Arahal, M.R., Cirre, C.M., Berenguel, M.: Serial grey-box model of a stratified thermal tank for hierarchical control of a solar plant. Sol. Energy **82**, 441–451 (2008)
20. ASHRAE: ASHRAE Handbook: HVAC Applications. ASHRAE, Atlanta (1999)
21. Åström, K.J., Hägglund, T.: PID Controllers: Theory, Design and Tuning. Instrument Society of America, Research Triangle Park (1995)
22. Åström, K.J., Hägglund, T.: PID control. In: Levine, S. (ed.) The Control Handbook. CRC Press/IEEE Press, Boca Raton/New York (1996)
23. Åström, K.J., Hägglund, T.: Advanced PID Control. ISA—The Instrumentation, Systems, and Automation Society, Research Triangle Park (2005)
24. Åström, K.J., Wittenmark, B.: Computer Controlled Systems. Theory and Design. Prentice Hall, New York (1984)
25. Åström, K.J., Wittenmark, B.: Adaptive Control. Addison-Wesley, Reading (1989)
26. Balas, G., Doyle, J., Glover, K., Packard, A., Smith, R.: μ-Analysis and Synthesis Toolbox User's Guide, 2nd edn. The Mathworks, Natick (1995)
27. Baltas, P., Tortoreli, M., Russel, P.: Evaluation of power output for fixed and step tracking photovoltaic arrays. Sol. Energy **37**(2), 147–163 (1986)
28. Barão, M.: Dynamics and nonlinear control of a solar collector field. PhD Thesis, Universidade Tecnica de Lisboa, Instituto Superior Tecnico, Lisbon, Portugal (2000)
29. Barão, M., Lemos, J.M., Silva, R.N.: Reduced complexity adaptive nonlinear control of a distributed collector solar field. J. Process Control **12**(1), 131–141 (2002)
30. Bazaraa, M.S., Shetty, C.M.: Nonlinear Programming. Wiley, New York (1979)
31. Bellecci, C., Conti, M.: Latent heat thermal storage for solar dynamic power generation. Sol. Energy **51**(3), 169–173 (1993)
32. Bemporad, A.: Hybrid Toolbox—User's Guide. http://www.dii.unisi.it/hybrid/toolbox (2004)
33. Bemporad, A., Morari, M.: Control of systems integrating logic, dynamics and constraints. Automatica **35**(3), 407–427 (1999)
34. Berenguel, M.: Contributions to the control of distributed solar collectors. PhD Thesis, Universidad de Sevilla, Spain (1996) (in Spanish)
35. Berenguel, M.: Some control applications to solar plants. In: Proc. of the Int. Workshop on Constrained Control Systems, DAS-CTC-UFSC-NECCOSYDE, Florianópolis, SC, Brazil, 1998
36. Berenguel, M., Camacho, E.F.: Frequency based adaptive control of systems with antiresonance modes. In: Proc. of the 5th IFAC Symp. of Adaptive Systems in Control and Signal Processing, ACASP'95, Budapest, Hungary, pp. 195–200 (1995)
37. Berenguel, M., Camacho, E.F.: Frequency based adaptive control of systems with antiresonance modes. Control Eng. Pract. **4**(5), 677–684 (1996)
38. Berenguel, M., Camacho, E.F., Rubio, F.R.: Simulation software package for the Acurex field. Internal Report, Dpto. de Ingeniería de Sistemas y Automática, ESI Sevilla, Spain. www.esi2.us.es/~rubio/libro2.html (1994)
39. Berenguel, M., Camacho, E.F., Rubio, F.R., Balsa, P.: Gain scheduling generalized predictive controller applied to the control of a parabolic trough solar collectors field. In: Proc. of the 8th Int. Symp. on Solar Thermal Concentrating Technologies, Cologne, Germany, vol. 2, pp. 685–703 (1996)
40. Berenguel, M., Arahal, M.R., Camacho, E.F.: Modeling free response of a solar plant for predictive control. In: Proc. of the 11th IFAC Symp. on Systems Identification, SYSID'97, Fukuoka, Japan, pp. 1291–1296 (1997)
41. Berenguel, M., Camacho, E.F., Rubio, F.R., Luk, P.C.K.: Incremental fuzzy PI control of a solar power plant. IEE Proc. Part D **144**(6), 596–604 (1997)
42. Berenguel, M., Arahal, M.R., Camacho, E.F.: Modeling free response of a solar plant for predictive control. Control Eng. Pract. **6**, 1257–1266 (1998)
43. Berenguel, M., Camacho, E.F., García-Martín, F., Rubio, F.R.: Temperature control of a solar furnace. IEEE Control Syst. Mag. **19**(1), 8–24 (1999)

44. Berenguel, M., Rubio, F.R., Camacho, E.F., Gordillo, F.: Techniques and applications of fuzzy logic control of solar power plants. In: Leondes, C.T. (ed.) Fuzzy Theory Systems Techniques and Applications, vol. 2. Academic Press, San Diego (1999) (Chap. 25)
45. Berenguel, M., Rubio, F.R., Valverde, A., Lara, P.J., Arahal, M.R., Camacho, E.F., López, M.: An artificial vision-based control system for automatic heliostat positioning offset correction in a central receiver solar power plant. Sol. Energy **76**, 563–575 (2004)
46. Berenguel, M., Cirre, C.M., Klempous, R., Maciejewski, H., Nikodem, J., Nikodem, M., Rudas, I., Valenzuela, L.: Hierarchical control of a distributed solar collector field. Lect. Notes Comput. Sci. **3643**, 614–620 (2005)
47. Berenguel, M., Klempous, R., Maciejewski, H., Nikodem, J., Nikodem, M., Valenzuela, L.: Explanatory analysis of data from a distributed solar collector field. Lect. Notes Comput. Sci. **3643**, 621–626 (2005)
48. Beschi, M., Berenguel, M., Visioli, A., Yebra, L.: Control strategies for disturbance rejection in a solar furnace. In: Proc. of the 18th IFAC World Congress, Milan, Italy, 2011
49. Bierman, G.: Factorization Methods for Discrete Estimation. Academic Press, New York (1977)
50. Biggs, F., Vittoe, C.N.: The Helios model for the optical behavior of reflecting solar concentrators. SAND 76-0347, Sandia National Laboratories, Albuquerque (1976)
51. Blackmon, J.B.: Development and performance of a digital image radiometer for heliostat evaluation at Solar One. In: Proc. of the ASME Solar Engineering Division 6th Annual Conf., Las Vegas, NV, USA, 1984
52. Blackmon, J.B., Edwards, D.K.: Apparatus and method for calculation of shape factor. US Patent 4,963,025, 16 October 1990
53. Blackmon, J.B., Stone, K.W.: Digital image system for determining relative position and motion of in-flight vehicles. US Patent 5,493,392, 20 February 1996
54. Blackmon, J.B., Caraway, M.J., Stone, K.W.: Digital image radiometer applications for solar concentrator optical evaluation. In: Proc. of the ISES World Congress, Jerusalem, Israel, 1999
55. Blanco, J., Malato, S.: Solar Detoxification. UNESCO, Paris (2004)
56. Blanco-Muriel, M., Alarcón-Padilla, D.C., López-Moratalla, T., Lara-Coira, M.: Computing the solar vector. Sol. Energy **70**(5), 431–441 (2001)
57. Bodson, M., Douglas, S.C.: Adaptive algorithms for the rejection of sinusoidal disturbances with unknown frequency. Automatica **33**(12), 2213–2221 (1997)
58. Bonilla, J., Roca, L., González, J., Yebra, L.J.: Modelling and real-time simulation of heliostat fields in central receiver plants. In: Proc. of the 6th Int. Conf. on Mathematical Modelling, Vienna, Austria, pp. 2576–2579 (2009)
59. Bonilla, J., Roca, L., Yebra, L.J., Dormido, S.: Real-time simulation of CESA-1 central receiver solar thermal power plant. In: Proc. of the 7th Int. Modelica Conf., Como, Italy, 2009
60. Bonilla, J., Yebra, L.J., Dormido, S.: A heuristic method to minimise the chattering problem in dynamic mathematical two-phase flow models. Math. Comput. Model. **54**(5–6), 1549–1560 (2011)
61. Borrelli, F., Bemporad, A., Fodor, M., Hrovat, D.: An MPC/hybrid system approach to traction control. IEEE Trans. Control Syst. Technol. **14**(3), 541–552 (2006)
62. Bosch, J.L., López, G., Batlles, F.J.: Daily solar irradiation estimation over a mountainous area using artificial neural networks. Renew. Energy **33**, 1622–1628 (2008)
63. Box, G.E.P., Luceno, A., Paniagua, M.C.: Statistical Control by Monitoring and Adjustment. Wiley, New York (2009)
64. Branicky, M.S., Borkar, V.S., Mitter, S.K.: A unified framework for hybrid control: model and optimal control theory. IEEE Trans. Autom. Control **43**(1), 31–45 (1998)
65. Braun, K.P.: HELIOS—simulation of the incident flux distribution on the TSA receiver. Internal Report, PSA (1996)
66. Braun, M.W., Ortiz-Mojica, R., Rivera, D.E.: Design of minimum crest factor multisinusoidal signals for plant-friendly identification of nonlinear process systems. Control Eng. Pract. **3**(3), 301–313 (2002)

67. Brdys, M.A., Tatjewski, P.: Iterative Algorithms for Multilayer Optimizing Control. Imperial College Press, London (2005)
68. Brockwell, P.J., Davis, R.A.: Introduction to Time Series and Forecasting. Springer, Berlin (2002)
69. Brosilow, C., Joseph, B.: Techniques of Model-Based Control. Prentice Hall, New York (2002)
70. Brus, L., Zambrano, D.: Black-box identification of solar collector dynamics with variant time delay. Control Eng. Pract. **18**, 1133–1146 (2010)
71. Brus, L., Wigren, T., Zambrano, D.: Feedforward model predictive control of a non-linear solar collector plant with varying delays. IET Control Theory Appl. **4**(8), 1421–1435 (2009)
72. Buie, D., Monger, A.G.: The effect of circumsolar radiation on a solar concentrating system. Sol. Energy **76**(1–3), 181–185 (2004)
73. Camacho, E.F.: Constrained generalized predictive control. IEEE Trans. Autom. Control **38**(2), 327–332 (1993)
74. Camacho, E.F., Berenguel, M.: Application of generalized predictive control to a solar power plant. In: Proc. of the EC Esprit/CIM CIDIC Conf. Advances in Model-Based Predictive Control, Oxford, UK, vol. 2, pp. 182–188 (1993)
75. Camacho, E.F., Berenguel, M.: Application of generalized predictive control to a solar power plant. In: Clarke, D.W. (ed.) Advances in Model-Based Predictive Control. Oxford University Press, London (1994)
76. Camacho, E.F., Berenguel, M.: Application of generalized predictive control to a solar power plant. In: Proc. of the Third IEEE Conf. on Control Applications, Glasgow, UK, pp. 1657–1662 (1994)
77. Camacho, E.F., Berenguel, M.: Robust adaptive model predictive control of a solar plant with bounded uncertainties. Int. J. Adapt. Control Signal Process. **11**(4), 311–325 (1997)
78. Camacho, E.F., Bordóns, C.: Simple implementation of generalized predictive self-tuning controllers for industrial processes. Int. J. Adapt. Control Signal Process. **7**, 63–73 (1993)
79. Camacho, E.F., Bordóns, C.: Model Predictive Control in the Process Industry. Springer, Berlin (1995)
80. Camacho, E.F., Bordóns, C.: Model Based Predictive Control. Springer, Berlin (2004)
81. Camacho, E.F., Rubio, F.R., Gutiérrez, J.A.: Modelling and simulation of a solar power plant with a distributed collector system. In: Proc. of the Int. IFAC Symp. on Power Systems Modelling and Control Applications, Brussels, Belgium, 1988
82. Camacho, E.F., Rubio, F.R., Hughes, F.M.: Self-tuning control of a solar power plant with a distributed collector field. IEEE Control Syst. Mag., 72–78 (1992)
83. Camacho, E.F., Berenguel, M., Bordóns, C.: Adaptive generalized predictive control of a distributed collector field. IEEE Trans. Control Syst. Technol. **2**(4), 462–467 (1994)
84. Camacho, E.F., Berenguel, M., Rubio, F.R.: Application of a gain scheduling generalized predictive controller to a solar power plant. Control Eng. Pract. **2**(2), 227–238 (1994)
85. Camacho, E.F., Berenguel, M., Rubio, F.R.: Advanced Control of Solar Plants. Springer, Berlin (1997)
86. Camacho, E.F., Rubio, F.R., Berenguel, M.: Application of fuzzy logic control to a solar power plant. In: Mielczarski, W. (ed.) Fuzzy Logic Techniques in Power Systems. Studies in Fuzziness and Soft Computing. Physica-Verlag, Heidelberg (1997)
87. Camacho, E.F., Rubio, F.R., Berenguel, M., Valenzuela, L.: A survey on control schemes for distributed solar collector fields. Part I: Modeling and basic control approaches. Sol. Energy **81**, 1240–1251 (2007)
88. Camacho, E.F., Rubio, F.R., Berenguel, M., Valenzuela, L.: A survey on control schemes for distributed solar collector fields. Part II: Advanced control approaches. Sol. Energy **81**, 1252–1272 (2007)
89. Camacho, E.F., Berenguel, M., Alvarado, I., Limón, D.: Control of solar power systems: a survey. In: Proc. of the 9th Int. Symp. on Dynamics and Control of Process Systems, DYCOPS 2010, Leuven, Belgium, 2010
90. Campo, P.J., Morari, M.: Robust model predictive control. In: Proc. of the American Control Conf., Minneapolis, MN, USA, 1987

91. Cao, J., Cao, S.: Forecast of solar irradiance using recurrent neural networks combined with wavelet analysis. Appl. Therm. Eng. **25**, 161–172 (2005)
92. Cao, J., Cao, S.: Study of forecasting solar irradiance using neural networks with preprocessing sample data by wavelet analysis. Energy **31**, 3435–3445 (2006)
93. Cao, Z., Ledwich, G.F.: Adaptive repetitive control to track variable periodic signals with fixed sampling rate. IEEE/ASME Trans. Mechatron. **7**(3), 374–384 (2002)
94. Cardoso, A.L., Henriques, J., Dourado, A.: Fuzzy supervisor and feedforward control of a solar power plant using accessible disturbances. In: Proc. of the European Control Conf., ECC'99, Karlsruhe, Germany, 1999
95. Cardoso, A.L., Gil, P., Henriques, J., Duarte-Ramos, H., Dourado, A.: A robust fault tolerant model-based control framework: application to a solar power plant. In: IASTED-ISC, Salzburg, Austria, 2003
96. Cardoso, A.L., Gil, P., Henriques, J., Carvalho, P., Duarte-Ramos, H., Dourado, A.: Experiments with a fault tolerant adaptive controller on a solar power plant. In: CONTROLO'04, 6th Portuguese Conf. on Automatic Control, Faro, Portugal, 2004
97. Carmona, R.: Modeling and control of a distributed solar collector field with a one-axis tracking system. PhD Thesis, University of Seville, Spain (1985) (in Spanish)
98. Carmona, R., Aranda, J.M., Silva, M., Andújar, J.M.: Regulation and automation of the SSPS-DCS ACUREX field of the PSA. Report No. R-15/87, PSA (1987)
99. Carotenuto, L., Cava, M.L., Raiconi, G.: Regular design for the bilinear distributed parameter of a solar power plant. Int. J. Syst. Sci. **16**, 885–900 (1985)
100. Carotenuto, L., Cava, M.L., Muraca, P., Raiconi, G.: Feedforward control for the distributed parameter model of a solar power plant. Large Scale Syst. **11**, 233–241 (1986)
101. Castro, M.A.: Simulation of solar energy plants. Application to energy management. PhD Thesis, ESII Madrid, Spain (1988) (in Spanish)
102. Chen, Y.T., Lim, B.H., Lim, C.S.: Sun tracking formula for heliostats with arbitrarily oriented axes. J. Sol. Energy Eng. **128**, 245–251 (2006)
103. Cirre, C.M.: Hierarchical control of the energy production using distributed solar collectors. PhD Thesis, University of Almería, Spain (2007) (in Spanish)
104. Cirre, C.M., Moreno, J.C., Berenguel, M.: Robust QFT control of a solar collectors field. In: Martínez, D. (ed.) IHP Programme. Research Results at PSA Within the Year 2002 Access Campaign. CIEMAT, Madrid (2003)
105. Cirre, C.M., Valenzuela, L., Berenguel, M., Camacho, E.F.: A control strategy integrating automatic setpoint generation and feedforward control for a distributed solar collector field. In: Proc. XIV Jornadas de Automática, León, Spain, 2004 (in Spanish)
106. Cirre, C.M., Valenzuela, L., Berenguel, M., Camacho, E.F.: Control de plantas solares con generación automática de consignas. Rev. Iberoam. Autom. Inform. Ind. **1**, 56–66 (2004)
107. Cirre, C.M., Valenzuela, L., Berenguel, M., Camacho, E.F.: Feedback linearization control for a distributed solar collector field. In: Proc. of the 16th IFAC World Congress, Prague, Czech Republic, 2005
108. Cirre, C.M., Valenzuela, L., Berenguel, M., Camacho, E.F., Zarza, E.: Fuzzy setpoint generator for a distributed collectors solar field. In: ICIIEM—1st Int. Congress on Energy and Environment Engineering and Management, Portalegre, Portugal, 2005
109. Cirre, C.M., Berenguel, M., Valenzuela, L., Camacho, E.F.: Feedback linearization control for a distributed solar collector field. Control Eng. Pract. **15**, 1533–1544 (2007)
110. Cirre, C.M., Berenguel, M., Valenzuela, L., Klempous, R.: Reference governor optimization and control of a distributed solar collector field. Eur. J. Oper. Res. **193**, 709–717 (2009)
111. Cirre, C.M., Moreno, J.C., Berenguel, M., Guzmán, J.L.: Robust control of solar plants with distributed collectors. In: Proc. of the 2010 IFAC Int. Symp. on Dynamics and Control of Process Systems, DYCOPS 2010, Leuven, Belgium, 2010, Paper ID: 103
112. Clarke, D.W., Mohtadi, C.: Properties of Generalized Predictive Control. Automatica **25**(6), 859–875 (1989)
113. Clarke, D.W., Mohtadi, C., Tuffs, P.S.: Generalized predictive control—Part I. The basic algorithm. Automatica **23**(2), 137–148 (1987)

114. Clifford, M.J., Eastwood, D.: Design of a novel passive solar tracker. Sol. Energy **77**, 269–280 (2004)
115. Cohen, W.C., Johnston, E.F.: Dynamic characteristics of double-pipe heat exchangers. Ind. Eng. Chem. **48**, 1031–1034 (1956)
116. Coito, F., Lemos, J.M., Rato, L.M., Silva, R.N.: Experiments in predictive control of a distributed collectors solar field. In: Proc. of the Portuguese Automatic Control Conf. CONTROLO'96, Porto, Portugal, 1996
117. Coito, F., Lemos, J.M., Silva, R.N., Mosca, E.: Adaptive control of a solar energy plant: exploiting acceptable disturbances. Int. J. Adapt. Control Signal Process. **11**(4), 327–342 (1997)
118. Collado, F.J.: One-point fitting of the flux density produced by a heliostat. Sol. Energy **84**(4), 673–684 (2010)
119. Corchero, M.A., Ortega, M.G., Rubio, F.R.: Robust H_∞ control applied to a solar plant. In: Proc. of the 16th IFAC World Congress, Prague, Czech Republic, 2005
120. Costa, B.A., Lemos, J.M., Guillot, E., Olalde, G., Rosa, L.G., Fernandes, J.C.: An adaptive temperature control law for a solar furnace. In: MED08, 16th Mediterranean Conf. on Control and Automation, Ajaccio, France, 2008
121. Costa, B.A., Lemos, J.M., Guillot, E., Olalde, G., Rosa, L.G., Fernandes, J.C.: Temperature control of a solar furnace for material testing. In: CONTROLO'08, 8th Portuguese Conf. on Automatic Control, Vila Real, Portugal, 2008
122. Costa-Castelló, R., Nebot, J., Griñó, R.: Demonstration of the internal model principle by digital repetitive control of an educational laboratory plant. IEEE Trans. Ed. **48**(1), 73–80 (2005)
123. Crispim, E.M., Ferreira, P.M., Ruano, A.E.: Solar radiation prediction using RBF neural networks and cloudiness indices. In: 2006 Int. Joint Conf. on Neural Networks, Vancouver, BC, Canada, 2006
124. Das, S., Dutta, T.K.: Mathematical modeling and experimental studies on solar energy storage in a phase change material. Sol. Energy **51**(5), 305–312 (2003)
125. Deshmukh, M.K., Deshmukh, S.S.: Modeling of hybrid renewable energy systems. Renew. Sustain. Energy Rev. **12**(1), 235–249 (2008)
126. Du, J., Song, C., Li, P.: Modeling and control of a continuous stirred tank reactor based on a mixed logical dynamical model. Chin. J. Chem. Eng. **15**(4), 533–538 (2007)
127. Dynasim, A.B.: Dymola 5.3 User Manual. http://www.dynasim.se (2004)
128. Eborn, J.: On model libraries for thermo-hydraulic applications. PhD Thesis, Department of Automatic Control, Lund Institute of Technology, Sweden (2001)
129. Eck, M., Eberl, M.: Controller design for injection mode driven direct solar steam generating parabolic trough collectors. In: Proc. of the ISES Solar World Congress 1999, Jerusalem, Israel, vol. I, pp. 247–257 (1999)
130. Eck, M., Zarza, E., Eickhoff, M., Rheilander, J., Valenzuela, L.: Applied research concerning the direct steam generation in parabolic troughs. Sol. Energy **74**, 341–351 (2003)
131. El-Nashar, A.: The economic feasibility of small solar med seawater desalination plants for remote arid areas. Desalination **134**, 173–186 (2001)
132. Ertekin, C., Evrendilek, F.: Spatio-temporal modeling of global solar radiation dynamics as a function of sunshine duration for Turkey. Agric. For. Meteorol. **145**, 36–47 (2007)
133. España, M.D., Rodríguez, V.L.: Approximate steady-state modeling of a solar trough collector. Sol. Energy **37**(6), 447–545 (1987)
134. Esram, T., Chapman, P.L.: Comparison of photovoltaic array maximum power point tracking techniques. IEEE Trans. Energy Convers. **22**(2), 439–449 (2007)
135. Farkas, I., Vajk, I.: Experiments with internal model-based controller for Acurex field. In: Martínez, D. (ed.) Proc. of the 2nd Users Group TMR Programme at PSA, CIEMAT. CIEMAT, Madrid (2002)
136. Farkas, I., Vajk, I.: Internal model-based controller for a solar plant. In: Proc. of the 15th IFAC World Congress, Barcelona, Spain, 2002
137. Farkas, I., Vajk, I.: Modeling and control of a distributed solar collector field. In: Proc. of the Energy and Environment Congress, Opatija, Croatia, 2002

138. Farkas, I., Vajk, I.: Experiments with robust internal model-based controller for Acurex field. In: Martínez, D. (ed.) IHP Programme. Research Results at PSA Within the Year 2002 Access Campaign. CIEMAT, Madrid (2003)
139. Fernández-García, A., Zarza, E., Valenzuela, L., Pérez, M.: Parabolic-trough solar collectors and their applications. Renew. Sustain. Energy Rev. **14**, 1695–1721 (2010)
140. Fernández-Reche, J., Cañadas, I., Sánchez, M., Ballestrín, J., Yebra, L., Monterreal, R., Rodríguez, J., García, G., Alonso, M., Chenlo, F.: PSA solar furnace: a facility for testing PV cells under concentrated solar radiation. Sol. Energy Mater. Sol. Cells **90**, 2480–2488 (2006)
141. Flamant, G., Ferriere, A., Laplaze, D., Monty, C.: Solar processing of materials: opportunities and new frontiers. Sol. Energy **66**(2), 117–132 (1999)
142. Flores, A., Saez, D., Araya, J., Berenguel, M., Cipriano, A.: Fuzzy predictive control of a solar power plant. IEEE Trans. Fuzzy Syst. **13**(1), 58–68 (2005)
143. Fortescue, T.R., Kershenbaum, L.S., Ydstie, B.E.: Implementation of self-tuning regulators with variable forgetting factors. Automatica **17**(6), 831–835 (1981)
144. Fosso, O.B., Gjelsvik, A., Haugstad, A., Mo, B., Wangensteen, I.: Generation scheduling in a deregulated system. The Norwegian case. IEEE Trans. Power Syst. **14**(1), 75–81 (1999)
145. Franklin, G.F., Powell, J.D.: Digital Control of Dynamic Systems. Addison-Wesley, London (1980)
146. Gálvez-Carrillo, M., De Keyser, R., Ionescu, C.: Nonlinear predictive control with dead-time compensator: application to a solar power plant. Sol. Energy **83**, 743–752 (2009)
147. Garcia, P., Ferriere, A., Bezian, J.J.: Codes for solar flux calculation dedicated to central receiver system applications: a comparative review. Sol. Energy **82**, 189–197 (2008)
148. Garcia-Gabin, W., Zambrano, D., Camacho, E.F.: Sliding mode predictive control of a solar air conditioning plant. Control Eng. Pract. **17**(6), 652–663 (2009)
149. García-Martín, F.J., Berenguel, M., Camacho, E.F., Rubio, F.R.: Automatic control of a solar furnace. Internal Report GAR 1996/06, U. Sevilla, Dept. Ing. Sistemas y Automática, ESI, Spain (1996)
150. García-Martín, F.J., Berenguel, M., Valverde, A., Camacho, E.F.: Heuristic knowledge-based heliostat field control for the optimization of the temperature distribution in a volumetric receiver. Sol. Energy **66**(5), 355–369 (1999)
151. German, R.M.: Sintering Theory and Practice. Wiley, New York (1996)
152. Ghezelayagh, H., Lee, K.Y.: Application of neuro-fuzzy identification in the predictive control of power plant. In: Proc. of the 15th IFAC World Congress, Barcelona, Spain, 2002
153. Gil, P., Henriques, J., Dourado, A.: Recurrent neural networks and feedback linearization for a solar power plant control. In: Proc. of EUNIT01, Tenerife, Spain, 2001
154. Gil, P., Henriques, J., Cardoso, A., Dourado, A.: Neural network in scheduling linear controllers with application to a solar power plant. In: Proc. of the 5th IASTED Int. Conf. on Control and Applications, Cancun, Mexico, 2002
155. Gil, P., Henriques, J., Carvalho, P., Duarte-Ramos, H., Dourado, A.: Adaptive neural model-based predictive controller of a solar power plant. In: Proc. of the IEEE Int. Joint Conf. on Neural Networks, IJCNN'02, Honolulu, USA, 2002
156. Gil, P., Cardoso, A., Henriques, J., Carvalho, P., Duarte-Ramos, H., Dourado, A.: Experiments with an adaptive neural model-based predictive controller applied to a distributed solar collector field: performance and fault tolerance assessment. In: Martínez, D. (ed.) IHP Programme. Research Results at PSA Within the Year 2002 Access Campaign. CIEMAT, Madrid (2003)
157. Giral, J., Rivoire, B., Robert, J.F.: A new advanced control and operating system for the heliostats of the French CNRS'1000 KW Solar Furnace. In: Proc. of the 8th Int. Symp. on Solar Thermal Concentrating Technologies, Cologne, Germany, vol. 2, pp. 1592–1608 (1996)
158. Glaser, P.E.: A solar furnace for use in applied research. Sol. Energy **1**(2–3), 63–67 (1957)
159. González, J., Yebra, L., Valverde, A., Berenguel, M., Peralta, M.: Real-time system applications in solar thermal plants. In: 13th SOLARPACES Int. Symp., Seville, Spain, 2006

160. Gordillo, F., Rubio, F.R., Camacho, E.F., Berenguel, M., Bonilla, J.P.: Genetic design of a fuzzy logic controller for a solar power plant. In: Proc. of the European Control Conf., ECC'97, Brussels, Belgium, p. 268 (1997)
161. Goswami, Y., Kreith, F., Kreider, J.F.: Principles of Solar Engineering. Taylor and Francis, London (2000)
162. Greco, C., Menga, G., Mosca, E., Zappa, G.: Performance improvements of self-tuning controllers by multistep horizons: the MUSMAR approach. Automatica **20**, 681–699 (1984)
163. Grena, R.: An algorithm for the computation of the solar position. Sol. Energy **82**, 462–470 (2008)
164. Grossmann, R.L., Nerode, A., Ravn, A.P., Rischel, H. (eds.): Hybrid Systems. Lecture Notes in Computer Science, vol. 736. Springer, Berlin (1993)
165. Gueymard, C.A.: Direct solar transmittance and irradiance predictions with broadband models. Part I: Detailed theoretical performance assessment. Sol. Energy **74**, 355–379 (2003)
166. Gueymard, C.A.: Direct solar transmittance and irradiance predictions with broadband models. Part II: Validation with high-quality measurements. Sol. Energy **74**, 381–395 (2003)
167. Gueymard, C.A.: Importance of atmospheric turbidity and associated uncertainties in solar radiation and luminous efficacy modeling. Energy **30**, 1603–1621 (2005)
168. Gueymard, C.A.: Prediction and validation of cloudless shortwave solar spectra incident on horizontal, tilted, or tracking surfaces. Sol. Energy **82**, 260–271 (2008)
169. Guzmán, J.L., Berenguel, M., Dormido, S.: Interactive teaching of constrained generalized predictive control. IEEE Control Syst. Mag. **25**(2), 52–66 (2005)
170. Haeger, M., Keller, L., Monterreal, R., Valverde, A.: Phoebus Technology Program Solar Air Receiver (TSA). CIEMAT, Madrid (1994)
171. Hägglund, T.: An industrial dead time compensating PI controller. Control Eng. Pract. (4), 749–756 (1996)
172. Hallet, R.W., Gervais, R.L.: Central Receiver Solar Thermal Power System, vol. III-book 2, SAN-1108-76-8, MDC G6776. McDonnell Douglas Corporation (1977)
173. Heemels, W.P.M.H., Schutter, B.D., Bemporad, A.: Equivalence of hybrid dynamical models. Automatica **37**(7), 1085–1091 (2001)
174. Heemels, W.P.M.H., Schutter, B.D., Bemporad, A.: On the equivalence of classes of hybrid dynamical models. In: 40th IEEE Conf. on Decision and Control, Orlando, FL, USA, pp. 364–369 (2001)
175. Heinemann, D., Lorenz, E., Girodo, M.: Forecasting of solar radiation. In: Dunlop, E.D., Wald, L., Suri, M. (eds.) Solar Energy Resource Management for Electricity Generation from Local Level to Global Scale. Nova Publishers, New York (2006) (Chap. 7)
176. Helwa, N.H., Bahgat, A.B.G., Shafee, A.M.R.E., Shanawy, E.T.E.: Maximum collectable solar energy by different solar tracking systems. Energy Sources **22**(1), 23–34 (2000)
177. Henning, H.: Solar assisted air conditioning of buildings—an overview. Appl. Therm. Eng. **27**(10), 1734–1749 (2007)
178. Henriques, J., Cardoso, A., Dourado, A.: Supervision and c-Means clustering of PID controllers for a solar power plant. Int. J. Approx. Reason. **22**(1–2), 73–91 (1999)
179. Henriques, J., Gil, P., Dourado, A.: Neural output regulation for a solar power plant. In: Proc. of the 15th IFAC World Congress, Barcelona, Spain, 2002
180. Hibbert, H., Pedreira, C., Souza, R.: Combining neural networks and ARIMA models for hourly temperature forecast. In: Proc. of the Int. Conf. on Neural Networks, IJCNN 2000, Como, Italy, pp. 414–419 (2000)
181. Hoffschmidt, B., Pitz-Paal, R., Böhmer, M.: Porous structures for volumetric receivers—comparison of experimental and numerical results. In: Proc. of the 8th Int. Symp. on Solar Thermal Concentrating Technologies, Cologne, Germany, pp. 567–587 (1996)
182. Horowitz, I.: Quantitative Feedback Design Theory (QFT). QFT Publications, Colorado (1993)
183. Hua, C., Lin, J.: A modified tracking algorithm for maximum power tracking of solar array. Energy Convers. Manag. **45**(2), 911–925 (2004)
184. Ibáñez, M., Rosell, J., Rosell Urrutia, J.: Tecnología Solar. Editorial MP, Madrid (2005)

185. Ibrahim, H., Ilincaa, A., Perron, J.: Energy storage systems. Characteristics and comparisons. Renew. Sustain. Energy Rev. **12**(5), 1221–1250 (2008)
186. Igreja, J.M., Lemos, J.M., Barão, M., Silva, R.N.: Adaptive nonlinear control of a distributed collector solar field. In: Proc. of the European Control Conf., ECC'03, Cambridge, UK, 2003
187. Ionescu, C., Wyns, B., Sbarciog, M., Boullart, L., De Keyser, R.: Comparison between physical modeling and neural network modeling of a solar power plant. In: Proc. of the IASTED Int. Conf. on Applied Simulation and Modeling, ASM'04, Rhodes, Greece, 2004
188. Iqbal, M.: An Introduction to Solar Radiation. Academic Press, Toronto (1983)
189. Isidori, A.: Nonlinear Control Systems. Springer, Berlin (1995)
190. ITET: The IEA/SSPS Solar Thermal Power Plants, vol. 2. Springer, Berlin (1986)
191. Jalili-Kharaajoo, M.: Predictive control of a solar power plant with neuro-fuzzy identification and evolutionary programming optimization. In: Proc. of the IEEE Conf. on Emerging Technologies and Factory Automation, ETFA'03, Lisbon, Portugal, vol. 2, pp. 173–176 (2004)
192. Jalili-Kharaajoo, M., Besharati, F.: Intelligent predictive control of a solar power plant with neuro-fuzzy identifier and evolutionary programming optimizer. In: Proc. of the IEEE Conf. on Emerging Technologies and Factory Automation, ETFA'03, Lisbon, Portugal, vol. 2, pp. 173–176 (2003)
193. Janjai, S., Pankaew, P., Laksanaboonsong, J.: A model for calculating hourly global solar radiation from satellite data in the tropics. Appl. Energy **86**, 1450–1457 (2009)
194. Jiang, J.: Optimal gain scheduling controller for a diesel engine. IEEE Control Syst. Mag., 42–48 (1994)
195. Johansen, T.A., Storaa, C.: An internal energy controller for distributed solar collector fields. In: Martínez, D. (ed.) Proc. of the 2nd Users Workshop IHP Programme, CIEMAT. CIEMAT, Madrid (2002)
196. Johansen, T.A., Storaa, C.: Energy-based control of a distributed solar collector field. Automatica **38**(7), 1191–1199 (2002)
197. Johansen, T.A., Hunt, K.J., Petersen, I.: Gain-scheduled control of a solar power plant. Control Eng. Pract. **8**(9), 1011–1022 (2000)
198. Juuso, E.K.: Fuzzy control in process industry. In: Verbruggen, H.B. et al. (eds.) Fuzzy Algorithms for Control. Kluwer Academic, Boston (1999)
199. Juuso, E.K., Valenzuela, L.: Adaptive intelligent control of a solar collector field. In: Proc. of the 3rd European Symp. on Intelligent Technologies, Hybrid Systems and Their Implementation on Smart Adaptive Systems, EUNITE 2003, Oulu, Finland, 2003
200. Juuso, E.K., Balsa, P., Leiviska, K.: Linguistic equation controller applied to a solar collectors field. In: Proc. of the European Control Conf., ECC'97, Brussels, Belgium, 1997
201. Juuso, E.K., Balsa, P., Valenzuela, L.: Multilevel linguistic equation controller applied to a 1MWh solar power plant. In: Proc. of the 1998 American Control Conf., ACC'98, Philadelphia, USA, vol. 6, pp. 3891–3895 (1998)
202. Juuso, E.K., Balsa, P., Valenzuela, L., Leiviska, K.: Robust intelligent control of a distributed solar collector field. In: Proc. of CONTROLO'98, the 3rd Portuguese Conf. on Automatic Control, Coimbra, Portugal, vol. 2, pp. 621–626 (1998)
203. Kalogirou, S.A.: Applications of artificial neural-networks for energy systems. Appl. Energy **67**, 17–35 (2000)
204. Kalogirou, S.A.: Artificial neural networks in renewable energy systems applications: a review. Renew. Sustain. Energy Rev. **5**, 373–401 (2001)
205. Kalogirou, S.A.: Solar thermal collectors and applications. Prog. Energy Combust. Sci. **30**, 231–295 (2004)
206. Kalt, A.: Distributed Collector System Plant Construction Report. IEA/SSPS Operating Agent DFVLR, Cologne (1982)
207. Kasabov, N.K.: Foundations of Neural Networks, Fuzzy Systems, and Knowledge Engineering. MIT Press, Cambridge (1995)
208. Ke, J.Y., Tang, K.S., Man, K.F., Luk, P.C.K.: Hierarchical genetic fuzzy controller for a solar power plant. In: Proc. of the IEEE Int. Symp. on Industrial Electronics, ISIE'98, South Africa, pp. 584–588 (1998)

209. Kim, D.S., Infante Ferreira, C.A.: Solar refrigeration options—a state-of-the-art review. Int. J. Refrig. **31**(1), 3–15 (2008)
210. King, D.L.: Beam Quality and Tracking Accuracy Evaluation of Second Generation and Barstow Production Heliostats, SAND82-0181. McDonnell Douglas Corporation (1982)
211. King, D.L., Arvizu, D.E.: Heliostat characterization at the central receiver test facility. Trans. ASME J. Sol. Energy Eng. **103**, 82–88 (1981)
212. Kingery, W.D., Bowen, H.K., Uhlmann, D.R.: Introduction to Ceramics, 2nd edn. Wiley, New York (1996)
213. Klein, A.A., Duffie, J.A., Beckman, W.A.: Transient considerations of flat-plate solar collectors. Trans. ASME J. Eng. Power **96A**, 109–110 (1974)
214. Kleissl, J., Harper, J., Dominguez, A.: A solar resource measurement network for solar intermittency at high spatio-temporal resolution. In: Proc. of the SOLAR 2010 Conf., Phoenix, Arizona, USA, 2010
215. Klempous, R., Maciejewski, H., Nikodem, M., Nikodem, J., Berenguel, M., Valenzuela, L.: Data driven methods and data analysis of a distributed solar collector field. In: Proc. of the 4th Int. Conf. on Applied Mathematics, APLIMAT 2005, Bratislava, Slovak, pp. 205–212 (2005)
216. Kolb, G.J., Scott, A.J., Donnelly, M.W., Gorman, D., Thomas, R., Davenport, R., Lumia, R.: Heliostat cost reduction study, SAND2007-3293. http://www.prod.sandia.gov/cgi-bin/techlib/access-control.pl/2007/073293.pdf (2007)
217. Kribus, A., Ries, H., Spirkl W.: Inherent limitations of volumetric solar receivers. J. Sol. Energy Eng. **118**, 151–155 (1996)
218. Kurzt, M.J., Henson, M.A.: Input–output linearizing control of constrained nonlinear processes. J. Process Control **7**, 3–17 (1996)
219. Kurzt, M.J., Henson, M.A.: Feedback linearizing control of discrete-time nonlinear systems with inputs constraints. Int. J. Control **70**(2), 603–616 (1998)
220. Lacasa, D., Berenguel, M., Cañadas, I., Yebra, L.: Modelling the thermal process of copper sintering in a solar furnace. In: Proc. of the 13th Solarpaces Int. Symp., Seville, Spain, pp. 2–9 (2006)
221. Lacasa, D., Berenguel, M., Yebra, L., Martínez, D.: Copper sintering in a solar furnace through fuzzy control. In: Proc. of the 2006 IEEE Int. Conf. on Control Applications, Munich, Germany, pp. 2144–2149 (2006)
222. Lahmidi, H., Mauran, S., Goetz, V.: Definition, test and simulation of a thermochemical storage process adapted to solar thermal systems. Sol. Energy **80**, 883–893 (2006)
223. Lee, C.C.: Fuzzy logic in control systems: fuzzy logic controller—Part I. IEEE Trans. Syst. Man Cybern. **20**(2), 404–418 (1990)
224. Lee, C.C.: Fuzzy logic in control systems: fuzzy logic controller—Part II. IEEE Trans. Syst. Man Cybern. **20**(2), 419–435 (1990)
225. Lemos, J.M., Rato, L.M., Mosca, E.: Integrating predictive and switching control: basic concepts and an experimental case study. In: Allgower, F., Zheng, A. (eds.) Nonlinear Model Predictive Control. Birkhäuser, Basel (2000)
226. León, J., Valenzuela, L.: DISS project. Results of three years operating a thermal solar plant with parabolic collectors for direct steam production. In: Proc. XI Congreso Ibérico–VI Congreso Iberoamericano de Energía Solar, Vilamoura-Algarve, Portugal, 2002 (in Spanish)
227. Liberzon, D.: Switching in Systems and Control. Springer, Boston (2003)
228. Limón, D., Alvarado, I., Álamo, T., Arahal, M.R., Camacho, E.F.: Robust control of the distributed solar collector field ACUREX using MPC for tracking. In: Proc. of the 17th World Congress of IFAC, Seoul, Korea, 2008
229. Ljung, L.: System Identification Toolbox. The Mathworks, Natick (1986)
230. Ljung, L.: System Identification, Theory for the User, 2nd edn. Prentice Hall, Englewood Cliffs (1999)
231. Lo, Y., Chiu, H.J., Lee, T.P., Purnama, I., Wang, J.M.: Analysis and design of a photovoltaic system DC connected to the utility with a power factor corrector. IEEE Trans. Ind. Electron. **56**(11), 4354–4362 (2009)

232. Loebis, D.: Fuzzy logic control of a solar power plant. Master's Dissertation, University of Sheffield, UK (2000)
233. Luk, P.C.K., Khoo, K.K., Berenguel, M.: Direct fuzzy logic control of a solar power plant using distributed collector fields. In: Proc. of the 2nd Int. ICSC Symp. on Soft Computing and Intelligent Industrial Automation, SOCO'97, Nimes, France, pp. 81–89 (1997)
234. Luk, P.C.K., Low, K.C., Sayiah, A.: GA-based fuzzy logic control of a solar power plant using distributed collector fields. Renew. Energy **16**(1–4), 765–768 (1999)
235. Maciejewski, H., Berenguel, M., Valenzuela, L., Cirre, C.M.: Data mining—applications and perspectives for solar plant control and monitoring. In: Martínez, D. (ed.) IHP Programme—Research Results at PSA Within the Year 2003 Access Campaign. CIEMAT, Madrid (2004)
236. Maciejewski, H., Valenzuela, L., Berenguel, M., Adamus, K.: Performing direct steam generation solar plant analysis through data mining. In: Proc. of the 13th Solarpaces Int. Symp., Seville, Spain, 2006
237. Mamdani, E.H.: Application of fuzzy algorithms for control of a simple dynamic plant. IEE Proc. Part D **121**, 1585–1588 (1974)
238. Mancini, T.: Heliostat daily centroid shift. Report No. III-1/99, IEA SolarPACES (1999)
239. Markou, H., Petropoulakis, L.: PID-type fuzzy control of the Acurex solar collector field. In: Martínez, D. (ed.) Proc. of the 2nd Users Group TMR Programme at PSA, CIEMAT. CIEMAT, Madrid (2002)
240. Martínez, D.: Solar furnace technologies. In: Solar Thermal Test Facilities. CIEMAT, Madrid (1996)
241. Meaburn, A.: Modeling and control of a distributed solar collector field. PhD Thesis, Department of Electrical Engineering and Electronics, UMIST, UK (1995)
242. Meaburn, A., Hughes, F.M.: Resonance characteristics of distributed solar collector fields. Sol. Energy **51**(3), 215–221 (1993)
243. Meaburn, A., Hughes, F.M.: A control technique for resonance cancellation. In: Proc. of the IEE Colloquium on Nonlinear Control Using Structural Knowledge of System Models, London, UK, 1993
244. Meaburn, A., Hughes, F.M.: Prescheduled adaptive control scheme for resonance cancellation of a distributed solar collector field. Sol. Energy **52**(2), 155–166 (1994)
245. Meaburn, A., Hughes, F.M.: A pre-scheduled adaptive control scheme based upon system knowledge. In: Proc. of the IEE Colloquium on Adaptive Controllers in Practice—Part One, London, UK, 1995
246. Meaburn, A., Hughes, F.M.: A simple predictive controller for use on large scale arrays of parabolic trough collectors. Sol. Energy **56**(6), 583–595 (1996)
247. Meaburn, A., Hughes, F.M.: Feedforward control of solar thermal power plants. Trans. ASME J. Sol. Energy Eng. **119**(1), 52–61 (1997)
248. Mechlouch, R.F., Brahim, A.B.: A global solar radiation model for the design of solar energy systems. Asian J. Sci. Res. **1**(3), 231–238 (2008)
249. Mellit, A., Kalogirou, S.A.: Artificial intelligence techniques for photovoltaic applications: a review. Prog. Energy Combust. Sci. **34**, 547–632 (2008)
250. Mellit, A., Massi-Pavan, A.: A 24-h forecast of solar irradiance using artificial neural network: application for performance prediction of a grid-connected PV plant at Trieste, Italy. Sol. Energy **84**, 807–821 (2010)
251. Menchinelli, P., Bemporad, A.: Hybrid model predictive control of a solar–air conditioning plant. Eur. J. Control **14**(6), 501–515 (2008)
252. Messaoud, H., Favier, G., Mendes, R.S.: Adaptive robust pole placement by connecting robust identification and control. In: Proc. of the IFAC Symp. on Adaptive Systems in Control and Signal Processing, ACASP'92, Grenoble, France, 1992
253. Mignone, D.: The REALLY big collection of logic propositions and linear inequalities. http://control.ee.ethz.ch/index.cgi?page=publications;action=details;id=377 (2002)
254. Mills, D.: Advances in solar thermal electricity technology. Sol. Energy **76**, 19–31 (2004)
255. Mittelmann, H.D., Pendse, G., Rivera, D.E., Lee, H.: Optimization-based design of plant-friendly multisine signals using geometric discrepancy criteria. Comput. Optim. Appl. **38**, 173–190 (2007)

256. Mo, S.H., Norton, J.P.: Fast and robust algorithm to compute exact polytop parameter bounds. Math. Comput. Simul. **32**, 481–493 (1990)
257. Modelica Association: Modelica, a unified object oriented language for physical systems modeling. Language specification 2.2. Technical Report. http://www.modelica.org (2005)
258. Modelica Association: Modelica standard library, version 2.2.1. http://www.modelica.org/libraries/Modelica/releases/2.2.1/ (2007)
259. Monterreal, R., Heller, P.: Large area heliostat comparison at PSA. Internal Report, PSA (1997)
260. Morari, M., Zafiriou, E.: Robust Process Control. Prentice Hall, New York (1989)
261. Moreno-Muñoz, A., de la Rosa, J.J.G., Posadillo, R., Bellido, F.: Very short term forecasting of solar radiation. In: Photovoltaic Specialists Conf., San Diego, CA, USA, 2008
262. Mosca, E.: Optimal, Predictive and Adaptive Control. Prentice Hall, New York (1995)
263. Mousazadeh, H., Keyhani, A., Javadi, A., Mobli, H., Abrinia, K., Sharifi, A.: A review of principle and sun-tracking methods for maximizing solar systems output. Renew. Sustain. Energy Rev. **13**, 1800–1818 (2009)
264. Nenciari, G., Mosca, E.: Supervised multicontrollers for temperature regulation of a distributed collector field. In: Martínez, D. (ed.) Proc. of the 1st Users Group TMR Programme at PSA, CIEMAT. CIEMAT, Madrid (1998)
265. Neumann, A., Groer, U.: Experimenting with concentrated sunlight using the DLR solar furnace. Sol. Energy **58**(4-6), 181–190 (1996)
266. Neumann, A., Witzke, A., Jones, S.A., Schimtt, G.: Representative terrestrial solar brightness profiles. J. Sol. Energy Eng. **124**(2), 198–204 (2002)
267. Neville, R.C.: Solar energy collector orientation and Sun tracking mode. Sol. Energy **20**(7), 7–11 (1978)
268. NIST: Engineering statistics handbook. Technical Report. http://www.itl.nist.gov/div898/handbook/ (2006)
269. Nordgren, R.E., Franchek, M.A.: New formulations for quantitative feedback theory. Int. J. Robust Nonlinear Control **4**, 47–64 (1994)
270. Normey-Rico, J.E., Camacho, E.F.: Control of Dead-Time Processes. Springer, Berlin (2007)
271. Normey-Rico, J.E., Camacho, E.F.: Dead-time compensators: a survey. Control Eng. Pract. **16**(4), 407–428 (2008)
272. Normey-Rico, J.E., Camacho, E.F.: Unified approach for robust dead-time compensator design. J. Process Control **19**(1), 38–47 (2008)
273. Normey-Rico, J.E., Bordóns, C., Camacho, E.F.: Improving the robustness of dead-time compensating PI controllers. Control Eng. Pract. (6), 801–810 (1997)
274. Normey-Rico, J.E., Bordóns, C., Berenguel, M., Camacho, E.F.: A robust adaptive dead-time compensator with application to a solar collector field. In: Proc. of the IFAC Linear Time Delay Systems Workshop, Grenoble, France, 1998
275. Nuñez-Reyes, A., Normey-Rico, J.E., Bordóns, C., Camacho, E.F.: A Smith predictive based MPC in a solar air conditioning plant. J. Process Control **15**, 1–10 (2005)
276. Ogata, K.: Modern Control Engineering, 5th edn. Pearson–Prentice Hall, Upper Saddle River–New York (2009)
277. Ogunnaike, B.A., Ray, W.H.: Process Dynamics, Modeling and Control. Academic Press, San Diego (1994)
278. Oksanen, P., Juuso, E.K.: Advanced control for solar systems at PSA. In: Proc. of TOOLMET99, Symp. Tool Environments and Development Methods for Intelligent Systems, Oulu, Finland, pp. 123–134 (1999)
279. Olfati-Saber, R.: Distributed Kalman filter with embedded consensus filters. in: 44th IEEE Conf. on Decision and Control and European Control Conf., CDC–ECC'05, Seville, Spain, pp. 8179–8184 (2005)
280. Orbach, A., Rorres, C., Fischl, R.: Optimal control of a solar collector loop using a distributed-lumped model. Automatica **27**(3), 535–539 (1981)
281. Ortega, M.G., Rubio, F.R.: Systematic design of weighting matrices for H_∞ mixed sensitivity problem. J. Process Control **14**(1), 89–98 (2004)

282. Ortega, M.G., Rubio, F.R., Berenguel, M.: An H_∞ controller for a solar power plant. In: Proc. of the IASTED Int. Conf. on Control, Cancún, México, pp. 122–125 (1997)
283. Paoli, C., Voyant, C., Muselli, M., Nivet, M.L.: Solar radiation forecasting using ad-hoc time series preprocessing and neural-network. Emerg. Intell. Comput. Technol. Appl. SpringerLink **5754**, 898–907 (2009)
284. Pasamontes, M., Álvarez, J.D., Guzmán, J.L., Berenguel, M.: Hybrid modeling of a solar cooling system. In: Proc. of the 3rd IFAC Conf. on Analysis and Design of Hybrid Systems, Zaragoza, Spain, 2009
285. Pasamontes, M., Álvarez, J.D., Guzmán, J.L., Lemos, J.M., Berenguel, M.: A switching control strategy applied to a solar collector field. Control Eng. Pract. **19**, 135–145 (2011)
286. Passino, K.M., Yurkovich, S.: Fuzzy Control. Addison-Wesley, Menlo Park (1998)
287. Patankar, S.V.: Numerical Heat Transfer and Fluid Flow. Series in Computational and Physical Processes in Mechanics and Thermal Sciences. Taylor & Francis, London (1980)
288. Pawlowski, A., Guzmán, J.L., Rodríguez, F., Berenguel, M., Sánchez, J.: Application of time-series methods to disturbance estimation in predictive control problems. In: Proc. of the 2010 IEEE Symp. on Industrial Electronics, Bari, Italy, 2010
289. Pereira, C., Dourado, A.: Application of a neuro-fuzzy network with support vector learning to a solar power plant. In: Martínez, D. (ed.) Proc. of the 2nd Users Workshop IHP Programme, CIEMAT. CIEMAT, Madrid (2002)
290. Pereira, C., Dourado, A.: Application of a neuro-fuzzy network with support vector learning to a solar power plant. In: Proc. of the 15th IFAC World Congress, Barcelona, Spain, 2002
291. Perez, R., Moore, K., Wilcox, S., Renne, D., Zelenka, A.: Forecasting solar radiation: preliminary evaluation of an approach based upon the national forecast database. Sol. Energy **81**, 809–812 (2007)
292. Pérez de la Parte, M., Cirre, C.M., Camacho, E.F., Berenguel, M.: Application of predictive sliding mode controllers to a solar plant. IEEE Trans. Control Syst. Technol. **16**(4), 819–825 (2008)
293. Petrasch, J., Osch, P., Steinfeld, A.: Dynamics and control of solar thermochemical reactors. Chem. Eng. J. **145**, 362–370 (2009)
294. Phipps, G.S.: Heliostat beam characterization system—calibration technique. In: Proc. ISA/79 Conf., Chicago, IL, USA, 1979
295. Pickhardt, R.: Application of adaptive controllers to a solar power plant using a multi-model description. In: Proc. of the American Control Conf., Albuquerque, NM, USA, 1998
296. Pickhardt, R.: Adaptive control of a solar power plant using a multi-model control. IEE Proc. Part D **147**(5), 493–500 (2000)
297. Pickhardt, R.: Nonlinear modeling and adaptive predictive control of a solar power plant. Control Eng. Pract. **8**(8), 937–947 (2000)
298. Pickhardt, R.: Results of the application of adaptive controllers to the Acurex field. In: Martínez, D. (ed.) Proc. of the 2nd Users Group TMR Programme at PSA, CIEMAT. CIEMAT, Madrid (2003)
299. Pickhardt, R., Silva, R.N.: Application of a nonlinear predictive controller to a solar power plant. In: Proc. of the 1998 IEEE Int. Conf. on Control Applications, Glasgow, UK, vol. 1, pp. 6–10 (1998)
300. Pilkington, S.I.: Status report on solar thermal power plants, Cologne, Germany. www.solarpaces.org/Library/docs/PiStaRep.pdf (1996)
301. Pin, G., Falchetta, M., Fenu, G.: Adaptive time-warped control of molten salt distributed collector solar fields. Control Eng. Pract. **16**, 813–823 (2008)
302. Pitz-Paal, R.: High temperature solar concentrators. In: Gálvez, J.B., Rodríguez, S.M. (eds.) Solar Energy Conversion and Photoenergy Systems. EOLSS Publishers, Oxford (1999)
303. Poulek, V., Libra, M.: A very simple solar tracker for space and terrestrial applications. Sol. Energy Mater. Sol. Cells **60**, 99–103 (2000)
304. Price, H., Kearney, D., Replogle, I.: Update on the performance and operation of SEGS III–VII. In: ASME Int. Solar Energy Conf., Miami, USA, 1990
305. Rabl, A.: Active Solar Collectors and Their Applications. Oxford University Press, New York (1985)

306. Ramos, F., Crespo, L.: A new powerful tool for heliostat field layout and receiver geometry optimization: NSPOC. In: Proc. of Concentrating Solar Power and Chemical Energy Systems SolarPACES, Berlin, Germany, 2009
307. Ratcliffe, J.D., Hatonen, J.J., Lewin, P.L., Rogers, E., Owens, D.H.: Repetitive control of synchronized operations for process applications. Int. J. Adapt. Control Signal Process. **21**, 300–325 (2007)
308. Rato, L., Borelli, D., Mosca, E., Lemos, J.M., Balsa, P.: MUSMAR based switching control of a solar collector field. In: Proc. of the European Control Conf., ECC'97, Brussels, Belgium, 1997
309. Rato, L., Silva, R.N., Lemos, J.M., Coito, F.: Multirate MUSMAR cascade control of a distributed solar field. In: Proc. of the European Control Conf., ECC'97, Brussels, Belgium, 1997
310. Rato, L., Silva, R.N., Lemos, J.M., Coito, F.J.: INESC research on adaptive control of ACUREX field. In: Martínez, D. (ed.) Proc. of the 1st Users Group TMR Programme at Plataforma Solar de Almería, CIEMAT. CIEMAT, Madrid (1998)
311. Reikard, G.: Predicting solar radiation at high resolutions: a comparison of time series forecasts. Sol. Energy **83**(3), 342–349 (2009)
312. Remund, J., Perez, R., Lorenz, E.: Comparison of solar radiation forecasts for the USA. In: 2008 European PV Conf., Valencia, Spain, 2008
313. Rhodes, C., Morari, M.: Determining the model order of nonlinear input/output systems directly from data. In: Proc. of the American Control Conf., Seattle, WA, pp. 1290–1295 (1995)
314. Riedmiller, M., Braun, H.: A direct adaptive method for faster backpropagation learning: the RPROP algorithm. In: Proc. of the European Symp. on Artificial Neural Networks, ESANN'93, Brussels, Belgium, pp. 917–922 (1993)
315. Rivera, D.E., Morari, M., Skogestad, S.: Internal model control. 4. PID controller design. Ind. Eng. Chem. Process Des. Dev. **25**, 252–265 (1986)
316. Rivera, D.E., Lee, H., Braun, M.W., Mittelmann, H.D.: Plant-friendly system identification: a challenge for the process industries. In: Proc. of the 13th IFAC Symp. on System Identification, Rotterdam, The Netherlands, pp. 97–104 (2003)
317. Rivera, D.E., Lee, H., Mittelmann, H.D., Braun, M.W.: Constrained multisine input signals for plant-friendly identification of chemical process systems. J. Process Control **19**(4), 623–635 (2009)
318. Roca, L., Berenguel, M., Yebra, L., Alarcón-Padilla, D.: Solar field control for desalination plants. Sol. Energy **82**, 772–786 (2008)
319. Roca, L., Guzmán, J.L., Normey-Rico, J.E., Berenguel, M., Yebra, L.: Robust constrained predictive feedback linearization controller in a solar desalination plant collector field. Control Eng. Pract. **17**, 1076–1088 (2009)
320. Rodríguez, D.P., López, V., Damborenea, J.J., Vázquez, A.J.: Surface transformation hardening on steels treated with solar energy in central tower and heliostats field. Sol. Energy Mater. Sol. Cells **37**, 1–12 (1995)
321. Rodríguez, J., Martínez, D., Shcheglov, V.: Materials treatments at the solar furnace of Plataforma Solar de Almería under EU's DGXII HCM programme. In: Proc. of the 8th Int. Symp. on Solar Thermal Concentrating Technologies, Cologne, Germany, vol. 2, pp. 1592–1608 (1996)
322. Rodríguez, M., de Prada, C., Capraro, F., Cristea, S.: Logic embedded NMPC of a solar air conditioning plant. Eur. J. Control **14**(6), 484–500 (2008)
323. Rohsenow, W.M., Hartnett, J.P., Cho, Y.I.: Handbook of Heat Transfer, 3rd edn. McGraw-Hill, New York (1998)
324. Romero, M., Buck, R., Pacheco, J.E.: An update on solar central receiver systems, projects, and technologies. J. Sol. Energy Eng. **124**(2), 98–109 (2002)
325. Rorres, C., Orbach, A., Fischl, R.: Optimal and suboptimal control policies for a solar collector system. IEEE Trans. Autom. Control **AC-25**, 1085–1091 (1980)
326. Rossiter, J.A.: Model-Based Predictive Control: A Practical Approach. CRC Press, Boca Raton (2003)

327. Roth, P., Georgiev, A., Boudinov, H.: Design and construction of a system for Sun-tracking. Renew. Energy **29**(3), 303–402 (2004)
328. Roth, P., Georgiev, A., Boudinov, H.: Cheap two axis Sun following device. Energy Convers. Manag. **46**, 1179–1192 (2005)
329. Rubio, F.R.: Adaptive control of industrial processes. Application to a solar plant. PhD Thesis, Universidad de Sevilla (1985) (in Spanish)
330. Rubio, F.R., López, M.J.: Control Adaptativo Y Robusto. Servicio de Publicaciones de la Universidad de Sevilla, Seville (1996)
331. Rubio, F.R., Carmona, R., Camacho, E.F.: Adaptive control of the Acurex field. In: Kesserlring, P., Selvage, C.S. (eds.) The IEA/SSPS Solar Thermal Power Plants, vol. 2. Springer, Berlin (1986)
332. Rubio, F.R., Hughes, F.M., Camacho, E.F.: Self-tuning PI control of a solar power plant. In: Prep. IFAC Symp. in Adaptive Systems in Control and Signal Processing, Glasgow, UK, pp. 335–340 (1989)
333. Rubio, F.R., Berenguel, M., Camacho, E.F.: Fuzzy logic control of a solar power plant. IEEE Trans. Fuzzy Syst. **3**(4), 459–468 (1995)
334. Rubio, F.R., Camacho, E.F., Berenguel, M.: Control de campos de colectores solares. Rev. Iberoam. Autom. Inform. Ind. **3**(4), 26–45 (2006)
335. Rubio, F.R., Ortega, M.G., Gordillo, F., López-Martinez, M.: Application of a new control strategy for Sun tracking. Energy Convers. Manag. **48**, 2174–2184 (2007)
336. Sala, G., Antón, I., Arborio, J.C., Luque, A., Camblor, E., Mera, E., Gasson, M., Cendagorta, M., Valera, P., Friend, M.P., Monedero, J., Gonzalez, S., Dobon, F., Luque, I.: The 480 kW_P EUCLIDES^{TM}-Thermie power plant: instalation, set-up and first results. In: Proceeding of the 16th European Photovoltaic Solar Energy Conf. and Exhibition, WIP—Stephens & Associates, Glasgow, Scotland, May 2000
337. Sarkka, S., Vehtari, A., Lampinen, J.: Time series prediction by Kalman smoother with cross-validated noise density. In: Proc. of the IEEE Int. Joint Conf. on Neural Networks, Budapest, Hungary, pp. 1615–1619 (2004)
338. Sayigh, A.M.: Solar Air Conditioning and Refrigeration. Pergamon, Elmsford (1992)
339. Sbarciog, M., Wyns, B., Ionescu, C., Keyser, R.D., Boullart, L.: Neural networks models for a solar plant. In: Proc. of the 2nd IASTED Conf. on Neural Networks and Computational Intelligence, NCI2004, Grindelwald, Switzerland, 2004
340. Seborg, D.E.: A perspective on advanced strategies for process control (revisited). In: Frank, P.M. (ed.) Advances in Control—Highlights of ECC'99. Springer, Berlin (1999)
341. Selvage, C.S.: The IEA/SSPS Solar Thermal Power Plants. Springer, Berlin (1986)
342. Sen, Z.: Solar energy in progress and future research trends. Prog. Energy Combust. Sci. **30**, 367–416 (2004)
343. Shahmaleki, P., Mahzoon, M.: GA modeling and ANFIS control design for a solar power plant. In: Proc. of the 2010 American Control Conf., Baltimore, MD, USA, pp. 3530–3535 (2010)
344. Silva, R.N.: Dual predictive control of processes with accessible disturbances. PhD Thesis, Universidade Técnica de Lisboa, Portugal (1999) (in Portuguese)
345. Silva, R.N.: Model based predictive control with time-scaling of a solar field. In: Martínez, D. (ed.) Proc. of the 2nd Users Group TMR Programme at Plataforma Solar de Almería. CIEMAT, Madrid (1999)
346. Silva, R.N.: Time scaled predictive controller of a solar power plant. In: Proc. of the European Control Conf. 99, Karlsruhe, Germany, 1999
347. Silva, R.N., Rato, L.M., Lemos, J.M., Coito, F.: Cascade control of a distributed collector solar field. J. Process Control **4**(2), 111–117 (1997)
348. Silva, R.N., Filatov, N., Lemos, J.M., Unbehauen, H.: Feedback/feedforward dual adaptive control of a solar collector field. In: Proc. of the IEEE Int. Conf. on Control Applications, Glasgow, UK, pp. 309–313 (1998)
349. Silva, R.N., Rato, L.M., Barão, L.M., Lemos, J.M.: A physical model based approach to distributed collector solar field control. In: Proc. of the American Control Conf., Anchorage, AK, USA, pp. 1822–3817 (2002)

350. Silva, R.N., Rato, L.M., Lemos, J.M.: Observer based time warped control of distributed collector solar fields. In: Martínez, D. (ed.) Proc. of the 2nd Users Workshop IHP Programme, CIEMAT. CIEMAT, Madrid (2002)
351. Silva, R.N., Lemos, J.M., Rato, L.M.: Variable sampling adaptive control of a distributed collector solar field. IEEE Trans. Control Syst. Technol. **11**(5), 765–772 (2003)
352. Silva, R.N., Rato, L.M., Lemos, J.M.: Time scaling internal state predictive control of a solar plant. Control Eng. Pract. **11**(12), 1459–1467 (2003)
353. Skogestad, S., Postlethwaite, I.: Multivariable Feedback Control. Analysis and Design. Wiley, New York (1996)
354. Slotine, J., Li, W.: Applied Nonlinear Control. Prentice Hall, New York (1991)
355. Sonntag, C., Ding, H., Engell, S.: Supervisory control of a solar air conditioning plant with hybrid dynamics. Eur. J. Control **14**(6), 451–463 (2008)
356. Spall, J.C.: Implementation of the simultaneous perturbation algorithm for stochastic optimization. IEEE Trans. Aerosp. Electron. Syst. **34**(3), 817–823 (1998)
357. Spirkl, W., Ries, H., Kribus, A.: Optimal parallel flow in solar collectors for nonuniform irradiance. Trans. ASME **119**, 156–159 (1997)
358. Steinbuch, M.: Repetitive control for system with uncertain period-time. Automatica **38**(12), 2103–2109 (2002)
359. Stine, W.B., Geyer, M.: Power from the Sun. http://www.powerfromthesun.net/book.html (2001)
360. Stirrup, R., Loebis, D., Chipperfield, A.J., Tang, K.S., Kwong, S., Man, K.F.: Gain-scheduled control of a solar power plant using a hierarchical MOGA-tuned fuzzy PI-controller. In: Proc. of ISIE 2001, the IEEE Int. Symp. on Industrial Electronics, Pusan, Korea, pp. 25–29 (2001)
361. Stone, K.W.: Automatic heliostat track alignment method. US Patent 4,564,275, 14 January 1986
362. Stone, K.W., Blackmon, J.B.: Alignment system and method for dish concentrators. US Patent 5,982,481, 9 November 1999
363. Stone, K.W., Lopez, C.W.: Evaluation of the solar one track alignment methodology. Trans. ASME J. Sol. Energy Eng. **S.1**, 521–526 (1995)
364. Strachan, J.W.: Revisiting the BCS, a measurement system for characterizing the optics of solar collectors. In: Proc. of the 39th Int. Symp. of Instrument Society of America, 1992
365. Strachan, J.W., Houser, R.: Testing and evaluation of large-area heliostat for solar thermal applications. SAND92-1381-UC-235, Sandia National Laboratories, USA (1993)
366. Stuetzle, T., Blair, N., Mitchell, J.W., Beckman, A.: Automatic control of a 30 MWe SEGS VI parabolic trough plant. Sol. Energy **76**, 187–193 (2004)
367. Sudkamp, T., Hammell, R.J. II: Interpolation, completion, and learning fuzzy rules. IEEE Trans. Syst. Man Cybern. **24**(2), 332–342 (1994)
368. Suresh, D., Rohatgi, P.K.: Melting and casting of alloys in a solar furnace. Sol. Energy **23**, 553–555 (1979)
369. Sweet, B.: Renewables ranked. IEEE Spectrum. http://spectrum.ieee.org/energywise/energy/renewables/renewables-ranked (2011)
370. Thalhammer, E.D.: Heliostat beam characterization system—update. In: Proc. ISA/79 Conf., Chicago, IL, USA, 1979
371. Thirugnanasambandam, M., Iniyan, S., Goic, R.: A review of solar thermal technologies. Renew. Sustain. Energy Rev. **14**, 312–322 (1995)
372. Torrico, B., Roca, L., Normey-Rico, J.E., Guzmán, J.L., Yebra, L.: Robust nonlinear predictive control applied to a solar collector field in a solar desalination plant. IEEE Trans. Control Syst. Technol. **18**(6), 1430–1439 (2010)
373. Torrisi, F.D., Bemporad, A.: HYSDEL—a tool for generating computational hybrid models for analysis and synthesis problems. IEEE Trans. Control Syst. Technol. **12**(2), 235–249 (2004)
374. Tovar-Pescador, J., Pozo-Vázquez, D., Ruiz-Arias, J.A., Batlles, J., López, G., Bosch, J.L.: On the use of the digital elevation model to estimate the solar radiation in areas of complex topography. Meteorol. Appl. **13**(3), 279–287 (2006)

375. Trombe, F.: Solar furnaces and their applications. Sol. Energy **1**(2–3), 9–15 (1957)
376. Tsuo, Y.S., Pitts, J.R., Landry, M.D., Bingham, C.E., Lewandowski, A., Ciszek, T.F.: High-flux solar furnace processing of silicon solar cells. In: Proc. of the 24th IEEE Photovoltaic Specialists Conf. on Photovoltaic Energy Conversion, Waikoloa, HI, USA, pp. 1307–1310 (1994)
377. Tummescheit, H.: Design and implementation of object-oriented model libraries using modelica. PhD Thesis, Department of Automatic Control, Lund Institute of Technology, Sweden (2002)
378. Tymvios, F.S., Jacovides, C.P., Michaelides, S.C., Scouteli, C.: Comparative study of Angström's and artificial neural networks' methodologies in estimating global solar radiation. Sol. Energy **78**, 752–762 (2005)
379. Tyner, C., Kolb, G., Prairie, M., Weinrebe, G., Valverde, A., Sánchez, M.: Solar power tower development: recent experiences. In: Proc. of the 8th Int. Symp. on Solar Thermal Concentrating Technologies, Cologne, Germany, pp. 193–216 (1996)
380. Tzafestas, S., Papanikolopoulos, N.P.: Incremental fuzzy expert PID control. IEEE Trans. Ind. Electron. **37**(5), 365–371 (1990)
381. Vadakkoot, R., Shah, M.D., Shrivastava, S.: Enhanced moving average computation. In: World Congress on Computer Science and Information Engineering. Los Angeles, USA, 2009
382. Valenzuela, L., Balsa, P.: Series and parallel feedforward control schemes to regulate the operation of a solar collector field. In: Martínez, D. (ed.) Proc. of the 2nd Users Workshop TMR Programme at PSA, CIEMAT. CIEMAT, Madrid (1998)
383. Valenzuela, L., Zarza, E., Berenguel, M., Camacho, E.F.: Control concepts for direct steam generation process in parabolic troughs. In: Proc. of the ISES Solar World Congress, Goteborg, Sweden, 2003
384. Valenzuela, L., Zarza, E., Berenguel, M., Camacho, E.F.: Direct steam generation in solar boilers. IEEE Control Syst. Mag. **24**(2), 15–29 (2004)
385. Valenzuela, L., Zarza, E., Berenguel, M., Camacho, E.F.: Control concepts for direct steam generation in parabolic troughs. Sol. Energy **78**, 301–311 (2005)
386. Valenzuela, L., Zarza, E., Berenguel, M., Camacho, E.F.: Control schemes for direct steam generation in parabolic solar collectors under recirculation operation mode. Sol. Energy **80**, 1–17 (2005)
387. Valverde, A., Weinrebe, G.: Implementation of an automatic aiming-point strategy for a volumetric receiver in the PSA's CESA-1 heliostat field. In: Proc. of the 8th Int. Symp. on Solar Thermal Concentrating Technologies, Cologne, Germany, pp. 1047–1066 (1996)
388. van Willigenburg, L.G., Bontsema, J., Koning, W.L.D., Valenzuela, L., Cirre, C.M.: Direct reduced-order digital control of a solar collector field. In: Martínez, D. (ed.) Proc. of the IHP Programme. Research Results at PSA Within the Year 2003 Access Campaign. CIEMAT. CIEMAT, Madrid (2004)
389. van Willigenburg, L.G., Bontsema, J., Koning, W.L.D., Valenzuela, L., Cirre, C.M.: Digital optimal reduced-order control of a solar power plant. In: Proc. of UKACC-IEE CONTROL 2004, University of Bath, UK, 2004
390. Vant-Hull, L.L., Izygon, M., Pitman, C.L.: Assessment of the real-time receiver excess-flux-density protection software at Solar Two. In: Proc. of the 8th Int. Symp. on Solar Thermal Concentrating Technologies, Cologne, Germany, pp. 951–970 (1996)
391. Vaz, F., Oliveira, R., Silva, R.N.: PID control of a solar plant with gain interpolation. In: Martínez, D. (ed.) Proc. of the 2nd Users Workshop TMR Programme at PSA, CIEMAT. CIEMAT, Madrid (1998)
392. Veres, S.M., Norton, J.P.: Predictive self-tuning control by parameter bounding and worst-case design. Automatica **29**(4), 911–928 (1993)
393. Wagner, W., Kruse, A.: Properties of Water and Steam. Springer, Berlin (1998)
394. Wang, Y.: Computer modeling and simulation of solid-state sintering: a phase field approach. Acta Mater. **54**(4), 953–961 (2006)
395. Wei, X., Lu, Z., Yu, W., Wang, Z.: A new code for the design and analysis of the heliostat field layout for power tower system. Sol. Energy **84**, 685–690 (2010)

396. Welch, G., Bishop, G.: An introduction to the Kalman filter. Technical Report, Chapel Hill, NC, USA (2006)
397. Wellstead, P.E., Prager, D., Zanker, P.: A pole assignment self-tuning regulator. IEE Proc. Part D **126-128**, 781–787 (1978)
398. Wettermark, G.: Performance of the SSPS solar power plants at Almería. J. Sol. Energy Eng. **110**, 235–246 (1988)
399. Wong, L.T., Chow, W.K.: Solar radiation model. Appl. Energy **69**, 191–224 (2001)
400. Wyns, B., Sbarciog, M., Ionescu, C., Boullart, L., De Keyser, R.: Neural network modeling versus physical modeling application to a solar power plant. In: Martínez, D. (ed.) Proc. of the IHP Programme. Research Results at PSA Within the Year 2003 Access Campaign, CIEMAT. CIEMAT, Madrid (2004)
401. Xiao, W., Lind, M.G.J., Dunford, W.G., Capel, A.: Real-time identification of optimal operating points in photovoltaic power systems. IEEE Trans. Ind. Electron. **53**(4), 1017–1026 (2006)
402. Yang, Z., Garimella, S.V.: Thermal analysis of solar thermal energy storage in a molten-salt thermocline. Sol. Energy **84**, 974–985 (2010)
403. Yang, Y., Torrance, A.A., Rodriguez, J.: The solar hardening of steels: experiments and predictions. Sol. Energy Mater. Sol. Cells **40**, 103–121 (1996)
404. Yebra, L.J., Berenguel, M., Dormido, S.: Extended moving boundary models for two-phase flows. In: Proc. of the 16th IFAC World Congress, Prague, Czech Republic, 2005
405. Yebra, L.J., Berenguel, M., Dormido, S., Romero, M.: Modelling and simulation of central receiver solar thermal power plants. In: Proc. of the 44th IEEE Conf. on Decision and Control, and the European Control Conf. 2005, Seville, Spain, 2005
406. Yebra, L.J., Berenguel, M., Dormido, S., Zarza, E.: Object oriented modelling and simulation of parabolic trough collectors with Modelica. Math. Comput. Model. Dyn. Syst. **14**(4), 361–375 (2008)
407. Yebra, L.J., Berenguel, M., Bonilla, J., Roca, L., Dormido, S., Zarza, E.: Object oriented modelling and simulation of Acurex solar thermal power plant. In: Proc. of MathMod09, Vienna, Austria, 2009
408. Yebra, L.J., Berenguel, M., Bonilla, J., Roca, L., Dormido, S., Zarza, E.: Object-oriented modelling and simulation of ACUREX solar thermal power plant. Math. Comput. Model. Dyn. Syst. **16**(3), 211–224 (2010)
409. Yona, A., Senjyu, T.: One-day-ahead 24-hours thermal energy collection forecasting based on time series analysis technique for solar heat energy utilization system. In: Proc. of IEEE T&D Asia 2009, Seoul, Korea, 2009
410. Zaharim, A., Razali, A.M., Gim, T.P., Sopian, K.: Time series analysis of solar radiation data in the tropics. Eur. J. Sci. Res. **25**(4), 672–678 (2009)
411. Zambrano, D., Garcia-Gabin, W.: Hierarchical control of a hybrid solar air conditioning plant. Eur. J. Control **14**(6), 464–483 (2008)
412. Zambrano, D., Bordóns, C., Garcia-Gabin, W., Camacho, E.F.: A solar cooling plant: a benchmark for hybrid systems control. In: The 2nd IFAC Conference on Analysis and Design of Hybrid Systems, pp. 199–204 (2006)
413. Zambrano, D., Bordóns, C., Garcia-Gabin, W., Camacho, E.F.: Model development and validation of a solar cooling plant. Int. J. Refrig. **31**, 315–327 (2008)
414. Zambrano, D., Garcia-Gabin, W., Camacho, E.F.: Application of a transition graph-based predictive algorithm to a solar air conditioning plant. IEEE Trans. Control Syst. Technol. **18**(5), 1162–1171 (2010)
415. Zarza, E., Ajona, J.I., León, J., Gregorzew, A., Genthner, K.: Solar thermal desalination project at the Plataforma Solar de Almeria. Sol. Energy Mater. **24**, 608–622 (1991)
416. Zarza, E., Valenzuela, L., León, J., Hennecke, K., Eck, M., Weyers, H.D., Eickhoff, M.: The DISS project: direct steam generation in parabolic troughs. Operation and maintenance experience & update on project status. In: Proc. of ASME Int. Solar Energy Conf.: Forum 2001, Washington, DC, USA, 2001

417. Zarza, E., Valenzuela, L., León, J., Hennecke, K., Weyers, H.D., Eickhoff, M.: Direct steam generation in parabolic troughs. Final results and conclusions of the DISS project. In: Proc. of the 11th SolarPaces Int. Symp. on Concentrated Solar Power and Chemical Energy Technologies, Zurich, Switzerland, 2002
418. Zarza, E., Valenzuela, L., León, J., Hennecke, K., Eck, M., Weyers, H.D., Eickhoff, M.: The DISS project: direct steam generation in parabolic trough systems. Operation & maintenance experience and update on project status. J. Sol. Energy Eng. Trans. ASME **124**, 126–133 (2004)
419. Zarza, E., Valenzuela, L., León, J., Weyers, H.D., Eickhoff, M., Eck, M., Hennecke, K.: Direct steam generation in parabolic troughs: final results and conclusions of the DISS project. Energy **29**, 635–644 (2004)
420. Zavala, V.M., Constantinescu, E.M., Krause, T., Anitescu, M.: On-line economic optimization of energy systems using weather forecast information. J. Process Control **19**, 1725–1736 (2009)
421. Zhang, G.P., Qi, M.: Neural network forecasting for seasonal and trend time series. Eur. J. Oper. Res. **160**, 501–514 (2005)
422. Zhang, L., Zhang, Y., Wang, D., Xu, D.: Multiple models generalized predictive control for superheated steam temperature based on MLD model. In: Proc. of 2007 IEEE Int. Conf. on Automation and Logistics, Jinan, Shandong, China, pp. 2740–2743 (2007)
423. Ziegler, J.G., Nichols, N.B.: Optimum settings for automatic controllers. Trans. ASME **64**, 759–768 (1942)

Index

E.F. Camacho et al., *Control of Solar Energy Systems*, Advances in Industrial Control,
DOI 10.1007/978-0-85729-916-1, © Springer-Verlag London Limited 2012